# Nanoparticle- and Microparticle-Based Delivery Systems

## Encapsulation, Protection and Release of Active Compounds

David Julian McClements

# Nanoparticle- and Microparticle-Based Delivery Systems

## Encapsulation, Protection and Release of Active Compounds

**CRC Press**
Taylor & Francis Group
Boca Raton London New York

CRC Press is an imprint of the
Taylor & Francis Group, an **informa** business

CRC Press
Taylor & Francis Group
6000 Broken Sound Parkway NW, Suite 300
Boca Raton, FL 33487-2742

First issued in paperback 2016

Version Date: 20140618

ISBN 13: 978-1-138-03403-7 (pbk)
ISBN 13: 978-1-4822-3315-5 (hbk)

### Library of Congress Cataloging-in-Publication Data

McClements, David Julian, author.
  Nanoparticle- and microparticle-based delivery systems : encapsulation, protection and release of active compounds / David Julian McClements.
    pages cm
  Includes bibliographical references and index.
  ISBN 978-1-4822-3315-5 (hardback)
  1. Food--Biotechnology. 2. Transport theory. 3. Colloids. 4. Microencapsulation. I. Title.

TP248.65.F66M365 2015
664--dc23
2014022773

**Visit the Taylor & Francis Web site at**
**http://www.taylorandfrancis.com**

**and the CRC Press Web site at**
**http://www.crcpress.com**

This book is dedicated to my wife Jayne and daughter Isobelle.

# Contents

# Foreword

Since the days when George Stainsby and I were writing our book *Colloids in Food* (Applied Science, London, 1982), there have been quite a lot of changes. Back in our innocent pre-"nano" and pre-"Internet" age, the phrase *delivery system* would most likely have immediately brought to mind the customized motor vehicle driven by our daily morning milkman when bringing his ubiquitous food colloid to our doorsteps in those old traditional pint-sized glass bottles. The intervening period has been witness to changes not only in the marketing and distribution of milk and other dairy-based products but also in the consumers' expectations of such foods in terms of shelf life, product variety, and lifestyle-related attributes. In particular, the growing interest within some parts of society toward the influence of diet on personal health and well-being has led the food research community to focus its attention on what happens to colloidal food systems during eating and within the gastrointestinal tract.

Independently, over the same period, we have seen quite relentless progress in the new field of nanoscience based on detailed fundamental physicochemical studies devoted to how specific systems of small and large molecules are organized into well-defined complex particles of various shapes and sizes within the colloidal domain. This improvement in the understanding of molecular self-assembly and organization, derived from a combination of sophisticated instrumentation and computer modeling/simulation, has provided the community of food scientists and technologists with enhanced confidence to explore novel delivery systems for encapsulating and releasing active ingredients, with the ultimate aim of enabling the food industry to improve diet-related human health. In the still brief history of this field, the publication record of one particular research group stands out well above all others. That is the group led by Professor D. J. (Julian) McClements. Hence, the appearance of a monograph collecting Julian's accumulated insight and understanding of this topical subject area is a very timely development.

An interesting paradox of the scientific enterprise is that the more a scientist makes his intellectual achievements widely available to others, the more it becomes securely identified as his own property. This statement seems especially applicable to Julian McClements: in his relatively short academic career (so far), he has established a recognizable scholarly style and methodological approach that are deservedly admired by his peers, both in industry as well as in academia, and in various geographical locations far and wide. The clarity of the written text and accompanying schematic diagrams within these pages will be immediately familiar to readers of his already classic textbook on *Food Emulsions* (2nd edition, CRC Press, 2005), as well as his many well-cited and readily accessible review articles. As we have come to expect for a McClements-authored work, the subject matter contained herein is built on solid physical chemistry principles. But, much more than this, the author's approach focuses strongly on providing the reader with some straightforward practical know-how in a field that is becoming increasingly diverse in scope and hence potentially confusing for general food scientists. Moreover, this is a book that clearly explains the "why" (and, often more importantly, the "why not") of any putative molecular design strategy that might be under consideration for delivery of a particular ingredient within some specific food product context. The authority of the author's writing comes from a solid base of knowledge derived from systematic testing of many different kinds of nanoscale and microscale delivery systems carried out in carefully designed experiments on model systems in the author's own laboratory at the University of Massachusetts in Amherst.

It is therefore indeed a special privilege for me to be able to introduce prospective readers to this new book. The field of food delivery systems is reaching a pivotal point in its development. For those starting to work in this exciting research area, some great opportunities and challenges surely lie ahead. Against such a heady background of activity, I confidently expect this new volume to become rapidly established as a respected source of basic information for those advancing the field of food colloids for the coming years.

**Eric Dickinson**
*Professor Emeritus*
*School of Food Science and Nutrition*
*University of Leeds*

# Preface

Nanoparticle and microparticle delivery systems are finding increasing application in the food, pharmaceutical, personal care, and other industries. These types of colloidal delivery systems were traditionally used in the food industry to encapsulate active ingredients designed to improve food quality and safety, such as flavors, colors, antioxidants, enzymes, and antimicrobials. More recently, they are finding increasing use to encapsulate bioactive components that may improve human health and well-being, such as vitamins, minerals, and nutraceuticals. A well-designed colloidal delivery system can be used to overcome many of the technical challenges normally associated with incorporation of these active ingredients into commercial food and beverage products.

The purposes of this book are to provide a comprehensive review of the various types of colloidal delivery systems available for encapsulating active ingredients in the food and other industries and to highlight their relative advantages and limitations. Initially, a discussion of the numerous kinds of active ingredients that the food industry is interested in incorporating into their products is given, and challenges associated with their encapsulation, protection, and delivery are discussed. The physicochemical and mechanical methods available for manufacturing colloidal particles are then discussed, and the importance of designing particles for specific applications is highlighted. Separate chapters are devoted to the three major types of colloidal delivery systems available for encapsulating active ingredients in the food industry: surfactant-based, emulsion-based, and biopolymer-based. The analytical tools available for characterizing the properties of colloidal delivery systems are then reviewed, and mathematical models for describing their properties are presented. Finally, the factors to consider when selecting an appropriate delivery system for a particular application are highlighted using a number of specific case studies.

I hope that this book will stimulate further developments in this important area and that it will provide the food industry with valuable information leading to the production of higher-quality and healthier foods.

Finally, I would like to thank my family, friends, colleagues, and students for all their help throughout the process of writing this book. In particular, I thank

Professor Eric Dickinson for introducing me to the field of colloid science during my undergraduate and graduate studies at the University of Leeds, and for serving as an excellent role model for how scientific research should be done in this area. I also thank all of the staff at Taylor & Francis for their help in bringing this book into its final form. Finally, I thank my wife Jayne and daughter Isobelle for always being so supportive.

# Author

 **David Julian McClements** is a professor in the Department of Food Science at the University of Massachusetts, and an adjunct professor in the Department of Biochemistry, Faculty of Science, King Abdulaziz University, Jeddah, Saudi Arabia. He specializes in the areas of food biopolymers and colloids, and in particular on the development of food-based structured delivery systems for bioactive components. Dr. McClements earned his PhD in food science (1989) at the University of Leeds (United Kingdom). He then did postdoctoral research at the University of Leeds, University of California (Davis), and University College Cork (Ireland). Dr. McClements is the sole author of the first and second editions of *Food Emulsions: Principles, Practice and Techniques*, coauthor of *Advances in Food Colloids* with Professor Eric Dickinson, and coeditor of *Developments in Acoustics and Ultrasonics, Understanding and Controlling the Microstructure of Complex Foods, Designing Functional Foods, Oxidation in Foods and Beverages* (Volumes 1 and 2), and *Encapsulation and Delivery Systems for Food Ingredients and Nutraceuticals*. In addition, he has published over 500 scientific articles in peer-reviewed journals (with an H-index over 60).

Dr. McClements has previously received awards from the American Chemical Society, American Oil Chemists Society, Society of Chemical Industry (UK), Institute of Food Technologists, and University of Massachusetts in recognition of his scientific achievements. His research has been funded by grants from the U.S. Department of Agriculture, National Science Foundation, U.S. Department of Commerce, Dairy Management Incorporated, and the food industry. He is a member of the editorial boards of a number of journals and has organized workshops, symposia, and conferences in the field of food colloids, food emulsions, and delivery systems.

# Chapter 1

# Background and Context

## 1.1 Introduction

This book is primarily concerned with the design, fabrication, and utilization of nanoparticle-based and microparticle-based delivery systems for application within the food and beverage industries. Nevertheless, much of the material covered in this book is also directly applicable to other applications, including pharmaceuticals, health care products, personal care products, cosmetics, and agrochemicals. Indeed, many of the same challenges exist in different application areas, and many of the same solutions to these challenges are applicable to different kinds of products. For the sake of clarity, the term *colloidal delivery systems* will often be used in this book to collectively refer to nanoparticle-based and microparticle-based delivery systems.

Food manufacturers often want to incorporate active ingredients with specific functional attributes into their products, for example, colors, flavors, nutraceuticals, antimicrobials, antioxidants, and preservatives (Table 1.1). Many of these active ingredients cannot simply be incorporated into these products because they are physically or chemically unstable, they are incompatible with the product matrix, or they lack the appropriate functional attributes. These challenges can often be overcome by incorporating the active ingredient into some kind of *delivery system* before it is introduced into the final product (Table 1.2). Delivery systems can be designed to have a number of potential benefits to the food industry: incorporation of active ingredients into food matrices without adversely affecting quality attributes such as appearance, texture, flavor, or stability; protection against chemical, physical, or biological degradation; off-flavor masking (such as bitterness or astringency); the ability to deliver active ingredients to particular sites-of-action thereby increasing their efficacy; improvement of product storage and handling;

1

**Table 1.1    Examples of Active Ingredients That May Need to Be Encapsulated before They Can Be Successfully Utilized within Foods and Beverages**

| Active Ingredients | Examples | Potential Advantages of Encapsulation |
|---|---|---|
| Flavors | Citrus oils<br>Natural extracts | Allow incorporation into aqueous medium<br>Facilitate storage and utilization<br>Retard chemical degradation<br>Control release profile |
| Antimicrobials | Essential oils | Improve matrix compatibility<br>Facilitate storage and utilization<br>Retard chemical degradation<br>Mask undesirable off-flavors<br>Increase potency |
| Antioxidants | Tocopherols<br>Carotenoids<br>Flavonoids<br>Phenolics | Allow incorporation into aqueous medium<br>Facilitate storage and utilization<br>Retard chemical degradation<br>Increase efficacy |
| Bioactive peptides | Milk peptides<br>Meat peptides<br>Plant peptides | Retard degradation in stomach<br>Reduce bitterness and astringency<br>Control release profile and bioactivity |
| Oligosaccharides and fibers | Prebiotics<br>Chitosan | Avoid adverse ingredient interactions<br>Improved product texture<br>Control delivery in the gastrointestinal tract |
| Minerals | Iron<br>Calcium | Avoid undesirable oxidative reactions<br>Prevent precipitation<br>Enhance bioavailability |

(*continued*)

**Table 1.1 (Continued) Examples of Active Ingredients That May Need to Be Encapsulated before They Can Be Successfully Utilized within Foods and Beverages**

| Active Ingredients | Examples | Potential Advantages of Encapsulation |
|---|---|---|
| Vitamins | Vitamins A, D, E<br>Vitamin C | Allow incorporation in aqueous medium<br>Improve ease of utilization<br>Prevent chemical degradation |
| Bioactive lipids | ω-3 fatty acids<br>Conjugated linoleic acid | Allow incorporation in aqueous medium<br>Improve ease of utilization<br>Avoid chemical degradation (oxidation)<br>Controlled delivery in GIT |
| Probiotics | Lactic acid bacteria | Improve cell viability in product<br>Avoid degradation in GIT |

extension of product shelf life (McClements et al. 2009a,b; Sagalowicz and Leser 2010; Sanguansri and Augustin 2006; Velikov and Pelan 2008). For example, a beverage cannot simply be formed by mixing flavor oil with water and other ingredients because the flavor oil is normally partly immiscible with water and would quickly separate. Instead, the flavor oil must first be converted into small particles (such as microemulsion or emulsion droplets) that remain stable within the final product (Given 2009; McClements 2011).

The purposes of this book are as follows: (i) to provide an overview of active ingredients that need to be encapsulated, and challenges associated with their successful incorporation into foods and beverages; (ii) to discuss the various approaches available for fabricating food-grade delivery systems and to highlight their relative advantages and disadvantages; and (iii) to demonstrate the potential utility of delivery systems in foods and beverages using specific examples. This book will primarily focus on colloidal delivery systems that can be constructed from food-grade ingredients (such as proteins, carbohydrates, lipids, and surfactants) using assembly principles largely derived from colloid or polymer science. In addition, it will be mainly concerned with those colloidal delivery systems that use water as the dispersing medium, such as surfactant-based, emulsion-based, biopolymer-based, and hybrid delivery systems (Table 1.2 and Figure 1.1).

**Table 1.2  Selected Examples of Colloidal Delivery Systems That Can Be Constructed from Food-Grade Ingredients**

| Delivery System Type | Characteristics |
|---|---|
| Surfactant based | |
| – Swollen micelles | $O_S/W$ |
| – Microemulsions | $O_S/W$ |
| – Liposomes | W/S/W |
| Emulsion based | |
| – Conventional emulsions | $O_S/W$ |
| – Nanoemulsions | $O_S/W$ |
| – Multiple emulsions | $(W_{S1}/O)_{S2}/W$ |
| – Multilayer emulsions | $O_{Snl}/W$ |
| – Solid lipid nanoparticles | $O'_S/W$ |
| – Microclusters | $(O_S)_N/W$ |
| Biopolymer based | |
| – Hydrogel particles | $W_1/W_2$ |
| – Coated hydrogel particles | $(W_1)_{nl}/W_2$ |
| – Biopolymer nanoparticles/microparticles | B/W |
| Hybrid systems | |
| – Filled hydrogel particles | $(O_S/W_1)/W_2$ |
| – Filled liposomes | $(O_S/W)/S/W$ |
| – Multilayer liposomes | $W/S_{nl}/W$ |
| – Colloidosomes | $[O_{S1}/(O_{S2})_N]/W$ |

*Note:* O, liquid oil; O′, solid fat; W, water; B, biopolymer; S, surfactant or emulsifier layer; nl, one or more interfacial coatings; N, two or more droplets. Only delivery systems with an aqueous continuous phase are shown here.

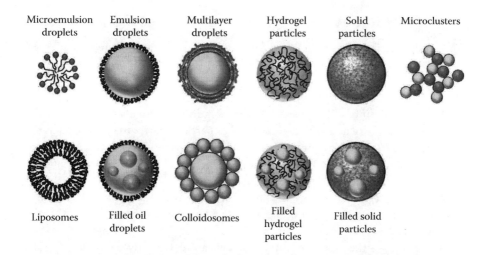

Microemulsion droplets    Emulsion droplets    Multilayer droplets    Hydrogel particles    Solid particles    Microclusters

Liposomes    Filled oil droplets    Colloidosomes    Filled hydrogel particles    Filled solid particles

**Figure 1.1** **Examples of colloidal delivery systems that can be used to encapsulate, protect, and deliver functional food ingredients (not drawn to scale).**

## CASE STUDY 1.1   FORTIFICATION OF FOODS WITH ω-3 FATTY ACIDS

Nutritional research has shown that consumption of appreciable quantities of ω-3 fatty acids (e.g., EPA and DHA) has a number of potential health benefits, including improved cognitive and psychological function (Luchtman and Song 2013), reduced inflammation and cancer (Laviano et al. 2013), and reduced risk of cardiovascular disease (Hu and Willett 2002; Nicholson et al. 2013). Nevertheless, polyunsaturated lipids are highly susceptible to chemical degradation owing to oxidation, which involves a complex series of reactions leading to the formation of rancid off-flavors and off-odors, as well as potentially toxic reaction products (McClements and Decker 2000; Shahidi and Zhong 2010; Taneja and Singh 2012). Consequently, the food industry would like to create products enriched with ω-3 fatty acids but is limited in doing this because of their poor chemical stability. This is a clear case where encapsulation is highly beneficial to the food industry. Delivery systems that do not adversely affect the physicochemical or sensory properties of food are needed to protect ω-3 fatty acids against chemical degradation during storage, while still ensuring that they are bioavailable after consumption. A number of delivery systems have been proposed to encapsulate, protect, and release ω-3 fatty acids including oil-in-water (O/W) emulsions (Lesmes et al. 2010; Lomova et al. 2010), microemulsions (Jakobsson and Sivik 1994), and filled hydrogel particles (Matalanis et al. 2012). A number of other delivery systems suitable for encapsulating ω-3 fatty acids will be discussed in this book.

## 1.2 Terminology

There are a number of terms associated with colloidal delivery systems that are used throughout this book and so it is useful to define them here to provide greater clarity and avoid confusion:

■ **Encapsulation**—The process of incorporating a particular component (the "active ingredient") within some kind of particle matrix (the "encapsulant"). This matrix may be composed of a single or multiple components, and it may have a homogeneous or heterogeneous structure, depending on the materials and procedures used to fabricate it (Figure 1.1).

■ **Retention**—The process of retaining the active ingredient within the encapsulating matrix. Typically, it is important to retain the active ingredient within the colloidal particles until they are released at the required site of action.

■ **Delivery**—The process of carrying an encapsulated active component to the required site of action, which may be the surface of a bacteria (in the case of an antimicrobial), the human mouth or nose (in the case of a flavor), the stomach or small intestine (in the case of a nutraceutical), or the colon (in the case of a probiotic or prebiotic). Once an active component has been encapsulated, it usually has to be retained by the delivery system for a certain period under specific environmental conditions before it is released.

■ **Delivery system**—A system designed to encapsulate, protect, and release one or more active components.

■ **Controlled release**—The process of releasing an encapsulated active ingredient with a specific concentration–time profile at the required site of action. The desired release profile may take different forms depending on the nature of the application (Aguzzi et al. 2010; Burey et al. 2008; Gibbs et al. 1999; Madene et al. 2006; Siepmann and Siepmann 2011), including the following:
  - *Burst release*: Rapid release of most of the active ingredient over a short period.
  - *Sustained release*: Prolonged release of the active ingredient at a relatively constant release rate.
  - *Triggered release*: Release of the active ingredient in response to a specific environmental trigger, such as dilution, pH, ionic strength, temperature, enzyme activity, or mechanical force.
  - *Targeted release*: Release of the active ingredient at a specific location (e.g., mouth, stomach, small intestine, colon). In the pharmaceutical industry, this is often achieved by attaching specific ligands to the surface of a colloidal particle that bind to target molecules or structures.

The nature of the delivery system selected for a particular application depends on the unique molecular characteristics and functional requirements of the active ingredient and food matrix under consideration.

# 1.3 Active Ingredients and the Need for Encapsulation

In this section, a brief overview of some of the most important active ingredients used in foods, as well as their physicochemical attributes and challenges associated with their incorporation into food and beverage products, is given. A more detailed discussion of active ingredients is given in Chapter 2.

The active ingredients used by the food industry to formulate commercial products may be isolated and purified from natural sources, or they may be chemically synthesized (Damodaran et al. 2008; Igoe 2011; Msagati 2013). Chemically synthesized ingredients may be "nature-identical" (i.e., they have the same structure and reactivity as molecules found in nature) or they may be entirely new molecules that do not normally occur in nature. This is an important consideration for food manufacturers when selecting active ingredients for particular products, as consumers often prefer natural rather than synthetic ingredients. In addition, any newly synthesized food ingredient has to undergo extensive testing and regulatory hurdles to ensure that it is safe to use, which is time consuming and expensive. Consequently, there is a trend toward the utilization of more natural ingredients within many food and beverage products.

In general, active ingredients vary in their molecular, physicochemical, and biological properties. At the molecular level, they can be characterized by their atomic composition, their molecular weight, their chemical formula, their three-dimensional structure, their flexibility, their polarity, and their electrical charge. The physiochemical properties of active ingredients are determined by their molecular characteristics and the environmental conditions. Some of the most important physicochemical properties of active ingredients are their physical state (e.g., gas, liquid, solid), their phase behavior (e.g., boiling point, melting point, glass transition temperature), their solubility in particular solvents (such as oil, water, or alcohol), their surface activity, their rheology (e.g., viscosity, elastic modulus, yield stress), their optical properties (e.g., color or opacity), and their chemical stability under particular conditions (e.g., oxidation or hydrolysis). At the biological level, active ingredients can be characterized by their interactions with microorganisms, animals, or humans (e.g., antimicrobial activity, aroma, taste, digestibility, or bioactivity). Some of the most important active ingredients that need to be encapsulated in foods and beverages are summarized in Table 1.1 and are briefly discussed here. A more detailed description of active ingredients is given in Chapter 2.

## 1.3.1 Flavors and Colors

Natural and artificial flavors are used to control the taste or aroma of food products (Jelen 2011; Lindsay 2008; Reineccius 2007). This category of active ingredients covers a wide variety of different molecular types, from nonvolatile water-soluble substances perceived in the mouth to volatile oil-soluble substances perceived in the nose. Natural and artificial colors are added to foods and beverages to control their appearance (Hutchings 1999; Schwartz et al. 2008). Flavors and colors are often encapsulated to prevent them from being lost during storage, to inhibit their chemical degradation, to improve their ease of handling and utilization, and to control their release profiles.

### 1.3.2 Antioxidants

Natural and synthetic antioxidants are widely used in the food industry to retard the rate and extent of lipid oxidation in foods and beverages (McClements and Decker 2000; Reische et al. 2008; Waraho et al. 2011). Antioxidants may be predominantly hydrophilic, lipophilic, or amphiphilic depending on their molecular structure. The polarity of an antioxidant determines its location and environment within a food or beverage system (e.g., the aqueous, oil, or interfacial phases), which in turn affects its ability to prevent oxidation by altering its proximity to the various participants in the oxidation reaction. Antioxidants are often encapsulated to improve their ease of handling, improve their compatibility with the food matrix, improve their stability during transport and storage, and enhance their activity.

### 1.3.3 Antimicrobials

Natural and synthetic antimicrobials are widely used by the food industry to kill or inhibit the growth of microorganisms, such as bacteria, molds, and yeasts (Brul and Coote 1999; Davidson et al. 2005). Food antimicrobials may operate through various mechanisms of action, including transition metal chelation, membrane disruption, and interference with essential biochemical pathways in microorganisms. Antimicrobials vary widely in their molecular characteristics and may be hydrophilic (e.g., sorbates and benzoates), lipophilic (e.g., essential oils), or amphiphilic (e.g., surfactants and proteins). Antimicrobial agents are often encapsulated to increase their compatibility with the food matrix, to increase their efficacy, to control release, to mask off-flavors, and to increase their ease of storage, transport, and utilization.

### 1.3.4 Bioactive Lipids

In general, the category of food lipids contains a broad spectrum of chemically diverse compounds that are soluble in organic solvents and (usually) insoluble in water (Akoh and Min 2008; McClements and Decker 2008). A variety of different classes of food molecules fall within this category, including monoacylglycerols, diacylglycerols, triacylglycerols, free fatty acids, phospholipids, carotenoids, phytosterols, and oil-soluble vitamins. Consumption of many of these lipophilic ingredients have been linked to specific health benefits, such as reduced incidences of obesity, coronary heart disease, diabetes, hypertension, cancer, brain disease, and eye disease. Consequently, there has been great interest in incorporating them into commercial food and beverage products (McClements et al. 2009a,b). Nevertheless, there are often challenges associated with successfully fortifying foods with these components. Most bioactive lipids have low water solubility and therefore must be encapsulated into a suitable delivery system before they can be incorporated into aqueous-based products (McClements and Decker 2000; McClements et al. 2009a,b; Waraho et al. 2011). In addition, many of them are highly susceptible to chemical degradation during storage (e.g., oxidation of polyunsaturated lipids), and therefore delivery systems must be specifically designed to protect them.

### *1.3.5 Bioactive Carbohydrates*

In general, carbohydrates can be categorized according to their molecular structures as monosaccharides, oligosaccharides, or polysaccharides (Biliaderis and Izydorczyk 2007; Cui 2005). They can also be further categorized depending on whether they are digestible or indigestible, with indigestible carbohydrates being considered to be dietary fibers (Redgwell and Fischer 2005; Wildman and Kelley 2007). Small carbohydrates (such as fructose, glucose, and sucrose) are often added to foods to provide sweetness, texture, and bulking properties, whereas large carbohydrates (such as starch, cellulose, and gums) are often used to modify texture and stability. There is currently considerable interest in developing delivery systems to control the release of sweeteners during eating so that reduced amounts of these high calorie ingredients can be used to provide the same flavor profile (Nehir El and Simsek 2012; Stieger and van de Velde 2013). There is also considerable interest in delivering dietary fibers, since their increased consumption has been linked to health benefits such as reduced constipation, cancer, and heart disease (Dikeman and Fahey 2006; Redgwell and Fischer 2005). One of the problems with simply adding dietary fibers to foods is that they may have undesirable effects on the texture or mouthfeel of the product. Some of these problems may be overcome by incorporating them into a suitable delivery system.

### *1.3.6 Bioactive Proteins*

As well as providing energy and essential nutrients to the diet, a number of proteins, peptides, and amino acids have also been claimed to have additional biological functions, such as acting as growth factors, antihypertensive agents, antimicrobial agents, antioxidants, food intake modifiers, and immune regulatory factors (Espitia et al. 2012; Meisel 1997; Playne et al. 2003; Udenigwe and Aluko 2012; Ward and German 2004). Consequently, there has been some interest in enriching food and beverage products with these bioactive protein-derived components. Again, there are often challenges associated with achieving this goal, such as physical instability in food and beverage products, undesirable sensory attributes (such as bitterness or astringency), and enzymatic degradation and loss of functionality in the gastrointestinal tract. Thus, delivery systems need to be specially designed to encapsulate, protect, and release bioactive proteins so that they can be successfully used in commercial applications.

### 1.3.7 Bioactive Minerals

Ingestion of certain types of minerals is essential for maintaining human health and wellness (White and Broadley 2005). Ingestion of calcium has been related to bone and teeth health, whereas ingestion of iron prevents anemia (Insel et al. 2010). Some individuals may not be consuming sufficient quantities of these minerals to maintain optimum health and wellness, and so there have been efforts to fortify foods with them (Salgueiro et al. 2002). Certain challenges have to be overcome before bioactive minerals can be successfully incorporated into commercial food and beverage products. Some minerals have high water solubility and can easily be dispersed in aqueous solutions (e.g., sodium or potassium chloride), whereas others have poor water solubility and are primarily present in a solid form (e.g., calcium phosphate). Some minerals promote physical instability in food and beverage products because of their ability to screen electrostatic interactions or bind to charged species, thereby promoting aggregation, precipitation, or gelation (McClements 2005). Some minerals have poor sensory attributes, such as astringency, bitterness, metallicity, or chalkiness (Hoehl et al. 2010; Lim and Lawless 2006; Yang and Lawless 2006). Transition metals, such as iron and copper, can reduce product quality by catalyzing undesirable chemical reactions, such as lipid oxidation (Kanner 2011; McClements and Decker 2000). Some minerals have relatively low oral bioavailability and therefore little is absorbed by the human body even if they are successfully added to foods (Davidsson and Haskell 2011; Glahn 2009; Rafferty et al. 2007). Encapsulation technologies can often be used to alleviate these problems, for example, by preventing mineral ions from interacting with other components in foods or in the mouth.

---

**CASE STUDY 1.2   INCORPORATION OF
FLAVOR OILS INTO BEVERAGES**

Flavor oils are usually extracted from natural products, such as the citrus fruits orange, lemon, lime, and grapefruit (Given 2009; McClements 2005). These oils are chemically complex substances that are widely used in beverages because they contain aromatic molecules with desirable flavor characteristics. Because flavor oils are primarily nonpolar, they have to be converted into an emulsified form before they can be incorporated into aqueous-based beverages. Nevertheless, there are a number of challenges that must be overcome to do this successfully. Many flavor oils are susceptible to chemical degradation during storage that leads to flavor loss or off-flavor production, and so environmental conditions may have to be controlled (pH, temperature, light, minerals) or preservatives may have to be added to improve their stability. Emulsified flavor oils may also be prone to physical instability, as a result of creaming, flocculation, coalescence, and Ostwald ripening, and so it may be necessary to carefully design a delivery system to avoid these processes. Finally, the impact of the emulsified flavor oils on the sensory or physicochemical properties of the beverage product, such as mouthfeel, flavor profile, or appearance, is important. For example, if the beverage is meant to be transparent, then it is important that the flavor oil droplets do not scatter light strongly.

## 1.4 Challenges to Incorporating Active Ingredients in Foods

The discussion in the previous section highlighted a number of challenges associated with particular types of active ingredient that need to be addressed when developing a suitable delivery system to encapsulate them. Some of the most common challenges that are generally encountered in the food and other industries are summarized in the following sections.

### 1.4.1 Low Solubility

A major challenge limiting the incorporation of some important active ingredients into food and beverage products is their low water solubility or oil solubility. Ingredients with poor water solubility (such as many bioactive lipids and some bioactive minerals) cannot readily be dispersed into aqueous-based products, such as many fortified waters, soft drinks, nutritional beverages, dressings, sauces, desserts, and yogurts. Conversely, ingredients with poor oil solubility cannot easily be incorporated into oil-based products, such as margarine and butter. Some active ingredients have low solubility in both oil and water phases (e.g., phytosterols), so they may have to be delivered in a crystalline form.

### 1.4.2 Inappropriate Physical State

Most food-grade active ingredients are either liquid or solid at ambient temperature, although some may by gaseous (e.g., carbon dioxide or nitrous oxide). The typical physical state of an active ingredient may not be appropriate for incorporation into a food product. For example, a crystalline version of an ingredient may give an undesirable appearance, texture, mouthfeel, or stability to a product. In this case, a crystalline ingredient may have to be heated above its melting point or dissolved below its saturation level to ensure that it is in a liquid state within a food product (McClements 2012a). Alternatively, a crystalline ingredient may have to be incorporated into a food product in a specific form, for example, as crystals with a particular size or shape.

### 1.4.3 Poor Physicochemical Stability

Some active ingredients are physically or chemically unstable during the manufacturing, transport, storage, or utilization processes of commercial products. A crystalline ingredient may grow, aggregate, or sediment during storage, thereby adversely affecting product quality (McClements 2012a). Similarly, some biopolymer-based ingredients have a tendency to aggregate or sediment during storage, for example, starch granules, protein particles, or polysaccharide aggregates. Some bio-active components are highly susceptible to chemical degradation within food or beverage matrices (Boon et al. 2010; McClements and Decker 2000; Waraho et al. 2011). For example, polyunsaturated lipids (such as ω-3 fatty acids and carotenoids) undergo a complex series of chemical transformations owing to oxidation reactions that lead to undesirable changes in product quality. These transformations are accelerated at elevated temperatures, upon exposure to light, at high oxygen levels, or in the presence of prooxidants, such as transition metals. Other examples are the chemical degradation of many natural colors and flavors during storage (Choi et al. 2009; Qian et al. 2012). Understanding the nature of the chemical degradation reactions involved in a particular food product and the factors that influence them are essential for developing effective strategies for improving their stability.

### 1.4.4 Poor Biochemical Stability

Many active ingredients in food and beverage products are physically or chemically transformed after they are ingested owing to exposure to different environments within the gastrointestinal tract (Golding et al. 2011; Lundin et al. 2008; McClements et al. 2009a,b; Singh and Ye 2013). Natural or artificial structures within foods are usually changed after ingestion because of alterations in temperature, mechanical forces, or solution conditions (such as dilution, pH, and ionic strength). Proteins, starches, and lipids are digested by proteases, amylases, and lipases in various regions of the gastrointestinal tract (e.g., mouth, stomach, and small intestine). Dietary fibers are broken down by the enzymes produced by the microorganisms that colonize the large intestine. In addition, various kinds of metabolic reactions may occur that alter the functional groups on an active ingredient, thereby changing its biological fate and activity.

### 1.4.5 Poor Flavor Profile

Some active ingredients have undesirable flavor profiles, such as bitterness or astringency, which normally limit their incorporation into foods and beverages. For example, some proteins lead to an astringent mouthfeel, whereas some bioactive peptides have a strong bitter taste. In this case, it is important to develop effective strategies to mask the undesirable flavor profile of the active ingredient.

### 1.4.6 Poor Handling Characteristics

Many active ingredients are normally in a form that is not easy to transport, store, or handle. For example, there may be large economic costs associated with transporting an active ingredient that is dissolved in a large quantity of solvent. Thus, it may be better to remove some or all of the solvent from the ingredient to reduce storage space and transport costs. In many situations, it is beneficial to have an active ingredient in a fluid or powdered form that can easily be incorporated into a product during manufacture (rather than having it present as a bulk solid).

## 1.5 Fabrication of Delivery Systems

Many of the challenges associated with the incorporation of active ingredients into food and beverage products can be overcome using well-designed delivery systems. These delivery systems can be fabricated from various food-grade materials (Table 1.3) using a variety of different processing operations (Table 1.4). Indeed, a wide range of food-grade delivery systems with different compositions, structures, and properties have been reported in the literature (Flanagan and Singh 2006; Gibbs et al. 1999; Madene et al. 2006; Matalanis et al. 2011; McClements 2011; McClements et al. 2009a,b; Müller et al. 2000; Sagalowicz and Leser 2010; Sanguansri and Augustin 2006; Velikov and Pelan 2008). In practice, only a limited number of these delivery systems are currently used in the food industry, although some of them are employed in other areas, such as the pharmaceutical, health care, or cosmetic industries. For successful application within commercial food and beverage products, the building blocks used to fabricate a delivery system should be generally recognized as safe, relatively inexpensive, easy to use, and readily available. In addition, the fabrication methods used to create the delivery system should be economical, reliable, reproducible, robust, and suitable for large-scale production.

In general, the methods used to fabricate colloidal delivery systems can be classified according to the physicochemical processes involved: top–down, bottom–up, or combination methods (McClements et al. 2009a,b; Velikov and Pelan 2008). In top–down methods, a bulk material or a suspension containing large particles is broken down into smaller particles (Figure 1.2). Examples of this approach include the following: (i) homogenization of oil and water phases to form tiny oil droplets dispersed within water, (ii) wet or dry grinding of a bulk solid into fine particles, and (iii) injecting or spraying a bulk liquid into another phase so as to form tiny particles (Chapter 4). These methods can be primarily characterized according to

**Table 1.3   Major Structural Components That Can Be Used to Fabricate Food-Grade Delivery Systems for Active Ingredients**

| Name | Important Characteristics | Examples |
|---|---|---|
| Lipids | Polarity<br><br>Rheology<br><br>Chemical stability<br><br>Physical state<br><br>Digestibility | *Animal fats*: beef, pork, chicken<br><br>*Fish oils*: cod liver, menhaden, salmon, tuna<br><br>*Plant oils*: palm, coconut, sunflower, safflower, corn, flax seed, soybean<br><br>*Flavor and essential oils*: lemon, lime, orange, thyme, clove<br><br>*Indigestible oils*: paraffin oils, olestra |
| Surfactants | Head group/tail group<br><br>Polarity and charge<br><br>Molecular geometry<br><br>Solubility<br><br>Surface activity and load<br><br>Adsorption kinetics | *Nonionic*: Tween, Span<br><br>*Anionic*: SLS, DATEM, CITREM<br><br>*Cationic*: lauric arginate<br><br>*Zwitterionic*: lecithin |
| Biopolymers | Molecular weight<br><br>Polarity and charge<br><br>Solubility<br><br>Conformation and flexibility<br><br>Surface activity and load<br><br>Adsorption kinetics | *Globular proteins*: whey, soy, egg<br><br>*Flexible proteins*: casein, gelatin<br><br>*Nonionic polysaccharides*: starch, dextran, agar, galactomannans, cellulose<br><br>*Anionic polysaccharides*: alginate, pectin, xanthan, carrageenan, gellan, gum arabic<br><br>*Cationic polysaccharides*: chitosan |

the processing operation used to carry out the size reduction, for example, homogenization, grinding, injection, or spraying. In bottom–up methods, delivery systems are prepared by assembling molecules or colloidal particles into larger particles (Figure 1.2). The assembly process may be either spontaneous (as in micelle or microemulsion formation) or directed (as in electrostatic deposition of charged polymers onto oppositely charged particles). Many of the methods developed to

**Table 1.4    Major Types of Mechanical Processing Operations That Can Be Used to Fabricate Delivery Systems for Active Components**

| Name | Important Characteristics | Examples |
|---|---|---|
| Shearing | Mechanical disruption of immiscible liquids | Emulsions Hydrogel particles |
| Homogenization | Particle size reduction of liquids by mechanical forces | Emulsions Nanoemulsions |
| Grinding | Particle size reduction of solids by mechanical forces | Powders Suspensions |
| Injection | Droplet formation by injecting liquid through small orifice | Hydrogel particles Oil droplets |
| Spraying | Droplet formation by rapidly forcing liquid through small orifice | Hydrogel particles Oil droplets |
| Dehydration or Desolvation | Removal of water or other solvents from materials | Powders Suspensions |

create delivery systems are actually combinations of the top–down and bottom–up approaches. For example, multilayer emulsions are formed by first blending bulk oil and water phases together in the presence of a charged emulsifier to form an emulsion using a homogenizer (top–down), and then electrostatic deposition is used to coat the resulting oil droplets with a biopolymer coating (bottom–up) (Guzey and McClements 2006). The different types of ingredients, processing operations, and mechanical devices used to prepare food-grade delivery systems are reviewed in later chapters.

In general, different delivery systems assembled from food-grade ingredients can be classified according to the major components used in their fabrication as surfactant-based, emulsion-based, biopolymer-based, or hybrid systems. A brief overview of these different kinds of delivery systems is given below, while a more detailed description is provided in later chapters.

## 1.5.1  Surfactant-Based Systems

The key building blocks for surfactant-based delivery systems are small-molecule surfactants and phospholipids (Chapter 5). These surface-active molecules have a polar head group and a nonpolar tail group. Different surface-active molecules vary according to the nature of their head and tail groups. The head group may vary in its dimensions, polarity, charge (positive, neutral, negative), and chemical

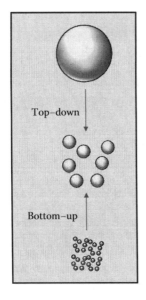 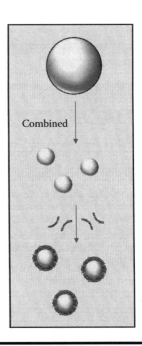

**Figure 1.2 Delivery systems can be fabricated from top–down or bottom–up approaches. Sometimes a combination of both approaches is used. In this case, an emulsion is formed and then coated by multilayers of biopolymer.**

reactivity. The tail group may vary depending on the number, length, unsaturation, and branching of the chains. The functionality of surface-active molecules depends on the nature of the head and tail groups, as well as environmental conditions. Consequently, it is possible to create surfactant-based delivery systems with a wide range of structures and functional properties by using different kinds of surface-active molecules to fabricate them. Some of these systems can be formed by simply mixing surfactant, oil, and water together, whereas others require additional processing, such as mixing, heating, chilling, evaporation, or homogenization. The main types of surfactant-based delivery systems are micelles, microemulsions, and liposomes.

*Micelles*: The particles in micelle solutions are thermodynamically stable structures that are normally formed from surfactants with polar head groups and nonpolar tail groups (Brodskaya 2012; Torchilin 2007). This type of surfactant spontaneously self-assembles into micelles when it is dispersed within water at levels that exceed the critical micelle concentration (CMC) (Leser et al. 2006). A surfactant exists primarily as monomers below the CMC, but as micelles above it. The main driving force for micelle formation is the hydrophobic effect, since this reduces the thermodynamically unfavorable contact area between nonpolar groups and water (Israelachvili 2011). Surfactants that form micelles in aqueous solutions tend to be water dispersible and have high hydrophilic-to-lipophilic balance (HLB) numbers (Israelachvili 2011; Leser et al. 2006). The nonpolar surfactant tails form a hydrophobic core inside the micelle interior, whereas the polar head groups form a hydrophilic shell that is in contact with the surrounding water (Figure 1.1). Micelles may be fabricated from a single surfactant type or from a mixture of surfactants and/or cosurfactants. Lipophilic active components can be incorporated into the hydrophobic core of surfactant micelles, and then the entire system is often referred to as a "swollen micelle" (Huang et al. 2010; Marze 2013). Typically, surfactant micelles have radii within the range of 2 to 10 nm depending on their composition and the environmental conditions (such as pH, ionic composition, and temperature).

*Microemulsions*: In general, microemulsions are thermodynamically stable systems that consist of surfactant, oil, and water molecules organized into particular structures, such as O/W, water-in-oil (W/O), bicontinuous, or liquid crystalline (Israelachvili 2011). O/W microemulsions are the most common form used for encapsulation and delivery of lipophilic active ingredients (Flanagan and Singh 2006; Spernath and Aserin 2006; Spernath et al. 2002). This type of microemulsion has a similar composition, structure, and properties as swollen micelles and so the terms *swollen micelle* and *microemulsion* are often used interchangeably. Microemulsions form spontaneously when certain types and combinations of surfactant, oil, and water are mixed together under appropriate environmental conditions (Israelachvili 2011). Sometimes, additional ingredients are needed to facilitate their formation, improve their stability, and tune their functional properties, such as cosurfactants and cosolvents (Flanagan and Singh 2006; Spernath and Aserin 2006; Spernath et al. 2002). Microemulsions have a similar structure to swollen micelles, consisting of a hydrophobic core composed of nonpolar surfactant tails and a hydrophilic shell composed of polar surfactant head groups (Figure 1.1). Lipophilic active ingredients are usually solubilized within the hydrophobic core; however, if they have any polar groups, these may protrude into the hydrophilic shell and water phase. The physical location of active components within a micelle or microemulsion may have an important impact on its chemical stability because this influences its accessibility to chemical reactions with components in the aqueous phase of the system. Typically, microemulsions have radii in the range of 5 to 100 nm depending on their composition and environmental conditions.

*Liposomes*: Liposomes consist of concentric bilayers, with each bilayer containing two layers of surfactant molecules with their nonpolar tail groups facing toward each other (Figure 1.1) (Sharma and Sharma 1997). A unilamellar liposome has a balloon-like structure consisting of a single bilayer, whereas a multilamellar liposome has an onion-like structure consisting of multiple concentric bilayer shells (Reineccius 1995; Taylor et al. 2005). The building blocks of liposomes are surface-active substances with intermediate HLB numbers and optimum curvatures close to zero, such as phospholipids (Israelachvili 2011). This kind of surface-active substance tends to spontaneously form bilayers when dispersed in aqueous solutions. However, appropriate preparation procedures are usually needed to ensure that the bilayer structures form liposomes of the appropriate characteristics, for example, dehydration/rehydration or homogenization (Maherani et al. 2011). Liposomes typically have radii in the approximately 10 to 10,000 nm range. Liposomes have a number of different hydrophilic and hydrophobic regions within their structures, and therefore they can be used to encapsulate different kinds of active ingredients in a single delivery system (Hashida et al. 2005; Maherani et al. 2011). For example, hydrophilic ingredients can be trapped in the internal aqueous phase, whereas hydrophobic ingredients can be trapped between the surfactant tails in the bilayers.

## 1.5.2 Emulsion-Based Systems

The key building blocks for emulsion-based delivery systems are oil droplets, which may vary in their size, composition, physical state, interfacial characteristics, and structural organization (McClements 2010b, 2012a; McClements and Li 2010). These systems are usually formed by mixing or blending emulsifier, oil, and water components together. The principal emulsion-based delivery systems are emulsions, nanoemulsions, and solid lipid nanoparticles (SLNs), but these systems can be used as building blocks to construct more complex structures, such as multilayer emulsions, colloidosomes, microclusters, and filled hydrogel particles (Figure 1.1).

*Emulsions and Nanoemulsions*: The particles in emulsions and nanoemulsions consist of a hydrophobic oil core and a hydrophilic surfactant shell (McClements 2011; McClements and Rao 2011) (Figure 1.1). Structurally, the main distinction between nanoemulsions ($r < 100$ nm) and emulsions ($r > 100$ nm) is the dimensions of the droplets. Unlike microemulsions, emulsions and nanoemulsions are thermodynamically *unstable* systems that break down through a variety of instability mechanisms during storage, such as flocculation, coalescence, partial coalescence, gravitational separation, Ostwald ripening, and phase separation (McClements 2005). The long-term kinetic stability of these systems can be improved by incorporating *stabilizers* such as emulsifiers, weighting agents, ripening inhibitors, and texture modifiers (McClements 2005, 2011). Emulsifiers are surface-active substances that adsorb to oil droplet surfaces during homogenization and provide a coating that facilitates droplet formation and stability. Weighting agents are dense nonpolar substances that are incorporated into the oil phase to increase its density so that it is closer to that of the surrounding aqueous phase, thereby reducing the driving force for gravitational separation. Ripening inhibitors are hydrophobic substances with very poor water solubility that are mixed with the oil phase to prevent Ostwald ripening. Texture modifiers, such as thickening and gelling agents, are usually proteins or polysaccharides that are added to the aqueous phase to increase the viscosity or form a gel that inhibits droplet movement.

Emulsions and nanoemulsions can be formed using various methods that can be broadly classified as either high-energy or low-energy methods (McClements

2011; McClements and Rao 2011). High-energy methods use mechanical devices ("homogenizers") capable of generating intense disruptive forces to break up and mingle the oil and aqueous phases, such as high-pressure valve homogenizers, microfluidizers, and sonicators (Koroleva and Yurtov 2012; McClements 2011; Silva et al. 2012). Low-energy methods rely on the spontaneous formation of oil droplets when certain combinations of surfactant, oil, and water are mixed together under controlled conditions, such as phase inversion temperature, spontaneous emulsification, and emulsion inversion point methods (Solans and Sole 2012). The food-grade emulsifiers that can be used to form emulsions and nanoemulsions include small-molecule surfactants, phospholipids, proteins, and polysaccharides.

*Multiple Emulsions*: Multiple emulsions are multiphase materials that have a structural organization that may have advantages for certain applications within the food industry (Figure 1.1). Water-in-oil-in-water (W/O/W) emulsions consist of water droplets trapped within oil droplets that are dispersed within an aqueous continuous phase (Garti 1997a,b; Garti and Benichou 2004; Garti and Bisperink 1998). For the sake of clarity, W/O/W emulsions should actually be designated as $W_1/O/W_2$ emulsions, with $W_1$ representing the inner water phase and $W_2$ representing the outer water phase, since these two regions may have different compositions and properties. There are two different interfaces in this type of emulsion: the $W_1$–O layer surrounding the inner water droplets and the O–$W_2$ layer surrounding the oil droplets. Consequently, it is usually necessary to use two different types of emulsifier to stabilize W/O/W emulsions: an oil-soluble emulsifier for the inner water droplets and a water-soluble emulsifier for the oil droplets. Multiple emulsions are prone to many of the same instability mechanisms as emulsions, such as creaming, flocculation, coalescence, and Ostwald ripening. However, there are also additional instability mechanisms associated with the inner water droplets, such as coalescence, expulsion, and shrinkage/growth (Benichou et al. 2004; Garti 1997a,b; Garti and Benichou 2004; Garti and Bisperink 1998). Multiple emulsions may have a number of potential advantages over conventional emulsions as delivery systems. Hydrophilic components (e.g., minerals, vitamins, enzymes, proteins, bioactive peptides, and fibers) can be trapped within the internal water phase, whereas hydrophobic components (e.g., bioactive lipids, antioxidants, and oil-soluble colors) can be trapped within the oil phase. As a result, it may be possible to deliver multiple active ingredients using a single delivery system. In addition, it may be possible to trap a hydrophilic ingredient inside the internal water phase, thereby preventing it from reacting with other hydrophilic ingredients in the external water phase or with biological surfaces (such as the tongue).

*Solid Lipid Nanoparticles*: SLNs and nanostructured lipid carriers (NLCs) are similar in structure to emulsions (Figure 1.1), but the lipid core is either fully or partially crystalline (Jores et al. 2004; Muller and Keck 2004; Weiss et al. 2008). SLNs and NLCs are typically formed from high-melting-point lipids using a two-step process. First, an emulsion or nanoemulsion is prepared at a temperature above the melting point of the lipid phase. Second, this system is cooled to a temperature below the melting point to promote crystallization of the lipid phase. SLNs and NLCs can also be formed using low-energy methods by spontaneously forming an emulsion or nanoemulsion at high temperatures and then cooling it to induce lipid phase crystallization. The morphology of the particles in SLNs and NLCs may change appreciably after the liquid-to-solid transition of the lipid phase has occurred owing to the physical constraints associated with packing lipid molecules within crystals (Bunjes 2011; Jores et al. 2004). The particles in SLN and NLC suspensions typically have radii in the 10 to 100 nm range if they are prepared from nanoemulsions and 100 nm to 100 μm if they are prepared from emulsions. For the sake of convenience, the term *SLN* will be used to refer to both fully and partially crystalline lipid nanoparticles and microparticles in this book.

*Multilayer Emulsions*: Multilayer O/W emulsions consist of oil droplets dispersed in an aqueous medium, with each oil droplet being surrounded by a nanolaminated shell that typically consists of an emulsifier layer plus one or more biopolymer layers (Figure 1.1). The possibility of precisely engineering the properties of the nanolaminated shells that coat the droplets provides great scope for improving their stability and functional performance (Guzey and McClements 2006; McClements 2010a). For example, thick and highly charged interfacial layers can be designed to increase the repulsive colloidal interactions between lipid droplets, thereby improving their stability to aggregation. In addition, the ability to control the properties of multilayer interfacial coatings enables one to engineer novel functional performance into emulsions, such as protection of encapsulated chemically labile components, or controlled release of encapsulated functional compounds. Multilayer emulsions are normally produced using a multistep process (Guzey and McClements 2006; McClements 2010a). Initially, an O/W emulsion containing electrically charged droplets is prepared. An oppositely charged biopolymer is then added to the system so that it adsorbs to the droplet surfaces and forms a biopolymer coating. This procedure can be repeated a number of times to form oil droplets coated by nanolaminated shells consisting of numerous biopolymer layers. Typically, the droplets in multilayer emulsions have radii in the range of 100 nm to 100 μm.

*Heteroaggregated Emulsions—Colloidosomes and Microclusters*: Colloidosomes and microclusters are formed by mixing two (or more) colloidal dispersions together that contain different kinds of particles that are attracted to each other (Mao and McClements 2011, 2012a,b). Colloidosomes consist of a central particle surrounded by a coating of different particles, whereas microclusters consist of a more disorganized arrangement of different particles (Figure 1.1). In principle, both of these systems can be formed from any kind of food-grade colloidal particle (such as droplets, bubbles, crystals, granules, or cells) and any kind of attractive interaction (such as electrostatic, hydrophobic, depletion, or hydrogen bonding). Recently, it has been shown that they can be formed by mixing together emulsions containing oppositely charged fat droplets (Mao and McClements 2011, 2012a). The structure of the heteroaggregates (colloidosomes or microclusters) formed depends on particle size, concentration and charge, solution pH and ionic strength, and mixing conditions. Heteroaggregated emulsions may be useful for encapsulating two different kinds of lipophilic components in a single delivery system.

## 1.5.3 Biopolymer-Based Systems

The key building blocks for biopolymer-based delivery systems are food biopolymers, such as proteins and polysaccharides (Burey et al. 2008; Jones and McClements 2010; Matalanis et al. 2011). These systems can be fabricated using a variety of different preparation methods depending on the biopolymers involved, and the desired functional performance. The most common biopolymer-based delivery systems are molecular complexes and hydrogel particles, but these systems can be used as building blocks for more complex structures, such as filled hydrogel particles.

*Molecular Complexes*: Active ingredients can be encapsulated within various kinds of molecular complexes fabricated from either individual or mixed biopolymers (Burey et al. 2008; Jones and McClements 2010; Matalanis et al. 2011). For example, casein molecules self-assemble in aqueous solutions to form micelle-like structures that can solubilize lipophilic molecules within their interior (Haham et al. 2012). Many globular proteins have nonpolar pockets on their surfaces that are capable of incorporating lipophilic molecules (Tavel et al. 2010). Protein–polysaccharide complexes that are capable of trapping active components within their structure can be formed (Schmitt and Turgeon 2011).

*Hydrogel Particles*: Hydrogel particles consist of nanometer- or micrometer-sized biopolymer beads dispersed within an aqueous medium. The hydrogel particles usually consist of physically, chemically, or enzymatically cross-linked biopolymer molecules that trap water molecules within the polymer network. These particles can be formed from single or mixed biopolymers using a variety of methods, including injection, molding, templating, and complexation methods (Burey et al. 2008; Matalanis et al. 2011). Both hydrophilic and lipophilic active ingredients can be trapped within hydrogel particles. The size and porosity of the hydrogel particles, as well as their response to changes in environmental conditions, can be controlled by selection of appropriate ingredients and preparation methods. Thus, it is possible to create delivery systems with different functional attributes.

*Filled-Hydrogel Particles*: Filled hydrogel particles consist of fat droplets trapped within biopolymer beads that are dispersed within an aqueous medium (Burey et al. 2008; Matalanis et al. 2011). These systems are usually formed from an emulsion with a biopolymer solution using methods similar to those used to fabricate hydrogel particles. Lipophilic active ingredients can be trapped within the fat droplets, whereas hydrophilic active ingredients can be trapped within the hydrogel matrix.

## 1.5.4 Hybrid Systems

The fundamental building blocks used in surfactant-, emulsion-, and biopolymer-based delivery systems can themselves be used to create more complicated structures that may have novel or improved functional performance. In general, the different approaches available for creating more complex structured delivery systems can be categorized into three main groups: coating, embedding, and clustering (Figure 1.3). For example, electrostatic deposition can be used to form biopolymer coatings around charged particles in liposomes, emulsions, SLNs, multiple emulsions, or hydrogel particles. The embedding approach can be used to trap microemulsions, emulsions, nanoemulsions, SLNs, or multiple emulsions within hydrogel particles. The clustering approach can be used to form colloidosomes or microclusters from nanoemulsions, emulsions, multiple emulsions, SLNs, or other types of particles. There is, therefore, great potential for the formation of a variety of structured delivery systems with different functional properties. Nevertheless, it is important to be aware that any structured delivery system formed should be economically viable and sufficiently robust to resist the harsh conditions food and beverage products typically experience during processing, storage, transport, and utilization.

## 1.5.5 Nature-Inspired Systems

Rather than constructing delivery systems from food-grade ingredients, it is possible to find structures in nature that can be used to encapsulate, protect, and release active ingredients.

Oil bodies are storage organelles naturally found in many plant seeds, such as those from soybeans or sunflowers (Murphy et al. 2000, 2001). They consist of a

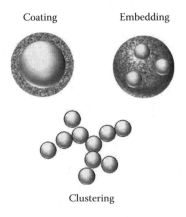

**Figure 1.3  General approaches for creating structure delivery systems: coating, embedding, and clustering.**

triacylglycerol oil core surrounded by a phospholipid-protein shell. The proteins are highly lipophilic molecules called oleosins that are embedded within the phospholipid layer of the oil bodies and are believed to naturally protect the oil bodies against physical and chemical degradation. A number of studies have shown that oil bodies may be suitable for encapsulating and delivering active ingredients, such as flavors, colors, and nutraceuticals (Bettini et al. 2013; Fisk et al. 2011).

Delivery systems based on spores isolated from certain types of plants, mosses, fungi, and algae have been developed (Barrier et al. 2010; Diego-Taboada et al. 2013; Hamad et al. 2011; Paunov et al. 2007). The spores are extracted from their natural source and are then processed to remove their interior (which is largely protein and polysaccharide), leaving the outer coating (which is largely lipid) as an encapsulating shell. The interior can be removed by treatment with suitable acids, bases, and enzymes. The spore coating is highly resistant to acid and alkali conditions and is therefore not readily degraded in the stomach or small intestine. Spores from a specific natural source are relatively monodisperse, but spores with a variety of sizes can be obtained by using different natural sources (e.g., *d* from 4 to 40 μm). The most commonly used spores appear to be those from sporopollenin, which is derived from a moss (*Lycopodium clavatum*). Hydrophilic active compounds (such as proteins, peptides, enzymes, vitamins, and probiotics) can be loaded into the shells by dissolving them in aqueous solutions and then mixing them with the spore shells. Hydrophobic active compounds (such as ω-3 fatty acids and other oils) can be dissolved in alcohol or alcohol–water solutions, which are mixed with the spore shells. Solvent carriers can then be removed by evaporation if required. Delivery systems based on these spores are commercially available and have been used not only to encapsulate various bioactive lipids, vitamins, and flavors, but also for taste masking (www.sporomex.co.uk). A fairly similar approach has been used to develop delivery systems from yeast cells (Soto and Ostroff 2010, 2012). There may also be many other types of structures present within plants and animals that could be isolated and used as delivery systems in the food industry.

# 1.6 Desirable Characteristics of Delivery Systems

There are a number of characteristics that any delivery system must have if it is going to be suitable for utilization within the food industry. A number of the most important of these characteristics are summarized in the following sections.

## 1.6.1 Food-Grade

The delivery system must be fabricated entirely from food ingredients and processing operations that have regulatory approval in the country where the food will be sold. Some common food-grade ingredients and processing operations that can be used to assemble delivery systems are listed in Tables 1.3 and 1.4.

## 1.6.2 Economic Production

The delivery system should be capable of being economically manufactured from cost-effective ingredients. The benefits gained from encapsulating an active ingredient within a delivery system (e.g., improved handling, shelf life, marketability, functionality, or bioavailability) should outweigh the additional costs associated with encapsulation.

## 1.6.3 Food Matrix Compatibility

The delivery system should be compatible with the food or beverage that it will be incorporated into; that is, it should not adversely affect the appearance, texture, flavor, or shelf life of the product. For example, if an active ingredient is going to be incorporated into a clear beverage, then it should be incorporated into a delivery system that does not increase the turbidity.

## 1.6.4 Protection against Chemical Degradation

A delivery system may have to be designed to protect an active ingredient against some form of chemical or biochemical degradation during storage after ingestion, for example, oxidation, hydrolysis, metabolism, and so on. Knowledge of the degradation mechanism and the major factors that affect it (e.g., oxygen, light, pH, heat, enzyme activity, etc.) will facilitate the design of a more effective delivery system.

## 1.6.5 Loading Capacity, Encapsulation Efficiency, and Retention

Ideally, a delivery system should be capable of encapsulating a relatively large amount of active ingredient per unit mass of carrier material and should efficiently retain the encapsulated ingredient until it needs to be delivered at the site of action. In addition, it is often important that a high percentage of the added active ingredient is actually encapsulated within the carrier material (rather than outside of it).

## 1.6.6 Delivery Mechanism

The delivery system may have to be designed so that it releases the active ingredient at a particular site of action, at a controlled rate, or in response to a specific environmental trigger (e.g., pH, ionic strength, temperature, or enzyme activity). This trigger could occur during food storage (e.g., release of an antimicrobial or antioxidant) or it could occur within the human body (e.g., release in the mouth, stomach, small intestine, or colon).

### 1.6.7 Bioavailability/Bioactivity

A delivery system may be designed to enhance (or at least not adversely affect) the bioavailability/bioactivity of an encapsulated ingredient.

## 1.7 Release Mechanisms

In many applications within the food and beverage industry, it is important to control the release of an active ingredient to achieve a specific effect, for example, a desirable flavor profile. An active ingredient may be released from a colloidal particle by a variety of different mechanisms (Liechty et al. 2010; Peppas et al. 2000) (Figure 1.4). The physicochemical basis and mathematical description of different release mechanisms are described in some detail in Chapter 10, and only a brief summary is given here.

### 1.7.1 Diffusion

In this case, the active ingredient (solute) is released from the particle by molecular diffusion through the particle matrix (Baker 1987). The particle matrix may remain largely unchanged throughout the diffusion process, or it may change because of

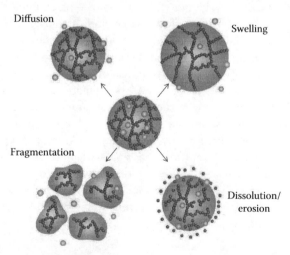

**Figure 1.4 Some common physicochemical mechanisms responsible for release of active ingredients: diffusion, swelling, dissolution/erosion, and fragmentation.**

swelling, shrinking, erosion, or fragmentation. The rate at which the active ingredient is released depends on many factors including solute properties (such as molecular weight and polarity), matrix properties (such as polarity, rheology, physical state, and interactions), particle characteristics (such as size, shape, and structure), solute concentration gradient between the particle and surrounding medium, and fluid flow conditions.

## 1.7.2 Dissolution

In this case, the active ingredient is released from the particle as a result of a dissolution process when the delivery system encounters specific solution or environmental conditions (Baker 1987). The active ingredient may make up the entirety of the particle and thereby be released as the particle dissolves. Alternatively, the active ingredient may be dispersed within a particle matrix and then be released when the particle matrix dissolves. The release rate is governed by the dissolution rate, which depends on the composition and structure of particle, as well as the magnitude and duration of the environmental factor responsible for dissolution (e.g., dilution, solvent type, pH, ionic strength, or temperature). Dissolution is usually assumed to start from the particle surface and work inward.

## 1.7.3 Erosion

In this case, the active ingredient is released as a result of particle erosion when the delivery system encounters specific conditions (Baker 1987). Erosion occurs owing to chemical degradation of the molecules within the particle matrix, which may occur throughout the entire particle (bulk erosion) or only at the particle exterior (surface erosion). The active ingredient is initially distributed within the particle matrix and is released when the particle erodes. Erosion may occur through a variety of physicochemical mechanisms, for example, physical (e.g., high temperatures), chemical (e.g., strong acids or bases), or enzymatic (e.g., lipases, proteases, or amylases). The release profile is governed by the rate at which erosion occurs, which depends on particle composition and structure, as well as the magnitude and duration of the factor responsible for erosion (e.g., temperature, pH, or enzyme activity). In addition, the release profile depends on whether erosion is a surface or bulk mechanism. Finally, the release profile also depends on the relative rates of erosion and diffusion. If erosion is much slower than diffusion, then release will be erosion limited. On the other hand, if diffusion is much slower than erosion, then release will be diffusion limited.

## 1.7.4 Fragmentation

In this case, the active ingredient is released from the particles when they are fragmented owing to physical, chemical, or enzymatic disruption. The rate of release depends on the fracture properties of the particle, such as the applied stress when fracture occurs, as well as the size and shape of the fragments formed. The active ingredient may still ultimately be released from the fragments by diffusion, dissolution, or erosion processes, but it will be released faster because of the increased surface area and reduced particle size.

## 1.7.5 Swelling

In this case, the active ingredient is released from the particles when they absorb solvent molecules and swell. For example, an active ingredient could initially be trapped within a hydrogel particle with a pore size sufficiently small to prevent its movement. Once the delivery system encounters particular solution or environmental conditions, the particles absorb solvent molecules from the surrounding medium and swell, thereby increasing the internal pore size. Release will occur when the pore size increases to a value similar to the molecular size of the active ingredient. In this case, the rate of release of the active ingredient will depend on the swelling rate and the time taken for the active ingredients to diffuse through the swollen particle matrix.

## 1.7.6 Designing Release Profiles

The design of an effective delivery system for an active ingredient often depends on establishing the desired release profile for the particular application involved, for example, burst, sustained, delayed, triggered, or targeted (Figure 1.5). An appropriate release mechanism can then be selected to obtain this release profile. For example, if one were designing a delivery system to rapidly release an active ingredient within the stomach, then one may select a carrier particle that quickly dissolved or eroded when it encountered highly acidic gastric conditions. Conversely, if one were designing a delivery system to release an active ingredient within the colon, then one would select a carrier particle that remained intact in the food product, mouth, stomach, and small intestine, but then dissolved or eroded within the large intestine (perhaps due to the activity of colonic bacteria).

A release profile is usually characterized in terms of the increase in concentration of the active ingredient at the site of action as a function of time (Figure 1.6). A number of parameters can be derived from such curves, such as the total area under the curve (AUC), the maximal concentration released ($C_{max}$), and the time to reach the maximum concentration ($t_{max}$) (Aguilera 2006). The release rate of encapsulated components from within delivery systems depends on many factors, including their equilibrium partition coefficients, their original location, the mass transfer coefficients of the components in the different phases, mechanical agitation, and the microstructure of the system, for example, particle size, particle degradation, and layer thickness. Mathematical models have been developed to predict the release of active ingredients from carrier particles owing to a variety of

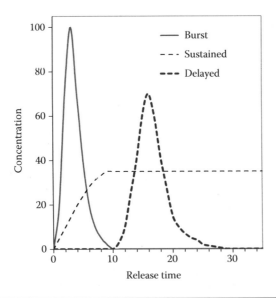

**Figure 1.5** **Schematic representation of different types of release profile that may be obtained for an active ingredient from a delivery system.**

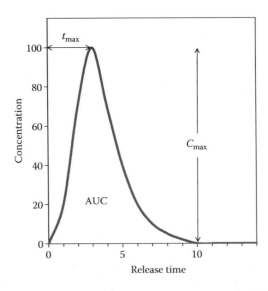

**Figure 1.6** **A typical release profile of an active ingredient from a delivery system can be characterized by the area under the curve (AUC), time to reach the maximum concentration ($t_{max}$), and the maximum concentration ($C_{max}$).**

different release mechanisms, for example, diffusion, dissolution, and erosion. A number of these models are discussed in more detail in Chapter 10. To model the release kinetics using mathematical models, it is necessary to have knowledge of various physicochemical properties of the system, for example, particle size, active ingredient concentrations in the particle and surrounding medium, and diffusion coefficients through the particle and surrounding medium. These empirical data can sometimes be found in the literature, or they may need to be measured for the specific system being studied. The rational development of colloidal delivery systems often depends on establishing the origin of the release mechanism in particular systems, quantifying the factors that influence the release rate, and utilizing appropriate mathematical relationships to describe these processes.

## 1.8 Summary

In this chapter, we have briefly discussed the need for delivery systems within the food industry to encapsulate, protect, and release active ingredients. Potential challenges to incorporating active ingredients into commercial food and beverage products were identified, and some of the delivery systems that may be suitable for encapsulating them were briefly highlighted. In the following chapters, we will examine in more detail how delivery systems can be designed, prepared, and characterized.

## References

Aguilera, J. M. (2006). "Food microstructure affects the bioavailability several nutrients." *Journal of Food Science* 72(2): R21–R32.

Aguzzi, C., P. Cerezo, I. Salcedo, R. Sanchez and C. Viseras (2010). "Mathematical models describing drug release from biopolymeric delivery systems." *Materials Technology* 25(3–4): 205–211.

Akoh, C. C. and D. B. Min (2008). *Food Lipids: Chemistry, Nutrition, and Biotechnology*. Boca Raton, FL, CRC Press.

Baker, R. W. (1987). *Controlled Release of Biologically Active Agents*. New York, John Wiley & Sons.

Barrier, S., A. S. Rigby, A. Diego-Taboada, M. J. Thomasson, G. Mackenzie and S. L. Atkin (2010). "Sporopollenin exines: A novel natural taste masking material." *Lwt-Food Science and Technology* 43(1): 73–76.

Benichou, A., A. Aserin and N. Garti (2004). "Double emulsions stabilized with hybrids of natural polymers for entrapment and slow release of active matters." *Advances in Colloid and Interface Science* 108–109: 29–41.

Bettini, S., D. Vergara, S. Bonsegna, L. Giotta, C. Toto, M. Chieppa, M. Maffia, G. Giovinazzo, L. Valli and A. Santino (2013). "Efficient stabilization of natural curcuminoids mediated by oil body encapsulation." *RSC Advances* 3(16): 5422–5429.

Biliaderis, C. G. and M. S. Izydorczyk (2007). *Functional Food Carbohydrates*. Boca Raton, FL, CRC Press.

Boon, C. S., D. J. McClements, J. Weiss and E. A. Decker (2010). "Factors influencing the chemical stability of carotenoids in foods." *Critical Reviews in Food Science and Nutrition* **50**(6): 515–532.

Brodskaya, E. N. (2012). "Computer simulations of micellar systems." *Colloid Journal* **74**(2): 154–171.

Brul, S. and P. Coote (1999). "Preservative agents in foods—Mode of action and microbial resistance mechanisms." *International Journal of Food Microbiology* **50**(1–2): 1–17.

Bunjes, H. (2011). "Structural properties of solid lipid based colloidal drug delivery systems." *Current Opinion in Colloid and Interface Science* **16**(5): 405–411.

Burey, P., B. R. Bhandari, T. Howes and M. J. Gidley (2008). "Hydrocolloid gel particles: Formation, characterization, and application." *Critical Reviews in Food Science and Nutrition* **48**(5): 361–377.

Choi, S. J., E. A. Decker, L. Henson, L. M. Popplewell and D. J. McClements (2009). "Stability of citral in oil-in-water emulsions prepared with medium-chain triacylglycerols and triacetin." *Journal of Agricultural and Food Chemistry* **57**(23): 11349–11353.

Cui, S. W. (2005). *Food Carbohydrates: Chemistry, Physical Properties and Applications*. Boca Raton, FL, Taylor & Francis.

Damodaran, S., K. L. Parkin and O. R. Fennema (2008). *Food Chemistry*, 4th Edition. Boca Raton, FL, CRC Press.

Davidson, P. M., J. N. Sofos and A. L. Branen (2005). *Antimicrobials in Food*. Boca Raton, FL, CRC Press.

Davidsson, L. and M. Haskell (2011). "Bioavailability of micronutrients: Stable isotope techniques to develop effective food-based strategies to combat micronutrient deficiencies." *Food and Nutrition Bulletin* **32**(1): S24–S30.

Diego-Taboada, A., L. Maillet, J. H. Banoub, M. Lorch, A. S. Rigby, A. N. Boa, S. L. Atkin and G. Mackenzie (2013). "Protein free microcapsules obtained from plant spores as a model for drug delivery: Ibuprofen encapsulation, release and taste masking." *Journal of Materials Chemistry B* **1**(5): 707–713.

Dikeman, C. L. and G. C. Fahey (2006). "Viscosity as related to dietary fiber: A review." *Critical Reviews in Food Science and Nutrition* **46**(8): 649–663.

Espitia, P. J. P., N. D. F. Soares, J. S. D. Coimbra, N. J. de Andrade, R. S. Cruz and E. A. A. Medeiros (2012). "Bioactive peptides: Synthesis, properties, and applications in the packaging and preservation of food." *Comprehensive Reviews in Food Science and Food Safety* **11**(2): 187–204.

Fisk, I. D., R. S. T. Linforth, A. J. Taylor and D. A. Gray (2011). "Aroma encapsulation and aroma delivery by oil body suspensions derived from sunflower seeds (Helianthus annus)." *European Food Research and Technology* **232**(5): 905–910.

Flanagan, J. and H. Singh (2006). "Microemulsions: A potential delivery system for bioactives in food." *Critical Reviews in Food Science and Nutrition* **46**(3): 221–237.

Garti, N. (1997a). "Double emulsions—Scope, limitations and new achievements." *Colloids and Surfaces A-Physicochemical and Engineering Aspects* **123**: 233–246.

Garti, N. (1997b). "Progress in stabilization and transport phenomena of double emulsions in food applications." *Food Science and Technology* **30**(3): 222–235.

Garti, N. and A. Benichou (2004). Recent developments in double emulsions for food applications. In *Food Emulsions*, S. E. Friberg, K. Larsson and J. Sjoblom (eds.). New York, Marcel Dekker.

Garti, N. and C. Bisperink (1998). "Double emulsions: Progress and applications." *Current Opinion in Colloid and Interface Science* **3**(6): 657–667.

Gibbs, B. F., S. Kermasha, I. Alli and C. N. Mulligan (1999). "Encapsulation in the food industry: A review." *International Journal of Food Sciences and Nutrition* **50**(3): 213–224.

Given, P. S. (2009). "Encapsulation of flavors in emulsions for beverages." *Current Opinion in Colloid and Interface Science* **14**(1): 43–47.

Glahn, R. (2009). The use of Caco-2 cells in defining nutrient bioavailability: Application to iron bioavailability of foods. In *Designing Functional Foods: Measuring and Controlling Food Structure Breakdown and Nutrient Absorption*, D. J. McClements and E. A. Decker (eds.) Woodhead Publishing, Cambridge, UK: 340–361.

Golding, M., T. J. Wooster, L. Day, M. Xu, L. Lundin, J. Keogh and P. Clifton (2011). "Impact of gastric structuring on the lipolysis of emulsified lipids." *Soft Matter* **7**(7): 3513–3523.

Guzey, D. and D. J. McClements (2006). "Formation, stability and properties of multi-layer emulsions for application in the food industry." *Advances in Colloid and Interface Science* **128**: 227–248.

Haham, M., S. Ish-Shalom, M. Nodelman, I. Duek, E. Segal, M. Kustanovich and Y. D. Livney (2012). "Stability and bioavailability of vitamin D nanoencapsulated in casein micelles." *Food and Function* **3**(7): 737–744.

Hamad, S. A., A. F. K. Dyab, S. D. Stoyanov and V. N. Paunov (2011). "Encapsulation of living cells into sporopollenin microcapsules." *Journal of Materials Chemistry* **21**(44): 18018–18023.

Hashida, M., S. Kawakami and F. Yamashita (2005). "Lipid carrier systems for targeted drug and gene delivery." *Chemical and Pharmaceutical Bulletin* **53**(8): 871–880.

Hoehl, K., G. U. Schoenberger and M. Busch-Stockfisch (2010). "Water quality and taste sensitivity for basic tastes and metallic sensation." *Food Quality and Preference* **21**(2): 243–249.

Hu, F. B. and W. C. Willett (2002). "Optimal diets for prevention of coronary heart disease." *JAMA-Journal of the American Medical Association* **288**(20): 2569–2578.

Huang, Q. R., H. L. Yu and Q. M. Ru (2010). "Bioavailability and delivery of nutraceuticals using nanotechnology." *Journal of Food Science* **75**(1): R50–R57.

Hutchings, J. B. (1999). *Food Color and Appearance*. New York, Springer Publishers.

Igoe, R. S. (2011). *Dictionary of Food Ingredients*. New York, Springer Scientific.

Insel, P., R. E. Turner and D. Ross (2010). *Discovering Nutrition*. Sudbury, MA, Jones and Bartlett Publishers.

Israelachvili, J. (2011). *Intermolecular and Surface Forces*, 3rd Edition. London, Academic Press.

Jakobsson, M. and B. Sivik (1994). "Oxidative stability of fish-oil included in a microemulsion." *Journal of Dispersion Science and Technology* **15**(5): 611–619.

Jelen, H. (2011). *Food Flavors: Chemical, Sensory and Technological Properties*. Boca Raton, FL, CRC Press.

Jones, O. G. and D. J. McClements (2010). "Functional biopolymer particles: Design, fabrication, and applications." *Comprehensive Reviews in Food Science and Food Safety* **9**(4): 374–397.

Jores, K., W. Mehnert, M. Drechsler, H. Bunjes, C. Johann and K. Mader (2004). "Investigations on the structure of solid lipid nanoparticles (SLN) and oil-loaded solid lipid nanoparticles by photon correlation spectroscopy, field-flow fractionation and transmission electron microscopy." *Journal of Controlled Release* **95**(2): 217–227.

Kanner, J. (2011). Metals and food oxidation. In *Oxidation in Foods and Beverages and Antioxidant Applications, Vol 1: Understanding Mechanisms of Oxidation and Antioxidant Activity*, E. A. Decker, R. J. Elias and D. J. McClements (eds.) Woodhead Publishing, Cambridge, UK: 36–56.

Koroleva, M. Y. and E. V. Yurtov (2012). "Nanoemulsions: The properties, methods of preparation and promising applications." *Russian Chemical Reviews* **81**(1): 21–43.

Laviano, A., S. Rianda, A. Molfino and F. R. Fanelli (2013). "Omega-3 fatty acids in cancer." *Current Opinion in Clinical Nutrition and Metabolic Care* **16**(2): 156–161.

Leser, M. E., L. Sagalowicz, M. Michel and H. J. Watzke (2006). "Self-assembly of polar food lipids." *Advances in Colloid and Interface Science* **123**: 125–136.

Lesmes, U., S. Sandra, E. A. Decker and D. J. McClements (2010). "Impact of surface deposition of lactoferrin on physical and chemical stability of omega-3 rich lipid droplets stabilised by caseinate." *Food Chemistry* **123**(1): 99–106.

Liechty, W. B., D. R. Kryscio, B. V. Slaughter and N. A. Peppas (2010). Polymers for drug delivery systems. In *Annual Review of Chemical and Biomolecular Engineering*, J. M. Prausnitz, M. F. Doherty and M. A. Segalman (eds.). Annual Reviews, Palo Alto, CA. **1**: 149–173.

Lim, J. and H. T. Lawless (2006). "Detection thresholds and taste qualities of iron salts." *Food Quality and Preference* **17**(6): 513–521.

Lindsay, R. C. (2008). Flavors. In *Food Chemistry*, 4th Edition, S. Damodaran, K. L. Parkin and O. R. Fennema (eds.). Boca Raton, FL, CRC Press: 639–688.

Lomova, M. V., G. B. Sukhorukov and M. N. Antipina (2010). "Antioxidant coating of micronsize droplets for prevention of lipid peroxidation in oil-in-water emulsion." *ACS Applied Materials and Interfaces* **2**(12): 3669–3676.

Luchtman, D. W. and C. Song (2013). "Cognitive enhancement by omega-3 fatty acids from child-hood to old age: Findings from animal and clinical studies." *Neuropharmacology* **64**: 550–565.

Lundin, L., M. Golding and T. J. Wooster (2008). "Understanding food structure and function in developing food for appetite control." *Nutrition and Dietetics* **65**: S79–S85.

Madene, A., M. Jacquot, J. Scher and S. Desobry (2006). "Flavour encapsulation and controlled release—A review." *International Journal of Food Science and Technology* **41**(1): 1–21.

Maherani, B., E. Arab-Tehrany, M. R. Mozafari, C. Gaiani and M. Linder (2011). "Liposomes: A review of manufacturing techniques and targeting strategies." *Current Nanoscience* **7**(3): 436–452.

Mao, Y. Y. and D. J. McClements (2011). "Modulation of bulk physicochemical properties of emulsions by hetero-aggregation of oppositely charged protein-coated lipid droplets." *Food Hydrocolloids* **25**(5): 1201–1209.

Mao, Y. Y. and D. J. McClements (2012a). "Influence of electrostatic heteroaggregation of lipid droplets on their stability and digestibility under simulated gastrointestinal conditions." *Food and Function* **3**(10): 1025–1034.

Mao, Y. Y. and D. J. McClements (2012b). "Fabrication of functional micro-clusters by hetero-aggregation of oppositely charged protein-coated lipid droplets." *Food Hydrocolloids* **27**(1): 80–90.

Marze, S. (2013). "Bioaccessibility of nutrients and micronutrients from dispersed food systems: Impact of the multiscale bulk and interfacial structures." *Critical Reviews in Food Science and Nutrition* **53**(1): 76–108.

Matalanis, A., E. A. Decker and D. J. McClements (2012). "Inhibition of lipid oxidation by encapsulation of emulsion droplets within hydrogel microspheres." *Food Chemistry* **132**(2): 766–772.

Matalanis, A., O. G. Jones and D. J. McClements (2011). "Structured biopolymer-based delivery systems for encapsulation, protection, and release of lipophilic compounds." *Food Hydrocolloids* **25**(8): 1865–1880.

McClements, D. J. (2005). *Food Emulsions: Principles, Practice, and Techniques*. Boca Raton, FL, CRC Press.

McClements, D. J. (2010a). "Design of nano-laminated coatings to control bioavailability of lipophilic food components." *Journal of Food Science* **75**(1): R30–R42.

McClements, D. J. (2010b). Emulsion design to improve the delivery of functional lipophilic components. In *Annual Review of Food Science and Technology*, M. P. Doyle and T. R. Klaenhammer (eds.) Annual Reviews, Palo Alto, CA. **1**: 241–269.

McClements, D. J. (2011). "Edible nanoemulsions: Fabrication, properties, and functional performance." *Soft Matter* **7**(6): 2297–2316.

McClements, D. J. (2012a). "Advances in fabrication of emulsions with enhanced functionality using structural design principles." *Current Opinion in Colloid and Interface Science* **17**(5): 235–245.

McClements, D.J. (2012b). "Crystals and crystallization in oil-in-water emulsions: Implications for emulsion-based delivery systems." *Advances in Colloid and Interface Science* **174**: 1–30.

McClements, D. J. and E. A. Decker (2000). "Lipid oxidation in oil-in-water emulsions: Impact of molecular environment on chemical reactions in heterogeneous food systems." *Journal of Food Science* **65**(8): 1270–1282.

McClements, D. J. and E. A. Decker (2008). Lipids. In *Food Chemistry*, 4th Edition, S. Damodaran, K. L. Parkin and O. R. Fennema (eds.). Boca Raton, FL, CRC Press: 155–216.

McClements, D. J., E. A. Decker and Y. Park (2009a). "Controlling lipid bioavailability through physicochemical and structural approaches." *Critical Reviews in Food Science and Nutrition* **49**(1): 48–67.

McClements, D. J., E. A. Decker, Y. Park and J. Weiss (2009b). "Structural design principles for delivery of bioactive components in nutraceuticals and functional foods." *Critical Reviews in Food Science and Nutrition* **49**(6): 577–606.

McClements, D. J. and Y. Li (2010). "Structured emulsion-based delivery systems: Controlling the digestion and release of lipophilic food components." *Advances in Colloid and Interface Science* **159**(2): 213–228.

McClements, D. J. and J. Rao (2011). "Food-grade nanoemulsions: formulation, fabrication, properties, performance, biological fate, and potential toxicity." *Critical Reviews in Food Science and Nutrition* **51**(4): 285–330.

Meisel, H. (1997). "Biochemical properties of bioactive peptides derived from milk proteins: Potential nutraceuticals for food and pharmaceutical applications." *Livestock Production Science* **50**(1–2): 125–138.

Msagati, T. A. M. (2013). *Chemistry of Food Additives and Preservatives*. Chichester, West Sussex, UK, Wiley-Blackwell.

Muller, R. H. and C. M. Keck (2004). "Challenges and solutions for the delivery of biotech drugs—A review of drug nanocrystal technology and lipid nanoparticles." *Journal of Biotechnology* **113**(1–3): 151–170.

Müller, R. H., K. Mader and S. Gohla (2000). "Solid lipid nanoparticles (SLN) for controlled drug delivery—A review of the state of the art." *European Journal of Pharmaceutics and Biopharmaceutics* **50**(1): 161–177.

Murphy, D. J., I. Hernandez-Pinzon and K. Patel (2001). "Role of lipid bodies and lipid-body proteins in seeds and other tissues." *Journal of Plant Physiology* **158**(4): 471–478.

Murphy, D. J., I. Hernendez-Pinzon, K. Patel, R. G. Hope and J. McLauchlan (2000). "New insights into the mechanisms of lipid-body biogenesis in plants and other organisms." *Biochemical Society Transactions* **28**: 710–711.

Nehir El, S. and S. Simsek (2012). "Food technological applications for optimal nutrition: An overview of opportunities for the food industry." *Comprehensive Reviews in Food Science and Food Safety* **11**(1): 2–12.

Nicholson, T., H. Khademi and M. H. Moghadasian (2013). "The role of marine n-3 fatty acids in improving cardiovascular health: A review." *Food and Function* **4**(3): 357–365.

Paunov, V. N., G. Mackenzie and S. D. Stoyanov (2007). "Sporopollenin micro-reactors for in-situ preparation, encapsulation and targeted delivery of active components." *Journal of Materials Chemistry* **17**(7): 609–612.

Peppas, N. A., P. Bures, W. Leobandung and H. Ichikawa (2000). "Hydrogels in pharmaceutical formulations." *European Journal of Pharmaceutics and Biopharmaceutics* **50**(1): 27–46.

Playne, M. J., L. E. Bennett and G. W. Smithers (2003). "Functional dairy foods and ingredients." *Australian Journal of Dairy Technology* **58**(3): 242–264.

Qian, C., E. A. Decker, H. Xiao and D. J. McClements (2012). "Inhibition of beta-carotene degradation in oil-in-water nanoemulsions: Influence of oil-soluble and water-soluble antioxidants." *Food Chemistry* **135**(3): 1036–1043.

Rafferty, K., G. Walters and R. P. Heaney (2007). "Calcium fortificants: Overview and strategies for improving calcium nutriture of the US population." *Journal of Food Science* **72**(9): R152–R158.

Redgwell, R. J. and M. Fischer (2005). "Dietary fiber as a versatile food component: An industrial perspective." *Molecular Nutrition and Food Research* **49**(6): 521–535.

Reineccius, G. (2007). *Flavor Chemistry and Technology*. Boca Raton, FL, CRC Press.

Reineccius, G. A. (1995). Liposomes for controlled-release in the food-industry. In *Encapsulation and Controlled Release of Food Ingredients*, S. J. Risch and G. A. Reineccius (eds.). **590**: 113–131.

Reische, D. W., D. A. Lillard and R. R. Eitenmiller (2008). Antioxidants. In *Food Lipids*, C. C. Akoh and D. B. Min (eds.). Boca Raton, FL, CRC Press: 409–434.

Sagalowicz, L. and M. E. Leser (2010). "Delivery systems for liquid food products." *Current Opinion in Colloid and Interface Science* **15**(1–2): 61–72.

Salgueiro, M. J., M. Zubillaga, A. Lysionek, R. Caro, R. Weill and J. Boccio (2002). "Fortification strategies to combat zinc and iron deficiency." *Nutrition Reviews* **60**(2): 52–58.

Sanguansri, P. and M. A. Augustin (2006). "Nanoscale materials development—A food industry perspective." *Trends in Food Science and Technology* **17**(10): 547–556.

Schmitt, C. and S. L. Turgeon (2011). "Protein/polysaccharide complexes and coacervates in food systems." *Advances in Colloid and Interface Science* **167**(1–2): 63–70.

Schwartz, S. J., J. H. Elbe and M. M. Giusti (2008). Colorants. In *Food Chemistry*, 4th Edition, S. Damodaran, K. L. Parkin and F. O.R. (eds.). Boca Raton, FL, CRC Press: 571–638.

Shahidi, F. and Y. Zhong (2010). "Lipid oxidation and improving the oxidative stability." *Chemical Society Reviews* **39**(11): 4067–4079.

Sharma, A. and U. S. Sharma (1997). "Liposomes in drug delivery: Progress and limitations." *International Journal of Pharmaceutics* **154**(2): 123–140.

Siepmann, J. and F. Siepmann (2011). "Mathematical modeling of drug release from lipid dosage forms." *International Journal of Pharmaceutics* **418**(1): 42–53.

Silva, H. D., M. A. Cerqueira and A. A. Vicente (2012). "Nanoemulsions for food applications: Development and characterization." *Food and Bioprocess Technology* **5**(3): 854–867.

Singh, H. and A. Q. Ye (2013). "Structural and biochemical factors affecting the digestion of protein-stabilized emulsions." *Current Opinion in Colloid and Interface Science* **18**(4): 360–370.

Solans, C. and I. Sole (2012). "Nano-emulsions: Formation by low-energy methods." *Current Opinion in Colloid and Interface Science* **17**(5): 246–254.

Soto, E. and G. Ostroff (2012). Glucan particles as carriers of nanoparticles for macrophage-targeted delivery. In *Nanomaterials for Biomedicine*, R. Nagarajan (ed). **1119**: 57–79.

Soto, E. R. and G. R. Ostroff (2010). "Chemical derivatization of glucan microparticles for targeted delivery." *Abstracts of Papers of the American Chemical Society* **240**.

Spernath, A. and A. Aserin (2006). "Microemulsions as carriers for drugs and nutraceuticals." *Advances in Colloid and Interface Science* **128**: 47–64.

Spernath, A., A. Yaghmur, A. Aserin, R. E. Hoffman and N. Garti (2002). "Food-grade microemulsions based on nonionic emulsifiers: Media to enhance lycopene solubilization." *Journal of Agricultural and Food Chemistry* **50**(23): 6917–6922.

Stieger, M. and F. van de Velde (2013). "Microstructure, texture and oral processing: New ways to reduce sugar and salt in foods." *Current Opinion in Colloid and Interface Science* **18**(4): 334–348.

Taneja, A. and H. Singh (2012). Challenges for the delivery of long-chain n-3 fatty acids in functional foods. In *Annual Review of Food Science and Technology*, M. P. Doyle and T. R. Klaenhammer (eds.). **3**: 105–123.

Tavel, L., C. Moreau, S. Bouhallab, E. C. Y. Li-Chan and E. Guichard (2010). "Interactions between aroma compounds and beta-lactoglobulin in the heat-induced molten globule state." *Food Chemistry* **119**(4): 1550–1556.

Taylor, T. M., P. M. Davidson, B. D. Bruce and J. Weiss (2005). "Liposomal nanocapsules in food science and agriculture." *Critical Reviews in Food Science and Nutrition* **45**(7–8): 587–605.

Torchilin, V. P. (2007). "Micellar nanocarriers: Pharmaceutical perspectives." *Pharmaceutical Research* **24**(1): 1–16.

Udenigwe, C. C. and R. E. Aluko (2012). "Food protein-derived bioactive peptides: Production, processing, and potential health benefits." *Journal of Food Science* **77**(1): R11–R24.

Velikov, K. P. and E. Pelan (2008). "Colloidal delivery systems for micronutrients and nutraceuticals." *Soft Matter* **4**(10): 1964–1980.

Waraho, T., D. J. McClements and E. A. Decker (2011). "Mechanisms of lipid oxidation in food dispersions." *Trends in Food Science and Technology* **22**(1): 3–13.

Ward, R. E. and J. B. German (2004). "Understanding milk's bioactive components: A goal for the Genomics toolbox." *Journal of Nutrition* **134**(4): 962S–967S.

Weiss, J., E. A. Decker, D. J. McClements, K. Kristbergsson, T. Helgason and T. Awad (2008). "Solid lipid nanoparticles as delivery systems for bioactive food components." *Food Biophysics* **3**(2): 146–154.

White, P. J. and M. R. Broadley (2005). "Biofortifying crops with essential mineral elements." *Trends in Plant Science* **10**(12): 586–593.

Wildman, R. E. C. and M. Kelley (2007). Nutraceuticals and functional foods. In *Handbook of Nutraceuticals and Functional Foods*, 2nd Edition, R. E. C. Wildman (ed.). Boca Raton, FL, CRC Press: 1–22.

Yang, H. H. L. and H. T. Lawless (2006). "Time-intensity characteristics of iron compounds." *Food Quality and Preference* **17**(5): 337–343.

# Chapter 2

# Active Ingredients

## 2.1 Introduction

There are a wide range of different active ingredients ("actives") used by the food industry that would benefit from being encapsulated within appropriate delivery systems, including flavors, colors, minerals, nutraceuticals, vitamins, antimicrobials, antioxidants, and preservatives (Chen et al. 2006; Shefer and Shefer 2003; Ubbink 2002; Ubbink and Kruger 2006). The molecular characteristics of different active ingredients vary considerably, for example, atomic composition, molecular weight, conformation, flexibility, polarity, and electrical charge. In turn, these differences in molecular characteristics lead to differences in physicochemical properties, such as physical state, solubility, partitioning, diffusion, interactions, optical characteristics, rheological properties, and stability. The unique molecular and physicochemical characteristics of active ingredients mean that no single delivery system is suitable for every situation. Instead, it is necessary to know the specific molecular and physicochemical characteristics of the active ingredient that needs to be encapsulated in order to select the most appropriate delivery system. The major types of active ingredients that need to be encapsulated are discussed in this chapter, along with a brief overview of their molecular and physicochemical properties, as well as possible challenges to their utilization within food and beverage products.

## 2.2 Lipid-Based Ingredients

The term *lipids* refers to a broad group of compounds that are usually soluble in organic solvents but insoluble in water (Akoh and Min 2008; McClements and Decker 2008). A number of molecular classes fall within this broad group, including triacylglycerols, diacylglycerols, monoacylglycerols, free fatty acids, phospholipids, carotenoids, phytosterols, oil-soluble vitamins, flavor oils, essential oils, and fat replacers. Some of the most important molecular and physicochemical properties of lipids that influence the type of delivery system needed to encapsulate them are shown in Tables 2.1 and 2.2. Examples of representative triacylglycerol, flavor, and essential oil molecules are shown in Figure 2.1.

**Table 2.1 Physicochemical Properties of Selected Triacylglycerol, Flavor, and Essential Oil Components**

| Compound | Molar Mass (g mol$^{-1}$) | $T_m$ (°C) | LogP | Molar Volume (cm$^3$ mol$^{-1}$) | $C_W^*$ (g L$^{-1}$) | p$K_a$ |
|---|---|---|---|---|---|---|
| **Triacylglycerols** | | | | | | |
| Tributyrin (4:0) | 302.4 | −75 | 3.27 | 287 | 0.28 | — |
| Tricaprylin (8:0) | 470.1 | 10 | 9.39 | 485 | $8.9 \times 10^{-5}$ | — |
| Tridedecanoin (12:0) | 554.8 | 31 | 12.44 | 584 | $2.9 \times 10^{-6}$ | — |
| Trilaurin (12:0) | 639.0 | 46 | 15.50 | 683 | $1.4 \times 10^{-7}$ | — |
| Trimyristin (14:0) | 723.2 | 57 | 18.56 | 782 | $1.0 \times 10^{-8}$ | — |
| Tripalmitin (16:0) | 807.3 | 66 | 21.61 | 881 | $1.1 \times 10^{-9}$ | — |
| Tristearin (18:0) | 891.5 | 73 | 24.67 | 980 | $1.8 \times 10^{-10}$ | — |
| Triolein (18:1) | 885.4 | 5 | 23.44 | 961 | $6.8 \times 10^{-10}$ | — |
| Trilinolein (18:2) | 897.4 | −43 | 22.23 | 942 | $2.6 \times 10^{-9}$ | — |
| Triarachin (20:0) | 975.6 | — | 27.73 | 1079 | $4.5 \times 10^{-11}$ | |
| **Flavor and Essential Oils** | | | | | | |
| Limonene | 136.2 | −95 | 4.55 | 163 | $3.4 \times 10^{-3}$ | — |
| Citral | 152.2 | <−10 | 3.13 | 178 | 1.7 | |
| Citral acetate | 196.3 | — | 3.90 | 217 | 0.57 | |
| Terpinolene | 136.2 | — | 4.21 | 160 | $7.1 \times 10^{-3}$ | |
| β-Bisabolene | 204.4 | — | 6.10 | 236 | $6.9 \times 10^{-6}$ | |

*Source:* SciFinder, American Chemical Society.

*Note:* The data are reported at pH 7 and 25°C unless otherwise stated. $T_m$, melting temperature; LogP, logarithm of oil–water partition coefficient (Log$_{10}$[K$_{OW}$]); $C_W^*$, water solubility.

**Table 2.2  Physicochemical Properties of Selected Lipophilic Components Claimed to Have Health Benefits When Used as Nutraceuticals**

| Compound | Molar Mass (g mol$^{-1}$) | $T_m$ (°C) | LogP | Molar Volume (cm$^3$ mol$^{-1}$) | $C_W^*$ (g L$^{-1}$) | $pK_a$ |
|---|---|---|---|---|---|---|
| **Phenolic Acids** | | | | | | |
| Curcumin | 368.4 | 183 | 3.07 | 288 | 0.052 | 8.1 |
| Capsaicin | 305.4 | 64 | 3.20 | 293 | 0.14 | 9.9 |
| Vanillin | 152.2 | 80 | 1.21 | 124 | 4.3 | 7.8 |
| **Flavonoids** | | | | | | |
| Catechin | 290.3 | 214 | 0.61 | 266 | 7.3 | 9.5 |
| Quercetin | 302.2 | 310 | 1.99 | 168 | 1.9 | 6.3 |
| Genistein | 270.2 | 300 | 3.11 | 175 | 0.7 | 6.5 |
| **Stilbenoids** | | | | | | |
| Resveratrol | 228.2 | 255 | 3.02 | 168 | 0.021 | 9.2 |
| **Carotenoids** | | | | | | |
| β-Carotene | 536.9 | 180 | 14.76 | 570 | $1.5 \times 10^{-9}$ | — |
| Lycopene | 536.9 | 175 | 14.5 | 604 | $3.7 \times 10^{-9}$ | — |
| Zeaxanthin | 568.9 | 205 | 10.9 | 564 | $8.0 \times 10^{-5}$ | 14.6 |
| Astaxanthin | 596.8 | 210 | 8.24 | 557 | $5.6 \times 10^{-7}$ | 12.3 |
| Lutein | 568.9 | 180 | 11.5 | 566 | $5.6 \times 10^{-5}$ | 14.6 |
| **Phytosterols** | | | | | | |
| β-Sitosterol | 414.7 | 138 | 10.5 | 424 | $8.7 \times 10^{-7}$ | 15.0 |
| Campesterol | 400.7 | 155 | 9.97 | 408 | $2.0 \times 10^{-6}$ | 15.0 |
| Stigmasterol | 412.7 | 150 | 10.07 | 418 | $1.6 \times 10^{-6}$ | 15.0 |
| **Oil-Soluble Vitamins** | | | | | | |
| Vitamin E (α-tocopherol) | 430.7 | 3 | 10.96 | 463 | $5.2 \times 10^{-5}$ | 11.4 |

(*continued*)

**Table 2.2 (Continued) Physicochemical Properties of Selected Lipophilic Components Claimed to Have Health Benefits When Used as Nutraceuticals**

| Compound | Molar Mass (g mol⁻¹) | $T_m$ (°C) | LogP | Molar Volume (cm³ mol⁻¹) | $C_w^*$ (g L⁻¹) | $pK_a$ |
|---|---|---|---|---|---|---|
| Vitamin E acetate (α-tocopherol acetate) | 472.7 | −27.5 | 10.69 | 502 | $2.0 \times 10^{-6}$ | — |
| Vitamin D₃ (cholecalciferol) | 384.6 | 85 | 9.09 | 397 | $6.5 \times 10^{-5}$ | 14.7 |
| Vitamin A (*trans*-retinol) | 286.5 | 63 | 6.08 | 300 | $7.7 \times 10^{-5}$ | 14.1 |
| **Miscellaneous** | | | | | | |
| Coenzyme-Q | 863.3 | 49 | 19.12 | 888.5 | $5.9 \times 10^{-8}$ | — |

*Source:* SciFinder, American Chemical Society; D. J. McClements, *Adv Colloid Interface*, 174, 1–30, 2012b.

*Note:* The data are reported at pH 7 and 25°C unless otherwise stated. $T_m$, melting temperature; LogP, logarithm of oil–water partition coefficient; and $C_w^*$, water solubility.

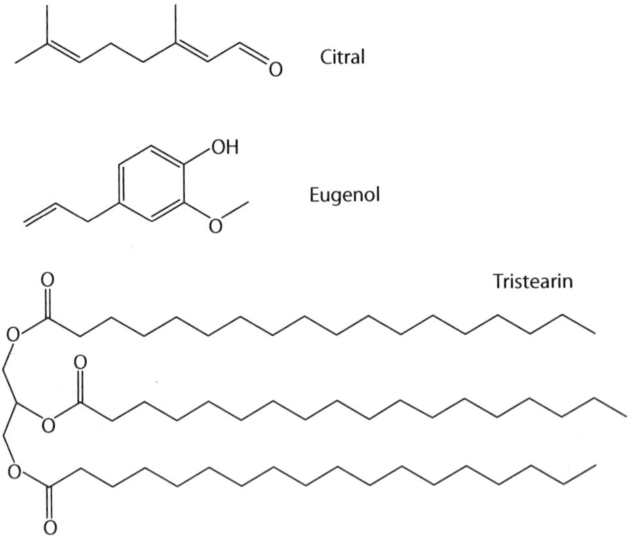

Citral

Eugenol

Tristearin

**Figure 2.1 Molecular structures of representative flavor oils (citral), essential oils (eugenol), and triacylglycerol oils (tristearin).**

## 2.2.1 Neutral Oils

Neutral oils can be considered to be lipids that have little flavor or color of their own. This type of lipid is added to many types of aqueous-based product to alter their physicochemical and sensory attributes in a way that makes them more palatable or desirable to consumers. For example, in emulsion-based products, neutral oil droplets provide a turbid or creamy appearance (because of light scattering), viscosity and mouthfeel enhancement (because of their impact on fluid flow), and flavor alterations (because of their ability to solubilize nonpolar flavors) (McClements 2005). In addition, neutral oils may also act as carriers for other nonpolar components in foods, such as oil-soluble pigments, preservatives, vitamins, and nutraceuticals. The neutral oils used for this purpose are usually high-molecular-weight molecules with extremely low water solubilities, such as triacylglycerols or terpene oils (Figure 2.1). Triacylglycerols are typically derived from natural sources such as animals (e.g., fish, meat, and milk), plants (e.g., canola, corn, rapeseed, olive, safflower, and sunflower), and microorganisms (e.g., algae) (Akoh and Min 2008), whereas terpenes are usually isolated by distillation of natural oils from plant sources, such as fruits, vegetables, leaves, herbs, and spices (Burt 2004). Because of their low water solubilities, these oils usually need to be encapsulated in a suitable form before they can be readily dispersed into aqueous-based food or beverage products. The most commonly used encapsulation technology for this type of lipid is oil-in-water emulsions (Given 2009), but other kinds of colloidal delivery systems can also be used to gain specific functional attributes, such as nanoemulsions, solid lipid nanoparticles, multilayer emulsions, or multiple emulsions (McClements 2012a). The formation and stability of colloidal delivery systems based on neutral oils depend on their polarity, phase behavior, viscosity, interfacial tension, and density (McClements 2005).

The viscosity and interfacial tension of lipids influence their ability to form small droplets during the homogenization process, with higher viscosities and interfacial tensions typically leading to larger droplets (Qian and McClements 2011; Wooster et al. 2008). The density of lipids determines the creaming velocity of the particles within emulsion-based delivery systems and may therefore influence their long-term stability (Chanamai and McClements 2000). The phase behavior of lipids is important for certain applications, such as the formation of solid lipid nanoparticles or the stability of emulsion-based delivery systems to partial coalescence (McClements 2012b; Rousseau 2000). The potential for crystallization to occur is particularly important in food or beverage products that are stored in refrigerators for prolonged periods. In these cases, it may be necessary to use oil that remains liquid at refrigerator temperatures (e.g., winterized or fractionated oil).

## 2.2.2 Flavor and Essential Oils

Flavor and essential oils are typically isolated from plant sources, such as fruits, vegetables, leaves, herbs, and spices, and consist of a chemically diverse group of nonpolar components (Bakkali et al. 2008; Burt 2004). Flavor oils, such as orange, lemon, and lime oils, are widely used in the food and beverage industries as flavorants, since they contain volatile constituents with characteristic aroma profiles (Given 2009). The composition of natural flavor oils depends on their biological origin, the extraction procedure used to isolate them, and any subsequent processing steps. For example, lemon oil is usually obtained from lemon peel by cold pressing, but it may then be further refined using steam distillation and other processing methods (Misharina et al. 2010). Flavor oils are commercially available in different forms ("folds") that differ in their chemical composition and physicochemical properties owing to differences in isolation and processing procedures. Flavor oils extracted by cold pressing are often referred to as single-fold (1×) oils, whereas those that have undergone further processing are referred to as higher-fold oils, for example, 3×, 5×, or 10× oils (Gamarra et al. 2006). The nature and intensity of the flavor profile depend on the type and amount of processing used during the manufacturing process, with higher-fold oils typically having different (often more intense) flavor profiles than lower-fold oils. Flavor oils with different chemical compositions have different physicochemical characteristics (such as water solubility, partitioning, density, viscosity, refractive index, and absorbance), which influences the formation, stability, and properties of colloidal delivery systems formed from them (Rao and McClements 2012a,b).

Essential oils also vary considerably in their compositions and physicochemical properties depending on isolation and processing conditions. Many essential oils have been claimed to have specific health benefits and functional attributes and are therefore used as nutraceuticals, antioxidants, or antimicrobials (Bakkali et al. 2008). There are a number of challenges associated with incorporation of flavor and essential oils into aqueous-based foods and beverages. First, even though they are hydrophobic, they do have some water solubility, which can cause instability problems owing to Ostwald ripening (McClements et al. 2012; Rao and McClements 2012a,b). Second, many flavor and essential oils are chemically unstable and degrade during storage (e.g., citral), which leads to the loss of desirable flavor notes or biological activity and possibly to the generation of off-flavor notes (Choi et al. 2009, 2010). Finally, these types of oils have relatively low molecular weights and are often fairly volatile so that they may be lost during processing, storage, or transport. Traditionally, flavor or essential oils are incorporated into aqueous-based products in the form of oil-in-water emulsions or microemulsions (Given 2009), but other systems can also be used, such as nanoemulsions, solid lipid nanoparticles, multilayer emulsions, or multiple emulsions (McClements 2012a).

## 2.2.3 Oil-Soluble Colorants

Many artificial natural colors added to foods and beverages are oil-soluble compounds (Hutchings 1999; Schwartz et al. 2008). These compounds have chemical structures that generate intense colors as a result of selective absorption of visible light in specific

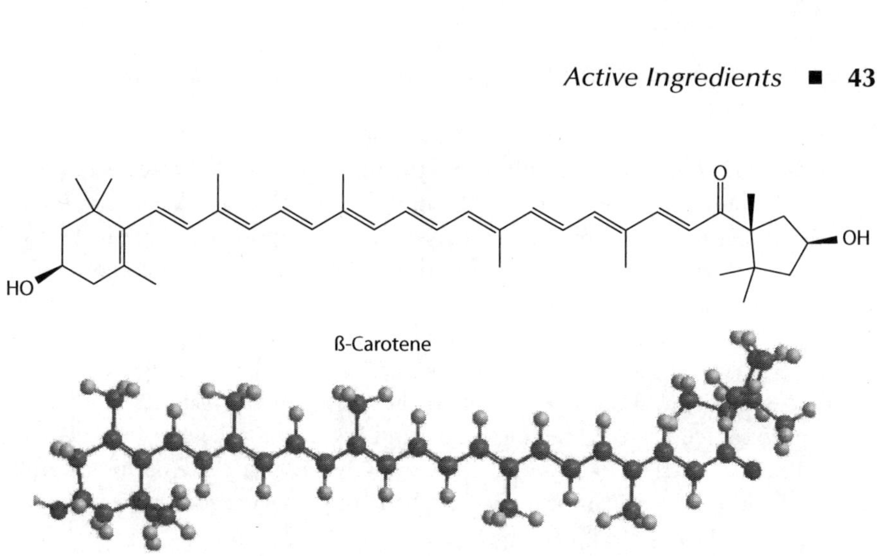

**Figure 2.2  Molecular structure of β-carotene—a nonpolar colorant and nutraceutical often used in food products.**

regions of the electromagnetic spectrum (Figure 2.2). Recently, there has been great interest in replacing artificial colors with natural ones, such as annatto, paprika, curcumin, lycopene, and β-carotene, which are all highly hydrophobic molecules (Gibbs et al. 1999; Kandansamy and Somasundaram 2012; Wrolstad and Culver 2012). Some of these compounds have also been reported to have beneficial biological activities and may therefore also be used as nutraceuticals (Ortiz-Moreno et al. 2011). Because of their very low water solubility, these compounds usually have to be incorporated into colloidal delivery systems before they can be introduced into aqueous-based food and beverage products (Velikov and Pelan 2008). In addition, many of these compounds are crystalline at ambient temperature and must therefore be either dissolved in a suitable carrier or introduced into the product in the form of a suspension of very fine crystals (McClements 2012b). Crystalline compounds can be dissolved in an oil phase and then homogenized to form a nanoemulsion or emulsion, or they can be solubilized directly within surfactant micelles or microemulsions. Many colorants from natural sources are highly unstable and may chemically degrade and therefore fade during storage, for example, carotenoids (Qian et al. 2012a,b). It is therefore important to establish the mechanism of chemical degradation for these types of colorant in specific food products and to determine the major factors affecting their degradation (such as pH, light, oxygen, and prooxidants) so that an effective colloidal delivery system can be designed to ensure appropriate product shelf life (Boon et al. 2010).

## 2.2.4 Lipophilic Nutraceuticals

There has been growing interest in the incorporation of various kinds of lipophilic nutraceuticals into commercial food and beverage products, such as polyunsaturated fats (particularly oils rich in omega-3 [ω-3] fatty acids), conjugated linoleic acid (CLA), carotenoids, phytosterols, and fat-soluble vitamins (McClements et al.

2009; Sagalowicz and Leser 2010; Spernath and Aserin 2006; Velikov and Pelan 2008). Lipophilic nutraceuticals may be mixed with other oil phase components prior to preparing a food product or they may be encapsulated in a suitable colloidal delivery system and then introduced into the product. The incorporation of lipophilic nutraceuticals into commercial food and beverage products is often challenging because they are prone to various physical or chemical degradation mechanisms. Each type of nutraceutical lipid has its own particular challenges depending on its physicochemical properties, such as chemical stability, water solubility, oil–water partition coefficient, melting point, and crystal morphology (McClements 2012a,b; McClements et al. 2009). Delivery systems must therefore be carefully designed taking into account the specific challenges associated with the particular type of lipophilic nutraceuticals present.

### 2.2.4.1 Polyunsaturated Lipids

There has been considerable interest in fortifying a wide range of food and beverage products with polyunsaturated fatty acids (PUFAs), and in particular ω-3 fatty acids, because of their potential health benefits (Jacobsen 2008; Ruxton et al. 2004, 2007; Siddiqui et al. 2004). Omega-3 fatty acids have a double bond that is three carbon atoms from the methyl end of the molecule (Figure 2.3). The three most common ω-3 fatty acids used to supplement food and beverage products are α-linolenic acid (ALA, 18:3), eicosapentaenoic acid (EPA, 20:5) and docosahexaenoic acid (DHA, 22:6), with EPA and DHA being the most bioactive forms (Larsen et al. 2011; Russo 2009). Omega-3 fatty acids are found in relatively high concentrations in a

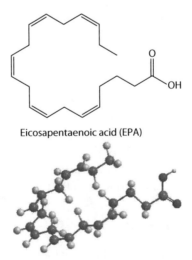

Eicosapentaenoic acid (EPA)

**Figure 2.3 Molecular structure of eicosapentaenoic acid (EPA)—an example of an ω-3 fatty acid.**

number of natural sources such as fish, algae, and flaxseed oils. Usually, PUFAs are introduced into foods as part of triacylglycerols, but they may also be introduced as part of monoacylglycerols, diacylglycerols, or phospholipids. The consumption of sufficiently high levels of ω-3 fatty acids has been linked to reduced risk of various types of chronic disease, including inflammation, cardiovascular disease, immune response disorders, mental disorders, and poor infant development (Orchard et al. 2012; Tur et al. 2012; Yashodhara et al. 2009). Consequently, many food manufacturers have attempted to incorporate ω-3 fatty acids into their products at levels sufficiently high to have a beneficial health effect (McClements and Decker 2000; Waraho et al. 2011). Nevertheless, there are a number of challenges associated with fortifying foods with polyunsaturated lipids. First, PUFAs are long-chain hydrophobic molecules that have low water solubility and therefore have to be mixed with an oil phase prior to food preparation or introduced in the form of a stable colloidal delivery system. Second, PUFAs are highly susceptible to degradation through lipid oxidation, which causes problems for their long-term storage (Arab-Tehrany al. 2012). Lipid oxidation involves a complex sequence of chemical reactions that eventually result in the generation of highly rancid off-flavors and potentially toxic reaction products. A combination of strategies is often needed to control lipid oxidation so as to achieve a sufficiently long shelf life, including control of initial ingredient quality, storage at reduced temperatures, avoidance of light exposure, addition of antioxidants, removal/deactivation of prooxidants (such as oxygen, peroxides, and transition metals), and interfacial engineering (McClements and Decker 2000; Waraho et al. 2011). A number of studies have shown that colloidal delivery systems can be designed to improve the oxidative stability of encapsulated PUFAs, including emulsions, multilayer emulsions, multiple emulsions, and filled hydrogel particles (Berton-Carabin et al. 2013; Djordjevic et al. 2007; Matalanis et al. 2012; Poyato et al. 2013; Salminen et al. 2013).

### 2.2.4.2 Fat-Soluble Antioxidants

Lipophilic antioxidants are often incorporated into the oil phase of food products to inhibit the oxidation of chemically labile substances (such as ω-3 oils or carotenoids), thereby extending product shelf life (Boon et al. 2010; McClements and Decker 2000; Waraho et al. 2011). Alternatively, they may be incorporated into a product as a functional ingredient so as to make an "antioxidant" claim on a label. Some of the most widely used lipophilic antioxidants currently used by the food industry are α-tocopherols, ascorbyl palmitate, butylhydroxytoluene, butylated hydroxyanisole, and rosemary extracts (Figure 2.4) (McClements and Decker 2008). The main physicochemical mechanism responsible for the antioxidant activity of these substances is their ability to act as free radical scavengers. This type of antioxidant absorbs free radicals from other substances (such as bioactive lipids), causing them to become more chemically stable. The physical location of an antioxidant within a food matrix is particularly important for determining

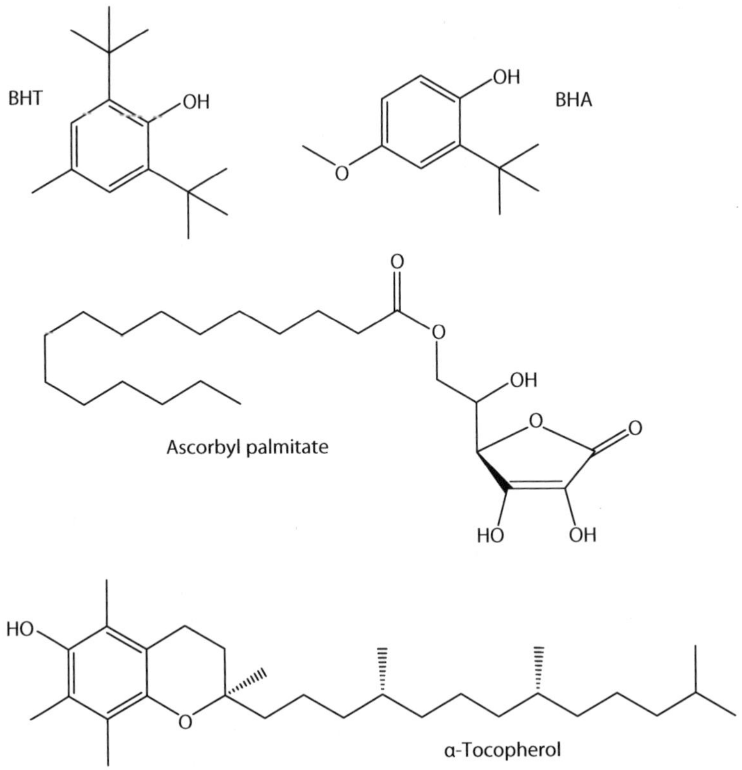

**Figure 2.4** Molecular structures of selected oil-soluble antioxidants: α-tocopherol, ascorbyl palmitate, butylhydroxytoluene (BHT), and butylated hydroxyanisole (BHA).

its effectiveness, as it should be present at the site where lipid oxidation primarily occurs, for example, within the oil, water, or interfacial regions (Figure 2.5). Appropriately designed delivery systems can be used to improve the efficacy of antioxidants in food and beverage products by delivering them to the appropriate site of action.

## 2.2.4.3 Fat-Soluble Vitamins

There is considerable interest in incorporating various types of fat-soluble vitamins into foods and beverages to improve the nutritional value of these products, for example, A, D, and E (Sagalowicz and Leser 2010; Velikov and Pelan 2008). Fat-soluble vitamins vary considerably in their molecular structures (Figure 2.6), but they are all highly hydrophobic substances with chemically reactive groups. Consequently, they usually have low water solubility and poor chemical stability.

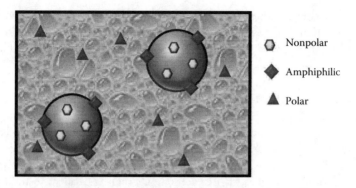

**Figure 2.5** **The location of an antioxidant within a colloidal delivery system plays an important role in determining its effectiveness.**

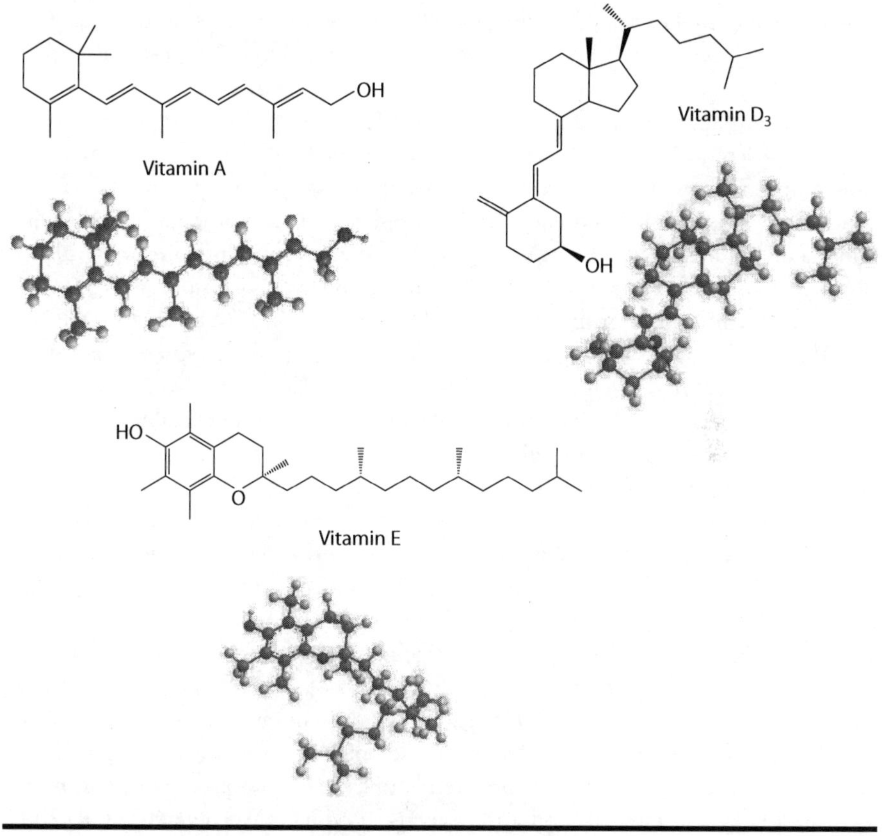

**Figure 2.6** **Molecular structures of major oil-soluble vitamins: vitamin A (*trans-*retinol), vitamin D₃ (cholecalciferol), and vitamin E (α-tocopherol).**

Vitamin A plays numerous roles in ensuring human health and well-being, including maintaining good vision, bone growth, reproduction, and skin health (Hark and Deen 2005; van den Berg et al. 2000). Vitamin A may also play other important roles in ensuring good health, for example, by acting as a natural antioxidant, protecting against heart disease, and inhibiting cancer development (Maiani et al. 2009). The two main sources of vitamin A in humans are *active forms* and *provitamins* (Hark and Deen 2005). Active forms are immediately available for utilization by the body and are mainly found in some animal products, for example, retinal and retinol. On the other hand, provitamin forms are converted into vitamin A within the human body after ingestion and mainly come from highly colored fruits and vegetables, that is, carotenoids such as β-carotene and lycopene. Provitamin forms of vitamin A therefore have many of the potential health benefits claimed for this class of nutraceutical compounds, as well as many of the challenges, such as low water solubility; poor stability to heat, oxygen, and light; high melting point; and low bioavailability (Boon et al. 2010; Rao and Rao 2007).

Vitamin D plays an essential role in bone, teeth, and cartilage development, which has mainly been attributed to its influence on the absorption of dietary calcium and phosphorus (Cranney et al. 2008; Hark and Deen 2005). In addition, it may also play important roles in preventing cancer, heart disease, and immune diseases (Haham et al. 2012; Holick 2004). It typically comes in two molecular forms: vitamin $D_2$ (ergocalciferol) and vitamin $D_3$ (cholecalciferol). Vitamin $D_2$ is naturally present in small amounts in a limited number of foods, whereas vitamin $D_3$ is usually synthesized in the human skin as a product of ultraviolet B irradiation of 7-dehydrocholesterol (Holick 2007; Lee et al. 2008). Significant segments of the population of many countries are vitamin D deficient owing to lack of exposure to the sun, the use of sun creams, or poor dietary intake (Haham et al. 2012; Holick 2004; Tsiaras and Weinstock 2011). Consequently, many food and beverage products (particularly dairy-based ones) have been fortified with vitamin D (Yang et al. 2013). Some of the challenges associated with vitamin D fortification are poor water solubility; chemical degradation when exposed to light, oxygen, or elevated temperatures; and variable oral bioavailability (Haham et al. 2012; Tsiaras and Weinstock 2011).

Vitamin E plays an essential role in maintaining the health and well-being of humans, which has been attributed to various antioxidant and nonantioxidant activities (Brigelius-Flohe and Galli 2010; Burton and Traber 1990; Zingg and Azzi 2004). However, the precise biological mechanisms by which vitamin E exerts its beneficial effects are still unclear (Brigelius-Flohe and Galli 2010; Golli and Azzi 2010). The term *vitamin E* actually refers to a group of nonpolar molecules with related chemical structures, physical properties, and biological activities (Chiu and Yang 1992; Rigotti 2007). The most biologically active form of vitamin E is α-tocopherol, and therefore this form tends to be the one that is most commonly used to fortify commercial food and beverage products (McClements et al. 2009; Yang and Huffman 2011). There are numerous challenges associated with incorporation of vitamin E into foods and beverages because of its chemical instability, poor water solubility, and variable bioavailability. α-Tocopherol is a polyunsaturated molecule that is prone to lipid oxidation during food processing, transport, and storage (Gawrysiak-Witulska et al. 2009; Yoon and Choe 2009). Consequently, the esterified form of vitamin E (α-tocopherol acetate) is often used in food products because of its higher chemical stability compared to the free form (Lauridsen et al. 2001). Another limitation of vitamin E is that it is a highly nonpolar molecule with poor water solubility, which means it cannot simply be dispersed into aqueous-based products (Sagalowicz et al. 2006; Velikov and Pelan 2008). Instead, it must be encapsulated within a suitable colloidal delivery system first (Flanagan and Singh 2006; Gonnet et al. 2010; McClements et al. 2009). Another challenge associated with utilization of vitamin E is its low and variable oral bioavailability (O'Callaghan and O'Brien 2010; Reboul et al. 2006). Research has therefore been carried out to overcome these challenges by incorporating vitamin E into colloidal delivery systems, such as microemulsions (Chiu and Yang 1992; Feng et al. 2009), nanoemulsions (Hatanaka et al. 2010; Li et al. 2011; Saberi et al. 2013a,b; Shukat et al. 2012), and emulsions (Chen and Wagner 2004; Gonnet et al. 2010). Encapsulation of vitamin E has been reported to improve its physicochemical stability during storage and its biological activity after consumption (Cortesi et al. 2002; Zuccari et al. 2005). Studies have also shown that the oral bioavailability of vitamin E may be increased when it is delivered in colloidal form rather than in bulk form (Feng et al. 2009).

### 2.2.4.4 Fat-Soluble Nutraceuticals

There are numerous fat-soluble compounds in foods that are claimed to have health benefits over and above their normal nutritional role (Espin et al. 2007; Ramaa et al. 2006; Wildman and Kelley 2007). Consumption of these nutraceutical lipids is claimed to promote human wellness and to reduce the risk of certain chronic diseases, such as coronary heart disease, cancer, diabetes, hypertension, eye disease, and brain disease (Arnoldi 2004; Wildman and Kelley 2007). Consequently, the food industry is interested in incorporating many of these lipophilic nutraceutical components into foods (McClements et al. 2009; Sagalowicz and Leser 2010; Velikov and Pelan 2008). Some of the most commonly studied nutraceuticals are briefly discussed below.

*Carotenoids*: Carotenoids consist of a diverse group of nonpolar compounds that contribute to the yellow, orange, and red colors of many foods, but that have also been claimed to have certain health benefits (Mayne 1996; Rao and Rao 2007). These molecules are distinguished by the fact that they contain multiple conjugated double bonds and carbon ring structures at one end or at both ends of the molecule (Figure 2.2). Carotenoids formed solely from carbon and hydrogen are known as carotenes (e.g., lycopene and β-carotene), whereas those containing carbon, hydrogen, and oxygen are known as xanthophylls (e.g., lutein and zeaxanthin). The carotenoids have been proposed to exhibit several health benefits:

β-carotene has provitamin A activity (Rao and Rao 2007); lutein and zeaxanthin may decrease age-related macular degeneration and cataracts (Stringham and Hammond 2005); lycopene may decrease the risk of prostate cancer (Basu and Imrhan 2007). There are a number of challenges associated with incorporating carotenoids into foods, including their low water solubility, low oil solubility, high melting points, poor chemical stability, and low and variable oral bioavailability (Boon et al. 2010). Carotenoids are relatively stable to chemical degradation when present in their natural environment in the tissues of fruits, vegetables, fish, and animals (Xianquan et al. 2005). However, they become highly unstable to degradation once they are isolated, becoming susceptible to light, oxygen, pH, and temperature (Qian et al. 2012a,b; Xianquan et al. 2005). As a result, incorporation of carotenoids into commercial food and beverage products can result in their rapid degradation (Heinonen et al. 1997; Ribeiro et al. 2003). Carotenoids are typically degraded by chemical or biochemical reactions that cause the loss of double bonds or scission of the molecule. In addition, the double bonds in carotenoids can undergo isomerization from the *trans* to the *cis* configuration (Xianquan et al. 2005). In some cases, isomerization is beneficial since *cis* isomers of carotenoids (such as lycopene) have been reported to be more bioavailable than *trans* isomers (Schieber and Carle 2005). A number of workers have shown that colloidal delivery systems can be used to encapsulate and protect various kinds of carotenoids, such as lycopene (Boon et al. 2009; Ribeiro et al. 2003), lutein (Losso et al. 2005), and β-carotene (Chu et al. 2007, 2008; Qian et al. 2012a,b; Silva et al. 2010).

*Phytosterols and Phytostanols*: Phytosterols and phytostanols are highly lipophilic molecules isolated from plant sources and include stigmasterol, β-sitosterol, and campesterol (Figure 2.7) (Fernandes and Cabral 2007; John et al. 2007; Rocha et al. 2011). Foods and beverages have been fortified with these types of nutraceutical because of their ability to reduce total and low-density lipoprotein (LDL) cholesterol in humans, which has been attributed to inhibition of dietary cholesterol

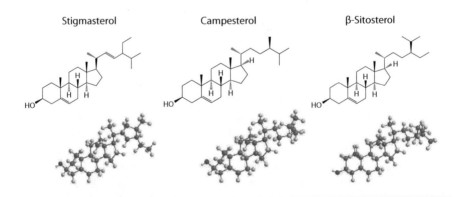

**Figure 2.7  Molecular structures of selected phytosterols.**

absorption (Ostlund 2004; Wong 2001). Consumption of 1.6 g of phytosterols per day has been reported to reduce LDL cholesterol in the blood by 10% (Hallikainen et al. 2000). Incorporation of phytosterols and phytostanols into aqueous-based food and beverage products is difficult because of their low water solubility, low oil solubility, high melting point, and poor chemical stability (Cercaci et al. 2007; Soupas et al. 2004). Some of these problems have been overcome by esterification of phytosterols with PUFAs. After consumption, the phytosterol esters are hydrolyzed by enzymes in the gastrointestinal tract (GIT) to produce free fatty acids and phytosterols. Initially, phytosterols were mainly incorporated into high-fat foods such as margarine and spreads by dispersing them within the oily continuous phase. The introduction of phytosterols into aqueous-based foods required that they be incorporated into a suitable colloidal delivery system first. This type of delivery system has been shown to be capable of overcoming some of the challenges associated with incorporating phytosterols and phytostanols into aqueous-based products (Engel and Schubert 2005; Leong et al. 2011; Rozner et al. 2007).

*Conjugated Linoleic Acid*: CLA is a fatty acid naturally found in the milk and body fat of certain animals, especially ruminants such as cows (Park and Pariza 2007). Research suggests that CLA has various biological activities in animals and humans, such as anticancer activity (Ha et al. 1988; Pariza et al. 1979), reduced atherosclerosis (Lee et al. 1994; Nicolosi et al. 1997), improved immune function (Cook et al. 1998; Miller et al. 1994), growth promotion (Chin et al. 1994), and reduction in body fat and increase in lean body mass (Park and Pariza 2007). CLA can exist in numerous isomeric forms, with the two most biologically active forms being the *cis*-9,*trans*-11 and *trans*-10,*cis*-12 isomers. The *cis*-9,*trans*-11 form is the major CLA isomer naturally occurring within foods, whereas the *trans*-10,*cis*-12 form is typically only a minor component (Kepler et al. 1966). On the other hand, both isomers are present at relatively high amounts in synthetic forms of CLA. Studies suggest that the wide range of biological activities exhibited by CLA is a result of interactions between these two CLA isomers (Park and Pariza 2007).

The levels of CLA in natural products are relatively low, and so foods must be fortified with CLA to reach levels where health benefits are expected. CLA may be incorporated into foods either as a free fatty acid, methyl ester form, or as part of a triacylglycerol molecule. Like other lipophilic compounds, there are numerous challenges associated with incorporating CLA into aqueous-based foods and beverages, such as low water solubility and susceptibility to oxidation. Consequently, effective delivery systems are needed to encapsulate and protect it (Liu et al. 2010; Lopes et al. 2011; Yang et al. 2009).

## 2.2.5 Challenges to Delivery of Lipid-Based Ingredients

In this section, the major challenges that need to be overcome when trying to incorporate lipid-based ingredients into commercial food and beverage products are summarized:

i. Most lipids are highly hydrophobic molecules with a low water solubility and so they must be incorporated into some kind of colloidal delivery system (e.g., microemulsion, nanoemulsion, solid lipid nanoparticle, emulsion, etc.) to make them readily dispersible in aqueous-based products.

ii. Some lipids are crystalline at storage temperatures in their pure form, for example, carotenoids and phytosterols. These lipids may have to be melted or dissolved within a suitable solvent before they can be incorporated into a colloidal delivery system. The crystalline nature of these lipids may provide challenges in the manufacture of certain types of food products (e.g., it may be necessary to use elevated temperatures during preparation), or they may adversely affect the long-term stability or organoleptic properties of the final product.

iii. The lipid and associated delivery system should be compatible with the food matrix; that is, they should not adversely affect the desirable appearance, stability, texture, mouthfeel, or flavor profile of the product. A lipid contained within an emulsion will cause a product to appear turbid or opaque because of light scattering by the droplets. On the other hand, a lipid contained within a microemulsion or nanoemulsion will appear transparent because the particles are so small that they do not scatter light strongly.

iv. Some lipids are highly susceptible to chemical degradation within food and beverage matrices, such as ω-3 oils, CLA, carotenoids, vitamin E, and some flavor oils. The rate of chemical degradation depends on factors such as light, oxygen, temperature, and the presence of prooxidants and antioxidants. These lipids can often be protected against degradation by incorporating them in suitable colloidal delivery systems.

v. Some lipids undergo biochemical changes after ingestion, thereby altering their biological activity, such as chemical or enzymatic hydrolysis, for example, by acids in the stomach or enzymes throughout the GIT.

vi. The oral bioavailability of many lipids is relatively low and variable because of their poor water solubility and food matrix effects. It may therefore be necessary to design colloidal delivery systems that increase their release, solubilization, and absorption within the GIT.

## 2.3 Protein-Based Ingredients

Proteins, peptides and amino acids are important nutrients because they provide an important source of energy and essential nutrients through the diet (Damodaran 2008). However, a number of proteins, peptides, and amino acids have also been claimed to have additional biological activities and can therefore be used as nutraceuticals, for example, acting as growth factors, antihypertensive agents, antimicrobial agents, antioxidants, food intake modifiers (satiation), and immune regulatory factors (Espitia et al. 2012; Huang et al. 2013; Marques et al. 2012; Meisel 1997; Playne et al. 2003; Ward and German 2004).

## 2.3.1 Proteins

Proteins are natural polymers that consist of amino acids linked by peptide bonds (Damodaran 2008; Hettiarachchy et al. 2012; Phillips and Williams 2011). Food proteins may form part of compositionally and structurally complicated ingredients (such as milk, flour, eggs, or meat) or they may be purified functional ingredients isolated from natural sources (such as gelatin, whey protein, caseinate, or soy protein). The type, number, and sequence of amino acids along the polypeptide chain of a protein molecule determine its molecular weight, conformation, electrical charge, flexibility, and hydrophobicity. These molecular characteristics then determine the functional and nutritional attributes of different proteins, for example, their ability to thicken solutions, form gels, hold water, adsorb to interfaces, stabilize emulsions and foams, catalyze reactions, bind specific molecules, and elicit biochemical responses. Proteins may have a variety of three-dimensional structures, including globular, fibrous, or random coil (Damodaran 2008). An example of the compact structure of a globular food protein from milk (β-lactoglobulin) is shown in Figure 2.8. Certain types of food proteins and their digestion products (peptides and amino acids) have been shown to exhibit biological activities, including soy, dairy, fish, and meat proteins (Chatterton et al. 2006; Kim and Mendis 2006; Kussmann and Affolter 2006; Wang and De Mejia 2005). These biological activities include antimicrobial activity, antioxidant activity, anticarcinogenic activity, inhibition of angiotensin-converting enzyme, reduced cholesterol and serum triglycerides, increased lean muscle mass, protection against pathogens,

β-Lactoglobulin

**Figure 2.8  Space-filling model of the molecular structure of a model globular protein (β-lactoglobulin) obtained from the Protein Database (NCBI).**

regulation of blood glucose levels, and satiety effects (Anderson and Moore 2004; Chatterton et al. 2006; Playne et al. 2003; Severin and Xia 2005; Seyler et al. 2007). Consequently, there has been considerable interest in fortifying foods and beverages with bioactive proteins.

## 2.3.2 Peptides and Amino Acids

Peptides and amino acids are naturally found in a range of natural sources, or they may be produced by hydrolysis of proteins (Damodaran 2008; Udenigwe and Aluko 2012). Certain peptides and amino acids have been shown to exhibit biological activities, such as mineral binding, antioxidant activity, antimicrobial activity, antihypertension, cancer prevention, and protection against heart disease (Espitia et al. 2012; Huang et al. 2013; Korhonen and Pihlanto 2003; Martinez-Maqueda et al. 2012; Meisel 2005; Muir 2005; Playne et al. 2003; Ricci-Cabello et al. 2012; Udenigwe and Aluko 2012). Bioactive peptides and amino acids are typically generated from food proteins by either controlled or uncontrolled hydrolysis. Protein hydrolysis may be carried out by a food or ingredient manufacturer (e.g., by chemical, enzymatic, or fermentation methods), or it may take place within the human GIT after food ingestion (e.g., by acids or digestive enzymes) (Lopez-Fandino et al. 2006; Meisel 1997; Muir 2005; Playne et al. 2003; Tavano 2013; Wang and De Mejia 2005). For example, casein phosphopeptides are calcium binding peptides that have been

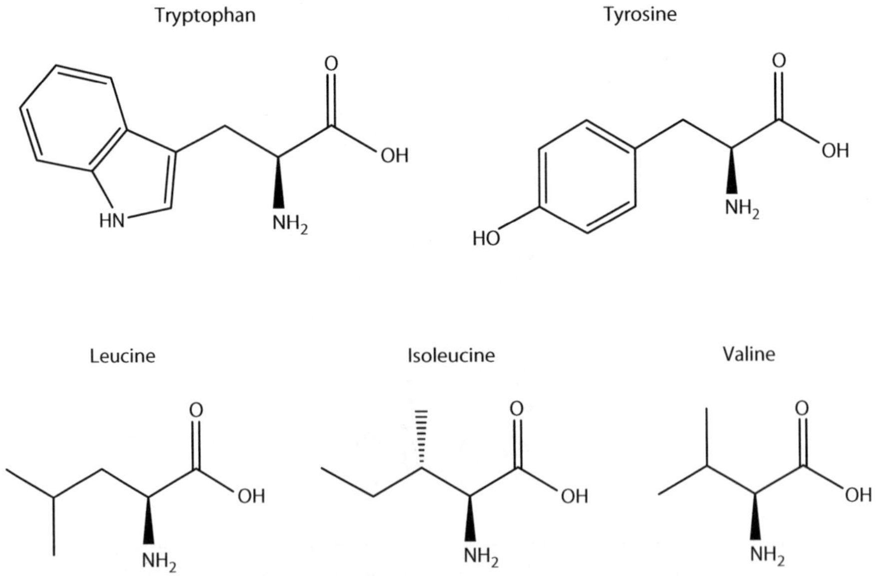

**Figure 2.9 Molecular structures of selected bioactive amino acids.**

claimed to exhibit a variety of biological activities, including protection against demineralization of teeth, antioxidant activity, antimicrobial activity, anticancer properties, and immuno-stimulation (Korhonen and Pihlanto 2003; Playne et al. 2003). Other peptides derived from milk, plants, and fish have been shown to be capable of reducing blood pressure (Martinez-Maqueda et al. 2012; Muir 2005; Ricci-Cabello et al. 2012).

Certain amino acids have also been demonstrated to have biological activities that make them suitable as nutraceutical food ingredients (McClements et al. 2009). Tryptophan is a precursor for brain serotonin synthesis whereas tyrosine is a precursor for dopamine synthesis (Figure 2.9). Both of these nonpolar amino acids have been linked to beneficial effects on human mood and stress (Banderet and Lieberman 1989; Chinevere et al. 2002; Young 1993). Leucine, isoleucine, and valine (Figure 2.9) have also been reported to have a variety of important biological functions in the brain (During et al. 1988; Fernstrom 2005; Harris et al. 2005; Russo et al. 2003).

## 2.3.3 Challenges to Delivery of Proteins, Peptides, and Amino acids

The major challenges to delivering proteins, peptides, and amino acids in food and beverage products are summarized below:

i. The solubility of proteins, peptides, and amino acids in aqueous solutions depends on their molecular structure, as well as environmental and solution conditions, such as temperature, pH, ionic strength, and ingredient interactions. It is important that the solubility characteristics of proteins are compatible with the food matrix that they are being incorporated into. In some products, proteins should remain water soluble (e.g., protein-enriched clear beverages), whereas in others, they may be either soluble or insoluble (e.g., desserts or yogurts).

ii. Some proteins have adverse effects on the mouthfeel or flavor profile of food products. For example, some proteins, peptides, and amino acids give a bitter or astringent mouthfeel that limits their application within certain foods (Cho et al. 2004; Pedrosa et al. 2006; Tripathi and Misra 2005). In addition, some proteins may bind volatile aroma molecules, thereby altering the overall flavor profile of the final product (Guichard 2002; Kuhn et al. 2006).

iii. Some proteins and peptides undergo alterations in their molecular structure during isolation, purification, or processing steps, which causes changes in their biological of functional activity, for example, unfolding, aggregation, or hydrolysis (Chatterton et al. 2006). For example, the structure and functionality of many globular proteins are altered by denaturation caused by changes in pH, ionic composition, or temperature (Damodaran 2008).

iv. Many proteins undergo alterations in their molecular structures and bio-activities within the human body after ingestion before they can reach the intended site of action (Moreno 2007; Wickham et al. 2009). For example, they may be hydrolyzed within the stomach or small intestine because of the presence of proteases. On the other hand, a certain amount of controlled hydrolysis within the GIT may be desirable since it releases specific bioactive peptides or amino acids (Meisel 1997, 2005).

Delivery systems may be developed to encapsulate proteins in a form where they can be easily dispersed into food products, without causing adverse effects on product appearance, texture, mouthfeel, or flavor, or undesirable losses in bioactivity (Balcao et al. 2013; Brandelli 2012; Kim 2012; Moutinho et al. 2012).

## 2.4 Carbohydrate-Based Ingredients

Carbohydrates are found in a wide variety of different biological species, with the most common food sources being plants, algae, and microorganisms (BeMiller and Huber 2008; Biliaderis and Izydorczyk 2007; Cui 2005). Carbohydrates are used as ingredients in foods for a number of reasons: an energy source (calories), colors (nonenzymatic browning), flavors (sweetness), water-activity control (humectants), texture modification and stabilization (thickeners and gelling agents), and emulsion and foam formation (emulsifiers and foaming agents). Carbohydrates also play an important role in human health (Biliaderis and Izydorczyk 2007). Digestible carbohydrates are a major source of calories and are therefore an important source of energy, although their overconsumption may lead to health problems such as over-weight, obesity, and diabetes (Johnson et al. 2009; Lustig et al. 2012). Consumption of certain types of indigestible carbohydrates (dietary fibers) has been linked to improving human health, for example, by promoting gut health, reducing choles-terol levels, and inhibiting chronic diseases, such as cancer, diabetes, and hypertension (Dhingra et al. 2012; Grabitske and Slavin 2009; Huang et al. 2013; Trigueros et al. 2013). Consequently, there is considerable interest in incorporating many of these health-promoting dietary fibers into foods.

There are a number ways of classifying carbohydrates (Cui 2005): (i) differences in molecular structures, such as type, number, bonding, and sequence of monosaccharides; (ii) differences in biological origin, such as wood, corn, wheat, apples, oranges, seaweed, and algae; (iii) differences in behavior within the human GI tract, for example, digestible or indigestible; (iv) differences in functionality in foods, for example, sweetener, humectant, thickener, gelling agent, bulking agent, water binder, emulsifier, or foaming agent. One of the most common ways of clas-sifying carbohydrates is according to the number of monomers they contain: *mono-saccharides* (1), *oligosaccharides* (2–20), and *polysaccharides* (>20).

## 2.4.1 Polysaccharides

### 2.4.1.1 Digestible Polysaccharides

Starches are the most commonly used digestible polysaccharide utilized in foods as functional ingredients (BeMiller and Huber 2008; Cui 2005). They are typically digested by amylases within the human GIT where they are eventually converted into glucose molecules suitable for absorption. Starches occur naturally in the roots, stems, seeds, and fruits of all green leaf plants, where their main function is to store energy. In its native state, starch is packed into plant cells as water-insoluble *granules* that vary in dimensions depending on their biological origin (typically 3–60 μm). In many processed foods, the native structure of starch is altered because of exposure of the granules to processing operations, such as heating, hydration, shearing, and enzymatic digestion. At the molecular level, starch consists of a mixture of two types of homo-polysaccharides that have glucose as the monosaccharide building block: *amylose*, which is primarily linear, and *amylopectin*, which is branched. Amylose typically contains 500 to 2000 glucose units mainly linked together into a linear chain by α-1,4 glycosidic bonds. Amylopectin typically contains greater than 1,000,000 glucose units linked together by α-1,4 bonds in linear segments and α-1,6 bonds at branch points. These two kinds of starch have different physiochemical and functional properties, and so it is often important to establish the *ratio* and *nature* of amylose and amylopectin present in the starch present within a particular food product.

In nature, amylose and amylopectin are organized into complex biological structures within starch granules that consist of *crystalline* regions separated by *amorphous* regions (Eliasson et al. 2013). When aqueous solutions of starch granules are heated above a critical temperature, they incorporate water and the crystalline regions are disrupted. The resultant *swelling* of the starch granule leads to an appreciable increase in solution viscosity ("gelatinization"). Upon further heating, a fraction of the starch leaches out of the granules and there is a subsequent decrease in viscosity. When the solution is cooled, linear regions of starch molecules associate with each other ("retrogradation") and there may be an increase in viscosity or even gelation. The rheological characteristics of a particular native starch depend on the structural organization of the molecules within the starch granule, the ratio of amylose to amylopectin, the precise molecular characteristics of each of these fractions, the solution composition (e.g., pH, ionic strength, sugar content), and environmental factors (e.g., shearing, temperature, pressure). The gels formed by native starch often have limited application in the food industry, because they do not have the desired solubility, textural, or stability characteristics. For this reason, starches are often physically, chemically, or enzymatically modified to improve their functional properties, for example, pregelatinization, limited hydrolysis, addition of side groups (polar, ionic, or hydrophobic) or controlled cross-linking. Many forms of ingested starch are broken in the mouth, stomach, and small intestine by amylases and other digestive enzymes to generate oligosaccharides and glucose (Singh et al. 2010). Nevertheless, certain types of starch may be more or less resistant to digestion within the GIT depending on their structure, for example, granule properties, degree of gelatinization, or extent of retrogradation (Parada and Aguilera 2011; Zhang and Hamaker 2009).

## 2.4.1.2 Indigestible Polysaccharides

Indigestible carbohydrates are not normally broken down by the enzymes secreted within the human GIT and therefore tend to reach the colon largely undigested. They form part of a larger group of indigestible polymeric food components called *dietary fibers*, which also includes lignin (Dikeman and Fahey 2006; Redgwell and Fischer 2005). Specific dietary fibers can be classified in a number of ways, including their biological origin, their molecular structure, their physicochemical properties, and their physiological effects (Redgwell and Fischer 2005; Wildman and Kelley 2007). In this book, we are primarily concerned with *soluble nondigestible polysaccharides* that have been isolated from their natural source and converted into functional food ingredients, rather than the fibers naturally present in whole foods, such as grains, fruits, and vegetables. This is because many isolated soluble fibers have been shown to be bioactive food components, which may need to be encapsulated in delivery systems before they can be successfully incorporated into foods.

Dietary fibers vary according to the type, number, distribution, and bonding of the monosaccharides they contain: they may be neutral, anionic, or cationic; they may have low, medium, or high molecular weights; they may be linear or branched; they may have ordered or disordered conformations; they vary in hydrophobicity; they may be homopolymers or heteropolymers; they may vary in chemical reactivity; they may vary in biological activity (Cui 2005, Biliaderis and Izydorczyk 2007). These differences in molecular characteristics cause differences in their physicochemical properties, for example, water solubility, viscosity enhancement capacity, gel formation, opacity, surface activity, and binding capacity (Cui 2005). In turn, these differences in molecular and physicochemical characteristics cause alterations in their biological activity after ingestion. The main bioactive functions that have been attributed to dietary fibers are cholesterol reduction, modulation of blood glucose levels, prevention of certain cancers, prevention of constipation, and prebiotic effects (Redgwell and Fischer 2005). Nevertheless, there is still a poor understanding of the molecular and physicochemical basis for the various bioactive functions that have been attributed to different kinds of dietary fibers.

At present, a large proportion of the population in many developed countries does not consume the recommended amount of dietary fiber (25 to 30 g per day) required for a healthy diet (Redgwell and Fischer 2005). Consequently, there is a need to increase the consumption of foods rich in dietary fibers to achieve the potential health benefits associated with them.

## 2.4.2 *Monosaccharides and Oligosaccharides*

Monosaccharides and oligosaccharides may be isolated from natural plant, animal, or microbial sources, or they may be produced by hydrolysis of polysaccharides (Barreteau et al. 2006). Monosaccharides and small oligosaccharides (such as glucose, fructose, lactose, and sucrose) are widely used in foods as sweeteners or bulking agents. However, there is considerable concern that overconsumption of sugars leads to rising overweight and obesity in developed countries and therefore there is a need to reduce their levels in foods (Lustig et al. 2012). This can be achieved by creating reduced-calorie artificial sweeteners to replace sugars (Lindsay 2008). Alternatively, delivery systems may prove useful for controlling the rate and extent of sugar release within the mouth, which may prove useful for fabricating reduced calorie foods that still have desirable sensory attributes.

Certain types of oligosaccharides (such as inulin, oligofructose, and oligochitosan) have been shown to have biological activity and can be utilized as nutraceutical agents in foods (Barreteau et al. 2006; Bosscher et al. 2009). Oligosaccharides in milk have been reported to have a range of biological activities, including protection against pathogens, modulation of the microbiome, improvement in immune response, enhancement of mineral absorption, and improvement in the formation of the brain and central nervous system (Al Mijan et al. 2011).

## 2.4.3 *Challenges to Delivery of Carbohydrates*

In this section, some of the major challenges commonly encountered when trying to incorporate carbohydrate-based functional ingredients into commercial food and beverage products are summarized:

i. Sugars are solid at ambient temperature but may melt upon heating or dissolve when they come into contact with water. It is often important to control the physical state of sugars in foods (e.g., liquid, glassy, or solid), as well as the morphology of any crystals formed (e.g., size, shape, and aggregation).

ii. Increasing the amount of dietary fiber present in foods to a level that has demonstrated bioactivity may have an adverse impact on product texture or mouthfeel. For example, many soluble dietary fibers cause a large increase in the viscosity of aqueous solutions when used at relatively low concentrations, which can impart an undesirable texture or mouthfeel to a product.

iii. Addition of dietary fibers to foods may cause product instability by promoting aggregation of other polymers or colloidal particles through depletion or bridging interactions (e.g., because of electrostatic or hydrophobic interactions).

iv. Dietary fibers may interact with other functional ingredients within food products and cause them to lose their desirable functional attributes; for example, anionic polysaccharides can bind cationic antimicrobials and cause them to lose their activity.

A well-designed colloidal delivery system may be able to increase the total amount of dietary fibers that can be incorporated into foods, without adversely affecting their organoleptic or physicochemical properties.

## 2.5 Mineral-Based Ingredients

Minerals are inorganic substances that may be added to foods as functional ingredients for a number of reasons, for example, flavor enhancement, food preservation, texture modification, or essential nutrients (Albarracin et al. 2011; Doyle and Glass 2010; Liem et al. 2011; White and Broadley 2005). There are health concerns associated with the overconsumption and underconsumption of certain types of mineral. For example, overconsumption of sodium may promote hypertension and related chronic diseases (Cook 2008; Mohan and Campbell 2009), whereas underconsumption of calcium may cause osteoporosis (Prentice 2004; Tang et al. 2007). There has therefore been considerable interest in reducing the levels of some minerals in foods, while increasing the levels of others. In general, there are numerous types of mineral that may be important to incorporate into food and beverage products (Miller 2008). In this section, only a limited number of minerals are discussed to highlight some of the issues involved.

### 2.5.1 Iron

Iron is an essential element for human health since it plays an important role in the functioning of various human proteins, including hemoglobin, myoglobin, and cytochromes (Bovell-Benjamin and Guinard 2003). Iron is important in biochemical processes such as oxygen transport, respiration, energy metabolism, and DNA synthesis (Miller 2008). The amount of iron absorbed from an ingested food depends on a number of factors, including meal composition, the physicochemical form of the iron, the nature of the food matrix, and the specific iron requirements of the individual (Bovell-Benjamin and Guinard 2003). Certain segments of the population do not absorb sufficient levels of iron from their diet, and therefore foods are often fortified with it. The main challenges to iron fortification are its relatively low and variable bioavailability, its propensity to promote undesirable chemical changes in other food components (e.g., lipid oxidation, protein oxidation, and color changes), and its undesirable taste profile (Bovell-Benjamin and Guinard 2003; Fairweather-Tait and Teucher 2002; Miller 2008). Various physicochemical forms of iron that are suitable for utilization within the food industry have been tabulated in the literature (Salgueiro et al. 2002). Water-soluble iron compounds are usually more bioavailable, but they have a tendency to promote oxidation reactions and have an undesirable metallic taste. Consequently, water-soluble iron compounds tend to be used mainly in low-moisture foods. Acid-soluble iron compounds are less chemically reactive in foods but still have a high bioavailability owing to their solubilization in gastric juices (Lynch et al. 2005). However, these compounds can only be used in limited food applications. Iron compounds that are insoluble in water over a wide pH range (such as elemental iron) have better chemical stability and taste characteristics, but their bioavailability is usually low (Salgueiro et al. 2002). Absorption of iron can be improved by chelation (e.g., with ascorbic acid or ethylenediaminetetraacetic acid) or by encapsulation in delivery systems (Fairweather-Tait and Teucher 2002; Lynch et al. 2005; Salgueiro et al. 2002).

## 2.5.2 Zinc

Zinc is another mineral that is essential for proper growth and development of humans because of its role in numerous metabolic processes, for example, maintaining the proper functioning of many enzymes and structural proteins within the human body (Miller 2008). Various physicochemical forms of zinc are available to fortify foods, with the most commonly used being zinc oxide, zinc sulfate, zinc chloride, zinc gluconate, and zinc stearate (Salgueiro et al. 2002). These compounds vary in their physicochemical and biological properties, which influences their suitability for incorporation into functional foods. Some of the main challenges to incorporating zinc into foods are its low and variable bioavailability and the potential for adverse ingredient interactions leading to precipitation.

## 2.5.3 Calcium

Calcium is a major constituent of human bones and it is therefore one of the most important minerals for ensuring proper bone health (Prentice 2004, Miller 2008). Dairy products are the major source of calcium in many countries, and therefore those who do not consume dairy products are at risk of inadequate calcium intake (Gao et al. 2006; Romanchik-Cerpovicz and McKemie 2007). Certain leafy vegetables are also good sources of this mineral, but the bioavailability of calcium is limited because of the presence of oxalate and phytic acid that bind to it in the GIT (White and Broadley 2005). Physicochemical forms of calcium suitable for fortification of foods have been tabulated elsewhere (Fairweather-Tait and Teucher 2002). Calcium fortification has been widely used in fruit juices, carbonated beverages, yeast breads, and breakfast cereals (Fairweather-Tait and Teucher 2002; Romanchik-Cerpovicz and McKemie 2007). Some of the major challenges associated with fortification of foods with calcium are as follows: it may promote physical instability owing to aggregation, precipitation, or gelation because of its interactions with other charged ingredients; it often has an undesirable mouthfeel; and it often has a low and variable bioavailability (Romanchik-Cerpovicz and McKemie 2007).

## 2.5.4 Challenges to Delivery of Minerals

There are a number of potential challenges associated with incorporating minerals into food and beverage products:

i. Mineral ions alter the electrostatic interactions between charged food components (e.g., many biopolymers and colloidal particles) owing to ion binding and electrostatic screening effects (McClements 2005). Consequently, they are capable of promoting aggregation and phase separation of charged biopolymers (e.g., proteins and polysaccharides) and colloidal particles (e.g., fat droplets, air bubbles, particulates) in some foods, which usually has a negative impact on appearance, texture, mouthfeel, and stability.

ii. Some mineral ions are capable of catalyzing the chemical degradation of certain food components. For example, transition metals, such as iron and copper, are capable of catalyzing the oxidative degradation of PUFAs and carotenoids (McClements and Decker 2000; Boon et al. 2009).

iii. Some forms of mineral have an inherently low oral bioavailability or their oral bioavailability is adversely affected by interactions with other food components (Intawongse and Dean 2006; Miller 2008). For example, the absorption of certain minerals is reduced because of their interactions with phytic acid and tannins from plants (Rimbach et al. 2008).

iv. Some minerals impart an undesirable taste or mouthfeel to food products, for example, high levels of soluble iron or calcium (Hong and Kim 2011).

v. Insoluble minerals may adversely affect the physicochemical attributes of food and beverage products because of their tendency to scatter light or sediment.

Colloidal delivery systems may be designed to encapsulate minerals in a form where they can be incorporated into foods without adversely affecting physicochemical, sensory, or nutritional properties.

# 2.6 Microorganisms

Consumption of certain kinds of nonpathogenic bacteria found in foods has been linked to beneficial health effects, such as reduction in irritable bowel syndrome, inflammatory bowel disease, colon cancer, and other chronic diseases (Boirivant and Strober 2007; Nair and Takeda 2011; Whelan and Quigley 2013). There has therefore been great interest in incorporating these beneficial bacteria ("probiotics") into foods. Nevertheless, the viability of these bacteria may be reduced either before or after ingestion. Prior to ingestion, bacterial viability may be compromised as a result of the harsh solution or environmental conditions they experience during food processing, transport, or storage, such as pH extremes, the presence of antimicrobial agents, ingredient interactions, elevated temperatures, shearing, or dehydration. After ingestion, bacterial viability may be reduced during their passage through the GIT because of the presence of highly acidic conditions in the stomach, bile salts in the small intestine, or interactions with other antimicrobial components in the GIT (Figure 2.10). There has therefore been considerable interest in developing effective delivery systems to protect probiotic bacteria within food

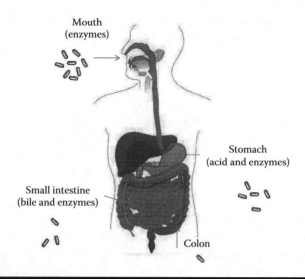

Mouth
(enzymes)

Stomach
(acid and enzymes)

Small intestine
(bile and enzymes)

Colon

**Figure 2.10    Probiotic bacteria must survive through the harsh conditions of the GIT before they reach the colon.**

products both before and after ingestion (Anal and Singh 2007; Cook et al. 2012; Heidebach et al. 2012; Islam et al. 2010; Teoh et al. 2011).

## 2.6.1  Probiotics

The nature of the microbial population within the human gut is believed to play a critical role in determining the health and wellness of human beings (Arora et al. 2013; Parvez et al. 2006; Teitelbaum and Walker 2002). Research has linked certain types of microbial populations to improved human health (Playne 2002). These bacterial populations have been claimed to have a variety of beneficial effects, including stimulating the immune system, improving gastrointestinal health, inhibition of harmful bacteria, improving lipid metabolism, synthesizing essential nutrients (such as vitamins), decreasing allergies, and reducing the incidences of certain kinds of cancer (Arora et al. 2013; Parvez et al. 2006; Teitelbaum and Walker 2002). Food manufacturers are therefore attempting to incorporate live beneficial bacteria (probiotics) into functional foods and beverages so as to increase their potential health benefits. Even though laboratory studies suggest that a particular type of bacteria is an effective probiotic, the viability and bioactivity of a probiotic may be lost before it reaches the colon owing to microbial degradation within the food or within the GIT (Figure 2.10). The viability of a probiotic may be increased by encapsulating it within an appropriate delivery system (Burgain et al. 2011; Cook et al. 2012; Heidebach et al. 2012; Islam et al. 2010).

There are a number of ways of improving the viability of probiotics in foods and beverages that do not involve using delivery systems. First, the viability of a probiotic depends on the strain of bacteria, with some strains having good resistance to the major stresses encountered within foods or within the GIT, such as high acidity and bile salts. Second, the nature of the food matrix surrounding the probiotics in a food or beverage can be optimized to improve their viability. For example, foods (like yogurt) that contain high levels of protein may be able to increase the pH of the stomach through their buffering capacity, thereby reducing the pH stress normally experienced by probiotics in the gastric environment. Third, the composition and handling of food or beverage products could be carefully controlled to prevent loss of probiotic viability, such as nutrient levels, pH, ionic strength, temperature, and oxygen. Nevertheless, nonencapsulated probiotics may still be susceptible to loss of viability in a food matrix or after ingestion, and therefore encapsulation may be required to ensure that they reach the large intestine in sufficient quantities. For example, probiotics could be encapsulated within the hydrophilic interior of hydrogel particles, liposomes, or W/O/W emulsions to protect them from hydrophilic components in the aqueous phases of foods or gastrointestinal fluids.

## 2.6.2 Challenges to Delivery of Probiotics

There are a number of challenges that the food industry faces when trying to incorporate sufficient levels of probiotics into foods and beverages.

i. The viability of probiotics may be decreased during the manufacture, storage, or utilization of food products, thereby reducing the number of beneficial bacteria consumed. In particular, thermal processing and pH extremes may severely reduce the number of beneficial bacteria present in food and beverage products.

ii. Probiotics usually have to reach the colon before they can exhibit their beneficial health effects, that is, by altering the composition of the microbial population colonizing the large intestine. However, the viability of probiotics may be decreased during passage through the upper GIT (e.g., because of the presence of high acidity, bile salts, or digestive enzymes), thereby reducing the number of beneficial bacteria reaching the colon.

iii. Probiotics may alter the desirable physicochemical and sensory attributes of foods, such as appearance, texture, flavor, or stability. Probiotic bacteria are relatively large ($d > 1$ μm) and therefore any delivery system should contain particles larger than this size. Consequently, they will tend to scatter light strongly and be prone to gravitational separation. As a result, probiotic delivery systems are most suitable for use in optically opaque foods that are highly viscous or gel-like (such as yogurts or desserts).

## 2.7 Summary

This chapter has highlighted a number of the functional ingredients that may need to be incorporated into food and beverage products and has discussed some of the potential challenges that need to be overcome. These functional ingredients include a diverse range of materials with different molecular, physicochemical, and biological properties. They also exhibit a wide range of different functional attributes, such as antioxidants, antimicrobials, colorants, flavorings, nutraceuticals, and probiotics. Consequently, different kinds of delivery systems are required for different applications. In the remainder of this book, various types of colloidal delivery systems that have been developed to encapsulate functional ingredients will be presented, and their relative advantages and disadvantages are discussed.

## References

Akoh, C. C. and D. B. Min (2008). *Food Lipids: Chemistry, Nutrition, and Biotechnology*. Boca Raton, FL, CRC Press.

Al Mijan, M., Y. K. Lee and H. S. Kwak (2011). "Classification, structure, and bioactive functions of oligosaccharides in milk." *Korean Journal for Food Science of Animal Resources* 31(5): 631–640.

Albarracin, W., I. C. Sanchez, R. Grau and J. M. Barat (2011). "Salt in food processing; usage and reduction: A review." *International Journal of Food Science and Technology* 46(7): 1329–1336.

Anal, A. K. and H. Singh (2007). "Recent advances in microencapsulation of probiotics for industrial applications and targeted delivery." *Trends in Food Science and Technology* 18(5): 240–251.

Anderson, G. H. and S. E. Moore (2004). "Dietary proteins in the regulation of food intake and body weight in humans." *Journal of Nutrition* 134(4): 974S–979S.

Arab-Tehrany, E., M. Jacquot, C. Gaiani, M. Imran, S. Desobry and M. Linder (2012). "Beneficial effects and oxidative stability of omega-3 long-chain polyunsaturated fatty acids." *Trends in Food Science and Technology* 25(1): 24–33.

Arnoldi, A. (2004). *Functional Foods, Cardiovascular Diease and Diabetes*. Boca Raton, FL, CRC Press.

Arora, T., S. Singh and R. K. Sharma (2013). "Probiotics: Interaction with gut microbiome and antiobesity potential." *Nutrition* 29(4): 591–596.

Bakkali, F., S. Averbeck, D. Averbeck and M. Waomar (2008). "Biological effects of essential oils—A review." *Food and Chemical Toxicology* 46(2): 446–475.

Balcao, V. M., C. I. Costa, C. M. Matos, C. G. Moutinho, M. Amorim, M. E. Pintado, A. P. Gomes, M. M. Vila and J. A. Teixeira (2013). "Nanoencapsulation of bovine lactoferrin for food and biopharmaceutical applications." *Food Hydrocolloids* 32(2): 425–431.

Banderet, L. E. and H. R. Lieberman (1989). "Treatment with tyrosine, a neurotransmitter precursor, reduces environmental-stress in humans." *Brain Research Bulletin* 22(4): 759–762.

Barreteau, H., C. Delattre and P. Michaud (2006). "Production of oligosaccharides as promising new food additive generation." *Food Technology and Biotechnology* **44**(3): 323–333.

Basu, A. and V. Imrhan (2007). "Tomatoes versus lycopene in oxidative stress and carcinogenesis: Conclusions from clinical trials." *European Journal of Clinical Nutrition* **61**(3): 295–303.

BeMiller, J. N. and K. C. Huber (2008). Carbohydrates. In *Food Chemistry*, S. Damodaran, K. L. Parkin and O. R. Fennema (eds.). Boca Raton, FL, CRC Press: 83–154.

Berton-Carabin, C., C. Genot, C. Gaillard, D. Guibert and M. H. Ropers (2013). "Design of interfacial films to control lipid oxidation in oil-in-water emulsions." *Food Hydrocolloids* **33**(1): 99–105.

Biliaderis, C. G. and M. S. Izydorczyk (2007). *Functional Food Carbohydrates*. Boca Raton, FL, CRC Press.

Boirivant, M. and W. Strober (2007). "The mechanism of action of probiotics." *Current Opinion in Gastroenterology* **23**(6): 679–692.

Boon, C. S., D. J. McClements, J. Weiss and E. A. Decker (2009). "Role of iron and hydroperoxides in the degradation of lycopene in oil-in-water emulsions." *Journal of Agricultural and Food Chemistry* **57**(7): 2993–2998.

Boon, C. S., D. J. McClements, J. Weiss and E. A. Decker (2010). "Factors influencing the chemical stability of carotenoids in foods." *Critical Reviews in Food Science and Nutrition* **50**(6): 515–532.

Bosscher, D., A. Breynaert, L. Pieters and N. Hermans (2009). "Food-based strategies to modulate the composition of the intestinal microbiota and their associated health effects." *Journal of Physiology and Pharmacology* **60**: 5–11.

Bovell-Benjamin, A. C. and J. X. Guinard (2003). "Novel approaches and application of contemporary sensory evaluation practices in iron fortification programs." *Critical Reviews in Food Science and Nutrition* **43**(4): 379–400.

Brandelli, A. (2012). "Nanostructures as promising tools for delivery of antimicrobial peptides." *Mini-Reviews in Medicinal Chemistry* **12**(8): 731–741.

Brigelius-Flohe, R. and F. Galli (2010). "Vitamin E: A vitamin still awaiting the detection of its biological function." *Molecular Nutrition and Food Research* **54**(5): 583–587.

Burgain, J., C. Gaiani, M. Linder and J. Scher (2011). "Encapsulation of probiotic living cells: From laboratory scale to industrial applications." *Journal of Food Engineering* **104**(4): 467–483.

Burt, S. (2004). "Essential oils: Their antibacterial properties and potential applications in foods—A review." *International Journal of Food Microbiology* **94**(3): 223–253.

Burton, G. W. and M. G. Traber (1990). "Vitamin E: Antioxidant activity, biokinetics, and bioavailability." *Annual Review of Nutrition* **10**(1): 357–382.

Cercaci, L., M. T. Rodriguez-Estrada, G. Lercker and E. A. Decker (2007). "Phytosterol oxidation in oil-in-water emulsions and bulk oil." *Food Chemistry* **102**(1): 161–167.

Chanamai, R. and D. J. McClements (2000). "Impact of weighting agents and sucrose on gravitational separation of beverage emulsions." *Journal of Agricultural and Food Chemistry* **48**(11): 5561–5565.

Chatterton, D. E. W., G. Smithers, P. Roupas and A. Brodkorb (2006). "Bioactivity of beta-lactoglobulin and alpha-lactalbumin—Technological implications for processing." *International Dairy Journal* **16**(11): 1229–1240.

Chen, C. C. and G. Wagner (2004). "Vitamin E nanoparticle for beverage applications." *Chemical Engineering Research and Design* **82**(A11): 1432–1437.

Chen, L. Y., G. E. Remondetto and M. Subirade (2006). "Food protein-based materials as nutraceutical delivery systems." *Trends in Food Science and Technology* **17**(5): 272–283.

Chin, S. F., J. M. Storkson, K. J. Albright and M. W. Pariza (1994). "Conjugated linoleic-acid (9,11-octadecadienoic and 10,12-octadecadienoic acid) is produced in conventional but not germ-free rats fed linoleic-acid." *Journal of Nutrition* **124**(5): 694–701.

Chinevere, T. D., R. D. Sawyer, A. R. Creer, R. K. Conlee and A. C. Parcell (2002). "Effects of L-tyrosine and carbohydrate ingestion on endurance exercise performance." *Journal of Applied Physiology* **93**(5): 1590–1597.

Chiu, Y. and W. Yang (1992). "Preparation of vitamin E microemulsion possessing high resistance to oxidation in air." *Colloids and Surfaces* **63**(3–4): 311–322.

Cho, M. J., N. Unklesbay, F. H. Hsieh and A. D. Clarke (2004). "Hydrophobicity of bitter peptides from soy protein hydrolysates." *Journal of Agricultural and Food Chemistry* **52**(19): 5895–5901.

Choi, S. J., E. A. Decker, L. Henson, L. M. Popplewell and D. J. McClements (2009). "Stability of citral in oil-in-water emulsions prepared with medium-chain triacylglycerols and triacetin." *Journal of Agricultural and Food Chemistry* **57**(23): 11349–11353.

Choi, S. J., E. A. Decker, L. Henson, L. M. Popplewell and D. J. McClements (2010). "Inhibition of citral degradation in model beverage emulsions using micelles and reverse micelles." *Food Chemistry* **122**(1): 111–116.

Chu, B., S. Ichikawa, S. Kanafusa and M. Nakajima (2007). "Preparation of protein-stabilized-carotene nanodispersions by emulsification–evaporation method." *Journal of the American Oil Chemists' Society* **84**(11): 1053–1062.

Chu, B. S., S. Ichikawa, S. Kanafusa and M. Nakajima (2008). "Stability of protein-stabilised beta-carotene nanodispersions against heating, salts and pH." *Journal of the Science of Food and Agriculture* **88**(10): 1764–1769.

Cook, M. E., D. L. Jerome, T. D. Crenshaw, D. R. Buege, M. W. Pariza, K. J. Albright, S. P. Schmidt, J. A. Scimeca, P. A. Lofgren and E. J. Hentges (1998). "Feeding conjugated linoleic acid improves feed efficiency and reduces carcass fat in pigs." *FASEB Journal* **12**(5): A836.

Cook, M. T., G. Tzortzis, D. Charalampopoulos and V. V. Khutoryanskiy (2012). "Microencapsulation of probiotics for gastrointestinal delivery." *Journal of Controlled Release* **162**(1): 56–67.

Cook, N. R. (2008). "Salt intake, blood pressure and clinical outcomes." *Current Opinion in Nephrology and Hypertension* **17**(3): 310–314.

Cortesi, R., E. Esposito, G. Luca and C. Nastruzzi (2002). "Production of lipospheres as carriers for bioactive compounds." *Biomaterials* **23**(11): 2283–2294.

Cranney, A., H. A. Weiler, S. O'Donnell and L. Puil (2008). "Summary of evidence-based review on vitamin D efficacy and safety in relation to bone health." *American Journal of Clinical Nutrition* **88**(2): 513S–519S.

Cui, S. W. (2005). *Food Carbohydrates: Chemistry, Physical Properties and Applications.* Boca Raton, FL, Taylor & Francis.

Damodaran, S. (2008). Amino acids, peptides and proteins. In *Food Chemistry*, S. Damodaran, K. L. Parkin and O. R. Fennema (eds.). Boca Raton, FL, CRC Press: 217–330.

Dhingra, D., M. Michael, H. Rajput and R. T. Patil (2012). "Dietary fibre in foods: A review." *Journal of Food Science and Technology-Mysore* **49**(3): 255–266.

Dikeman, C. L. and G. C. Fahey (2006). "Viscosity as related to dietary fiber: A review." *Critical Reviews in Food Science and Nutrition* **46**(8): 649–663.

Djordjevic, D., L. Cercaci, J. Alamed, D. J. McClements and E. A. Decker (2007). "Chemical and physical stability of citral and limonene in sodium dodecyl sulfate-chitosan and gum arabic-stabilized oil-in-water emulsions." *Journal of Agricultural and Food Chemistry* **55**(9): 3585–3591.

Doyle, M. E. and K. A. Glass (2010). "Sodium reduction and its effect on food safety, food quality, and human health." *Comprehensive Reviews in Food Science and Food Safety* **9**(1): 44–56.

During, M. J., I. N. Acworth and R. J. Wurtman (1988). "Effects of systemic L-tyrosine on dopamine release from rat corpus striatum and nucleus accumbens." *Brain Research* **452**(1–2): 378–380.

Eliasson, A. C., B. Bergenstahl, L. Nilsson and M. Sjoo (2013). "From molecules to products: Some aspects of structure-function relationships in cereal starches." *Cereal Chemistry* **90**(4): 326–334.

Engel, R. and H. Schubert (2005). "Formulation of phytosterols in emulsions for increased dose response in functional foods." *Innovative Food Science and Emerging Technologies* **6**(2): 233–237.

Espin, J. C., M. T. Garcia-Conesa and F. A. Tomas-Barberan (2007). "Nutraceuticals: Facts and fiction." *Phytochemistry* **68**(22–24): 2986–3008.

Espitia, P. J. P., N. D. F. Soares, J. S. D. Coimbra, N. J. de Andrade, R. S. Cruz and E. A. A. Medeiros (2012). "Bioactive peptides: Synthesis, properties, and applications in the packaging and preservation of food." *Comprehensive Reviews in Food Science and Food Safety* **11**(2): 187–204.

Fairweather-Tait, S. J. and B. Teucher (2002). "Iron and calcium bioavailability of fortified foods and dietary supplements." *Nutrition Reviews* **60**(11): 360–367.

Feng, J.-L., Z.-W. Wang, J. Zhang, Z.-N. Wang and F. Liu (2009). "Study on food-grade vitamin E microemulsions based on nonionic emulsifiers." *Colloids and Surfaces A: Physicochemical and Engineering Aspects* **339**(1–3): 1–6.

Fernandes, P. and J. M. S. Cabral (2007). "Phytosterols: Applications and recovery methods." *Bioresource Technology* **98**(12): 2335–2350.

Fernstrom, J. D. (2005). "Branched-chain amino acids and brain function." *Journal of Nutrition* **135**(6): 1539S–1546S.

Flanagan, J. and H. Singh (2006). "Microemulsions: A potential delivery system for bioactives in food." *Critical Reviews in Food Science and Nutrition* **46**(3): 221–237.

Gamarra, F. M. C., L. S. Sakanaka, E. B. Tambourgi and F. A. Cabral (2006). "Influence on the quality of essential lemon (Citrus aurantifolia) oil by distillation process." *Brazilian Journal of Chemical Engineering* **23**(1): 147–151.

Gao, X., P. E. Wilde, A. H. Lichtenstein and K. L. Tucker (2006). "Meeting adequate intake for dietary calcium without dairy foods in adolescents aged 9 to 18 years (National Health and Nutrition Examination Survey 2001–2002)." *Journal of the American Dietetic Association* **106**(11): 1759–1765.

Gawrysiak-Witulska, M., A. Siger and M. Nogala-Kalucka (2009). "Degradation of tocopherols during near-ambient rapeseed drying." *Journal of Food Lipids* **16**(4): 524–539.

Gibbs, B. F., S. Kermasha, I. Alli and C. N. Mulligan (1999). "Encapsulation in the food industry: A review." *International Journal of Food Sciences and Nutrition* **50**(3): 213–224.

Given, P. S. (2009). "Encapsulation of flavors in emulsions for beverages." *Current Opinion in Colloid and Interface Science* **14**(1): 43–47.

Golli, F. and A. Azzi (2010). "Present trends in vitamin E research." *Biofactors* **36**(1): 33–42.

Gonnet, M., L. Lethuaut and F. Boury (2010). "New trends in encapsulation of liposoluble vitamins." *Journal of Controlled Release* **146**(3): 276–290.

Grabitske, H. and J. Slavin (2009). "Gastrointestinal effects of low-digestible carbohydrates." *Critical Reviews in Food Science and Nutrition* **49**(4): 327–360.

Guichard, E. (2002). "Interactions between flavor compounds and food ingredients and their influence on flavor perception." *Food Reviews International* **18**(1): 49–70.

Ha, Y. L., N. K. Grimm and M. W. Pariza (1988). "Anticarcinogenic fatty-acids from fried ground-beef." *FASEB Journal* **2**(5): A1192.

Haham, M., S. Ish-Shalom, M. Nodelman, I. Duek, E. Segal, M. Kustanovich and Y. D. Livney (2012). "Stability and bioavailability of vitamin D nanoencapsulated in casein micelles." *Food and Function* **3**(7): 737–744.

Hallikainen, M. A., E. S. Sarkkinen and M. I. J. Uusitupa (2000). "Plant stanol eaters affect serum cholesterol concentrations of hypercholesterolemic men and women in a dose-dependent manner." *Journal of Nutrition* **130**(4): 767–776.

Hark, L. and D. Deen (2005). *Nutrition for Life*. London, UK, Dorling Kindersley.

Harris, R. A., M. Joshi, N. H. Jeoung and M. Obayashi (2005). "Overview of the molecular and biochemical basis of branched-chain amino acid catabolism." *Journal of Nutrition* **135**(6): 1527S–1530S.

Hatanaka, J., H. Chikamori, H. Sato, S. Uchida, K. Debari, S. Onoue and S. Yamada (2010). "Physicochemical and pharmacological characterization of alpha-tocopherol-loaded nano-emulsion system." *International Journal of Pharmaceutics* **396**(1–2): 188–193.

Heidebach, T., P. Forst and U. Kulozik (2012). "Microencapsulation of probiotic cells for food applications." *Critical Reviews in Food Science and Nutrition* **52**(4): 291–311.

Heinonen, M., K. Haila, A. M. Lampi and V. Piironen (1997). "Inhibition of oxidation in 10% oil-in-water emulsions by beta-carotene with alpha- and gamma-tocopherols." *Journal of the American Oil Chemists Society* **74**(9): 1047–1052.

Hettiarachchy, N. S., K. Sato, M. R. Marshall and A. Kannan (2012). *Food Proteins and Peptides: Chemistry, Functionality, Interactions, and Commercialization*. Boca Raton, FL, CRC Press.

Holick, M. F. (2004). "Vitamin D: Importance in the prevention of cancers, type 1 diabetes, heart disease, and osteoporosis." *American Journal of Clinical Nutrition* **79**(3): 362–371.

Holick, M. F. (2007). "Vitamin D deficiency." *New England Journal of Medicine* **357**(3): 266–281.

Hong, J. H. and K. O. Kim (2011). "Operationally defined solubilization of copper and iron in human saliva and implications for metallic flavor perception." *European Food Research and Technology* **233**(6): 973–983.

Huang, W. Y., S. T. Davidge and J. P. Wu (2013). "Bioactive natural constituents from food sources—Potential use in hypertension prevention and treatment." *Critical Reviews in Food Science and Nutrition* **53**(6): 615–630.

Hutchings, J. B. (1999). *Food Color and Appearance*. New York, Springer Publishers.

Intawongse, M. and J. R. Dean (2006). "In-vitro testing for assessing oral bioaccessibility of trace metals in soil and food samples." *Trac-Trends in Analytical Chemistry* **25**(9): 876–886.

Islam, M. A., C. H. Yun, Y. J. Choi and C. S. Cho (2010). "Microencapsulation of live probiotic bacteria." *Journal of Microbiology and Biotechnology* **20**(10): 1367–1377.

Jacobsen, C. (2008). "Omega-3s in food emulsions: Overview and case studies." *Agro Food Industry Hi-Tech* **19**(5): 9–12.

John, S., A. V. Sorokin and P. D. Thompson (2007). "Phytosterols and vascular disease." *Current Opinion in Lipidology* **18**(1): 35–40.

Johnson, R. K., L. J. Appel, M. Brands, B. V. Howard, M. Lefevre, R. H. Lustig, F. Sacks, L. M. Steffen, J. Wylie-Rosett, M. Council Nutr Phys Activity, E. Council and Prevention (2009). "Dietary sugars intake and cardiovascular health a scientific statement from the American Heart Association." *Circulation* **120**(11): 1011–1020.

Kandansamy, K. and P. D. Somasundaram (2012). "Microencapsulation of colors by spray drying—A review." *International Journal of Food Engineering* **8**(2). doi: 10.1515/1556-3758.2647.

Kepler, C. R., K. P. Hirons, J. J. McNeill and S. B. Tove (1966). "Intermediates and products of biohydrogenation of linoleic acid by butyrivibrio fibrisolvens." *Journal of Biological Chemistry* **241**(6): 1350–1354.

Kim, H. (2012). "Natural products to improve quality of life targeting for colon drug delivery." *Current Drug Delivery* **9**(2): 132–147.

Kim, S. K. and E. Mendis (2006). "Bioactive compounds from marine processing byproducts—A review." *Food Research International* **39**(4): 383–393.

Korhonen, H. and A. Pihlanto (2003). "Bioactive peptides: New challenges and opportunities for the dairy industry." *Australian Journal of Dairy Technology* **58**(2): 129–134.

Kuhn, J., T. Considine and H. Singh (2006). "Interactions of milk proteins and volatile flavor compounds: Implications in the development of protein foods." *Journal of Food Science* **71**(5): R72–R82.

Kussmann, M. and M. Affolter (2006). "Proteomic methods in nutrition." *Current Opinion in Clinical Nutrition and Metabolic Care* **9**(5): 575–583.

Larsen, R., K. E. Eilertsen and E. O. Elvevoll (2011). "Health benefits of marine foods and ingredients." *Biotechnology Advances* **29**(5): 508–518.

Lauridsen, C., M. S. Hedemann and S. K. Jensen (2001). "Hydrolysis of tocopheryl and retinyl esters by porcine carboxyl ester hydrolase is affected by their carboxylate moiety and bile acids." *The Journal of Nutritional Biochemistry* **12**(4): 219–224.

Lee, J. H., J. H. O'Keefe, D. Bell, D. D. Hensrud and M. F. Holick (2008). "Vitamin D deficiency: An important, common, and easily treatable cardiovascular risk factor?" *Journal of the American College of Cardiology* **52**(24): 1949–1956.

Lee, K. N., D. Kritchevsky and M. W. Pariza (1994). "Conjugated linoleic-acid and atherosclerosis in rabbits." *Atherosclerosis* **108**(1): 19–25.

Leong, W. F., O. M. Lai, K. Long, Y. B. C. Man, M. Misran and C. P. Tan (2011). "Preparation and characterisation of water-soluble phytosterol nanodispersions." *Food Chemistry* **129**(1): 77–83.

Li, X., N. Anton, M. C. T. Thi, M. J. Zhao, N. Messaddeq and T. F. Vandamme (2011). "Microencapsulation of nanoemulsions: Novel Trojan particles for bioactive lipid molecule delivery." *International Journal of Nanomedicine* **6**: 1313–1325.

Liem, D. G., F. Miremadi and R. S. J. Keast (2011). "Reducing sodium in foods: The effect on flavor." *Nutrients* **3**(6): 694–711.

Lindsay, R. C. (2008). Food additives. In *Food Chemistry*, 4th Edition, S. Damodaran, K. L. Parkin and O. R. Fennema (eds). Boca Raton, FL, CRC Press: 690–749.

Liu, F., R. Q. Zhong and Z. W. Wang (2010). "Formulation of conjugated linoleic acid microemulsion using mixed nonionic surfactants." *Journal of Dispersion Science and Technology* **31**(6): 715–721.

Lopes, D., M. P. C. Silvestre, H. Chiarini-Garcia, E. S. Garcia, H. A. Morais and M. R. Silva (2011). "Evaluation of conjugated linoleic acid addition to a chocolate milk drink." *International Journal of Food Engineering* **7**(2). doi: 10.2202/1556-3758.1842.

Lopez-Fandino, R., J. Otte and J. van Camp (2006). "Physiological, chemical and technological aspects of milk-protein-derived peptides with antihypertensive and ACE-inhibitory activity." *International Dairy Journal* **16**(11): 1277–1293.

Losso, J. N., A. Khachatryan, M. Ogawa, J. S. Godber and F. Shih (2005). "Random centroid optimization of phosphatidylglycerol stabilized lutein-enriched oil-in-water emulsions at acidic pH." *Food Chemistry* **92**(4): 737–744.

Lustig, R. H., L. A. Schmidt and C. D. Brindis (2012). "The toxic truth about sugar." *Nature* **482**(7383): 27–29.

Lynch, J. M., A. L. Lock, D. A. Dwyer, R. Noorbakhsh, D. M. Barbano and D. E. Bauman (2005). "Flavor and stability of pasteurized milk with elevated levels of conjugated linoleic acid and vaccenic acid." *Journal of Dairy Science* **88**(2): 489–498.

Maiani, G., M. J. P. Caston, G. Catasta, E. Toti, I. G. Cambrodon, A. Bysted, F. Granado-Lorencio et al. (2009). "Carotenoids: Actual knowledge on food sources, intakes, stability and bioavailability and their protective role in humans." *Molecular Nutrition and Food Research* **53**: S194–S218.

Marques, C., M. M. Amorim, J. O. Pereira, M. E. Pintado, D. Moura, C. Calhau and H. Pinheiro (2012). "Bioactive peptides—Are there more antihypertensive mechanisms beyond ace inhibition?" *Current Pharmaceutical Design* **18**(30): 4706–4713.

Martinez-Maqueda, D., B. Miralles, I. Recio and B. Hernandez-Ledesma (2012). "Antihypertensive peptides from food proteins: A review." *Food and Function* **3**(4): 350–361.

Matalanis, A., E. A. Decker and D. J. McClements (2012). "Inhibition of lipid oxidation by encapsulation of emulsion droplets within hydrogel microspheres." *Food Chemistry* **132**(2): 766–772.

Mayne, S. T. (1996). "Beta-carotene, carotenoids, and disease prevention in humans." *FASEB Journal* **10**(7): 690–701.

McClements, D. J. (2005). *Food Emulsions: Principles, Practice, and Techniques*. Boca Raton, FL, CRC Press.

McClements, D. J. (2012a). "Advances in fabrication of emulsions with enhanced functionality using structural design principles." *Current Opinion in Colloid and Interface Science* **17**(5): 235–245.

McClements, D. J. (2012b). "Crystals and crystallization in oil-in-water emulsions: Implications for emulsion-based delivery systems." *Advances in Colloid and Interface Science* **174**: 1–30.

McClements, D. J. and E. A. Decker (2000). "Lipid oxidation in oil-in-water emulsions: Impact of molecular environment on chemical reactions in heterogeneous food systems." *Journal of Food Science* **65**(8): 1270–1282.

McClements, D. J. and E. A. Decker (2008). Lipids. In *Food Chemistry*, 4th Edition, S. Damodaran, K. L. Parkin and O. R. Fennema (eds.). Boca Raton, FL, CRC Press: 155–216.

McClements, D. J., E. A. Decker, Y. Park and J. Weiss (2009). "Structural design principles for delivery of bioactive components in nutraceuticals and functional foods." *Critical Reviews in Food Science and Nutrition* **49**(6): 577–606.

McClements, D. J., L. Henson, L. M. Popplewell, E. A. Decker and S. J. Choi (2012). "Inhibition of ostwald ripening in model beverage emulsions by addition of poorly water soluble triglyceride oils." *Journal of Food Science* **77**(1): C33–C38.

Meisel, H. (1997). "Biochemical properties of bioactive peptides derived from milk proteins: Potential nutraceuticals for food and pharmaceutical applications." *Livestock Production Science* **50**(1–2): 125–138.

Meisel, H. (2005). "Biochemical properties of peptides encrypted in bovine milk proteins." *Current Medicinal Chemistry* **12**(16): 1905–1919.

Miller, C. C., Y. Park, M. W. Pariza and M. E. Cook (1994). "Feeding conjugated linoleic-acid to animals partially overcomes catabolic responses due to endotoxin injection." *Biochemical and Biophysical Research Communications* **198**(3): 1107–1112.

Miller, D. D. (2008). Minerals. In *Food Chemistry*, 4th Edition, S. Damodaran, K. L. Parkin and O. R. Fennema (eds.). Boca Raton, FL, CRC Press: 523–569.

Misharina, T. A., M. B. Terenina, N. I. Krikunova and I. B. Medvedeva (2010). "Autooxidation of a mixture of lemon essential oils, methyl linolenoate, and methyl oleinate." *Applied Biochemistry and Microbiology* **46**(5): 551–556.

Mohan, S. and N. R. C. Campbell (2009). "Salt and high blood pressure." *Clinical Science* **117**(1–2): 1–11.

Moreno, F. J. (2007). "Gastrointestinal digestion of food allergens: Effect on their allergenicity." *Biomedicine and Pharmacotherapy* **61**(1): 50–60.

Moutinho, C. G., C. M. Matos, J. A. Teixeira and V. M. Balcao (2012). "Nanocarrier possibilities for functional targeting of bioactive peptides and proteins: State-of-the-art." *Journal of Drug Targeting* **20**(2): 114–141.

Muir, A. D. (2005). "Natural peptides in blood pressure control. A review." *Agro Food Industry Hi-Tech* **16**(5): 15–17.

Nair, G. B. and Y. Takeda (2011). *Probiotic Foods in Health and Disease*. Boca Raton, FL, CRC Press.

Nicolosi, R. J., E. J. Rogers, D. Kritchevsky, J. A. Scimeca and P. J. Huth (1997). "Dietary conjugated linoleic acid reduces plasma lipoproteins and early aortic atherosclerosis in hypercholesterolemic hamsters." *Artery* **22**(5): 266–277.

O'Callaghan, Y. and N. O'Brien (2010). "Bioaccessibility, cellular uptake and transepithelial transport of alpha-tocopherol and retinol from a range of supplemented foodstuffs assessed using the caco-2 cell model." *International Journal of Food Science and Technology* **45**(7): 1436–1442.

Orchard, T. S., X. L. Pan, F. Cheek, S. W. Ing and R. D. Jackson (2012). "A systematic review of omega-3 fatty acids and osteoporosis." *British Journal of Nutrition* **107**: S253–S260.

Ortiz-Moreno, A., L. Dorantes-Alvarez, M. M. Hernandez-Ortega, G. Chamorro-Cevallos and L. Garduno-Siciliano (2011). Bioactive compounds related to fats and oils. In *Nutraceuticals and Functional Foods: Conventional and Non-conventional Sources*, M. E. Jaramillo-Flores, E. C. Lugo-Cervantes and L. Chel-Guerrero (eds.). Houston, TX, Studium Press LLC: 203–228.

Ostlund, R. E. (2004). "Phytosterols and cholesterol metabolism." *Current Opinion in Lipidology* **15**(1): 37–41.

Parada, J. and J. M. Aguilera (2011). "Review: Starch matrices and the glycemic response." *Food Science and Technology International* **17**(3): 187–204.

Pariza, M. W., S. H. Ashoor, F. S. Chu and D. B. Lund (1979). "Effects of temperature and time on mutagen formation in pan-fried hamburger." *Cancer Letters* **7**(2–3): 63–69.

Park, Y. and M. W. Pariza (2007). "Mechanisms of body fat modulation by conjugated linoleic acid (CLA)." *Food Research International* **40**(3): 311–323.

Parvez, S., K. A. Malik, S. A. Kang and H. Y. Kim (2006). "Probiotics and their fermented food products are beneficial for health." *Journal of Applied Microbiology* **100**(6): 1171–1185.

Pedrosa, M., C. Y. Pascual, J. I. Larco and M. M. Esteban (2006). "Palatability of hydrolysates and other substitution formulas for cow's milk-allergic children: A comparative study of taste, smell, and texture evaluated by healthy volunteers." *Journal of Investigational Allergology and Clinical Immunology* **16**(6): 351–356.

Phillips, G. O. and P. A. Williams (2011). *Handbook of Food Proteins.* Oxford, UK, Woodhead Publishing.

Playne, M. J. (2002). "The health benefits of probiotics." *Food Australia* **54**(3): 71–74.

Playne, M. J., L. E. Bennett and G. W. Smithers (2003). "Functional dairy foods and ingredients." *Australian Journal of Dairy Technology* **58**(3): 242–264.

Poyato, C., I. Navarro-Blasco, M. I. Calvo, R. Y. Cavero, I. Astiasaran and D. Ansorena (2013). "Oxidative stability of O/W and W/O/W emulsions: Effect of lipid composition and antioxidant polarity." *Food Research International* **51**(1): 132–140.

Prentice, A. (2004). "Diet, nutrition and the prevention of osteoporosis." *Public Health Nutrition* **7**(1A): 227–243.

Qian, C., E. A. Decker, H. Xiao and D. J. McClements (2012a). "Inhibition of beta-carotene degradation in oil-in-water nanoemulsions: Influence of oil-soluble and water-soluble antioxidants." *Food Chemistry* **135**(3): 1036–1043.

Qian, C., E. A. Decker, H. Xiao and D. J. McClements (2012b). "Physical and chemical stability of beta-carotene-enriched nanoemulsions: Influence of pH, ionic strength, temperature, and emulsifier type." *Food Chemistry* **132**(3): 1221–1229.

Qian, C. and D. McClements (2011). "Formation of nanoemulsions stabilized by model food-grade emulsifiers using high pressure homogenization: Factors affecting particle size." *Food Hydrocolloids* **25**(5): 1000–1008.

Ramaa, C. S., A. R. Shirode, A. S. Mundada and V. J. Kadam (2006). "Nutraceuticals—An emerging era in the treatment and prevention of cardiovascular diseases." *Current Pharmaceutical Biotechnology* **7**(1): 15–23.

Rao, A. V. and L. G. Rao (2007). "Carotenoids and human health." *Pharmacological Research* **55**(3): 207–216.

Rao, J. and D. J. McClements (2012a). "Food-grade microemulsions and nanoemulsions: Role of oil phase composition on formation and stability." *Food Hydrocolloids* **29**(2): 326–334.

Rao, J. J. and D. J. McClements (2012b). "Lemon oil solubilization in mixed surfactant solutions: Rationalizing microemulsion and nanoemulsion formation." *Food Hydrocolloids* **26**(1): 268–276.

Reboul, E., M. Richelle, E. Perrot, C. Desmoulins-Malezet, V. Pirisi and P. Borel (2006). "Bioaccessibility of carotenoids and vitamin E from their main dietary sources." *Journal of Agricultural and Food Chemistry* **54**(23): 8749–8755.

Redgwell, R. J. and M. Fischer (2005). "Dietary fiber as a versatile food component: An industrial perspective." *Molecular Nutrition and Food Research* **49**(6): 521–535.

Ribeiro, H. S., K. Ax and H. Schubert (2003). "Stability of lycopene emulsions in food systems." *Journal of Food Science* **68**(9): 2730–2734.

Ricci-Cabello, I., M. O. Herrera and R. Artacho (2012). "Possible role of milk-derived bioactive peptides in the treatment and prevention of metabolic syndrome." *Nutrition Reviews* **70**(4): 241–255.

Rigotti, A. (2007). "Absorption, transport, and tissue delivery of vitamin E." *Molecular Aspects of Medicine* **28**(5–6): 423–436.

Rimbach, G., J. Pallauf, J. Moehring, K. Kraemer and A. M. Minihane (2008). "Effect of dietary phytate and microbial phytase on mineral and trace element bioavailability—A literature review." *Current Topics in Nutraceutical Research* **6**(3): 131–144.

Rocha, M., C. Banuls, L. Bellod, A. Jover, V. M. Victor and A. Hernandez-Mijares (2011). "A review on the role of phytosterols: New insights into cardiovascular risk." *Current Pharmaceutical Design* **17**(36): 4061–4075.

Romanchik-Cerpovicz, J. E. and R. J. McKemie (2007). "Research and professional briefs—Fortification of all-purpose wheat-flour tortillas with calcium lactate, calcium carbonate, or calcium citrate is acceptable." *Journal of the American Dietetic Association* **107**(3): 506–509.

Rousseau, D. (2000). "Fat crystals and emulsion stability—A review." *Food Research International* **33**(1): 3–14.

Rozner, S., A. Aserin, E. J. Wachtel and N. Garti (2007). "Competitive solubilization of cholesterol and phytosterols in nonionic microemulsions." *Journal of Colloid and Interface Science* **314**(2): 718–726.

Russo, G. L. (2009). "Dietary n-6 and n-3 polyunsaturated fatty acids: From biochemistry to clinical implications in cardiovascular prevention." *Biochemical Pharmacology* **77**(6): 937–946.

Russo, S., I. P. Kema, R. Fokkema, J. C. Boon, P. H. B. Willemse, E. G. E. de Vries, J. A. den Boer and J. Korf (2003). "Tryptophan as a link between psychopathology and somatic states." *Psychosomatic Medicine* **65**(4): 665–671.

Ruxton, C. H. S., S. C. Reed, M. J. A. Simpson and K. J. Millington (2004). "The health benefits of omega-3 polyunsaturated fatty acids: A review of the evidence." *Journal of Human Nutrition and Dietetics* **17**(5): 449–459.

Ruxton, C. H. S., S. C. Reed, M. J. A. Simpson and K. J. Millington (2007). "The health benefits of omega-3 polyunsaturated fatty acids: A review of the evidence." *Journal of Human Nutrition and Dietetics* **20**(3): 275–285.

Saberi, A. H., Y. Fang and D. J. McClements (2013a). "Effect of glycerol on formation, stability, and properties of vitamin-E enriched nanoemulsions produced using spontaneous emulsification." *Journal of Colloid and Interface Science* **411**: 105–113.

Saberi, A. H., Y. Fang and D. J. McClements (2013b). "Fabrication of vitamin E-enriched nanoemulsions: Factors affecting particle size using spontaneous emulsification." *Journal of Colloid and Interface Science* **391**: 95–102.

Sagalowicz, L., M. Leser, H. Watzke and M. Michel (2006). "Monoglyceride self-assembly structures as delivery vehicles." *Trends in Food Science and Technology* **17**(5): 204–214.

Sagalowicz, L. and M. E. Leser (2010). "Delivery systems for liquid food products." *Current Opinion in Colloid and Interface Science* **15**(1–2): 61–72.

Salgueiro, M. J., M. Zubillaga, A. Lysionek, R. Caro, R. Weill and J. Boccio (2002). "Fortification strategies to combat zinc and iron deficiency." *Nutrition Reviews* **60**(2): 52–58.

Salminen, H., K. Herrmann and J. Weiss (2013). "Oil-in-water emulsions as a delivery system for n-3 fatty acids in meat products." *Meat Science* **93**(3): 659–667.

Schieber, A. and R. Carle (2005). "Occurrence of carotenoid cis-isomers in food: Technological, analytical, and nutritional implications." *Trends in Food Science and Technology* **16**(9): 416–422.

Schwartz, S. J., J. H. Elbe and M. M. Giusti (2008). Colorants. In *Food Chemistry*, 4th Edition, S. Damodaran, K. L. Parkin and O. R. Fennema (eds.). Boca Raton, FL, CRC Press: 571–638.

Severin, S. and W. S. Xia (2005). "Milk biologically active components as nutraceuticals: Review." *Critical Reviews in Food Science and Nutrition* **45**(7–8): 645–656.

Seyler, J. E., R. E. C. Wildman and D. K. Layman (2007). Protein as a functional food ingredient for weight loss and maintaining body composition. In *Handbook of Nutraceuticals and Functional Foods*, 2nd Edition, R. E. C. Wildman (ed.). Boca Raton, FL, CRC Press: 391–407.

Shefer, A. and S. Shefer (2003). "Novel encapsulation system provides controlled release of ingredients." *Food Technology* 57: 40–43.

Shukat, R., C. Bourgaux and P. Relkin (2012). "Crystallisation behaviour of palm oil nanoemulsions carrying vitamin E." *Journal of Thermal Analysis and Calorimetry* 108(1): 153–161.

Siddiqui, R. A., S. R. Shaikh, L. A. Sech, H. R. Yount, W. Stillwell and G. P. Zaloga (2004). "Omega 3-fatty acids: Health benefits and cellular mechanisms of action." *Mini-Reviews in Medicinal Chemistry* 4(8): 859–871.

Silva, H., M. Cerqueira, B. Souza, C. Ribeiro, M. Avides, M. Quintas, J. Coimbra, M. Carneiro-da-Cunha and A. Vicente (2010). "Nanoemulsions of [beta]-carotene using a high-energy emulsification-evaporation technique." *Journal of Food Engineering* 102(2): 130–135.

Singh, J., A. Dartois and L. Kaur (2010). "Starch digestibility in food matrix: A review." *Trends in Food Science and Technology* 21(4): 168–180.

Soupas, L., L. Juntunen, A. M. Lampi and V. Piironen (2004). "Effects of sterol structure, temperature, and lipid medium on phytosterol oxidation." *Journal of Agricultural and Food Chemistry* 52(21): 6485–6491.

Spernath, A. and A. Aserin (2006). "Microemulsions as carriers for drugs and nutraceuticals." *Advances in Colloid and Interface Science* 128: 47–64.

Stringham, J. M. and B. R. Hammond (2005). "Dietary lutein and zeaxanthin: Possible effects on visual function." *Nutrition Reviews* 63(2): 59–64.

Tang, B. M. P., G. D. Eslick, C. Nowson, C. Smith and A. Bensoussan (2007). "Use of calcium or calcium in combination with vitamin D supplementation to prevent fractures and bone loss in people aged 50 years and older: A meta-analysis." *Lancet* 370(9588): 657–666.

Tavano, O. L. (2013). "Protein hydrolysis using proteases: An important tool for food biotechnology." *Journal of Molecular Catalysis B-Enzymatic* 90: 1–11.

Teitelbaum, J. E. and W. A. Walker (2002). "Nutritional impact of pre- and probiotics as protective gastrointestinal organisms." *Annual Review of Nutrition* 22: 107–138.

Teoh, P. L., H. Mirhosseini, S. Mustafa, A. S. M. Hussin and M. Y. A. Manap (2011). "Recent approaches in the development of encapsulated delivery systems for probiotics." *Food Biotechnology* 25(1): 77–101.

Trigueros, L., S. Pena, A. V. Ugidos, E. Sayas-Barbera, J. A. Perez-Alvarez and E. Sendra (2013). "Food ingredients as anti-obesity agents: A review." *Critical Reviews in Food Science and Nutrition* 53(9): 929–942.

Tripathi, A. K. and A. K. Misra (2005). "Soybean—A consummate functional food: A review." *Journal of Food Science and Technology-Mysore* 42(2): 111–119.

Tsiaras, W. G. and M. A. Weinstock (2011). "Factors influencing vitamin D status." *Acta Dermato-Venereologica* 91(2): 115–124.

Tur, J. A., M. M. Bibiloni, A. Sureda and A. Pons (2012). "Dietary sources of omega 3 fatty acids: Public health risks and benefits." *British Journal of Nutrition* 107: S23–S52.

Ubbink, J. (2002). "Flavor delivery systems: Trends, technologies and applications." *Abstracts of Papers of the American Chemical Society* 223: U34.

Ubbink, J. and J. Kruger (2006). "Physical approaches for the delivery of active ingredients in foods." *Trends in Food Science and Technology* 17(5): 244–254.

Udenigwe, C. C. and R. E. Aluko (2012). "Food protein-derived bioactive peptides: Production, processing, and potential health benefits." *Journal of Food Science* 77(1): R11–R24.

van den Berg, H., R. Faulks, H. F. Granado, J. Hirschberg, B. Olmedilla, G. Sandmann, S. Southon and W. Stahl (2000). "The potential for the improvement of carotenoid levels in foods and the likely systemic effects." *Journal of the Science of Food and Agriculture* 80(7): 880–912.

Velikov, K. P. and E. Pelan (2008). "Colloidal delivery systems for micronutrients and nutraceuticals." *Soft Matter* 4(10): 1964–1980.

Wang, W. Y. and E. G. De Mejia (2005). "A new frontier in soy bioactive peptides that may prevent age-related chronic diseases." *Comprehensive Reviews in Food Science and Food Safety* 4(4): 63–78.

Waraho, T., D. J. McClements and E. A. Decker (2011). "Mechanisms of lipid oxidation in food dispersions." *Trends in Food Science and Technology* 22(1): 3–13.

Ward, R. E. and J. B. German (2004). "Understanding milk's bioactive components: A goal for the genomics toolbox." *Journal of Nutrition* 134(4): 962S–967S.

Whelan, K. and E. M. M. Quigley (2013). "Probiotics in the management of irritable bowel syndrome and inflammatory bowel disease." *Current Opinion in Gastroenterology* 29(2): 184–189.

White, P. J. and M. R. Broadley (2005). "Biofortifying crops with essential mineral elements." *Trends in Plant Science* 10(12): 586–593.

Wickham, M., R. Faulks and C. Mills (2009). "In vitro digestion methods for assessing the effect of food structure on allergen breakdown." *Molecular Nutrition and Food Research* 53(8): 952–958.

Wildman, R. E. C. and M. Kelley (2007). Nutraceuticals and functional foods. In *Handbook of Nutraceuticals and Functional Foods*, 2nd Edition, R. E. C. Wildman (ed.). Boca Raton, FL, CRC Press: 1–22.

Wong, N. C. W. (2001). "The beneficial effects of plant sterols on serum cholesterol." *Canadian Journal of Cardiology* 17(6): 715–721.

Wooster, T. J., M. Golding and P. Sanguansri (2008). "Impact of oil type on nanoemulsion formation and ostwald ripening stability." *Langmuir* 24(22): 12758–12765.

Wrolstad, R. E. and C. A. Culver (2012). Alternatives to those artificial FD&C food colorants. In *Annual Review of Food Science and Technology*, M. P. Doyle and T. R. Klaenhammer (eds.). Palo Alto, CA, Annual Reviews: 59–77.

Xianquan, S., J. Shi, Y. Kakuda and J. Yueming (2005). "Stability of lycopene during food processing and storage." *Journal of Medicinal Food* 8(4): 413–422.

Yang, Y., Z. B. Gu and G. Y. Zhang (2009). "Delivery of bioactive conjugated linoleic acid with self-assembled amylose-CLA complex." *Journal of Agricultural and Food Chemistry* 57(15): 7125–7130.

Yang, Z., A. Laillou, G. Smith, D. Schofield and R. Moench-Pfanner (2013). "A review of vitamin D fortification: Implications for nutrition programming in Southeast Asia." *Food and Nutrition Bulletin* 34(2): S81–S89.

Yang, Z. Y. and S. L. Huffman (2011). "Review of fortified food and beverage products for pregnant and lactating women and their impact on nutritional status." *Maternal and Child Nutrition* 7: 19–43.

Yashodhara, B. M., S. Umakanth, J. M. Pappachan, S. K. Bhat, R. Kamath and B. H. Choo (2009). "Omega-3 fatty acids: A comprehensive review of their role in health and disease." *Postgraduate Medical Journal* 85(1000): 84–90.

Yoon, Y. and E. Choe (2009). "Lipid oxidation and stability of tocopherols and phospholipids in soy-added fried products during storage in the dark." *Food Science and Biotechnology* **18**(2): 356–361.

Young, S. N. (1993). "The use of diet and dietary-components in the study of factors controlling affect in humans—A review." *Journal of Psychiatry and Neuroscience* **18**(5): 235–244.

Zhang, G. Y. and B. R. Hamaker (2009). "Slowly digestible starch: Concept, mechanism, and proposed extended glycemic index." *Critical Reviews in Food Science and Nutrition* **49**(10): 852–867.

Zingg, J. M. and A. Azzi (2004). "Non-antioxidant activities of vitamin E." *Current Medicinal Chemistry* **11**(9): 1113–1133.

Zuccari, G., R. Carosio, A. Fini, P. Montaldo and I. Orienti (2005). "Modified polyvinyl-alcohol for encapsulation of all-trans-retinoic acid in polymeric micelles." *Journal of Controlled Release* **103**(2): 369–380.

*Chapter 3*

# Particle Characteristics and Their Impact on Physicochemical Properties of Delivery Systems

## 3.1 Introduction

The physicochemical, sensory, and biological properties of colloidal delivery systems are mainly determined by the characteristics of the carrier particles they contain, for example, their composition, structure, dimensions, charge, physical state, and interactions (Jones and McClements 2010; Lesmes and McClements 2009). Food or ingredient manufacturers can therefore design colloidal delivery systems with specific functional performances by carefully controlling particle attributes during the production process. In this chapter, the most important particle characteristics are highlighted, and the relationships between particle attributes and the physicochemical properties of delivery systems are discussed. Particular emphasis is given to establishing quantitative structure–function relationships between particle attributes and the physicochemical properties of colloidal delivery systems, as these relationships are useful for the rational design of delivery systems with specific functional attributes.

## 3.2 Particle Building Blocks

### 3.2.1 Molecules, Particles, and Phases

As discussed in Chapter 1, the nanoparticles or microparticles in a colloidal delivery system may be fabricated using top–down (mechanical), bottom–up (physicochemical), or combined approaches. For top–down approaches, the building blocks for the formation of colloidal particles are either bulk liquid or solid phases (such as oil, water, or crystalline solids) or large liquid or solid particles (such as large oil droplets, water droplets, or crystals). For bottom–up approaches, the building blocks are usually molecules or fine particles, such as proteins, polysaccharides, lipids, surfactants, phospholipids, water, or minerals. The characteristics of the building blocks used to fabricate the particles in a colloidal delivery system ultimately determine its functional attributes. It is therefore important to have good knowledge of the molecules, particles, or phases used to assemble a delivery system so as to better control its performance.

**Table 3.1   Summary of Major Interactions between Pairs of Molecules or Particles**

| Interaction | Sign | Range | Magnitude | Features |
|---|---|---|---|---|
| Van der Waals | Attractive | Medium | Medium | Acts between all particle types |
| Steric | Repulsive | Short | Strong | Depends on thickness, packing, and chemistry of interfacial layers |
| Electrostatic | Attractive or repulsive | Short to long | Weak to strong | Depends on surface charge, pH, and ionic strength |
| Depletion | Attractive | Medium | Weak to strong | Depends on concentration and size of excluded particles |
| Hydrophobic | Attractive | Medium | Weak to strong | Depends on surface hydrophobicity |
| Hydrogen bonds | Attractive | Medium | Medium | Depends on type and location of polar groups |

*Note:* For the sake of brevity, the term *particles* is used to refer to both molecules and particles.

### 3.2.2 Molecular and Colloidal Interactions

The creation of colloidal particles by the breakdown of larger particles (top–down) or by the assembly of smaller particles (bottom–up) depends on the nature of the forces holding the various components together. A variety of forces may act between molecules and colloidal particles, including van der Waals, electrostatic, steric, hydrophobic, hydrogen bonding, and depletion interactions (Israelachvili 2011). Each of these forces can be characterized by its sign (attractive or repulsive), range (short to long), and magnitude (weak to strong). A brief summary of the characteristics of the major interactions between molecules and colloidal particles is shown in Table 3.1. Knowledge of the types of interactions involved and the factors that influence their sign, range, and magnitude is often important for the rational design of colloidal delivery systems. A number of the most important types of these interactions are discussed in later chapters on specific types of surfactant-based, emulsion-based, or biopolymer-based delivery systems.

## 3.3 Particle Characteristics

### 3.3.1 Composition

The particles in food-grade colloidal delivery systems may be fabricated from a variety of food ingredients, including water, proteins, carbohydrates, lipids, surfactants, phospholipids, and minerals (Jones and McClements 2010; Matalanis et al. 2011; McClements 2005). Selection of an appropriate particle composition is important for the design of food-grade delivery systems with specific functional attributes. Economically, the type and amount of ingredients used to fabricate colloidal particles determine the overall cost of a delivery system, which is an important consideration for the food industry where profit margins are often tight. The nature of the ingredients used also has important implications for the way final products are labeled and marketed, for example, natural, vegan, vegetarian, Kosher, Halal, or allergy status. As examples, Tween 80 cannot be used in "all natural" foods, gelatin cannot be used in vegetarian foods, and milk proteins cannot be used in vegan foods. Particle composition also influences how a delivery system responds to specific environmental conditions that it may experience during its lifetime, such as dilution, pH, ionic strength, temperature, and enzyme activity. For example, triacylglycerols, starches, and proteins are digested by specific enzymes within the upper gastrointestinal tract (i.e., lipases, amylases, and proteases), whereas dietary fibers are not. Instead, dietary fibers tend to be degraded by microbial fermentation in the lower gastrointestinal tract. This knowledge is important for designing delivery systems that release active ingredients at specific points within the gastrointestinal tract. Particle composition also influences the potential impact of colloidal particles on the physicochemical properties and sensory attributes of delivery systems (such as their appearance, stability, rheology, and flavor profile), since it determines the electrical characteristics, surface hydrophobicity, physical state, physicochemical properties, chemical stability, and morphology of the particles.

A brief overview of the major food ingredients typically used to fabricate food-grade colloidal delivery systems is given in Table 3.2. A summary of potential release mechanisms for active ingredients encapsulated within carrier particles with different compositions is given in Table 3.3. A more detailed description of specific delivery systems is provided in the later chapters on surfactant-based, emulsion-based, and biopolymer-based delivery systems (Chapters 5 through 7).

The physical state of a component used to construct a particle depends on its phase behavior (e.g., melting, crystallization, polymorphic transitions, boiling, glass

**Table 3.2  Compositions, Morphologies, and Dimensions of Particles That Could Be Used to Fabricate Food-Based Delivery Systems**

| Particle Type | Typical Composition | Morphology | Typical Dimensions |
|---|---|---|---|
| Micelle | Surfactants, cosurfactants, or cosolvents | Core–shell | 5–20 nm |
| Microemulsion | Surfactants, cosurfactants, or cosolvents | Core–shell | 10–100 nm |
| Liposome | Phospholipids | Core–shell | 100 nm–100 μm |
| Emulsion | Lipids, emulsifiers | Core–shell | 100 nm–100 μm |
| Nanoemulsion | Lipids, emulsifiers | Core–shell | 30–100 nm |
| Solid lipid nanoparticle | Lipids, emulsifiers | Dispersion–shell | 30 nm–100 μm |
| Multilayer emulsion | Lipids, emulsifiers, biopolymers | Core–shell | 100 nm–100 μm |
| Multiple emulsion | Lipids, water, emulsifiers | Dispersion–shell | 500 nm–100 μm |
| Colloidosome | Lipids, emulsifiers | Mixed | 100 nm–100 μm |
| Hydrogel particle | Biopolymers, water | Dispersion | 100 nm–100 μm |
| Filled hydrogel particle | Biopolymers, water, lipids, emulsifiers | Mixed | 100 nm–100 μm |

**Table 3.3   Influence of Particle Composition on Potential Release Mechanisms of Active Ingredients from Colloidal Delivery Systems**

| Composition | Potential Release Mechanisms | Examples |
|---|---|---|
| **Lipid** | | |
| • Digestible | Diffusion; phase transitions (melting/crystallization); enzyme degradation (lipases in stomach and small intestine) | Triacylglycerols: canola, corn, fish, soy, sunflower oils |
| • Indigestible | Diffusion | Mineral oils, flavor oils, fat substitutes (Olestra) |
| **Carbohydrate** | | |
| • Digestible | Diffusion; hydrogen bond weakening (temperature); enzyme degradation (amylase in mouth, stomach, and small intestine); swelling | Starch |
| • Indigestible | Diffusion; swelling; hydrogen bond weakening (temperature); electrostatic screening (pH and ionic strength); electrostatic bridge disruption (chelating agents); bacterial fermentation (colon) | Dietary fibers: cellulose, pectin, xanthan, carrageenan, agar, alginate |
| **Protein** | | |
| • Digestible | Diffusion; thermal denaturation (heat); acid denaturation (acid in stomach); enzyme degradation (proteases in stomach and small intestine); acid hydrolysis (acid in stomach); electrostatic bridge disruption (chelating agents); electrostatic screening (pH and ionic strength); swelling | Whey protein, caseinate, soy protein, gelatin, egg protein, myosin, wheat protein, zein |

transitions, thermal denaturation, helix–coil transitions, or sol–gel transitions), as well as environmental conditions (i.e., thermal–mechanical history). Thus, two colloidal particles may have the same overall composition but different functionalities depending on the physical state of the components they are constructed from, for example, solid versus liquid lipid droplets.

## 3.3.2 *Morphology*

The structure of the colloidal particles in food-grade delivery systems varies widely depending on the ingredients and methods used to fabricate them (Figure 1.1). Micelles and microemulsions consist of small molecular assemblies of surfactant molecules with the polar head groups directed toward the aqueous phase and the nonpolar tails directed away from the aqueous phase where they form a hydrophobic interior. Oil-in-water (O/W) emulsions or nanoemulsions contain lipid droplets coated by a layer of surface-active (emulsifier) molecules, with the polar part of the emulsifier facing the aqueous phase and the nonpolar part facing the lipid phase. Multilayer emulsions consist of lipid droplets coated by multiple layers of biopolymers (or other charged substances). Hydrogel particles contain a network of aggregated biopolymer molecules that traps water and possibly other components (such as lipid droplets). Conceptually, it is often useful to group the structure of colloidal particles into different categories, since this influences the type of physical theory or mathematical model that can be used to describe their physicochemical properties and functional performance (Figure 3.1):

- *Homogeneous.* These particles consist of a single phase (which may be composed of one or more types of material intimately mixed with each other).
- *Core–shell.* These particles consist of an inner core of one phase surrounded by an outer shell of another phase. The shell may be composed of a single layer or a number of concentric layers.
- *Dispersion.* These particles consist of one phase dispersed as small particles within another phase.
- *Cluster.* These particles consist of two or more smaller particles that are associated with each other.
- *Combination.* These particles are some combination of the above categories, for example, a dispersion trapped within a core–shell structure.

The structures of various types of particle have been classified according to the scheme in Table 3.2. For example, micelles, microemulsions, nanoemulsions, and

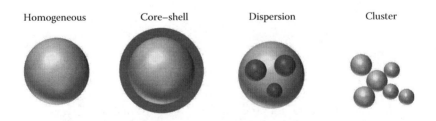

Homogeneous    Core–shell    Dispersion    Cluster

**Figure 3.1 Examples of different kinds of structural organizations of the particles possible in delivery systems.**

emulsions have a core–shell structure with the polar head groups being the shell and the nonpolar tail groups or lipids forming the core. Nevertheless, this classification scheme does have considerable limitations when applied to many of types of colloidal particles used in food-based delivery systems, since the structures are too complex to fall into a single category. For example, a multiple emulsion droplet consists of *core–shell* particles (water droplets coated with a lipophilic surfactant) *dispersed* within larger *core–shell* particles (oil droplets coated with a hydrophilic surfactant).

Particles may also be classified according to their shape, for example, as either spherical or nonspherical (Figure 3.2). The most common particle morphology is the sphere (e.g., micelles, liposomes, and emulsion droplets), but other shapes may also be present, such as spheroids (e.g., some hydrogel particles formed by shearing), disk-like (e.g., some solid lipid nanoparticles), rod-like (e.g., some biopolymer fibers or crystals), or irregular shaped (e.g., bacteria, spores, viruses, or aggregated systems). Many particulate delivery systems can be approximated as spheres or spheroids, which may be either prolate (rod-like) or oblate (disk-like). A spheroid is characterized by its axis ratio $r_p = a/b$, where $a$ is the major axis and $b$ is the minor axis. For a sphere, $a = b$; for a prolate spheroid, $a > b$; and for an oblate spheroid, $a < b$ (Hunter 1986). Particle shape has a major impact on the rheology, appearance, and stability of particulate delivery systems. For example, the rheology of a particle suspension is larger for nonspherical than spherical particles of the same volume fraction (Genovese et al. 2007; Hunter 1986). The morphology of particles may be conveniently determined using visualization methods such as optical or electron microscopy, with the choice of method depending on the resolution of the microscope compared to the dimensions of the particles, as well as the sensitivity of the particles to disruption during sample preparation (McClements 2007).

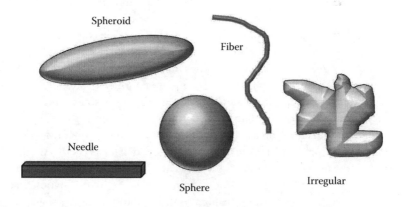

Spheroid

Fiber

Needle

Sphere

Irregular

**Figure 3.2    Examples of different shapes of the particles possible in food-grade delivery systems.**

### 3.3.3 Concentration

The concentration of the colloidal particles within a delivery system plays an important role in determining its overall functional performance, as well as on the impact it has on the final properties of the food product that it is incorporated into. It is therefore important to clearly specify and reliably report the particle concentration of colloidal delivery systems. Particle concentration may be specified in a number of ways depending on the feature of the particles that is of most interest to the researcher (McClements 2005):

- *Number–volume concentration* ($c_{NV}$): The total number of particles per unit volume of the system ($m^{-3}$).
- *Surface–volume concentration* ($c_{SV}$): The total surface area of particles per unit volume of the system ($m^2\ m^{-3}$ or $m^{-1}$).
- *Volume–volume concentration* ($c_{VV}$): The total volume of particles per unit volume of the system ($m^3\ m^{-3}$ or dimensionless). This parameter is often referred to as the disperse phase volume fraction ($\varphi$).
- *Mass–mass concentration* ($c_{mm}$): The total mass of particles per unit mass of the system ($kg\ kg^{-1}$ or dimensionless). This parameter is often referred to as the disperse phase mass fraction ($\varphi_m$), and is often reported as a weight percentage (wt%).
- *Mass–volume concentration* ($c_{mV}$): The total mass of particles per unit volume of the system ($kg\ m^{-3}$ or $g\ mL^{-1}$).

These particle concentration parameters can also be specified in a variety of other ways. For example, they can be expressed per unit mass of the system rather than per unit volume, for example, the surface area of particles per unit mass of the system. Alternatively, they can be expressed per unit amount of particles (volume or mass), rather than per unit amount of the overall system, for example, the number of particles per unit mass of particles, or the surface area per unit volume of particles. Knowledge of the relationship between particle dimensions, surface area ($A$), and volume ($V$) is useful for converting one type of particle concentration parameter into another. For spheres: $A = 4\pi r^2$ and $V = (4/3)\pi r^3$; for cylinders: $A = 2\pi r^2 + 2\pi rh$ and $V = \pi r^2 h$, where $r$ is particle radius and $h$ is particle length or height. It is important to carefully select the most appropriate parameter to represent the particle concentration in a specific system and always to clearly state which parameter has been used when reporting data.

The particle concentration can be controlled by varying the quantities of materials or manufacturing conditions used to prepare the delivery system. Alternatively, delivery systems may be prepared with a particular particle concentration and then be diluted (e.g., by adding solvent) or concentrated (e.g., by gravitational separation, filtration, centrifugation, or evaporation). It should be noted that changing the particle size will also change the number and surface area of the particles present, even if their total mass or volume remains constant.

## 3.3.4 Particle Physicochemical Properties

There are a number of physicochemical properties of particles that are important in determining their functional performance, including their density, refractive index, and rheology. Knowledge of these parameters is often necessary to quantitatively relate particle characteristics to the functional performance of colloidal delivery systems.

### 3.3.4.1 Density

The density of a particle determines how it behaves within a gravitational or centrifugal field, which will influence the stability of delivery systems to creaming/sedimentation. In addition, particle density influences the ability to separate colloidal particles from the surrounding medium (e.g., by gravitational settling or centrifugation), which may be important in industrial production of food-grade delivery systems. The density of a carrier particle is determined by the type, amounts, and densities of the various components present within it. To a first approximation, the density of a multicomponent particle is given by the following relationship, which assumes ideal mixing:

$$\rho_{particle} = \sum_i \varphi_i \rho_i, \tag{3.1}$$

where $\varphi_i$ and $\rho_i$ are the volume fraction and density of the $i$th component. Thus, for a two component particle:

$$\rho_{particle} = \varphi_i \rho_2 + (1 - \varphi_i)\rho_1. \tag{3.2}$$

Equation 3.2 is suitable for systems that can be approximated by ideal mixing conditions, such as mixed oils inside lipid droplets (e.g., emulsions and nanoemulsions), proteins or polysaccharides dispersed in water (e.g., hydrogel particles), mixtures of solid and liquid fat (e.g., solid lipid nanoparticles), or particulate dispersions (e.g., filled hydrogel particles or multiple emulsion droplets). Different models may be required when the system deviates appreciably from ideal mixing conditions, for example, when minerals or sugars are dissolved in an aqueous solution. The densities of some of the major components used to construct food-grade colloidal particles are summarized in Table 3.4.

### 3.3.4.2 Rheology

The rheological characteristics of particles are important in many applications of colloidal delivery systems, for example, the mechanical response of a particle to an

**Table 3.4 Physicochemical Properties of the Major Components Used to Construct Food-Grade Colloidal Particles**

| Material | Density (kg m⁻³) | Refractive Index | Viscosity (mPa s) |
|---|---|---|---|
| Water | 1000 | 1.33 | 1 |
| Triglyceride oils | 920 | 1.43 | ≈50 |
| Flavor oils | 800 | 1.44 | ≈10 |
| Protein | 1470 | 1.5 | Solid |
| Polysaccharides | 1500 | 1.5 | Solid |

*Note:* There are actually considerable variations between different ingredient types, and these values are only representative.

applied stress or the diffusion of active ingredients within particles. The material within particles may be fluid-like, solid-like, or some combination (e.g., plastic or viscoelastic). The rheology of liquids is usually characterized in terms of their shear viscosity, which is a measure of their resistance to flow when a stress is applied (Genovese et al. 2007; Hunter 1986). The viscosity of the material within a fluid particle may vary from being relatively low (like water) to relatively high (like honey). The viscosity of the particles in food-grade delivery systems may play an important role in both their fabrication and functional performance. For example, the viscosity of the internal phase in emulsions determines how easy it is to disrupt droplets within a homogenizer and therefore determines the smallest size of particles that can be formed (Walstra 1993; Wooster et al. 2008). It also determines the rate at which components diffuse in or out of the particles, which is important for long-term stability and controlled release applications (Baker 1987; Li and Jastic 2005).

The rheology of solid and semisolid materials is usually characterized in terms of their elastic modulus and fracture properties (Foegeding et al. 2011). The elastic modulus is a measure of the ability of a material to resist deformation when a stress is applied, which may vary from soft (easily deformed) to hard (difficult to deform). The fracture properties describe the tendency for a material to break when a stress is applied, for example, fracture stress (the stress required to cause fracture) and fracture strain (the extent to which a material deforms before fracture occurs). The fracture stress ($\sigma_F$) and strain ($\gamma_F$) determine how likely a particle is to break up when external forces are applied, which may be important for a number of reasons: the formation of particle suspensions by grinding, the stability of particles during storage and transport, and the release of active components within the mouth or stomach owing to fragmentation. For a particle to break up, the applied shear stress acting across it should be greater than the fracture stress (Walstra 2003). The rheological properties of some of the major components used to construct food-grade colloidal particles are summarized in Table 3.4.

### 3.3.4.3 Refractive Index

The refractive index of a colloidal particle determines how it interacts with electromagnetic radiation (Hunter 1986). In particular, the (complex) refractive index influences the scattering and absorption of light by particles, which determines the optical properties and appearance of delivery systems and the food matrices into which they are incorporated. For relatively homogeneous particles, such as oil or water droplets, it is possible to find tabulated values for the refractive index of the material that makes up the particles (e.g., *CRC Handbook of Chemistry and Physics*). For multicomponent particles, the refractive index depends on the concentration, structural organization, and optical properties of the various components present. Theoretical models have been developed to calculate the effective refractive index of multicomponent particles. To a first approximation, the refractive index of a multicomponent particle is given by the following simple relationship, which involves weighting the contribution from different components (Liu and Daum 2008):

$$n_{particle} = \sum_i \varphi_i n_i, \tag{3.3}$$

where $n_i$ is the refractive index of the *i*th component. Thus, for a two component particle:

$$n_{particle} = \varphi n_2 + (1 - \varphi) n_1. \tag{3.4}$$

These expressions can be used to estimate the refractive index of multicomponent colloidal particles, such as hydrogel particles (biopolymers + water). The refractive indices of some of the major components used to construct food-grade colloidal particles are summarized in Table 3.4. Nevertheless, these simple equations do not fully describe the complex light scattering behavior of multicomponent particles with complicated structures (e.g., dispersion or core–shell), where scattering may occur from different phases. In this case, more complex theories are needed to calculate the effective particle refractive index (Aden and Kerker 1951; Bhandari 1985). In addition, particle size analyzers that rely on knowledge of particle refractive index (such as light scattering) may give incorrect results for colloidal particles with more complicated structures, which should be taken into account when interpreting data from this kind of system.

### 3.3.5 Particle Dimensions

The dimensions of the particles within a colloidal delivery system is one of the most important factors determining its performance (e.g., release rate and bioaccessibility) and its impact on the properties of food matrices (e.g., optical properties, stability, and mouthfeel). If all the particles in a system have the same size, they are referred to as being *monodisperse*, but if they have a range of sizes, they are referred to as being *polydisperse*. The particle size ($x$) of a monodisperse particulate system can be completely characterized by a single number, such as the particle diameter ($d$) or radius ($r$). Monodisperse systems are rare, although there are some examples of delivery systems that have approximately monodisperse particles, such as micelles, microemulsions, and emulsions produced by membrane homogenization. Nevertheless, most other delivery systems typically contain a range of particles with different dimensions, which is characterized by the *particle size distribution* (PSD), that is, the fraction of particles in different size classes. For polydisperse systems, it is usually important to measure and report the full PSD as this provides important information about the potential stability and functional performance of the overall system. Nevertheless, it is often convenient to summarize the dimensions of a polydisperse system in terms of a mean size or a mean size plus a measure of the spread of the distribution (McClements 2005; Walstra 2003).

### 3.3.5.1 Particle Size Distributions

A PSD is usually represented by specifying the fraction of particles within a series of different discrete size classes (Walstra 2003). The upper and lower size limits of a PSD are selected so as to cover the entire range of particles present within the system, while the number and width of each size class are chosen to give an accurate representation of the distribution. PSD data are typically presented in either a tabular or a graphical form (Figure 3.3). The two most important parameters to

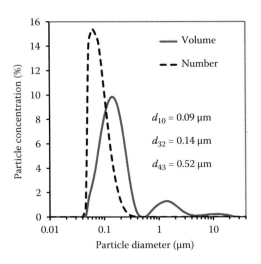

**Figure 3.3   Particle size distribution measured by light scattering for a polydisperse food emulsion. The particle concentration is represented as either volume-weighted or number-weighted, which gives distinctly different profiles.**

consider when presenting particle size data are (i) the way of representing the particle concentration in each class (e.g., number, area, volume, or mass) and (ii) the way
of representing the particle size (e.g., radius or diameter). Particle concentration is
typically presented as either a fraction or a percentage. It is important to note that
the shape of a PSD changes appreciably depending on the particle concentration
parameter used to represent it. The volume of a droplet is proportional to the cube
of its dimensions ($V \propto x^3$), and so a volume distribution is skewed more toward larger
droplets, whereas a number distribution is skewed more toward the smaller droplets, since one large droplet may equal many smaller droplets (Figure 3.3).

The size class that contains the largest concentration of particles is called the
*mode*, whereas the size class where 50% of the particles are smaller and 50% are
larger is called the *median* (Hunter 1986). The numerical values of the median and
the modal particle sizes depend on the way that the particle concentration in each
size class is expressed, for example, number or volume. Hence, there are number
and volume median sizes and number and volume modal sizes of a distribution.

The PSD of colloidal delivery systems can be conveniently measured using a
variety of commercially available particle size analyzers based on different physical
principles, such as static light scattering, dynamic light scattering, and electrical
pulse counting (McClements 2007). PSDs can also be determined from images of
delivery systems acquired using various kinds of microscopes, with the most common being light and electron microscopy (McClements 2007). To obtain reliable
results, it is important to ensure that the equipment is operated correctly, the appropriate input parameters are used to interpret the data (such as refractive indices of
the different phases for light scattering), and sample preparation is carried out carefully to avoid generating artifacts (e.g., dilution and stirring).

### 3.3.5.2 Mean and Standard Deviation

It is often more convenient to represent the size characteristics of the particles
within a polydisperse particle suspension by one or two parameters, rather than as
a full PSD (Hunter 1986). The most useful parameters are the mean particle size,
$\bar{x}$, which is a measure of the *central tendency* of the distribution, and the standard
deviation, $\sigma$, which is a measure of the *width* of the distribution. The mean and
standard deviation of a PSD can be calculated using the following equations:

$$\bar{x} = \frac{\sum_{i=1} n_i x_i}{N} \tag{3.5}$$

$$\sigma = \sqrt{\frac{\sum_{i=1} n_i (x_i - \bar{x})^2}{N}} \tag{3.6}$$

Here, $N$ represents the total number of particles present in the system. These summations are carried out over the total number of size classes used to represent the distribution. Nevertheless, this is only one way of expressing the mean and standard deviation, and there are a number of other ways that emphasize different physical characteristics of the PSD that are often more appropriate (Walstra 2003). A number of the most commonly used mean particle sizes are presented in Table 3.5.

The mean particle size of colloidal dispersions is often expressed as the surface area-weighted mean diameter ($d_{32}$), which is related to the average surface area of droplets exposed to the continuous phase per unit volume of the colloidal dispersion, $A_N$:

$$A_N = \frac{6\varphi}{d_{32}}.$$

(3.7)

This relationship is particularly useful for calculating the total surface area of particles in a delivery system from knowledge of the mean particle diameter and concentration (McClements 2007). Another commonly used method of expressing the mean particle size of a polydisperse system is the volume-weighted mean diameter ($d_{43}$). It should be noted that $d_{43}$ is more sensitive to the presence of any large particles in a PSD than $d_{32}$, and hence it is often more indicative of a phenomenon such as particle aggregation. It should also be stressed that there may be large differences between the values of different mean sizes depending on the method used to calculate them, and so it is always important to clearly establish the type of mean that is being used. In general, mean particle sizes can be considered to be represented by a *weighting factor* multiplied by some *quantity being averaged*, after a suitable conversion has been made to obtain the appropriate units for length, that is, meters. The standard deviation ($\sigma$) can also be expressed in a number of different ways depending on how it is defined, and the most appropriate one for each mean should be used (McClements 2005). Finally, as mentioned earlier, the median and mode of a distribution can also be represented in different ways and so it is important to be sure that the appropriate value is used.

### 3.3.5.3 Internal Dimensions

In particles with more complicated internal structures, the dimensions of the various phases within them often play an important role in determining their overall functionality, for example, shell thickness, core diameter, or the size of any internal inclusions. For example, the diffusion rate of an active ingredient out of a core–shell particle depends on the thickness of the shell as well as the diameter of the core (Baker 1987). Measurement of the internal dimensions of colloidal particles is more difficult than measurement of their external dimensions. Typically, microscopy

**Table 3.5  Different Ways of Expressing the Mean Particle Diameter of Polydisperse Emulsions**

| Name of Mean Diameter | Symbol | Definition | Quantity Averaged | Weighting Factor |
|---|---|---|---|---|
| Number–length mean diameter | $d_{NL}$ or $d_{10}$ | $d_{10} = \left(\sum n_i d_i\right)\big/\left(\sum n_i\right)$ | Diameter ($\propto$ L) | Number in class |
| Number–area mean diameter | $d_{NA}$ or $d_{20}$ | $d_{20} = \left[\left(\sum n_i d_i^2\right)\big/\left(\sum n_i\right)\right]^{1/2}$ | Diameter squared ($\propto$ A) | Number in class |
| Number–volume mean diameter | $d_{NV}$ or $d_{30}$ | $d_{30} = \left[\left(\sum n_i d_i^3\right)\big/\left(\sum n_i\right)\right]^{1/3}$ | Diameter cubed ($\propto$ V) | Number in class |
| Area–volume mean diameter | $d_{AV}$ or $d_{32}$ | $d_{32} = \left(\sum n_i d_i^3\right)\big/\left(\sum n_i d_i^2\right)$ | Diameter ($\propto$ L) | Area in class |
| Volume–length mean diameter | $d_{VL}$ or $d_{43}$ | $d_{43} = \left(\sum n_i d_i^4\right)\big/\left(\sum n_i d_i^3\right)$ | Diameter ($\propto$ L) | Volume in class |

*Source:* Adapted from Hunter (1986) and Walstra (2003).

methods such as optical or electron microscopy are required in combination with appropriate sample preparation methods that allow discrimination of the different phases within the particles, such as sectioning, staining, or dyeing.

### 3.3.6 Particle Charge

Knowledge of the electrical charge on a colloidal particle within a delivery system is important because it affects many of its important functional attributes, such as stability to aggregation, binding to biological surfaces, or interaction with other charged species within a food matrix (McClements 2005, 2007). The electrical properties of a particle are usually characterized in terms of its surface electrical potential ($\psi_0$), surface charge density ($\sigma$) and/or $\zeta$-potential ($\zeta$) (Hunter 1986; Israelachvili 2011). The surface charge density is the number of electrical charges per unit area and depends on the type and amount of ionized groups present at the particle surface. The surface electrical potential is the free energy required to increase the surface charge density from zero to $\psi_0$. The surface electrical potential not only is related to the surface charge density but also depends on the ionic composition of the surrounding medium owing to electrostatic screening effects. The magnitude of the surface electrical potential decreases as the ionic strength of the aqueous phase increases since it takes less free energy to bring a new charge to a similarly charged surface in the presence of salt. The $\zeta$-potential is the electrical potential at the "shear plane," which is defined as the distance away from the particle surface below which any counterions remain strongly attached to the particle surface as it moves within an electrical field. Practically, the $\zeta$-potential is often a better representation of the electrical characteristics of a particle because it inherently accounts for the adsorption of any charged counterions. The $\zeta$-potential is also much easier to measure than the electrical potential or surface charge density. The electrical characteristics of particles can be controlled by careful selection of system composition, for example, emulsifier type, biopolymer type, pH, and ionic strength. Typically, the electrical characteristics of a particular kind of particle are conveniently represented by the $\zeta$-potential versus pH profile, which can be measured using commercial instruments based on electrophoresis or electroacoustics (McClements 2007).

### 3.3.7 Particle Interactions

The bulk physicochemical properties, sensory attributes, and biological fate of colloidal delivery systems are affected by particle interactions (McClements 2005). These interactions may operate between different particles (particle–particle), between particles and surfaces (particle–surface), or between particles and molecular species (e.g., particle–ion or particle–polymer). There are many different kinds of interactions that may involve particles, including van der Waals, electrostatic, steric, depletion, hydrogen bonding, and hydrophobic interactions (Israelachvili

2011). These interactions vary in their sign (attractive or repulsive), magnitude (strong to weak), and range (long to short). The overall characteristics of the interactions in a particular delivery system are determined by the relative contribution of the different kinds of interactions operating (Figure 3.4), which depends on composition, microstructure, and environment.

*Particle–particle* interactions play a major role in determining the aggregation stability of colloidal delivery systems, which in turn affects their shelf life, rheology, and functional performance. Particles tend to associate with each other when attractive forces dominate but remain as individual entities when repulsive forces dominate (Dickinson 2010; Israelachvili 2011). To prevent particle aggregation, it is often necessary to design a delivery system so that there is a relatively strong repulsion between the particles under the specific environmental conditions operating in that system. This is usually achieved by increasing the electrostatic or steric repulsion between the particles and avoiding bridging or depletion effects (McClements 2005). Theoretical models are available for calculating the sign, magnitude, and range of the most important colloidal interactions operating between particles (Hunter 1986; Israelachvili 2011; McClements 2005), and these can be used to predict the stability of particles to aggregation. Extensive particle aggregation causes an increase in the viscosity of colloidal dispersions and may lead to gel formation (McClements 2005). In dilute particle suspensions,

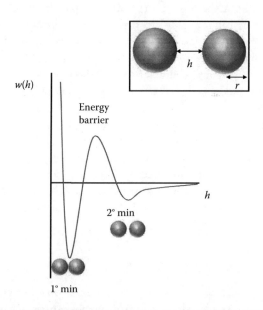

**Figure 3.4**  **Schematic representation of the interaction potential between two emulsion droplets showing the primary minimum (1° min), secondary minimum (2° min) and energy barrier.**

aggregation promotes gravitational separation because of the increase in mean particle size, but in concentrated suspensions, it may inhibit gravitational separation by leading to the formation of a three-dimensional network of aggregated particles whose movement is hindered.

*Particle–surface* interactions are also important for a number of potential applications of food-based colloidal delivery systems. In targeted release applications, the ability of a particle to adhere to a specific biological surface within the gastrointestinal tract (e.g., mouth, stomach, small intestine, or large intestine) can be used to increase the efficiency of delivery of an active ingredient and reduce unwanted side effects (Ensign et al. 2012). For example, if a colloidal particle adheres to the mucus layer lining the small intestine, it will increase the retention time of the particle within the gastrointestinal tract, thereby enhancing the release of bioactive components. Alternatively, if a particle adheres to the tongue, it may delay the release of flavor molecules and provide a prolonged flavor profile (Dresselhuis et al. 2008; van Aken et al. 2007).

*Particle–molecular species* interactions are important in determining the chemical and physical stability of many types of delivery systems, as well as the compatibility of a delivery system with a food matrix. For example, particles may bind specific mineral ions (such as calcium) or polyelectrolytes (such as pectin) that reduce the magnitude of the particle–particle electrostatic repulsion, thereby promoting aggregation (Dickinson and Golding 1998; Guzey et al. 2004; Tippetts and Martini 2012). On the other hand, anionic lipid droplets may bind cationic transition metal ions (such as iron) to their surfaces through electrostatic attraction, which promotes oxidation of encapsulated polyunsaturated lipids due to the fact that the prooxidant metal ions come into proximity (Mei et al. 1998, 1999). Knowledge of the nature and characteristics of the various types of particle interactions that can occur in colloidal delivery systems is therefore important when designing them for specific applications.

Particle interactions are also important in the formation of certain types of colloidal delivery systems: multilayer emulsions can be built by depositing charged biopolymer molecules onto oppositely charged oil droplets (Guzey and McClements 2006); colloidosomes can be fabricated by depositing small charged particles onto the surfaces of larger, oppositely charged droplets (Gu et al. 2007); microclusters can be assembled by mixing oppositely charged particles together (Mao and McClements 2012a,b, 2013).

### 3.3.8 Loading Characteristics

Practically, it is important to consider the potential loading characteristics of a delivery system, for example, what is the loading mechanism, what is the maximum amount of active that can be loaded, and how much of the active material remains encapsulated? The process of loading active components into colloidal delivery systems varies widely depending on the nature of the active ingredient and of the delivery system. Some examples of loading methods for different types of carrier particles are briefly listed below:

- *Micelles and microemulsions*: Lipophilic active ingredients can often simply be mixed with surfactant and water to spontaneously form swollen micelles or microemulsions whose hydrophobic interior is loaded with the active ingredient (Chapter 5). There is usually a maximum amount of the lipophilic ingredient that can be loaded into this system, which is referred to as the solubilization capacity.
- *Liposomes*: A number of different methods are available to load hydrophilic active ingredients into liposomes (Chapter 5). Hydrophilic components can be trapped in the internal aqueous core of liposomes by dissolving them in the aqueous phase before liposome formation or by homogenization after liposome formation, although the encapsulation efficiency is often rather low. For ionizable hydrophilic components, the encapsulation efficiency can sometimes be improved by altering the pH of the external aqueous phase (so that the active component loses its charge and diffuses into the internal aqueous phase where it becomes charged again and remains trapped). Lipid-soluble components can simply be mixed with the phospholipid phase prior to or after liposome formation.
- *Emulsions and nanoemulsions*: For emulsions and nanoemulsions, lipophilic active ingredients are usually dissolved in the oil phase while hydrophilic active ingredients are dissolved in the aqueous phase (Chapter 6). This process is often carried out before homogenization, but it can also sometimes be carried out after homogenization.
- *Multiple emulsions*: For $W_1/O/W_2$ emulsions, water-soluble active ingredients are usually dissolved in the internal water phase ($W_1$) that forms the $W_1/O$ emulsion prior to homogenization (Chapter 6). The resulting emulsion is then homogenized with an external water phase ($W_2$), which may contain other types of water-soluble active ingredients.
- *Hydrogel particles*: Lipophilic components can be dissolved in an oil phase and then homogenized with water and emulsifier to form an O/W emulsion (Chapter 7). The lipid droplets can then be introduced into the hydrogel particles by a number of different mechanisms, for example, injection, coacervation, phase separation, and molding.

More details about loading methods for specific systems are given in the chapters on specific types of food-grade delivery systems (Chapters 5 through 7). The amount and efficiency of loading of active ingredients into colloidal particles is an important consideration in designing, selecting, and fabricating food-grade delivery systems. The *loading capacity* (LC) can be defined as the amount of active ingredient loaded per unit mass of carrier particle:

$$LC = 100 \times m_A/m_P. \tag{3.8}$$

Here, $m_A$ is the mass of the active ingredient and $m_P$ is the total mass of the particle (active ingredient + carrier material). This value may vary from around 0% (low) to 100% (high) depending on the nature of the active ingredient and delivery system. For example, the loading capacity of a lipophilic active ingredient may be <5% in a micelle system but >90% in an emulsion system.

The *encapsulation efficiency* (EE) can be defined as the percentage of active ingredient added to a system that is actually trapped within the carrier particles:

$$EE = 100 \times m_{A,E}/m_{A,T}. \tag{3.9}$$

Here, $m_{A,E}$ is the mass of the encapsulated active ingredient and $m_{A,T}$ is the total mass of the active ingredient in the system (encapsulated + nonencapsulated). This value can range from 0% (poor) to 100% (good) depending on how well the particles are able to trap the active ingredient relative to the surrounding medium.

The *retention efficiency* (RE) can be defined as the percentage of active material initially present within the carrier particles that remains inside them after some specified storage period or treatment:

$$RE = 100 \times m_{A,t}/m_{A,0}. \tag{3.10}$$

Here, $m_{A,t}$ and $m_{A,0}$ are the masses of the encapsulated active ingredient within the particles at time $t$ and zero, respectively. The retention efficiency may range from close to 0% (all released) to 100% (all retained) depending on the system.

An ideal colloidal delivery system would have a high loading capacity, a high encapsulation efficiency, and a high retention efficiency.

### 3.3.9 Release Characteristics

A delivery system is usually designed so that it releases the active component at the required site of action, which may be the mouth, stomach, small intestine, or colon. As discussed in Chapter 1, an active ingredient may be released from carrier particles into the surrounding medium through a variety of different physicochemical mechanisms depending on the nature of the active ingredient and carrier particle, including diffusion, dissolution, erosion, swelling, and fragmentation (Figure 1.4). Mathematical models have been developed to describe many of these release mechanisms (Aguzzi et al. 2010; Arifin et al. 2006; Grassi and Grassi 2005; Kaunisto et al. 2011; Peppas 2013; Sackett and Narasimhan 2011), which are discussed in more detail in Chapter 10. The rate and extent of the different release mechanisms can often be controlled by careful selection of the ingredients and processing operations used to fabricate a delivery system. Consequently, it is possible to design delivery systems with release profiles appropriate for the specific application, such as burst, sustained, delayed, triggered, or targeted (Figure 1.5).

## 3.4 Impact of Particle Properties on Physicochemical Properties

A delivery system should be compatible with the food matrix that it is going to be incorporated into; that is, it should not adversely affect the appearance, texture, stability, or flavor (McClements 2010). It is therefore important to understand the relationship between particle properties and the bulk physicochemical and sensory characteristics of delivery systems.

### 3.4.1 Rheology and Texture

The incorporation of a colloidal delivery system into a food matrix may cause a pronounced alteration in its rheological properties and perceived texture. The impact of a delivery system on product rheology depends on the concentration, interactions, properties, and size of the carrier particles (Genovese et al. 2007; Pal 2011; Tadros 1994). The potential impact of the carrier particles in a delivery system on the rheology of fluid foods can be demonstrated by considering the factors that affect the viscosity of colloidal dispersions. The viscosity of a colloidal dispersion increases with increasing particle concentration, gradually at first but then steeply

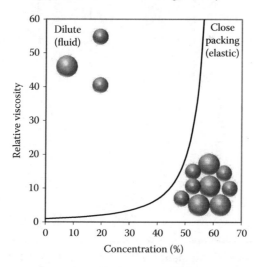

**Figure 3.5** Influence of particle concentration on the *relative* viscosity of particle suspensions.

as the particles become more closely packed (Figure 3.5). Around and above the particle concentration where close packing occurs (typically around 50%–70% for a nonflocculated system), the suspension exhibits elastic-like characteristics, such as viscoelasticity and plasticity (McClements 2005; Tadros 1994). The precise value of the particle concentration where this steep increase in viscosity is observed depends on the nature of the particle interactions in the system, decreasing for either strong attractive or strong repulsive interactions (Quemada and Berli 2002). The viscosity of a colloidal dispersion tends to increase when the particles are flocculated because the effective particle concentration is increased owing to the continuous phase trapped within the floc structure (Figure 3.6). In addition, shear thinning behavior is observed in flocculated systems because of deformation and breakdown of the floc structure as shear stresses increase. The impact of particle characteristics on the overall rheology of a colloidal dispersion may be an important consideration when designing a colloidal delivery system for a particular food application. Some food systems have a relatively low viscosity (such as beverages) and therefore the delivery system itself should not significantly increase the viscosity. Other food systems are highly viscous or gel-like (e.g., dressings, dips, desserts), and in these cases, the delivery system should not decrease the viscosity or disrupt the gel network.

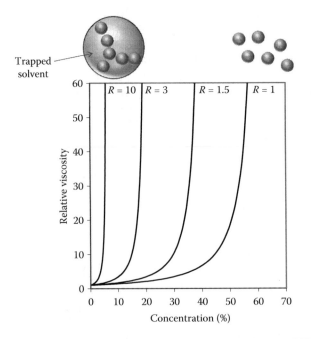

**Figure 3.6** **Influence of volume ratio (*R*) on the *relative* viscosity of particle suspensions as a function of particle concentration. The effective volume of a particle may be much larger than that of the individual constituents because of the presence of an entrapped solvent.**

Mathematical theories have been established to relate the viscosity of colloidal dispersions to particle characteristics for many types of model system (Larson 1999; Pal 2011). These theories can be used to construct structure–function relationships to guide delivery system development (Lesmes and McClements 2009). For example, Einstein developed the following equation to describe the viscosity of dilute suspensions of rigid spherical particles:

$$\eta = \eta_1(1 + 2.5\varphi). \tag{3.11}$$

Here, $\eta_1$ is the viscosity of the liquid surrounding the droplets and $\varphi$ is the dispersed phase volume fraction. Equation 3.11 assumes that the liquid surrounding the particles is Newtonian, the particles are rigid and spherical, there are no particle–particle interactions, there is no slip at the particle–liquid interface, and Brownian motion effects are unimportant (Larson 1999). The Einstein equation predicts that the viscosity of a dilute suspension of rigid spherical particles increases linearly with particle volume fraction and is independent of particle size and shear rate. It gives excellent agreement with experimental measurements up to particle concentrations of approximately 5% for colloidal dispersions that meet the above criteria (Larson 1999). Equation 3.11 also shows that the viscosity of dilute colloidal dispersions is primarily governed by the viscosity of the continuous phase, rather than by the particle characteristics. Equation 3.11 can be extended so that it can be used to model a wide range of colloidal delivery systems, for example, by including fluid particles, nonspherical particles, particle–particle interactions, and a non-Newtonian continuous phase (McClements 2005).

In most applications, the carrier particles incorporated into a food product will only be present at relatively low concentrations (<5%), and therefore they should not cause an appreciable increase in the viscosity (<12.5%) of the product. Nevertheless, the particle concentration within the delivery system itself may be much higher, and so it is important to understand the relationship between particle concentration and rheology for more concentrated systems. For concentrated colloidal dispersions containing rigid spherical particles suspended in an ideal liquid, the following semiempirical equation can be used to relate the viscosity to the particle concentration:

$$\frac{\eta}{\eta_1} = \left(1 - \frac{\varphi}{\varphi_c}\right)^{-2}. \tag{3.12}$$

Here, $\varphi_c$ is the critical packing parameter for a suspension of rigid spherical particles ($\approx 0.65$). The critical packing parameter represents the concentration where the particles become jammed together and the suspension gains solid-like characteristics (Tadros 1994). There is a large increase in viscosity when the particle concentration is increased toward the critical packing parameter (Figure 3.5).

It should be noted that the effective volume fraction ($\varphi_{eff}$) of particles within a delivery system may be considerably larger than the actual volume fraction ($\varphi$) of the constituents used to fabricate it. This occurs when some of the continuous phase is trapped within the particles, which may occur for nonspherical particles that rotate in solution, flocculated particles, multiple emulsions, or filled hydrogel particles. The above equations can be modified to take into account these effects by replacing $\varphi$ with $\varphi_{eff}$. The effective volume fraction can be related to the *actual* volume fraction by the following expression: $\varphi_{eff}$ ($=R\varphi$), where $R$ is the *volume ratio*, that is, the total volume occupied by the particle divided by the volume of the constituents used to fabricate it. Colloidal dispersions with a high volume ratio ($R \gg 1$) have a more pronounced impact on the viscosity than those with a low volume ratio (Figure 3.6). To highlight the importance of this effect, consider the impact of hydrogel particles on the viscosity of an aqueous solution. The composition of a hydrogel particle may be 20% biopolymer molecules and 80% water so that the volume of the particle is five times greater than the actual volume of the biopolymer used (i.e., $R = 5$). Thus, for an aqueous hydrogel particle dispersion containing 5% biopolymer, the effective particle volume fraction would be 25%, which would mean the viscosity was considerably higher than expected.

The equations relating the viscosity of a colloidal dispersion to the particle concentration presented above suggest that particle size does not play a major role in determining the rheological properties of these systems. This is normally true for relatively dilute nonaggregated systems (<20%–30% particles). However, there may be a considerable increase in viscosity in more concentrated nonaggregated systems with decreasing particle size owing to a change in the relative importance of Brownian motion and hydrodynamic effects (McClements 2005). In addition, the viscosity of flocculated systems containing smaller particles is usually much greater than that of similar systems containing larger particles (Mao and McClements 2012a,b). Consequently, varying the size of the particles in a colloidal delivery system may influence its rheological properties or those of any food matrix it is incorporated into.

The size of the particles in a colloidal delivery system may also influence the perceived mouthfeel of food and beverage products. If the particles within a product are detected as discrete entities within the mouth, then the product may be perceived as being either "grainy" or "gritty," which are undesirable sensory attributes (Burey et al. 2008; Engelen et al. 2005b). There appears to be a critical particle size below which particles are not detected as discrete entities within the mouth (although they still contribute to the overall mouthfeel through their impact on viscosity and lubrication). This critical particle size has been reported to depend on characteristics of the colloidal particles and of the surrounding medium, such as particle rheology, particle shape, and continuous phase rheology (Engelen et al. 2005a,b; Imai et al. 1997). "Hard" particles (such as solid particles) are detected as individual entities by the tongue at smaller diameters than "soft" particles (such as emulsion droplets or hydrogel particles). For example, alumina particles can be detected as separate entities at sizes as low as 10 μm, whereas the particles in margarine can only be detected when particle size exceeds approximately 22 μm. Studies have shown that humans can even rank the size of solid particles in the 2 to 200 μm range (Engelen et al. 2005a,b). Irregular shaped particles are usually perceived more readily than rounder shaped ones (Engelen et al. 2005b; Imai et al. 1999). It also becomes easier to detect particles as separate entities as the viscosity of the surrounding medium decreases (Engelen et al. 2005b; Imai et al. 1997). Particle size is therefore an important consideration when designing colloidal delivery systems that will not adversely affect the desirable texture and mouthfeel of food products.

## 3.4.2 Optical Properties and Appearance

The characteristics of the particles within a colloidal delivery system determine how it interacts with light waves, which affect the optical properties and appearance of food products into which it is incorporated. The most important optical properties affected by particle characteristics are opacity and color, which can be quantitatively described using tristimulus color coordinates, such as the $L^*a^*b^*$ system (Danviriyakul et al. 2002; Nuchi et al. 2002). In this color system, $L^*$ represents lightness, and $a^*$ and $b^*$ are color coordinates, where $+a^*$ is the red direction, $-a^*$ is the green direction, $+b^*$ is the yellow direction, $-b^*$ is the blue direction, low $L^*$ is dark, and high $L^*$ is light. The opacity of a food is characterized by its lightness ($L^*$), while its color intensity can be characterized by its chroma: $C = (a^{*2} + b^{*2})^{1/2}$. The optical properties of colloidal dispersions are mainly determined by the relative refractive index, the particle concentration, and the particle size (McClements 2002a,b, 2005). Theoretical relationships have been developed to mathematically relate particle characteristics to the optical properties of colloidal dispersions (Chantrapornchai et al. 1999a,b; McClements 2002a,b). These expressions can be presented simply in the following form:

$$\tau = f(\varphi, r, \Delta n, \alpha(\lambda), \lambda) \tag{3.13}$$

$$L^*, a^*, b^* = f(\varphi, r, \Delta n, \alpha(\lambda)) \tag{3.14}$$

Here, $\tau$ is the turbidity, $L^*a^*b^*$ are the tristimulus color coordinates, $\varphi$ is the particle concentration, $r$ is the particle radius, $\Delta n$ is the refractive index difference between particles and surrounding medium, $\alpha(\lambda)$ depends on the visible absorption spectrum of the particles and surrounding medium, $\lambda$ is the wavelength of light, and $f()$ means "a function of." A full description of the mathematical approach used to describe the relationship between particle properties and optical properties is given elsewhere (McClements 2002a,b). To quantitatively predict the impact of particle characteristics on the optical properties of delivery systems, it is necessary to have information about the structure, composition, and properties of the system, for example, PSD, particle morphology, particle concentration, and the complex refractive indices (real and imaginary part) of the particles and surrounding medium.

Theoretical and experimental studies indicate that the lightness ($L^*$) of colloidal dispersions increases with increasing particle concentration and refractive index contrast and has a maximum value at an intermediate particle size close to the wavelength of light (Chantrapornchai et al. 1998, 1999a,b, 2000, 2001). As an example, the lightness of O/W emulsions increases steeply as the droplet concentration increases from around 0 to 5 wt% but then increases more gradually as the droplet concentration is further increased (Figure 3.7). The impact of particle size on the predicted turbidity of colloidal dispersions containing monodisperse oil droplets dispersed in water is shown in Figure 3.8. The turbidity has a relatively low value for small particles, increases with increasing particle diameter to a maximum value, and then decreases when the particle size is increased further. Colloidal dispersions with particle radius less than approximately 25 nm appear transparent (Wooster et al. 2008). In principle, the impact of particles on the optical properties of a colloidal dispersion can be controlled by manipulating the refractive index of the particles or surrounding medium (Chantrapornchai et al. 2001). In practice, this is often difficult because of the limited number of food-grade materials available to control refractive index or because of the large amounts of materials that must be added (e.g., >50% sorbitol needs to be added to an emulsion to make it transparent by contrast index matching). Colloidal particles that give relatively high refractive indices may cause a large increase in the turbidity or opacity of a food or beverage product that they are incorporated into, unless the particles are very small compared to the wavelength of light.

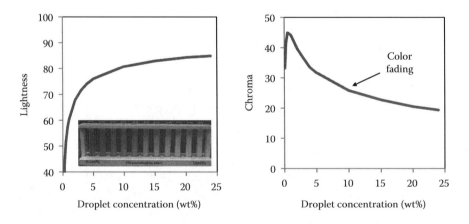

**Figure 3.7** **Influence of droplet concentration on the lightness and chroma of hexadecane O/W emulsions ($d$ = 300 nm) containing a fixed concentration of red dye (0.1%).**

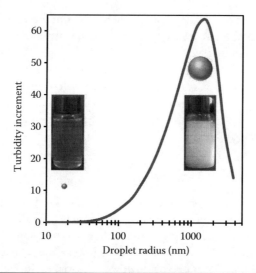

**Figure 3.8    Influence of particle size on the turbidity increase per unit increase in particle concentration. Colloidal dispersions appear transparent when the particle size falls below some critical value. Photographs are of orange O/W emulsions with different droplet sizes.**

The impact of particle characteristics on the optical properties of foods and beverages is an important consideration when designing colloidal delivery systems for commercial products. Some products should be optically clear or only slightly turbid (e.g., flavored waters, soft drinks, and fruit beverages) and so the delivery system should be designed not to cause a large increase in turbidity or opacity. Other food products are optically opaque (e.g., dressings, sauces, desserts, and mayonnaise) and therefore the light scattering characteristic of the particles within the delivery system may be less important.

The standard mathematical models for relating particle characteristics to optical properties assume that the particles are homogeneous spheres. In reality, the particles in many colloidal delivery systems are nonspherical or inhomogeneous and therefore it is important to be aware of the impact of particle composition and morphology on their optical properties. More sophisticated theoretical models have been developed to account for the interaction of light waves with more complex particles, for example, nonspherical, core–shell, and dispersion (Bohren and Huffman 1983).

### 3.4.3  Stability and Shelf Life

Colloidal delivery systems may be thermodynamically stable or unstable depending on their composition and structure. Micelles and microemulsions are thermodynamically stable systems under particular environmental conditions, such as composition, temperature, and pressure (Chapter 5). On the other hand, liposomes, emulsions, and hydrogel particles are thermodynamically unstable systems that have a tendency to break down over time (Chapters 5 through 7). Thermodynamically unstable colloidal dispersions may break down through a variety of physicochemical mechanisms, including gravitational separation (creaming or sedimentation), aggregation (flocculation, coalescence, and partial coalescence) and ripening (compositional and Ostwald ripening) of the particles (Dickinson 1992; Friberg et al. 2004; McClements 2005).

### 3.4.3.1 Gravitational Separation

Gravitational separation is one of the most common forms of instability in foods and may take the form of either creaming or sedimentation depending on the relative densities of the particles and surrounding liquid. Creaming is the upward movement of particles that occurs when they have a lower density than the surrounding liquid, whereas sedimentation is the downward movement of particles that occurs when they have a higher density than the surrounding liquid. Liquid flavor or triglyceride oils normally have lower densities than water and so lipid droplets tend to cream in aqueous systems. On the other hand, proteins, polysaccharides, phospholipids, and solid triglyceride oils usually have higher densities than water and so are prone to sedimentation. Many particulate delivery systems contain multicomponent particles and so their densities depend on the type and concentration of components present (Matalanis et al. 2011; Matalanis and McClements 2013). In principle, gravitational separation can be retarded by preparing particles that have similar densities to the surrounding medium, for example, by mixing appropriate amounts of a high- and low-density material. This approach is widely used in the beverage industry where weighting agents (e.g., brominated vegetable oil [BVO], sucrose acetoisobutyrate [SAIB], or ester gum) are added to the oil phase to match the droplet density to the surrounding aqueous phase (Chanamai and McClements 2000; McClements 2005). The creaming rate of an isolated spherical particle in a liquid is given by the following expression, which is known as Stokes' law:

$$v = -\frac{2gr^2(\rho_2 - \rho_1)}{9\eta_1} \tag{3.15}$$

Here, $v$ is the creaming velocity, $g$ is the acceleration caused by gravity, $r$ is the radius of the particle, $\rho$ is the density, $\eta$ is the shear viscosity, and the subscripts 1 and 2 refer to the continuous phase and particles, respectively. The sign of $v$ determines whether the particle moves upward (+) (creaming) or downward (−) (sedimentation). Equation 3.15 shows that the creaming rate increases with increasing particle size, increasing density contrast, and decreasing continuous phase viscosity. Equation 3.15 is only strictly applicable to dilute solutions of noninteracting rigid spherical particles dispersed in an ideal (Newtonian) fluid. More sophisticated mathematical models are available that take into account polydispersity, nonspherical particles, particle fluidity, particle–particle interactions, and non-Newtonian fluids (McClements 2005). To use this equation to predict the stability of a colloidal delivery system within a food or beverage product, it is necessary to know the size distribution and density of the particles and the density and viscosity of the surrounding medium. Stokes' law can be used to predict the stability of structured colloidal delivery systems containing multicomponent particles to gravitational separation, such as multiple emulsions or filled hydrogel particles. However, an effective particle density should be used. For example, for a particle containing two phases (I and II) with different densities, the following expression can be used: $\rho_2 = \varphi_{II}\rho_{II} + (1 - \varphi_{II})\rho_I$.

Knowledge of the factors that influence the movement of colloidal particles in centrifugal fields is also important, since it is often necessary to separate particles from the surrounding medium or to concentrate particles by centrifugation. In this case, the driving force for separation is the centrifugal force ($\omega R$) rather than gravity ($g$), and hence Stokes' equation can still be used by substituting $\omega R$ for $g$, where $\omega$ is the angular velocity of rotation and $R$ is the distance of the particle from the center of rotation.

## 3.4.3.2 Particle Aggregation

The particles in colloidal delivery systems are often subject to forces that cause them to collide with their neighbors, such as thermal energy, gravity, or applied mechanical stresses (Friberg et al. 2004; McClements 2005). After an encounter, the colloidal particles may move apart or remain aggregated, depending on the magnitude and range of the attractive and repulsive interactions between them (Israelachvili 2011). Particles tend to remain together (aggregate) when the attractive forces dominate the repulsive forces. The major attractive forces that may act between particles in colloidal dispersions are van der Waals, depletion, and hydrophobic attraction, as well as bridging effects (McClements 2005). The major repulsive forces that are used to inhibit particle aggregation in colloidal dispersions are electrostatic and steric repulsion. Electrostatic repulsion tends to increase as the magnitude of the charge on the particle surfaces increases and the ionic strength of the surrounding medium decreases (Israelachvili 2011; McClements 2005). The steric repulsion tends to increase as the thickness and hydrophilicity of the interfacial layer increases. In general, the type, sign, magnitude, and range of the colloidal interactions operating in a particular system depends on the composition and properties of the system, for example, particle size, particle composition, surface charge, surface hydrophobicity, the thickness and density of any adsorbed polymer layers, the ionic strength and pH of the aqueous solution, and the concentration and type of nonadsorbed polymers (McClements 2005). There are a number of different types of particle aggregation that can occur in colloidal delivery systems: flocculation, coalescence, and partial coalescence.

*Flocculation*: Flocculation is the process whereby two or more particles come together to form an aggregate ("floc") in which the particles retain their individual integrity (Figure 3.9). Flocculation is usually detrimental to the stability of colloidal delivery systems because it increases the rate of gravitational separation and may promote thickening or gel formation. Flocculation may occur through a variety of different physicochemical mechanisms that either increase the attractive forces or decrease the repulsive interactions operating between the particles:

■ *Reduced electrostatic repulsion.* Electrostatically stabilized colloidal dispersions tend to flocculate when the electrostatic repulsion between the particles is reduced below a critical level. A reduction in electrostatic repulsion may occur for a variety of reasons: the pH is altered so that the magnitude of the electrical charge on the colloidal particles is reduced; counterions that bind to the particle surfaces and reduce their net charge are added; the ionic strength of the surrounding medium is increased to screen the electrostatic interactions.
■ *Reduced steric repulsion.* Sterically stabilized colloidal dispersions tend to flocculate when the steric repulsion between the particles is reduced below a

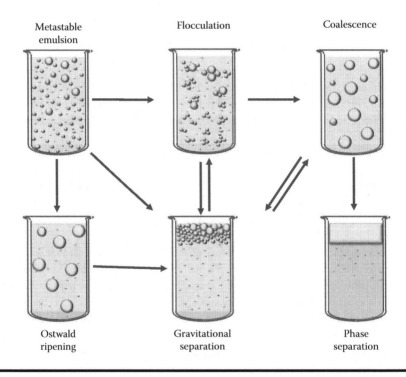

**Figure 3.9 Schematic diagram of most common instability mechanisms that occur in colloidal delivery systems: gravitational separation, flocculation, coalescence, Ostwald ripening, and phase inversion.**

critical level. A reduction in steric repulsion may occur through a variety of processes that alter the properties of adsorbed polymer layers: solvent quality may be altered so that polymer–polymer interactions are favored over polymer–solvent interactions; pH or ionic strength may be altered so that the electrostatic interactions between charged polymers are reduced, thereby leading to a reduction in layer thickness; the polymer layer may be degraded by chemicals or enzymes; the polymer layer may be detached from the particle surfaces by competitive adsorption.

■ *Increased depletion attraction.* The carrier particles in colloidal delivery systems are often suspended in a liquid that also contains other types of relatively small nonadsorbing colloidal substances, such as surfactant micelles, biopolymer molecules, or biopolymer aggregates. These nonadsorbing substances increase the attractive force between larger particles owing to an osmotic effect associated with their exclusion from a narrow region surrounding each particle. This attractive force increases as the concentration of nonadsorbing substances increases, until eventually it may become large enough to overcome the repulsive interactions between the particles and cause them to flocculate. This type of particle aggregation is usually referred to as *depletion flocculation.* Depletion flocculation depends on the concentration and size of the nonabsorbing substances within the continuous phase of a colloidal dispersion and can be avoided by ensuring that their concentration remains below some critical level.

■ *Increased hydrophobic interactions.* This type of interaction is important in delivery systems that contain colloidal particles that have some nonpolar regions exposed to the surrounding aqueous phase. There is then a relatively strong hydrophobic attraction between nonpolar patches on the surfaces of different particles that promotes aggregation. This may occur if hydrophobic particles are not fully coated with emulsifier molecules or if the emulsifier itself has some hydrophobic character (such as globular proteins). Hydrophobic-driven aggregation can be avoided by ensuring that the colloidal particles have hydrophilic surfaces, for example, by adding sufficient emulsifier to completely cover all nonpolar patches or by avoiding processes that increase surface hydrophobicity (such as heating of particles coated by globular proteins).

■ *Bridge formation.* Some substances in foods are able to promote flocculation in colloidal delivery systems by forming bridges between two or more particles. These substances may be biopolymers (such as proteins or polysaccharides), multivalent ions (such as calcium or polyphosphates), or small particles (such as micelles). To be able to link colloidal particles together, there must be a sufficiently strong attractive interaction between these substances and the particle surfaces. The most common types of interaction that operate in food colloids are electrostatic and hydrophobic. For example, a negatively charged biopolymer might adsorb to the surface of two positively charged colloidal particles, causing them to flocculate.

A number of mathematical models have been developed to calculate the interactions between colloidal particles and predict the stability of colloidal dispersions to particle aggregation (Israelachvili 2011; McClements 2005). These models can often be used to predict the impact of specific particle characteristics (such as size, concentration, charge, hydrophobicity, and interfacial thickness) and solution conditions (such as pH, ionic strength, and nonabsorbing substances) on the stability of colloidal delivery systems. The development of a suitable strategy to stabilize a particular colloidal delivery system depends on identification of the physicochemical origin of particle flocculation in that system. In general, flocculation can be prevented by ensuring that the repulsive forces are stronger than the attractive forces (Israelachvili 2011).

*Coalescence*: Coalescence is the process whereby two or more liquid droplets encounter each other and then fuse together to form a single larger droplet (Figure 3.9). Coalescence is therefore only important in systems that contain fluid carrier particles, such as nanoemulsions, emulsions, multiple emulsions, and some types of soft hydrogel particles. Coalescence tends to cause particles to cream or sediment more rapidly because of the increase in their dimensions (Stokes' law). In O/W emulsions, coalescence eventually leads to the formation of a layer of oil on top of the sample, which is commonly referred to as *oiling off*. In water-in-oil emulsions, it leads to the accumulation of water at the bottom of the material. In general, the susceptibility of particles to coalescence is determined by the nature of the forces that act on and between them (i.e., gravitational, colloidal, hydrodynamic, and mechanical forces), as well as the resistance of any interfacial layers to disruption. To improve the stability of particles to coalescence, it is necessary to either prevent them from coming into proximity for extended periods or ensure that the interfacial coating has a high resistance to disruption.

*Partial coalescence*: Partial coalescence occurs when two or more partially crystalline particles come into contact and form an irregularly shaped aggregate (Figure 3.10). It is initiated when a solid crystal from one particle penetrates into

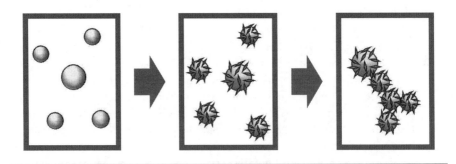

**Figure 3.10  Schematic diagram of partial coalescence, an instability mechanism that may occur in colloidal delivery systems containing partially crystalline particles.**

the liquid portion of another particle and acts as a bridge between them (Fredrick et al. 2010). Partial coalescence is most important in systems that contain a mixture of solid and liquid fat, such as emulsion-based systems and solid lipid nanoparticles. Normally, protruding fat crystals would be surrounded by the aqueous phase, but when they penetrate into another droplet, they are surrounded by liquid oil. This causes the droplets to remain aggregated because it is thermodynamically more favorable for a fat crystal to be wetted by oil molecules than by water molecules. Over time, the partially crystalline particles merge together because this reduces the surface area of oil exposed to water. Nevertheless, the cluster formed partly retains the shape of the original droplets from which it was formed because the fat crystal network within the droplets prevents them from completely merging together.

### 3.4.3.3 Ostwald Ripening

Ostwald ripening is the process whereby large particles grow at the expense of smaller ones owing to mass transport of dispersed phase from one particle to another through the intervening continuous phase (Figure 3.11). This process is not important in colloidal dispersions where the solubility of the dispersed phase is negligible within the surrounding medium (e.g., triacylglycerol or mineral oils in water), because the mass transport rates are then insignificant (Kabalnov and Shchukin 1992). Nevertheless, it is important in systems where the disperse phase does have an appreciable solubility in the surrounding medium (e.g., flavor oils, essential oils, or short-chain triglycerides in water) (Li et al. 2009; Lim et al. 2011). In this type of delivery system, one has to consider methods for retarding the rate of Ostwald ripening.

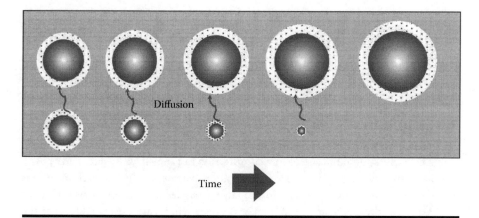

Figure 3.11 Schematic diagram of physicochemical processes involved in Ostwald ripening. Solute molecules move from smaller to larger droplets, leading to a net increase in particle size.

Ostwald ripening occurs because the solubility of the material in a spherical particle increases as the size of the particle decreases (Kabalnov and Shchukin 1992):

$$S(r) = S(\infty) \exp\left(\frac{2\gamma V_m}{RTr}\right). \tag{3.16}$$

Here, $V_m$ is the molar volume of the dispersed phase, $\gamma$ is the interfacial tension at the particle boundary, and $S(\infty)$ and $S(r)$ are the solubilities of the dispersed phase in the continuous phase for particles with infinite curvature (planar boundary) and radius $r$, respectively. The increase in solubility with decreasing particle radius means that there is a higher local concentration of solubilized material around a smaller particle than around a larger one (Figure 3.11). The solubilized disperse phase molecules therefore move from the smaller particles to the larger ones because of this concentration gradient. Once steady state has been achieved, the rate of Ostwald ripening is given by (Kabalnov and Shchukin 1992)

$$\frac{d\langle r\rangle^3}{dt} = \frac{8\gamma V_m S(\infty) D}{9RT}. \tag{3.17}$$

Here, $D$ is the translation diffusion coefficient of the dispersed phase molecules through the continuous phase. Equation 3.17 indicates that particle growth becomes more rapid as the solubility of the dispersed phase in the continuous phase increases. The rate of Ostwald ripening depends on the initial PSD, with the ripening rate increasing as the width of the distribution increases. Hence, tight control over the initial PSD is often important to create delivery systems that are more stable to Ostwald ripening.

The most effective method of inhibiting Ostwald ripening in colloidal delivery systems is to use ripening inhibitors, which are molecules that have a high solubility in the dispersed phase, but a very low solubility in the continuous phase. In O/W emulsions, these are typically highly hydrophobic molecules, such as long-chain triacylglycerols (Li et al. 2009; Wooster et al. 2008). Ripening inhibitors work through an entropy of mixing effect, usually referred to as "compositional ripening," which opposes droplet growth through Ostwald ripening. Initially, all of the particles in the system have the same composition but may have different sizes. As a response, disperse phase molecules with an appreciable solubility in the continuous phase will diffuse from the small to the large particles because of Ostwald ripening. After this process occurs, the smaller droplets will have an elevated level of the ripening inhibitor since it cannot diffuse out of the droplets, whereas the larger droplets will have a reduced level of the ripening inhibitor since it is diluted with the disperse phase entering these droplets. Consequently, there is a concentration gradient setup in the system, which is thermodynamically unfavorable owing to an entropy of mixing effect that opposes further transport of the disperse phase between the droplets and thereby inhibits Ostwald ripening. The amount of ripening inhibitor that needs to be added to the disperse phase to effectively eliminate Ostwald ripening depends on its molecular weight and solubility and can be calculated using mathematical models (Kabalnov and Shchukin 1992; Wooster et al. 2008).

## 3.4.4 Molecular Partitioning and Transport

When a colloidal delivery system is incorporated into a food or beverage product, there may be a redistribution of the various types of active ingredient present among the different phases (e.g., carrier particle, food matrix, and headspace), which is governed by their equilibrium partition coefficients and the kinetics of molecular motion (McClements 2005). For example, if a delivery system was incorporated into a food product, then some of the encapsulated active ingredients may move from the carrier particles into the food matrix. Conversely, some of the substances initially present within the food matrix may move into the delivery system particles. These molecular transport processes may have an adverse, neutral, or beneficial impact on the functionality of a delivery system and should be considered when designing colloidal delivery systems for specific food matrices.

### 3.4.4.1 Molecular Partitioning

The location of the active ingredients within a colloidal delivery system plays an important role in its functional performance. For example, the physical environment of an active ingredient may influence its physicochemical stability, sensory perception, or biological activity. In this section, the importance of molecular partitioning of active ingredients is demonstrated by focusing on flavor molecules, since their physical location influences the sensory perception of food and beverage products (McClements 2005). The aroma of a food is sensed by receptors within the nasal cavity owing to the presence of volatile flavor molecules released from a food before, during, or after mastication. Aroma perception therefore depends on the type and amount of flavor molecules in the gas (headspace) above a food. The taste of a food is perceived by specific receptors within the oral cavity because of certain nonvolatile molecules that elicit a taste response, for example, sweet, bitter, sour, salty, and umami flavors. Taste perception therefore depends on the type and concentration of flavor molecules released from a food into the surrounding saliva (aqueous phase), which then reach the receptors on the tongue and inside of the mouth. In general, the distribution of flavor molecules among the various environments within a food or beverage product (e.g., headspace, disperse phase, continuous phase) depends on the characteristics of the flavor molecules themselves, as well as on the nature and proportions of the various phases. The incorporation of a colloidal delivery system into a food or beverage product may alter the levels of volatile and nonvolatile molecules in the headspace (aroma) and aqueous phase (taste), thereby altering its overall flavor profile. It is therefore important to establish the potential impact of delivery systems on flavor partitioning. This is also true for other types of active ingredients that may be utilized as functional agents in foods and beverages, such as antimicrobials, antioxidants, or nutraceuticals.

The partitioning of active ingredients between the different phases in a food material can be quantified in terms of *equilibrium partition coefficients*: $K_{12} = c_1/c_2$, where $c_1$ and $c_2$ are the concentrations of the active ingredients in phases 1 and 2, respectively (Chapter 10). These phases may be the particles in the colloidal delivery system, as well as various discrete phases that may be present within the original food product, for example, fat, water, solids, or air. The tendency for a particular active ingredient to partition into one phase rather than another depends on its molecular characteristics (e.g., molecular weight and polarity), relative to that of the other phases. Thus, for an oil–water system, nonpolar active ingredients will tend to partition more into the oil phase whereas polar active ingredients will tend to partition more into the water phase. This phenomenon has important consequences when a delivery system is added to a food matrix. For example, a delivery system containing hydrophobic particles (such as micelles, lipid droplets, or liposomes) may absorb some of the nonpolar flavor molecules from the food, thereby altering its flavor profile. The concentration of active ingredients present in the headspace above a food depends on the relative amounts of dispersed and continuous phase present:

$$\Phi_m = \left(1 + \frac{V_F}{V_G}\left(\frac{\varphi_D K_{DC}}{K_{GC}} + \frac{(1-\varphi_D)}{K_{GC}}\right)\right)^{-1}. \tag{3.18}$$

Here, $\Phi_m$ is the mass fraction of active ingredient (usually a flavor) in the gas phase (headspace) above the food, $V_F$ is the volume of the food, $V_G$ is the volume of the gaseous headspace, $\varphi_D$ is the volume fraction of the disperse phase (carrier particles) within the food, $K_{DC}$ is the equilibrium partition coefficient of the active ingredient between the dispersed and continuous phase, and $K_{GC}$ is the equivalent value between the gas phase and the continuous phase.

Equation 3.18 can be used to predict the potential impact of the carrier particles within a delivery system on the flavor profile of foods and beverages. For example, adding an emulsion-based delivery system to a food will alter the total amount of oil phase present. If the total oil phase concentration is increased, then the concentration of nonpolar flavor molecules in the headspace decreases (Figure 3.12). This has important implications when incorporating delivery systems containing nonpolar phases (such as microemulsions, nanoemulsions, and emulsions) into foods. The type and amount of flavor molecules present may have to be altered to achieve the desired flavor profile.

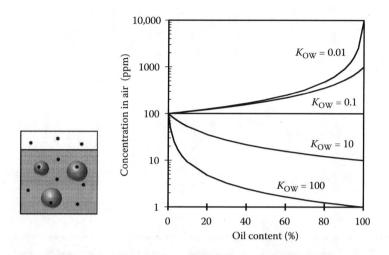

**Figure 3.12** **Influence of oil content and partition coefficient on the concentration of an active ingredient in the air above an oil–water system.**

It is also possible to predict the influence of the overall oil content of a mixed system on the concentration of active ingredients within the aqueous phase, which has important implications for determining the potential impact of delivery systems on taste perception. In general, the mass fraction of an active ingredient within the continuous phase ($\Phi_{m,C}$) of a mixed system depends on its partition coefficient and system composition:

$$\Phi_{m,C} = \left(1 + \frac{\varphi_D K_{DC}}{1-\varphi_D}\right)^{-1}. \tag{3.19}$$

For an oil–water system, $K_{DC} = K_{OW}$ and $\Phi_{m,C} = \Phi_{m,W}$. The value of $K_{OW}$ depends on the chemical structure of the active ingredient, as well as the properties of the oil and aqueous phases. Knowledge of $\Phi_{m,W}$ is often important for the development of effective delivery systems since the chemical degradation of some active ingredients occurs more rapidly in the aqueous phase than in the oil phase (e.g., citral). The aqueous phase concentration of highly nonpolar active ingredients ($K_{OW} \gg 1$) decreases as the oil content increases (Figure 3.13).

Another factor that may influence the flavor profile of a food system into which a delivery system is incorporated is the potential for flavor binding (McClements 2005). Many proteins and carbohydrates are capable of binding flavor molecules and therefore altering their distribution within a multiphase system. Flavor binding can therefore cause a significant alteration in the flavor profile of a food. This alteration is often detrimental to food quality because it changes the characteristic flavor profile, but it can also be beneficial when the bound molecules are off-flavors.

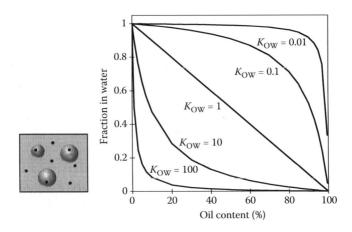

**Figure 3.13    Influence of oil content and partition coefficient on the fraction of an active ingredient present within the water phase of an oil–water system.**

Mathematical models have been developed to describe the impact of flavor binding on the flavor profile of colloidal food products (McClements 2005).

### 3.4.4.2 Mass Transport Processes

The movement of active ingredients from one location to another is important for many applications of colloidal delivery systems. For example, an active ingredient may have to move from its original location within carrier particles to its intended site of action, which may be the nose (aroma), the mouth (flavor), bacterial surfaces (antimicrobials), the small intestine (nutraceuticals), or the colon (prebiotics or probiotics). An understanding of mass transport processes is therefore important for the design of many kinds of colloidal delivery systems. The physical and mathematical basis of mass transport processes associated with release of active ingredients from colloidal particles is discussed in Chapter 10. In this section, only a brief overview of the importance of molecular transport processes is given, with special emphasis on flavor release since this is one of the most important potential applications of delivery systems in the food industry.

Flavor release is the process whereby flavor molecules move out of a food product and into the surrounding saliva or vapor phase during mastication (McClements 2005). The release of the flavors from a food material occurs under extremely complex and dynamic conditions (Liu and Daum 2008). A food usually spends a relatively short period (typically 1 to 30 s) in the oral cavity before being swallowed. During this period, it is diluted with saliva, experiences temperature changes, and is subjected to a variety of complex mechanical forces and flow profiles. Mastication may therefore cause dramatic changes in the composition, physicochemical properties, and structural characteristics of colloidal delivery systems and food matrices.

During mastication, nonvolatile flavor molecules must move from within the food, through the saliva to the taste receptors on the tongue and oral cavity, whereas volatile flavor molecules must move from the food, through the saliva and into the gas phase, where they are carried to the aroma receptors in the nasal cavity. The two major factors determining the rate at which these processes occur are the equilibrium partition coefficients (because this determines the initial flavor concentration gradients at the various phase boundaries) and mass transport coefficients (because this determines the speed at which molecules move through each phase). A variety of mathematical models have been developed to describe the release of flavor molecules from colloidal dispersions (McClements 2005). The incorporation of a delivery system into a food product may change its flavor profile by altering the initial partitioning of the flavor molecules and by modifying the rate at which these molecules travel from one location to another. For example, flavor molecules may initially be trapped inside colloidal particles and may only be released in response to some specific trigger within the mouth (such as dilution, dissolution, temperature changes, or enzyme activity).

## 3.5 Summary

This chapter has shown that particle characteristics, such as composition, size, concentration, charge, interactions, and physical state, ultimately determine the physicochemical properties and functional performance of colloidal delivery systems, for example, their optical properties, rheology, stability, retention, and release characteristics. When designing colloidal delivery systems, it is important to ensure that the carrier particles have suitable functional characteristics, but that they do not adversely affect the quality attributes of the food or beverage products that they are going to be utilized in. For example, lipophilic active ingredients may need to be incorporated into aqueous-based products that are required to have high optical clarity and low viscosity, such as fortified waters, soft drinks, and some fruit juices. In these cases, it is important to use a colloidal delivery system that contains very small carrier particles that do not scatter light strongly and that are stable to gravitational separation. On the other hand, lipophilic active ingredients may need to be incorporated into aqueous-based products that are highly viscous and optically opaque, for example, dressings, dips, sauces, and desserts. In these cases, the size of the particles in the colloidal delivery system may be less critical since optical clarity and gravitational separation are not major problems. An important factor to be aware of when designing a colloidal delivery system is that changing the particle characteristics to achieve one goal (e.g., reducing gravitational separation) may have an impact on all of the other quality attributes of the final product (e.g., appearance, rheology, flavor, and release characteristics). Consequently, one needs to take an integrated approach to design colloidal delivery systems to meet all of the required quality specifications.

# References

Aden, A. L. and M. Kerker (1951). "Scattering of electromagnetic waves from 2 concentric spheres." *Journal of Applied Physics* **22**(10): 1242–1246.

Aguzzi, C., P. Cerezo, I. Salcedo, R. Sanchez and C. Viseras (2010). "Mathematical models describing drug release from biopolymeric delivery systems." *Materials Technology* **25**(3–4): 205–211.

Arifin, D. Y., L. Y. Lee and C.-H. Wang (2006). "Mathematical modeling and simulation of drug release from microspheres: Implications to drug delivery systems." *Advanced Drug Delivery Reviews* **58**(12–13): 1274–1325.

Baker, R. W. (1987). *Controlled Release of Biologically Active Agents*. New York, John Wiley and Sons.

Bhandari, R. (1985). "Scattering coefficients for a multilayered sphere—Analytic expressions and algorithms." *Applied Optics* **24**(13): 1960–1967.

Bohren, C. F. and D. R. Huffman (1983). *Absorption and Scattering of Light by Small Particles*. New York, John Wiley and Sons, Inc.

Burey, P., B. R. Bhandari, T. Howes and M. J. Gidley (2008). "Hydrocolloid gel particles: Formation, characterization, and application." *Critical Reviews in Food Science and Nutrition* **48**(5): 361–377.

Chanamai, R. and D. J. McClements (2000). "Impact of weighting agents and sucrose on gravitational separation of beverage emulsions." *Journal of Agricultural and Food Chemistry* **48**(11): 5561–5565.

Chantrapornchai, W., F. Clydesdale and D. J. McClements (1998). "Influence of droplet size and concentration on the color of oil-in-water emulsions." *Journal of Agricultural and Food Chemistry* **46**(8): 2914–2920.

Chantrapornchai, W., F. Clydesdale and D. J. McClements (1999a). "Influence of droplet characteristics on the optical properties of colored oil-in-water emulsions." *Colloids and Surfaces A-Physicochemical and Engineering Aspects* **155**(2–3): 373–382.

Chantrapornchai, W., F. Clydesdale and D. J. McClements (1999b). "Theoretical and experimental study of spectral reflectance and color of concentrated oil-in-water emulsions." *Journal of Colloid and Interface Science* **218**(1): 324–330.

Chantrapornchai, W., F. M. Clydesdale and D. J. McClements (2000). "Optical properties of oil-in-water emulsions containing titanium dioxide particles." *Colloids and Surfaces A-Physicochemical and Engineering Aspects* **166**(1–3): 123–131.

Chantrapornchai, W., F. M. Clydesdale and D. J. McClements (2001). "Influence of relative refractive index on optical properties of emulsions." *Food Research International* **34**(9): 827–835.

Danviriyakul, S., D. J. McClements, E. Decker, W. W. Nawar and P. Chinachoti (2002). "Physical stability of spray-dried milk fat emulsion as affected by emulsifiers and processing conditions." *Journal of Food Science* **67**(6): 2183–2189.

Dickinson, E. (1992). *An Introduction to Food Colloids*. Oxford, UK, Oxford University Press.

Dickinson, E. (2010). "Flocculation of protein-stabilized oil-in-water emulsions." *Colloids and Surfaces B-Biointerfaces* **81**(1): 130–140.

Dickinson, E. and M. Golding (1998). "Influence of calcium ions on creaming and rheology of emulsions containing sodium caseinate." *Colloids and Surfaces A-Physicochemical and Engineering Aspects* **144**(1–3): 167–177.

Dresselhuis, D. M., M. A. C. Stuart, G. A. van Aken, R. G. Schipper and E. H. A. de Hoog (2008). "Fat retention at the tongue and the role of saliva: Adhesion and spreading of 'protein-poor' versus 'protein-rich' emulsions." *Journal of Colloid and Interface Science* **321**(1): 21–29.

Engelen, L., R. A. de Wijk, A. van der Bilt, J. F. Prinz, A. M. Janssen and F. Bosman (2005a). "Relating particles and texture perception." *Physiology and Behavior* **86**(1–2): 111–117.

Engelen, L., A. Van der Bilt, M. Schipper and F. Bosman (2005b). "Oral size perception of particles: Effect of size, type, viscosity and method." *Journal of Texture Studies* **36**(4): 373–386.

Ensign, L. M., R. Cone and J. Hanes (2012). "Oral drug delivery with polymeric nanoparticles: The gastrointestinal mucus barriers." *Advanced Drug Delivery Reviews* **64**(6): 557–570.

Foegeding, E. A., C. R. Daubert, M. A. Drake, G. Essick, M. Trulsson, C. J. Vinyard and F. Van de Velde (2011). "A comprehensive approach to understanding textural properties of semi- and soft-solid foods." *Journal of Texture Studies* **42**(2): 103–129.

Fredrick, E., P. Walstra and K. Dewettinck (2010). "Factors governing partial coalescence in oil-in-water emulsions." *Advances in Colloid and Interface Science* **153**(1–2): 30–42.

Friberg, S., K. Larsson and J. Sjoblom (2004). *Food Emulsions*. New York, Marcel Dekker.

Genovese, D. B., J. E. Lozano and M. A. Rao (2007). "The rheology of colloidal and noncolloidal food dispersions." *Journal of Food Science* **72**(2): R11–R20.

Grassi, M. and G. Grassi (2005). "Mathematical modelling and controlled drug delivery: Matrix systems." *Current Drug Delivery* **2**(1): 97–116.

Gu, Y. S., E. A. Decker and D. J. McClements (2007). "Formation of colloidosomes by adsorption of small charged oil droplets onto the surface of large oppositely charged oil droplets." *Food Hydrocolloids* **21**(4): 516–526.

Guzey, D., H. J. Kim and D. J. McClements (2004). "Factors influencing the production of o/w emulsions stabilized by beta-lactoglobulin-pectin membranes." *Food Hydrocolloids* **18**(6): 967–975.

Guzey, D. and D. J. McClements (2006). "Formation, stability and properties of multilayer emulsions for application in the food industry." *Advances in Colloid and Interface Science* **128**: 227–248.

Hunter, R. J. (1986). *Foundations of Colloid Science: Volume 1*. Oxford, UK, Oxford Science Publications.

Imai, E., K. Saito, M. Hatakeyama, K. Hatae and A. Shimada (1999). "Effect of physical properties of food particles on the degree of graininess perceived in the mouth." *Journal of Texture Studies* **30**(1): 59–88.

Imai, E., Y. Shimichi, I. Maruyama, A. Inoue, S. Ogawa, K. Hatae and A. Shimada (1997). "Perception of grittiness in an oil-in-water emulsion." *Journal of Texture Studies* **28**(3): 257–272.

Israelachvili, J. (2011). *Intermolecular and Surface Forces*, 3rd Edition. London, Academic Press.

Jones, O. G. and D. J. McClements (2010). "Functional biopolymer particles: Design, fabrication, and applications." *Comprehensive Reviews in Food Science and Food Safety* **9**(4): 374–397.

Kabalnov, A. S. and E. D. Shchukin (1992). "Ostwald ripening theory—Applications to fluorocarbon emulsion stability." *Advances in Colloid and Interface Science* **38**: 69–97.

Kaunisto, E., M. Marucci, P. Borgquist and A. Axelsson (2011). "Mechanistic modelling of drug release from polymer-coated and swelling and dissolving polymer matrix systems." *International Journal of Pharmaceutics* **418**(1): 54–77.

Larson, R. G. (1999). *The Structure and Rheology of Complex Fluids.* Oxford, UK, Oxford University Press.

Lesmes, U. and D. J. McClements (2009). "Structure-function relationships to guide rational design and fabrication of particulate food delivery systems." *Trends in Food Science and Technology* **20**(10): 448–457.

Li, X. and B. R. Jastic (2005). *Design of Controlled Release Drug Delivery Systems.* New York, McGraw-Hill Professional.

Li, Y., S. Le Maux, H. Xiao and D. J. McClements (2009). "Emulsion-based delivery systems for tributyrin, a potential colon cancer preventative agent." *Journal of Agricultural and Food Chemistry* **57**(19): 9243–9249.

Lim, S. S., M. Y. Baik, E. A. Decker, L. Henson, L. M. Popplewell, D. J. McClements and S. J. Choi (2011). "Stabilization of orange oil-in-water emulsions: A new role for ester gum as an Ostwald ripening inhibitor." *Food Chemistry* **128**(4): 1023–1028.

Liu, Y. G. and P. H. Daum (2008). "Relationship of refractive index to mass density and self-consistency of mixing rules for multicomponent mixtures like ambient aerosols." *Journal of Aerosol Science* **39**(11): 974–986.

Mao, Y. Y. and D. J. McClements (2012a). "Fabrication of functional micro-clusters by heteroaggregation of oppositely charged protein-coated lipid droplets." *Food Hydrocolloids* **27**(1): 80–90.

Mao, Y. Y. and D. J. McClements (2012b). "Modulation of emulsion rheology through electrostatic heteroaggregation of oppositely charged lipid droplets: Influence of particle size and emulsifier content." *Journal of Colloid and Interface Science* **380**: 60–66.

Mao, Y. Y. and D. J. McClements (2013). "Modification of emulsion properties by heteroaggregation of oppositely charged starch-coated and protein-coated fat droplets." *Food Hydrocolloids* **33**(2): 320–326.

Matalanis, A., O. G. Jones and D. J. McClements (2011). "Structured biopolymer-based delivery systems for encapsulation, protection, and release of lipophilic compounds." *Food Hydrocolloids* **25**(8): 1865–1880.

Matalanis, A. and D. J. McClements (2013). "Hydrogel microspheres for encapsulation of lipophilic components: Optimization of fabrication and performance." *Food Hydrocolloids* **31**(1): 15–25.

McClements, D. J. (2002a). "Colloidal basis of emulsion color." *Current Opinion in Colloid and Interface Science* **7**(5–6): 451–455.

McClements, D. J. (2002b). "Theoretical prediction of emulsion color." *Advances in Colloid and Interface Science* **97**(1–3): 63–89.

McClements, D. J. (2005). *Food Emulsions: Principles, Practice, and Techniques.* Boca Raton, FL, CRC Press.

McClements, D. J. (2007). "Critical review of techniques and methodologies for characterization of emulsion stability." *Critical Reviews in Food Science and Nutrition* **47**(7): 611–649.

McClements, D. J. (2010). "Emulsion design to improve the delivery of functional lipophilic components." *Annual Review of Food Science and Technology* **1**(1): 241–269.

Mei, L. Y., D. J. McClements and E. A. Decker (1999). "Lipid oxidation in emulsions as affected by charge status of antioxidants and emulsion droplets." *Journal of Agricultural and Food Chemistry* **47**(6): 2267–2273.

Mei, L. Y., D. J. McClements, J. N. Wu and E. A. Decker (1998). "Iron-catalyzed lipid oxidation in emulsion as affected by surfactant, pH and NaCl." *Food Chemistry* **61**(3): 307–312.

Nuchi, C. D., P. Hernandez, D. J. McClements and E. A. Decker (2002). "Ability of lipid hydroperoxides to partition into surfactant micelles and alter lipid oxidation rates in emulsions." *Journal of Agricultural and Food Chemistry* **50**(19): 5445–5449.

Pal, R. (2011). "Rheology of simple and multiple emulsions." *Current Opinion in Colloid and Interface Science* **16**(1): 41–60.

Peppas, N. A. (2013). "Historical perspective on advanced drug delivery: How engineering design and mathematical modeling helped the field mature." *Advanced Drug Delivery Reviews* **65**(1): 5–9.

Quemada, D. and C. Berli (2002). "Energy of interaction in colloids and its implications in rheological modeling." *Advances in Colloid and Interface Science* **98**(1): 51–85.

Sackett, C. K. and B. Narasimhan (2011). "Mathematical modeling of polymer erosion: Consequences for drug delivery." *International Journal of Pharmaceutics* **418**(1): 104–114.

Tadros, T. F. (1994). "Fundamental principles of emulsion rheology and their applications." *Colloids and Surfaces A-Physicochemical and Engineering Aspects* **91**: 39–55.

Tippetts, M. and S. Martini (2012). "Influence of iota-carrageenan, pectin, and gelatin on the physicochemical properties and stability of milk protein-stabilized emulsions." *Journal of Food Science* **77**(2): C253–C260.

van Aken, G. A., M. H. Vingerhoeds and E. H. A. de Hoog (2007). "Food colloids under oral conditions." *Current Opinion in Colloid and Interface Science* **12**(4–5): 251–262.

Walstra, P. (1993). "Principles of emulsion formation." *Chemical Engineering Science* **48**: 333.

Walstra, P. (2003). *Physical Chemistry of Foods*. New York, Marcel Decker.

Wooster, T. J., M. Golding and P. Sanguansri (2008). "Impact of oil type on nanoemulsion formation and ostwald ripening stability." *Langmuir* **24**(22): 12758–12765.

# Chapter 4

# Mechanical Particle Fabrication Methods

## 4.1 Introduction

In general, particle fabrication methods are usually divided into two main categories: *physicochemical methods* (bottom–up) and *mechanical methods* (top–down). Physicochemical methods are primarily based on spontaneous or directed assembly of molecules or colloidal particles into larger particles (Figure 4.1). Particle formation is driven by a reduction in the free energy of the system and is mainly caused by alterations in molecular interactions when the molecules or particles reorganize themselves, such as electrostatic, hydrogen bonding, or hydrophobic forces (McClements et al. 2009). On the other hand, mechanical methods are primarily based on the use of processing equipment specifically designed to create small particles from larger materials, such as homogenization, grinding, spraying, or drying methods (Muller et al. 2011). In practice, many particle formation methods are a combination of these two different approaches; for example, multilayer emulsions are formed by creating small lipid droplets using mechanical methods (homogenizers) and then coating them using physicochemical methods (electrostatic deposition). Physicochemical methods of particle formation are covered in the later chapters concerned with surfactant-based, emulsion-based, and biopolymer-based systems (Chapters 5 through 7). In this chapter, we focus on various mechanical devices that are used to fabricate particles suitable for use as colloidal delivery systems in the food industry.

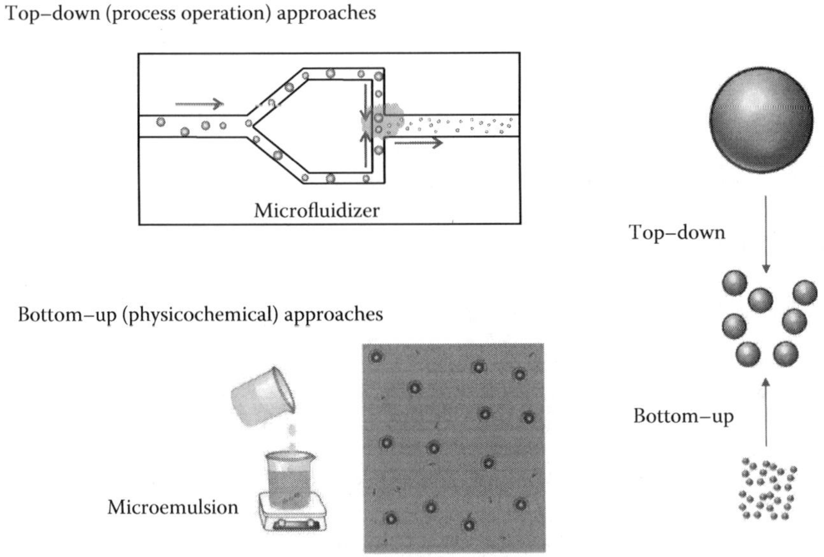

Top–down (process operation) approaches

Bottom–up (physicochemical) approaches

**Figure 4.1** **Particle fabrication methods can be characterized as top–down (process operation) methods or bottom–up (physicochemical) approaches.**

## 4.2 Homogenization Methods

Initially, mechanical devices have been developed to create colloidal dispersions from two immiscible liquids or to reduce the size of the particles in existing colloidal dispersions (Lee et al. 2013; McClements 2005; Santana et al. 2013). These devices are most commonly used to prepare emulsions, but they can also be used to create other types of colloidal delivery systems, for example, wet milling of solid particle suspensions (Muller et al. 2011). A number of different types of homogenizer have been developed for creating colloidal dispersions, with each type having its own advantages and disadvantages for particular applications. In this section, a brief overview of some of the homogenization devices most commonly used to prepare colloidal dispersions is given, whereas more detailed information is provided elsewhere (Lee et al. 2013; McClements 2005; Santana et al. 2013; Walstra 1993, 2003). The main focus in this section will be on the production of emulsions, but as mentioned earlier, many of these mechanical devices can also be employed during the production of other types of colloidal delivery systems.

### 4.2.1 Higher-Shear Mixers

High-shear mixers are one of the most widely used methods for homogenizing oil and water phases together to form coarse emulsions containing relatively large droplets (McClements 2005; Walstra 1993). They may also be used to create colloidal dispersions from other starting materials, such as phase-separated biopolymer mixtures, that is, so-called water-in-water (W/W) or oil-in-water-in-water (O/W/W) systems (Matalanis and McClements 2013; Wolf et al. 2001a,b). A high-shear mixer typically consists of a container and a mixing head capable of revolving at high speeds. To produce an emulsion using a batch process, the oil, water, and other ingredients to be homogenized are placed in the container, which may be relatively small for laboratory use (a few cubic centimeters) or relatively large for industrial use (several cubic meters). Intense shearing forces are generated by rapidly rotating the mixing head, which causes the oil and water phases to break up and intermingle with each other. The efficiency of homogenization is often improved by having baffles fixed to the inside walls of the mixing vessel. The design of the mixing head also determines the efficiency of homogenization, and a number of different designs are available for different applications, for example, blades, propellers, and turbines (Walstra 1993). Different mixing heads are designed to generate more intense and evenly distributed disruptive forces, so as to create smaller droplets, reduce homogenization times, improve energy efficiency, or ensure more uniform mixing. Many industrial high-shear mixers are capable of in-line (rather than batch) operation so that products can be continuously homogenized. High-shear mixing often leads to an increase in product temperature because of frictional losses. If any of the active ingredients within a delivery system are heat sensitive, it may therefore be necessary to control the temperature of the mixing vessel during homogenization (e.g., using cooling coils).

High-shear mixers are most suitable for preparing emulsions from low- or intermediate-viscosity fluids but are less effective for homogenizing very high viscosity fluids. The size of the droplets produced tends to decrease as the homogenization time or mixing head speed increases, until a lower limit is reached that depends on the type and concentration of ingredients used and the maximum power density of the mixing device. Typically, the mean diameter of the droplets produced by high-shear mixers is around 2 to 10 μm. Industrial mixers are often designed to avoid excessive incorporation of air bubbles during homogenization, because this can have an adverse effect on the subsequent processing and stability of products.

## 4.2.2 Colloid Mills

Colloid mills are widely used to create emulsions from intermediate- and high-viscosity liquids (McClements 2005; Walstra 1993). A colloid mill typically contains two disks with a narrow gap between them: a *rotor* (a rotating disk) and a *stator* (a static disk) (Figure 4.2). The liquids to be homogenized are fed into the center of the colloid mill in the form of a preformed coarse emulsion, rather than as separate oil and aqueous phases, because colloid mills are more efficient at reducing the size of existing particles than at breaking up two separate phases. Coarse emulsions are typically prepared by blending oil, water, and other ingredients together

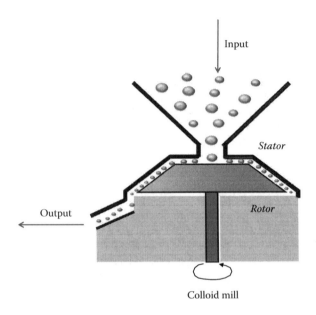

Colloid mill

**Figure 4.2 Small droplets can be prepared from larger ones using a colloidal mill. In this case, a coarse emulsion is passed through a narrow gap between two rotating disks that generate intense disruptive forces.**

using a high-shear mixer (Section 4.2.1). The rapid rotation of the rotor generates an intense shear stress in the narrow gap between the rotor and stator, which causes large droplets to be broken down into smaller ones. In addition, rotation generates a centrifugal force that causes the fluids to move from the center to the edge of the disks where they are collected or transported to a subsequent processing operation. The intensity of the shear stresses can be modulated by varying the thickness of the gap between the rotor and stator, the rotation speed, or the types of disks used (e.g., smooth, roughened, or toothed). Droplet disruption can also be increased by extending the period the emulsion spends within the colloid mill, for example, by decreasing the flow rate or by passing the emulsion through the device multiple times. Colloid mills often have cooling devices attached to prevent excessive temperature increases caused by frictional losses. Colloid mills are most suitable for homogenizing intermediate- and high-viscosity fluids and typically produce relatively large droplets (1 to 5 μm). In addition to forming emulsions, they can also be used to reduce the size of solid particles in suspensions.

## 4.2.3 High-Pressure Valve Homogenizers

High-pressure valve homogenizers (HPVHs) are one of the widely used methods of producing emulsions containing small droplets (Lee et al. 2013; McClements 2005; Walstra 1993). Like colloid mills, HPVHs are more effective at reducing the particle size of existing emulsions than at combining two separate liquids. Typically, a coarse emulsion is produced using a high-shear mixer and this is then fed into the inlet of the HPVH. An HPVH typically has a piston that pulls the coarse emulsion into a chamber on its backstroke and then forces it through a narrow valve at the end of the chamber on its forward stroke (Figure 4.3). As the coarse emulsion passes through this valve, it is subjected to intense disruptive forces that break larger droplets into smaller ones. Various valve designs have been created to alter the contribution from different types of droplet disruption forces (shear, turbulence, and cavitation) and so increase the efficiency of droplet breakup. The homogenization pressure can be manipulated by adjusting the dimensions of the valve gap through which the emulsion passes. Typically, the droplet size can be decreased by increasing homogenization pressure, increasing number of passes, and increasing emulsifier concentration (Qian and McClements 2011). The droplet size produced also depends on the oil–water interfacial tension, emulsifier type, and the ratio of the viscosities of the dispersed and continuous phases (Lee et al. 2013; Wooster et al. 2008).

Many commercial HPVHs utilize a "two-stage" homogenization process, in which the emulsion is forced through two different valves in series. The first valve is often set at a high pressure to break down the droplets, while the second valve is set at a lower pressure to dissociate any "flocs" formed during the first stage. HPVHs are most suitable for formation of emulsions from low- and intermediate-viscosity materials, particularly when small droplet sizes are required. It is often possible to create emulsions with small particles ($d < 500$ nm) using a single pass through the homogenizer. If smaller droplets are required, it is usually necessary to pass the emulsion through the homogenizer a number of times. Emulsions containing lipid

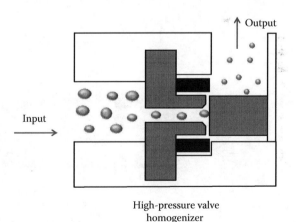

High-pressure valve
homogenizer

**Figure 4.3  Small droplets can be prepared from larger ones using a high-pressure valve homogenizer. In this case, a coarse emulsion is passed through a narrow valve at high pressure that generates intense disruptive forces.**

droplets with diameters <100 nm can be produced using this method, provided there is sufficient emulsifier present to completely cover the oil–water interface formed, the emulsifier adsorbs rapidly enough to prevent droplet coalescence, and the viscosity ratio is in an appropriate range (Wooster et al. 2008).

The temperature rise in an HPVH is typically quite small if an emulsion is only passed through it once, but it may become appreciable if the emulsion is recirculated or if particularly high pressures are used. In these cases, it may be necessary to cool the device using a water-jacketed chamber during homogenization. Conversely, in other circumstances, it may be necessary to keep the homogenizer warm during the homogenization process, for example, to prevent crystallization of a lipid phase during the formation of solid lipid nanoparticles. As well as forming emulsions and nanoemulsions, HPVHs may also be utilized for the fabrication of other types of colloidal delivery system; for example, they can be used to produce nanocrystal suspensions by breaking down large crystals into smaller ones (Muller et al. 2011).

### 4.2.4 Ultrasonic Homogenizers

Ultrasonic homogenizers utilize high-intensity, rapidly fluctuating pressure waves to break large droplets into smaller ones, primarily through cavitation and turbulent effects (Jafari et al. 2007b; Kentish et al. 2008; Leong et al. 2009; McClements 2005; Walstra 1993). The two most common types of ultrasonic homogenizer are (i) the ultrasonic probe homogenizer and (ii) the ultrasonic jet homogenizer (Figure 4.4). Bench-top ultrasonic probe homogenizers are widely used in research

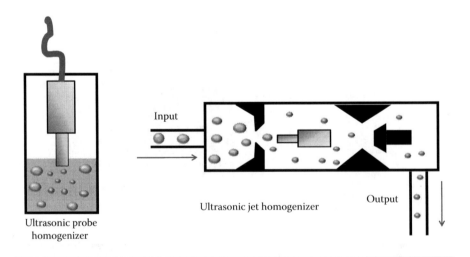

**Figure 4.4 Small droplets can be prepared using ultrasonic probe or jet homogenizers.**

laboratories because they are relatively inexpensive, easy to operate, and can prepare small volumes of emulsions (a few cubic centimeters to hundreds of cubic centimeters). They contain an ultrasonic transducer contained within a protective casing that converts applied electrical waves into ultrasonic (pressure) waves. These pressure waves radiate from the tip of the ultrasonic transducer into the surrounding fluids where they generate intense pressure and shear gradients (mainly due to cavitation). The intense disruptive forces generated cause oil and water phases to be disrupted and intermingled with one another. The fact that the disruptive forces are typically focused on a small volume of sample near the tip of the ultrasonic transducer means that it is important to have good agitation within the sample container. In small vessels, the fluid flow induced by the ultrasonic field itself is usually sufficient to ensure thorough mixing, but in large vessels, it may be necessary to employ additional mixing to ensure effective homogenization of the entire sample (Kentish et al. 2008). In order to prepare an emulsion, ultrasonic waves are typically applied to a sample from periods ranging from a few seconds to a few minutes. Application of continuous ultrasonic energy to a sample can cause appreciable heating, and so it is often applied in a number of short bursts. Ultrasonic probe homogenizers are typically used for batch preparation of emulsions, but in-line flow-through versions have been developed for continuous production (Kentish et al. 2008; Leong et al. 2009). The droplet size typically decreases with increasing homogenization time and intensity, provided that none of the functional ingredients (such as emulsifiers) are damaged by the high ultrasonic energies produced. Ultrasonic probe homogenizers are most suitable for producing emulsions from low- and intermediate-viscosity liquids.

Ultrasonic jet homogenizers are typically used for the continuous production of emulsions in industrial applications. The fluids to be homogenized are usually converted into a coarse emulsion first using a high-shear mixer. This coarse emulsion is then pumped into the inlet of the ultrasonic homogenizer and forced through an orifice at high pressure. The stream of fluid passing through the orifice impinges upon a sharp-edged blade, causing it to break up owing to a combination of cavitation, shear, and turbulence. Some of the major advantages of ultrasonic jet homogenizers are that they can be used for the continuous emulsion production, they generate very small droplets, and they are often more energy efficient than HPVHs. On the other hand, the vibrating blade may be prone to erosion and may need to be replaced frequently.

The principal factors determining the efficiency of ultrasonic homogenizers are the frequency, intensity, and duration of the ultrasonic waves (Jafari et al. 2007; Kentish et al. 2008; Leong et al. 2009). The size of the droplets produced by ultrasonic homogenization can be decreased by increasing the intensity or duration of the applied ultrasonic waves. The efficiency of ultrasonic homogenization usually deceases as the viscosity of the component phases increases. As well as producing emulsions, ultrasonic homogenizers are useful for the production of other kinds of colloidal delivery systems. For example, they can be used to facilitate the formation of small particles in some antisolvent precipitation methods of producing particles (Hatkar and Gogate 2012; Jiang et al. 2012). They can also be used to break down large solid particles or agglomerates into smaller particles.

## 4.2.5 Microfluidization

Microfluidization is one of the most effective methods of fabricating emulsions with very fine droplet sizes (Jafari et al. 2007a,b; McClements 2005; Qian and McClements 2011; Walstra 1993; Wooster et al. 2008). This type of homogenizer consists of a fluid inlet, some kind of pumping device, and an interaction chamber containing two channels through which the fluids are made to flow and interact with each other (Figure 4.5). Typically, a preformed coarse emulsion is formed by blending oil, water, and emulsifier together using a high-shear mixer. This coarse emulsion is then pumped into the inlet of the microfluidizer and passes through a channel that splits the fluid into two narrower channels. The arrangement of these two channels is such that two streams of fluid are made to influence each other at high velocity. Intense disruptive forces are generated when the two fluid streams collide, which causes larger droplets to be broken down. The size of the droplets produced by a microfluidizer can be decreased by increasing the homogenization pressure, increasing the number of passes, and increasing the emulsifier concentration (Qian and McClements 2011; Wooster et al. 2008). The droplet size also depends on the oil–water interfacial tension, the viscosity ratio of the oil and water phases, the adsorption kinetics of the emulsifier, and the ability of the emulsifier to prevent coalescence within the homogenizer (Jafari et al. 2008; Lee et al. 2013). A variety of different channel types have been designed to increase the efficiency of droplet disruption, for example, straight or zigzag channels. Microfluidizers are

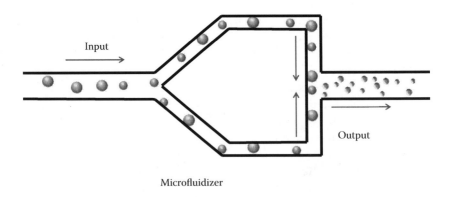

Microfluidizer

**Figure 4.5 Small droplets can be prepared from larger ones using a microfluidizer. In this case, a coarse emulsion is passed through a series of narrow channels that generates intense disruptive forces when the fluids collide.**

available as small-scale bench-top, pilot plant, and full-scale production versions. The minimum volumes that can be produced using laboratory-scale microfluidizers may be as small as a few milliliters, which is convenient for preparation of delivery systems using ingredients that are costly or scarce. Microfluidizers are often the most efficient method of producing emulsions containing very small droplets (e.g., $d < 100$ nm). Typically, microfluidizers are most suitable for homogenizing low- and intermediate-viscosity fluids.

As well as producing emulsions or nanoemulsions, microfluidizers can also be used to facilitate the preparation of other kinds of colloidal delivery systems. For example, they can be used to break down lipophilic crystalline materials into small nanocrystals (Muller et al. 2011) or to break large biopolymer particles into smaller ones (Jong 2013). In addition, microfluidizers are available where two separate fluid streams can be brought into contact in the reaction chamber, which means that they can be used to produce fine crystals using the antisolvent precipitation method, or they can homogenize oil and water phases without the need to form a coarse emulsion premix (Panagiotou and Fisher 2008; Panagiotou et al. 2008, 2009).

## 4.2.6 Membrane and Microchannel Homogenizers

Membrane and microchannel homogenizers are particularly useful for producing particles with well-defined dimensions and internal structures (Nazir et al. 2010; Nisisako 2008; Vladisavljevic and Williams 2005). They are widely used for research purposes but are less commonly used for industrial purposes because of their relatively low throughput. Membrane homogenizers can be used in a number of different ways; for example, they can be used to form an emulsion directly from separate oil and water phases, or they can be used to reduce the size of the droplets in a coarse emulsion. In direct membrane homogenization, an emulsion is formed when one immiscible liquid (the disperse phase) is forced into another immiscible liquid (the continuous phase) through a solid membrane that contains small pores (Figure 4.6). The continuous phase usually contains an emulsifier that adsorbs to the droplet surfaces and stabilizes them against aggregation. The size of the droplets in the emulsion formed depends on the membrane pore size, the oil–water interfacial tension, the applied pressure, the flow profile of the continuous phase, and the type and amount of emulsifier used. Membranes can be manufactured with different pore sizes so that emulsions with different droplet sizes can be produced. Some membranes contain pores with well-defined sizes and shapes, whereas others have less well-defined ones, which has an impact on the polydispersity of the particles produced. The membrane polarity should also be carefully selected as it determines the type of emulsions that can be created: hydrophobic membranes are needed to produce water-in-oil (W/O) emulsions, whereas hydrophilic membranes are needed to produce oil-in-water emulsions. Membrane homogenization can be carried out as either a batch or a continuous process. In the batch process, droplets are formed by forcing the dispersed phase through a membrane in contact with the

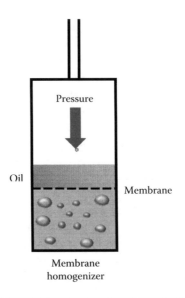

Membrane
homogenizer

---

**Figure 4.6 Droplets with well-defined structures and dimensions can be produced using membrane homogenizers or microchannel homogenizers.**

continuous phase (Figure 4.6). In the continuous process, the continuous phase is made to flow through a cylindrical inner membrane, while the disperse phase is made to flow through an external outer cylinder. The dispersed phase is pressurized so that it is forced through the inner membrane, where it forms small droplets in the continuous phase in the outer membrane. Membrane homogenizers can also be used to reduce the size of droplets in preexisting coarse emulsions. In this approach, a coarse emulsion is formed first (e.g., by simple stirring) and is then forced through the membrane to reduce the droplet size. The major advantages of this approach are that higher fluxes can be achieved and emulsions with higher droplet concentrations can be produced more easily.

Microchannel homogenizers work on a fairly similar principle to membrane homogenizers, except that the disperse phase is forced through microchannels with well-defined geometries to form droplets (Neethirajan et al. 2011; Vladisavljevic et al. 2012). This technique has proved to be particularly useful for producing droplets with very narrow particle size distributions and with well-defined internal structures, such as core–shell or dispersion (Chapter 1). More elaborate particle structures are typically produced by using concentric microchannels and by controlling the relative flow rates of the different fluids involved. These kinds of structured particles are suitable for highly specialized commercial applications or for research purposes but are usually unsuitable for large-scale commercial applications in the food industry because they are too expensive and time consuming to make. Emulsions with mean droplet diameters as low as 300 nm can be produced using these types of homogenizers, but typically they are more suitable for producing larger droplets (1 to 100 μm).

Although membrane and microchannel homogenizers are mainly used for producing emulsions, they can also be used to form other kinds of colloidal particles, such as hydrogel beads (Zhou et al. 2008). Hydrogel beads can be formed by preparing a W/O emulsion containing a gelling agent in the water phase. After formation of the W/O emulsion, the system conditions are altered so that the water phase is gelled, and then the hydrogel beads formed are separated from the oil phase (e.g., by centrifugation followed by washing with organic solvent).

## 4.3 Atomization Methods

This group of methods consists of mechanical devices that create particles through an atomization process, which typically involves breaking a bulk liquid into a series of tiny droplets (Oxley 2012b; Peltonen et al. 2010). Some of the homogenization methods discussed in the previous section could be considered to be atomization methods, such as certain forms of membrane or microchannel homogenization. As these devices have already been discussed, they will not be considered further in this section.

### 4.3.1 Spray Drying

Spray drying methods have been used industrially for many years to produce powdered forms of active ingredients from fluid solutions and suspensions (Desai and Park 2005; Fang and Bhandari 2012; Gharsallaoui et al. 2007; Madene et al. 2006; Okuyama et al. 2006; Paudel et al. 2013; Rokka and Rantamaki 2010; Vega and Roos 2006). This method is suitable for economical large-scale production of food ingredients, and the equipment is already widely utilized within the food industry. Spray drying has been used to encapsulate many kinds of active ingredients, including flavors, colors, vitamins, nutraceuticals, and probiotics (Desai and Park 2005; Fang and Bhandari 2012). There are a number of advantages to converting a fluid form of an active ingredient into a powdered form. First, the storage stability of labile active ingredients can often be increased because the wall material that surrounds the active ingredient forms an impermeable solid matrix that retards molecular diffusion processes, thereby inhibiting the active ingredients from interacting with other molecular species that might promote their degradation. Second, the formation of an impermeable solid matrix can also improve the retention of active ingredients during storage by inhibiting their evaporation or expulsion. Third, conversion of a fluid form of an active ingredient into a powder form can improve its handling and utilization, as well as reduce transport and storage costs. Fourth, the spray drying process is actually quite mild and can be used to encapsulate ingredients that are heat sensitive (such as some flavors, colors, bioactive lipids, and probiotics). This is because drying occurs very rapidly and the temperature rise is relatively low because of the endothermic latent heat associated with solvent evaporation from the particles during drying (Fang and Bhandari 2012).

The initial fluid form of the ingredient used in the spray drying process could be a simple solution (such as protein or carbohydrate dissolved in solvent) or it could be a colloidal dispersion (such as an emulsion droplet, solid lipid nanoparticle, or hydrogel particle suspended in solvent). The most common solvent used in the food industry is water, but other solvents can also be used (such as organic solvents) for specialist applications. For colloidal dispersions, the final particles produced after spray drying have a complex internal structure, consisting of small colloidal particles embedded within larger solid particles formed during the drying process. Furthermore, the particles formed by spray drying may be agglomerated or coated afterward, creating even more complex structures. This ability provides great flexibility in producing colloidal delivery systems with different functional characteristics.

The basic principle of the spray drying process is illustrated schematically in Figure 4.7. Initially, the active ingredient is dissolved or dispersed in a *feed fluid*, which is then pumped through a pipe until it reaches an *atomizer*. The atomizer

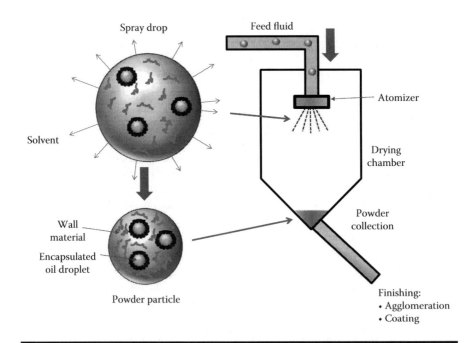

**Figure 4.7 Spray driers can be used to produce powdered forms of delivery systems. In this highly schematic diagram, an emulsion is spray dried to form a powder.**

converts the bulk fluid into a fine mist of fluid drops that is sprayed into a *drying chamber*. The temperature within the drying chamber is kept sufficiently high to cause the solvent (usually water) to evaporate from these *spray drops*, leaving behind small powder particles, which typically have diameters around 10 to 1000 μm. These powder particles are separated from the gas phase and collected, for example, using cyclones or filter bags. They may then be subjected to further processing operations to improve their handling, storage stability, or functional properties, for example, by agglomeration or coating (Fang and Bhandari 2012).

A number of factors must be considered to obtain a final powdered product with the desired properties, for example, particle size, packing density, flowability, stability, encapsulation efficiency, retention efficiency, and dispersibility (Fang and Bhandari 2012; Paudel et al. 2013; Peltonen et al. 2010; Vega and Roos 2006). A brief outline of some of the most important factors to consider is given here:

*Composition and Structure of Feed Fluid*: The composition and structure of the initial fluid material fed into the spray drier have a major impact on the final properties of the powder produced. The active ingredient is dissolved in a solution or dispersed in a suspension, which also contains a substance that eventually becomes the wall material after drying (typically carbohydrate or protein). A water-soluble active ingredient is usually dissolved in an aqueous phase, whereas a water-insoluble one is usually incorporated into some kind of emulsion or suspension. The type and concentration of the substances that eventually become the wall material largely determine the physicochemical properties of the solid matrix surrounding the active ingredients after drying, such as the rheology, molecular mobility, permeability, and dissolution properties. It is therefore important to choose wall materials that will give the desired physicochemical attributes required for the particular application.

*Spray Drying Conditions*: The design and operation conditions of a spray dryer must also be optimized to produce a high-quality powdered product with the desired properties. Spray dryers are available for bench-top, pilot plant, and industrial applications, which vary in the type and quantity of materials that can be processed, the versatility of materials that can be processed, the nature of the powders produced, and the capital and operating costs. In commercial applications, it is important to ensure that there is a good correlation between the properties of powders prepared using bench-top or pilot plant equipment, and those made using industrial equipment when scaling up industrial processes. One of the main factors affecting the performance of a spray dryer is the device used to atomize the samples, which is typically either a rotary or a spray nozzle atomizer. A variety of atomizer designs are commercially available to increase the efficiency of the spray drying process or to improve the quality of the powders produced. Once a suitable design of equipment has been selected, it is important to optimize operating parameters, such as flow rate, inlet temperature, and outlet temperature.

*Postdying Processing*: After a powder has been produced, it is possible to modify its properties further so as to improve its handling or functional properties. For example, powder particles can be agglomerated or coated using fluid bed methods after spray drying (Meiners 2012). Agglomeration is usually carried out to produce dustless powders that are easier to handle, flow better, and rehydrate more rapidly. Coating may be carried out to further improve the chemical stability, retention, handling, or release characteristics of active ingredients (Desai and Park 2005).

## 4.3.2 Spray Chilling

The equipment used to produce particles using spray chilling is fairly similar in design and operation to that used for spray drying (Gamboa et al. 2011; Gibbs et al. 1999; Madene et al. 2006; Oxley 2012b). Indeed, it is often possible to use the same equipment to produce particles using both approaches. In spray chilling, a hot liquid is sprayed into a cold chamber, which causes the small liquid drops formed to solidify into solid particles, which are then collected as they fall to the bottom of the vessel. Like spray drying, it is possible to use different kinds of atomizers to form a fine mist of liquid drops, such as rotary or spray nozzle atomizers. When the hot liquid drops first contact the cold air in the chamber, they are cooled to a temperature where they begin to solidify and their temperature then remains constant at the solidification temperature because of the release of heat associated with the liquid-to-solid exothermic phase transition. When they are fully solidified, the temperature of the solid particles then decreases. This approach can be used to create delivery systems from low melting carrier lipids, for example, those with melting points in the range from around 30°C to 70°C. The substance to be encapsulated is dispersed within a lipid phase held at a temperature above its melting point so that it is liquid. The active ingredient and carrier lipid are then forced through an atomizer to form a mist of fine lipid droplets. When these droplets encounter the surrounding cold air, they solidify. Knowledge of the melting behavior of the lipid phase is important to establish the optimum operating conditions for the spray chilling process, for example, atomizer type, operating temperatures, flow rates, and cooling chamber size. Lipids may crystallize over a narrow or broad range of temperature depending on their composition. They may also exhibit a large degree of supercooling so that the crystallization temperature may be well below the melting temperature.

Spray chillers are also useful for producing certain kinds of hydrogel particles, that is, those containing cold-setting biopolymers (Oxley 2012b). The substance to be encapsulated is dispersed within an aqueous biopolymer solution maintained at a temperature above its gelation temperature. When the liquid droplets encounter the cold air, they form gelled particles that can be collected. Materials that form gelled or glassy states (rather than crystalline ones) do not have a large heat of crystallization, and therefore they will not maintain at a constant temperature during the cooling process. After solidification, the hydrogel particles are usually separated from the surrounding gas phase using cyclones or filter bags. Spray chillers are often used in combination with fluidized beds that use cold air to carry out a second stage of cooling. As with spray drying, the particles formed by spray chilling may be further processed to alter their functional properties, for example, agglomeration or coating.

### 4.3.3 Rotary Disk Atomization

Rotary disk atomization methods involve directing a fluid stream of the material to be converted into particles onto a rapidly spinning disk (Oxley 2012b). The centrifugal forces generated by the spinning disk cause the fluid to move to the periphery of the disk where it forms thin filaments that leave the disk surface. After leaving, these thin filaments break up into a series of tiny drops. These drops can be solidified by cooling them down (e.g., cooling below the melting point of a lipid or gelation point of a biopolymer solution) or by evaporating the solvent (e.g., using heat). Rotary disk atomization may be an integral part of conventional spray drying or spray chilling equipment, or it may be used in its own right to form particles.

### 4.3.4 Electrospraying

Colloidal particles suitable for application as delivery systems can be formed using electrospray methods (Jaworek 2008; Zhang et al. 2012). Research has shown that both nanoparticles and microparticles may be fabricated using electrospraying. In this case, the material to be converted into powdered form is dissolved in a suitable solvent and then placed into an injection device ("nozzle") that is placed some distance away from a collection plate. A high electrical potential is then applied between the nozzle and collection plate, which pulls a stream of the solution out of the nozzle owing to electrical effects. Under appropriate conditions, the liquid stream breaks up into a series of tiny electrically charged liquid droplets that are rapidly converted into solid particles as they travel through the air and the solvent evaporates. The resulting solid particles accumulate on the collection plate, where they can be brushed or scraped off and collected. The size of the particles produced can be controlled by adjusting the fluid flow rate, applied voltage, and solution conditions. Electrospraying devices are relatively easy to assemble and operate, but they have relatively low throughputs, which may limit their widespread commercial application.

## 4.4 Milling Methods

Some colloidal delivery systems are formed by breaking up solid bulk materials or large solid particles using specialized mechanical milling devices (Hu et al. 2004; Muller et al. 2011; Peltonen and Hirvonen 2010; Van Eerdenbrugh et al. 2008). Milling devices are available to produce small particles from dry materials (such as bulk phases or powders) or from wet materials (such as suspensions of solid particles in liquids). Milling methods have some similarities to homogenization methods because both involve size reduction, but the materials to be disrupted are usually solid in milling but liquid in homogenization. Depending on their design, the mechanical devices used in milling may generate a combination of shear, attrition, compression, or impact forces to reduce the size of the particles in a material. Different milling methods vary in the type of materials they can be applied to (such as wet/dry, hard/soft, brittle/rubbery), the smallest particle size they can produce, equipment and operating costs, ease of use, and amount of material that can be processed.

For dry milling, the starting material may already be a solid or it may be a liquid that is converted into a solid form, for example, by air or freeze drying (Fang and Bhandari 2012). For wet milling, the starting material may be particles already suspended in a fluid or it may be a powder that is dispersed into a fluid. A wide variety of milling methods are available depending on the nature of the material to be reduced in size, for example, impact, hammer, cutting, grinding, and ball mills (Burmeister and Kwade 2013). Pearl (bead) mills are particularly effective at producing suspensions containing very fine solid particles (Muller et al. 2011). The initial particle suspension is placed within a disruption chamber containing a number of hard beads (typically 200 to 600 µm in diameter) that are made to roll against each other under high force, thereby disrupting any particles in the narrow gap between them. Some of the same equipment can be used in wet milling as that used in homogenization, for example, high-pressure homogenizers, microfluidizers, and colloid mills (Muller et al. 2011; Nagarwal et al. 2011). In this case, a suspension of solid particles is passed through the homogenizer and the particles are fragmented by the intense disruptive forces generated within the device (similar to the disruption of large liquid droplets). As with homogenization, it is usually necessary to have a stabilizer (i.e., emulsifier) present during this process to coat the new surfaces formed during particle disruption and thereby prevent particle aggregation.

## 4.5 Extrusion Methods

A number of methods used to form colloidal particles suitable for encapsulation and delivery of food ingredients involve extruding a fluid through one or more small orifices and then trapping the particles formed in a stable form, for example, by coating or solidifying them (Hu et al. 2004; Oxley 2012a). The membrane and microchannel homogenizers discussed in Section 4.2 can be considered to be specialized forms of extrusion methods. As the principles behind these two methods have been outlined already, they will not be considered further here.

A simple example of an extrusion method is demonstrated schematically in Figure 4.8, which shows the extrusion of a biopolymer solution into another solution that promotes gelation. Gelation may be induced by incorporating an appropriate gelling agent in the second solution, such as cross-linking mineral ions,

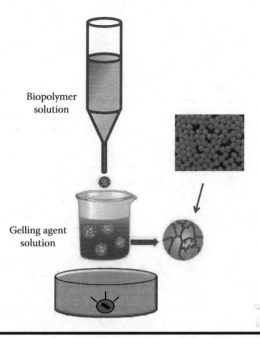

**Figure 4.8  Schematic representation of the injection method used to produce filled hydrogel beads. A biopolymer solution is injected into another aqueous solution containing a gelling agent or different environmental conditions.**

gluteraldehyde, enzymes, or alcohol. Alternatively, the second solution could be heated or chilled to promote heat-set or cold-set gelation of thermal-setting biopolymers. Extrusion methods can be used to form hydrogel particles with different characteristics by using different kinds of biopolymers and gelling mechanisms. The extrusion method can also be used to produce lipid particles by extruding a molten lipid into a cold aqueous solution to promote solidification of the particle. Extrusion methods can be adopted to form core–shell structures by having a coaxial extrusion system with one fluid on the inside and another on the outside (Oxley 2012a). In addition, they can be designed to increase the number of particles formed by having many extruders in parallel (as in some microchannel homogenizers) or by using a solid mesh that contains a number of pores through which the fluid is passed (as in membrane homogenization).

Microfluidic methods are a specialized type of extrusion method that are particularly versatile for creating particles with well-defined dimensions and internal structures (Desmarais et al. 2012; Helgeson et al. 2011; Selimovic et al. 2012). Microfluidic devices can be designed to create particles with different sizes, shapes, and internal structures by carefully designing the nature of the microchannels through which the fluids are made to flow. A highly schematic diagram of a microfluidic device capable of forming biopolymer particles is shown in Figure 4.9. Microfluidic methods are particularly suitable for fundamental research because of

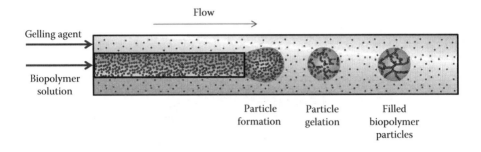

**Figure 4.9** **Schematic diagram of the microfluidic method for forming biopolymer particles using a coaxial device. A biopolymer solution is passed through an inner cylinder, while a gelling agent solution is passed through an outer cylinder. Hydrogel particles are formed at the end of the inner cylinder, which are gelled by the gelling agent.**

the control one has over the characteristics of the particles produced. Nevertheless, they are less suitable for widespread commercial applications within the food industry owing to scale up and economic problems. The microchannel homogenizers discussed in Section 4.2.6 are an example of microfluidic methods specifically designed to produce emulsions.

## 4.6 Coating Methods

Colloidal particles are often covered with a coating to alter their ability to protect and release encapsulated active ingredients (Gibbs et al. 1999; Meiners 2012). As mentioned in previous sections, a coating is automatically created around colloidal particles during the particle formation process for certain methods: a thin layer of emulsifier is formed around colloidal particles in most homogenization and some milling methods; a layer of coating material is formed around particles in coextrusion methods. Colloidal particles can also be coated after they have been formed. In this section, we focus on fluid bed coating methods, since these are widely used in industry for encapsulation of food ingredients in colloidal delivery systems. Other coating methods (such as pan and drum coating) used for creating coatings around larger objects (such as candies and tablets) in the food and pharmaceutical industry will not be considered.

Spray coating involves spraying a liquid material onto the surface of particles and then hardening the coating (e.g., by drying or cooling). The liquid material may be a solution, a suspension, or a hot melt. In the case of a solution or suspension, the coating is usually hardened by evaporating the solvent using hot air, whereas in the case of a hot melt (such as a solidifying lipid or gelling biopolymer), the coating is hardened using cold air (Gibbs et al. 1999; Oxley 2012b). Core particles with diameters in the range of approximately 50 to 500 μm are commonly used in spray coating. Typically, the particles are fluidized using a gas stream (e.g., air) that causes them to become suspended. A liquid material is then sprayed onto the surface of the particles, and the liquid coating is hardened, for example, by solvent evaporation or a temperature change (Figure 4.10). The operation conditions have to be carefully controlled to ensure that each particle is covered with a uniform coating and that particle aggregation does not occur (Oxley 2012a,b). These conditions include the particle concentration, gas flow rate, operating temperature, relative humidity, chamber dimensions, and nozzle design and operation. Spray coating may be performed as either a batch or a continuous process. Equipment manufacturers have designed a variety of different approaches to carry out the spray coating procedure, each with its own advantages and disadvantages for particular applications, for example, top, bottom, or tangential sprayers; batch or continuous sprayers (Oxley 2012a,b).

In the batch methods, the particles to be coated are introduced into a vessel that consists of a perforated base plate and a spray nozzle. In the *top coating* method, hot air is forced upward through the perforated base plate, which fluidizes the particles, and liquid coating is sprayed onto the particles from a nozzle

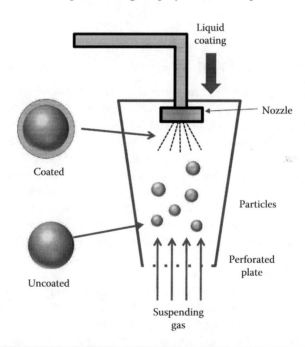

**Figure 4.10** **Highly schematic diagram of a fluidized bed coating device based on a top spray. A gas is forced through a perforated membrane at the bottom of the device, which causes the particles to be suspended in the gas. The coating material is then sprayed from the top of the device, which causes a coating to form.**

located at the top of the vessel. To ensure even coating, it is important that the coating has a low viscosity and that it forms tiny droplets during spraying. In the *bottom coating* method, hot air is also forced upward through the perforated base plate to fluidize the particles, but in this case, the coating is sprayed from a nozzle facing upward in the same direction as the hot air. As the upward-moving particles pass through the spray cone formed by the nozzle, they are coated with liquid, which is then dried by the hot air. The design of the instrument ensures that the particles move through the spray cone a number of times, ensuring a more even coating. In the *tangential coating* method, the base plate rotates and hot air is forced upward, which causes the particles to be set into a rotating motion. The coating liquid is then sprayed from the side onto the swirling particles. This method is particularly useful for forming thick coatings on particulate materials.

In continuous methods, the particles are carried along through the equipment by an air flow. The equipment may be divided into a number of sections that have different spraying nozzles and temperatures, so as to have different coating and drying regimes. This method may therefore be used to coat particles with a number of different layers of coating material to create more sophisticated protection or release systems.

## 4.7 Supercritical Fluid Methods

A number of supercritical fluid methods have been developed for encapsulating active ingredients into colloidal delivery systems (Cocero et al. 2009; Kalani and Yunus 2011; Sheth et al. 2012). In general, a supercritical fluid is a material whose pressure and temperature are above its critical pressure and temperature. Under these conditions, the material exists as a single-phase fluid that has some characteristics of a gas (such as low viscosity) and some characteristics of a liquid (such as high solubilization capacity), which makes it particularly suitable as a solvent. The physicochemical properties of a supercritical fluid can be controlled by manipulating the temperature and pressure of the system. Typically, supercritical carbon dioxide ($CO_2$) is used as a solvent in the fabrication of colloidal delivery systems because its critical temperature and pressure are in a convenient range, and it is relatively safe and inexpensive to use.

Colloidal particles suitable for encapsulating active ingredients can be produced in a number of ways using supercritical fluids as either solvents or antisolvents (Cocero et al. 2009). For example, supercritical methods can be utilized to produce powder particles in a similar manner to spray drying, but using a supercritical fluid (rather than water) as the solvent for the material to be converted into powder form. In this case, the active ingredient and a carrier material (that will form the wall material) are dissolved in a supercritical fluid at high pressure and temperature. The resulting mixture is then sprayed through a nozzle into a low-pressure chamber, which causes the carbon dioxide to evaporate, leading to the formation of small solid powder particles that can be collected. Alternatively, supercritical fluid methods can be used to create particles using an antisolvent precipitation method (Cocero et al. 2009; Kalani and Yunus 2011). In this case, the active ingredient and a carrier material are dissolved in a conventional solvent first, and then this mixture is brought into contact with a supercritical fluid. The supercritical fluid dissolves the conventional solvent, which causes the active ingredient and carrier material to precipitate and form particles. This approach can also be used to form core–shell-type particles by using antisolvent precipitation to deposit a shell of a coating material (e.g., biopolymers) around preexisting particles (e.g., fat droplets) (Cocero et al. 2009). A number of the major factors that influence the formation of colloidal delivery systems using supercritical fluids have been reviewed, including polymer type, solvent type, and processing conditions (Cocero et al. 2009).

## 4.8 Summary

There have been considerable advances in the development of mechanical devices capable of producing small particles with different kinds of properties, such as composition, size, shape, physical state, and interfacial properties. Many of these mechanical devices are therefore useful for producing food-grade colloidal delivery systems suitable for the encapsulation, protection, and release of active ingredients. This chapter has given a brief overview of some of the mechanical devices that can be used to fabricate food-grade delivery systems, and a number of other devices will be mentioned in later chapters on specific types of colloidal systems. The selection of an appropriate particle fabrication method is one of the most important factors determining the functional performance of a colloidal delivery system. Typically, one should first carefully define the functional requirements of a colloidal delivery system for a particular application: optical properties; rheological characteristics; stability; and protection, retention, and release profile. These functional requirements will then determine the characteristics of the carrier particles that are being used to encapsulate the active ingredient, such as composition, size, shape, physical state, and interfacial properties. This information can then be used to select the most appropriate bottom–up or top–down method to fabricate particles with the required properties.

## References

Burmeister, C. F. and A. Kwade (2013). "Process engineering with planetary ball mills." *Chemical Society Reviews* **42**(18): 7660–7667.

Cocero, M. J., A. Martin, F. Mattea and S. Varona (2009). "Encapsulation and co-precipitation processes with supercritical fluids: Fundamentals and applications." *Journal of Supercritical Fluids* **47**(3): 546–555.

Desai, K. G. H. and H. J. Park (2005). "Recent developments in microencapsulation of food ingredients." *Drying Technology* **23**(7): 1361–1394.

Desmarais, S. M., H. P. Haagsman and A. E. Barron (2012). "Microfabricated devices for biomolecule encapsulation." *Electrophoresis* **33**(17): 2639–2649.

Fang, Z. and B. Bhandari (2012). Spray drying, freeze drying, and related processes for food ingredient and nutraceutical encapsulation. In *Encapsulation Technologies and Delivery Systems for Food Ingredients and Nutraceuticals*, N. Garti and D. J. McClements (eds.). Oxford, UK, Woodhead Publishing: 73–108.

Gamboa, O. D., L. G. Goncalves and C. F. Grosso (2011). Microencapsulation of tocopherols in lipid matrix by spray chilling method. In *11th International Congress on Engineering and Food*, G. Saravacos, P. Taoukis, M. Krokida et al. (eds.) Procedia Food Science, Elsevier. **1**: 1732–1739.

Gharsallaoui, A., G. Roudaut, O. Chambin, A. Voilley and R. Saurel (2007). "Applications of spray-drying in microencapsulation of food ingredients: An overview." *Food Research International* **40**(9): 1107–1121.

Gibbs, B. F., S. Kermasha, I. Alli and C. N. Mulligan (1999). "Encapsulation in the food industry: A review." *International Journal of Food Sciences and Nutrition* **50**(3): 213–224.

Hatkar, U. N. and P. R. Gogate (2012). "Process intensification of anti-solvent crystallization of salicylic acid using ultrasonic irradiations." *Chemical Engineering and Processing* **57–58**: 16–24.

Helgeson, M. E., S. C. Chapin and P. S. Doyle (2011). "Hydrogel microparticles from lithographic processes: Novel materials for fundamental and applied colloid science." *Current Opinion in Colloid and Interface Science* **16**(2): 106–117.

Hu, J. H., K. P. Johnston and R. O. Williams (2004). "Nanoparticle engineering processes for enhancing the dissolution rates of poorly water soluble drugs." *Drug Development and Industrial Pharmacy* **30**(3): 233–245.

Jafari, S. M., Y. H. He and B. Bhandari (2007a). "Optimization of nano-emulsions production by microfluidization." *European Food Research and Technology* **225**(5): 733–741.

Jafari, S. M., Y. H. He and B. Bhandari (2007b). "Production of sub-micron emulsions by ultrasound and microfluidization techniques." *Journal of Food Engineering* **82**(4): 478–488.

Jafari, S. M., E. Assadpoor, Y. H. He and B. Bhandari (2008). "Re-coalescence of emulsion droplets during high-energy emulsification." *Food Hydrocolloids* **22**(7): 1191–1202.

Jaworek, A. (2008). "Electrostatic micro- and nanoencapsulation and electroemulsification: A brief review." *Journal of Microencapsulation* **25**(7): 443–468.

Jiang, T. Y., N. Han, B. W. Zhao, Y. L. Xie and S. L. Wang (2012). "Enhanced dissolution rate and oral bioavailability of simvastatin nanocrystal prepared by sonoprecipitation." *Drug Development and Industrial Pharmacy* **38**(10): 1230–1239.

Jong, L. (2013). "Characterization of soy protein nanoparticles prepared by high shear microfluidization." *Journal of Dispersion Science and Technology* **34**(4): 469–475.

Kalani, M. and R. Yunus (2011). "Application of supercritical antisolvent method in drug encapsulation: A review." *International Journal of Nanomedicine* **6**: 1429–1442.

Kentish, S., T. Wooster, M. Ashokkumar, S. Balachandran, R. Mawson and L. Simons (2008). "The use of ultrasonics for nanoemulsion preparation." *Innovative Food Science and Emerging Technologies* **9**(2): 170–175.

Lee, L. L., N. Niknafs, R. D. Hancocks and I. T. Norton (2013). "Emulsification: Mechanistic understanding." *Trends in Food Science and Technology* **31**(1): 72–78.

Leong, T., T. Wooster, S. Kentish and M. Ashokkumar (2009). "Minimising oil droplet size using ultrasonic emulsification." *Ultrasonics Sonochemistry* **16**(6): 721–727.

Madene, A., M. Jacquot, J. Scher and S. Desobry (2006). "Flavour encapsulation and controlled release—A review." *International Journal of Food Science and Technology* **41**(1): 1–21.

Matalanis, A. and D. J. McClements (2013). "Hydrogel microspheres for encapsulation of lipophilic components: Optimization of fabrication and performance." *Food Hydrocolloids* **31**(1): 15–25.

McClements, D. J. (2005). *Food Emulsions: Principles, Practice, and Techniques.* Boca Raton, FL, CRC Press.

McClements, D. J., E. A. Decker, Y. Park and J. Weiss (2009). "Structural design principles for delivery of bioactive components in nutraceuticals and functional foods." *Critical Reviews in Food Science and Nutrition* **49**(6): 577–606.

Meiners, J. A. (2012). Fluid bed microencapsulation and other coating methods for food ingredient and nutraceutical bioactive compounds. In *Encapsulation Technologies and Delivery Systems for Food Ingredients and Nutraceuticals*, N. Garti and D. J. McClements (eds.). Oxford, UK, Woodhead Publishing: 151–176.

Muller, R. H., S. Gohla and C. M. Keck (2011). "State of the art of nanocrystals—Special features, production, nanotoxicology aspects and intracellular delivery." *European Journal of Pharmaceutics and Biopharmaceutics* **78**(1): 1–9.

Nagarwal, R. C., R. Kumar, M. Dhanawat, N. Das and J. K. Pandit (2011). "Nanocrystal technology in the delivery of poorly soluble drugs: An overview." *Current Drug Delivery* **8**(4): 398–406.

Nazir, A., K. Schroen and R. Boom (2010). "Premix emulsification: A review." *Journal of Membrane Science* **362**(1–2): 1–11.

Neethirajan, S., I. Kobayashi, M. Nakajima, D. Wu, S. Nandagopal and F. Lin (2011). "Microfluidics for food, agriculture and biosystems industries." *Lab on a Chip* **11**(9): 1574–1586.

Nisisako, T. (2008). "Microstructured devices for preparing controlled multiple emulsions." *Chemical Engineering and Technology* **31**(8): 1091–1098.

Okuyama, K., M. Abdullah, I. W. Lenggoro and F. Iskandar (2006). "Preparation of functional nanostructured particles by spray drying." *Advanced Powder Technology* **17**(6): 587–611.

Oxley, J. D. (2012a). Coextrusion for food ingredients and nutraceutical encapsulation: Principles and technologies. In *Encapsulation Technologies and Delivery Systems for Food Ingredients and Nutraceuticals*, N. Garti and D. J. McClements (eds.). Oxford, UK, Woodhead Publishing: 131–149.

Oxley, J. D. (2012b). Spray cooling and spray chilling for food ingredient and nutraceutical encapsulation. In *Encapsulation Technologies and Delivery Systems for Food Ingredients and Nutraceuticals*, N. Garti and D. J. McClements (eds.). Oxford, UK, Woodhead Publishing: 110–130.

Panagiotou, T. and R. J. Fisher (2008). "Form nanoparticles via controlled crystallization." *Chemical Engineering Progress* **104**(10): 33–39.

Panagiotou, T., S. V. Mesite, J. M. Bernard, K. J. Chomistek and R. J. Fisher (2008). *Production of Polymer Nanosuspensions Using Microfluidizer (R) Processor Based Technologies.* NSTI-Nanotech, **1**: 688–691.

Panagiotou, T., S. V. Mesite and R. J. Fisher (2009). "Production of norfloxacin nanosuspensions using microfluidics reaction technology through solvent/antisolvent crystallization." *Industrial and Engineering Chemistry Research* **48**(4): 1761–1771.

Paudel, A., Z. A. Worku, J. M. S. Guns, S. Guns and G. Van den Mooter (2013). "Manufacturing of solid dispersions of poorly water soluble drugs by spray drying: Formulation and process considerations." *International Journal of Pharmaceutics* **453**(1): 253–284.

Peltonen, L. and J. Hirvonen (2010). "Pharmaceutical nanocrystals by nanomilling: Critical process parameters, particle fracturing and stabilization methods." *Journal of Pharmacy and Pharmacology* **62**(11): 1569–1579.

Peltonen, L., H. Valo, R. Kolakovic, T. Laaksonen and J. Hirvonen (2010). "Electrospraying, spray drying and related techniques for production and formulation of drug nanoparticles." *Expert Opinion on Drug Delivery* **7**(6): 705–719.

Qian, C. and D. J. McClements (2011). "Formation of nanoemulsions stabilized by model food-grade emulsifiers using high-pressure homogenization: Factors affecting particle size." *Food Hydrocolloids* **25**(5): 1000–1008.

Rokka, S. and P. Rantamaki (2010). "Protecting probiotic bacteria by microencapsulation: Challenges for industrial applications." *European Food Research and Technology* **231**(1): 1–12.

Santana, R. C., F. A. Perrechil and R. L. Cunha (2013). "High- and low-energy emulsifications for food applications: A focus on process parameters." *Food Engineering Reviews* **5**(2): 107–122.

Selimovic, S., J. Oh, H. Bae, M. Dokmeci and A. Khademhosseini (2012). "Microscale strategies for generating cell-encapsulating hydrogels." *Polymers* **4**(3): 1554–1579.

Sheth, P., H. Sandhu, D. Singhal, W. Malick, N. Shah and M. S. Kislalioglu (2012). "Nanoparticles in the pharmaceutical industry and the use of supercritical fluid technologies for nanoparticle production." *Current Drug Delivery* **9**(3): 269–284.

Van Eerdenbrugh, B., G. Van den Mooter and P. Augustijns (2008). "Top–down production of drug nanocrystals: Nanosuspension stabilization, miniaturization and transformation into solid products." *International Journal of Pharmaceutics* **364**(1): 64–75.

Vega, C. and Y. H. Roos (2006). "Invited review: Spray-dried dairy and dairy-like—Emulsions compositional considerations." *Journal of Dairy Science* **89**(2): 383–401.

Vladisavljevic, G. T., I. Kobayashi and M. Nakajima (2012). "Production of uniform droplets using membrane, microchannel and microfluidic emulsification devices." *Microfluidics and Nanofluidics* **13**(1): 151–178.

Vladisavljevic, G. T. and R. A. Williams (2005). "Recent developments in manufacturing emulsions and particulate products using membranes." *Advances in Colloid and Interface Science* **113**(1): 1–20.

Walstra, P. (1993). "Principles of emulsion formation." *Chemical Engineering Science* **48**: 333.

Walstra, P. (2003). *Physical Chemistry of Foods*. New York, Marcel Decker.

Wolf, B., W. J. Frith and I. T. Norton (2001a). "Influence of gelation on particle shape in sheared biopolymer blends." *Journal of Rheology* **45**(5): 1141–1157.

Wolf, B., W. J. Frith, S. Singleton, M. Tassieri and I. T. Norton (2001b). "Shear behaviour of biopolymer suspensions with spheroidal and cylindrical particles." *Rheologica Acta* **40**(3): 238–247.

Wooster, T., M. Golding and P. Sanguansri (2008). "Impact of oil type on nanoemulsion formation and Ostwald ripening stability." *Langmuir* **24**(22): 12758–12765.

Zhang, L. L., J. W. Huang, T. Si and R. X. Xu (2012). "Coaxial electrospray of microparticles and nanoparticles for biomedical applications." *Expert Review of Medical Devices* **9**(6): 595–612.

Zhou, Q. Z., L. Y. Wang, G. H. Ma and Z. G. Su (2008). "Multi-stage premix membrane emulsification for preparation of agarose microbeads with uniform size." *Journal of Membrane Science* **322**(1): 98–104.

# Chapter 5

# Surfactant-Based Delivery Systems

## 5.1 Introduction

Surfactants are the key building blocks of a number of colloidal delivery systems that can be utilized in the food industry, for example, swollen micelles, microemulsions, and liposomes. The main driving force for the formation of this type of delivery system is usually self-assembly of the surfactants based on the hydrophobic effect, and so they are primarily fabricated using physicochemical (rather than mechanical) means. Nevertheless, mechanical methods are often employed to facilitate their initial formation or to alter the structures formed, for example, mixing of surfactant, oil, and water together to form microemulsions or homogenization of liposomes to reduce their particle size. This chapter begins by discussing the molecular and physicochemical characteristics of surfactants and then discusses each of the major delivery systems that can be fabricated from them. Finally, the potential applications of these kinds of delivery systems in the food industry are discussed.

## 5.2 Building Blocks: Surfactants

As mentioned earlier, the key building blocks of this type of colloidal delivery system are surfactants. Nevertheless, other components may also be used to facilitate their formation or stability, such as cosurfactants, cosolvents, or carrier oils.

**149**

## 5.2.1 Molecular Characteristics

Surfactants are amphiphilic surface-active molecules that consist of a hydrophilic head group and a lipophilic tail group. The head group has a high affinity for water, while the tail group has a high affinity for oil. A wide variety of surfactants are available for utilization within food and beverage products (Hasenhuettl and Hartel 2010; Kralova and Sjoblom 2009; Stauffer 1999), and some of the most commonly used are listed in Table 5.1. These surfactants can be represented by the formula $R - X$, where $X$ represents the hydrophilic head group and $R$ represents the lipophilic tail group (McClements 2005). The physicochemical characteristics and functional performance of a surfactant are primarily governed by the nature of its head and tail groups. Thus, surfactants with different functional properties can be created by using different types of head and tail groups. Head groups vary in their chemical compositions, dimensions, polarities, and charges, for example, nonionic, anionic, cationic, or zwitterionic. Tail groups may also vary considerably depending on their molecular characteristics. Food-grade surfactants typically have one or more hydrocarbon chains, with between 10 and 20 carbon atoms per chain. These chains may be saturated or unsaturated, linear or branched, and aliphatic or aromatic. However, most common food surfactants have either one or two linear aliphatic chains, which are either saturated or unsaturated. The functional properties of a particular surfactant are determined by its unique molecular structure (e.g., head group and tail group), solution composition (e.g., pH, ionic strength, and cosolvents), and environmental conditions (such as temperature). The choice of a surfactant for a particular application depends not only on its functional performance but also on factors such as the price, usage level, legal status, naturalness, taste characteristics, food matrix compatibility, reliability of source, and ease of utilization. Consequently, there is no single surfactant that can be used to fabricate every kind of colloidal delivery system. Instead, it is necessary to select the most appropriate surfactant for each particular application depending on the type of active ingredient to be encapsulated, the nature of the food matrix, the environmental conditions that the product will experience, and the required functional attributes (e.g., protection, retention, or release profile).

Even though many food-grade surfactants are designated by a particular commercial or chemical name (such as Tween 20 or Polyoxyethylene [20] sorbitan monolaurate), it is important to note that they are actually compositionally complex mixtures of various kinds of molecules. The main reason for this heterogeneity is that food-grade surfactants are produced industrially by chemical processes utilizing a variety of different raw materials, such as fats, oils, glycerol, organic acids, sugars, and polyols (Kralova and Sjoblom 2009). This compositional heterogeneity can have a large impact on their functional performance in food products and on their ability to form colloidal delivery systems. In addition, there may be considerable batch-to-batch variations in the composition of commercial surfactants and they may chemically degrade during storage (e.g., because of oxidation or hydrolysis).

**Table 5.1 Summary of the Properties of Small-Molecule Surfactants That Can Be Used to Formulate Colloidal Delivery Systems**

| Chemical Name | Abbreviation | Solubility | HLB Number | Parameter |
|---|---|---|---|---|
| **Ionic** | | | | **Charge** |
| Lecithin | — | O and W | 2–8 | Zwitterionic |
| Lysolecithin | — | W | 8–11 | Zwitterionic |
| Fatty acid salts | FA | O and W | 1–3 | Negative |
| Sodium stearoyl lactylate | SSL | W | 11 | Negative |
| Calcium stearoyl lactylate | CSL | O | 7–9 | Negative |
| Citric acid esters of MG | CITREM | W | | Negative |
| Diacetyl tartaric acid esters of MG | DATEM | W | 9.2 | Negative |
| Lauryl arginate | LAE | W | | Positive |
| **Nonionic** | | | | **Cloud Point** |
| Mono- and diglycerides | MDG | O | 2–5 | |
| Acetyl esters of MG | ACETEM | O | 2.5–3.5 | |
| Lactyl esters of MG | LACTEM | O | 3–4 | |
| Succinic acid esters of MG | SMG | O | 5.3 | |
| Polyglycerol esters of FA | PGE | O and W | 2–13 | |
| Propylene glycol esters of FA | PGMS | O | 1–3 | |
| Sucrose monooleate | SMO | W | | |
| Sucrose monostearate | SMS | W | 15 | |
| Sucrose monopalmitate | SMP | W | 16 | |
| Sucrose monolaurate | SML | W | 15 | |
| Sucrose distearate | SDS | O | 6 | |
| Sorbitan monooleate | Span 80 | O | 4.3 | — |
| Sorbitan monostearate | Span 60 | O | 4.7 | — |

(*continued*)

**Table 5.1 (Continued) Summary of the Properties of Small-Molecule Surfactants That Can Be Used to Formulate Colloidal Delivery Systems**

| Chemical Name | Abbreviation | Solubility | HLB Number | Parameter |
|---|---|---|---|---|
| Sorbitan monopalmitate | Span 40 | O | 6 | — |
| Sorbitan monolaurate | Span 20 | O | 8.6 | — |
| Sorbitan tristearate | Span 65 | O | 2.2 | — |
| POE sorbitan monooleate | Tween 80 | W | 15 | 65 |
| POE sorbitan monostearate | Tween 60 | W | 14.9 | 80 |
| POE sorbitan monopalmitate | Tween 40 | W | 15.6 | 73 |
| POE sorbitan monolaurate | Tween 20 | W | 16.7 | 76 |
| POE sorbitan tristearate | Tween 65 | W | 10.5 | |
| Polyglycerol polyricinoleate | PGPR | O | 1.5 | — |

*Source:* McClements, D. J., *Food Emulsions: Principles, Practice, and Techniques.* Boca Raton, CRC Press, 2005; and commercial surfactant suppliers.

*Note:* The solubility of some classes of surfactants depends on the relative lengths of their hydrophilic and hydrophobic parts. MG, monoglyceride; POE, polyoxyethylene.

Consequently, it is often important to determine the type and concentration of different chemical species within a commercial surfactant ingredient to ensure that it performs reliably in a particular application (Hasenhuettl and Hartel 2010).

Commercially, surfactants are often used in combinations with other types of ingredients (such as cosurfactants and cosolvents) to improve their ability to form and stabilize colloidal delivery systems (Flanagan and Singh 2006). The surfactants used to form delivery systems come in a variety of different forms, including liquids, pastes, solids, powders, and beads. During the preparation of a colloidal delivery system, a surfactant is usually combined with other ingredients such as actives, oil, water, cosurfactants, cosolvents, or water. The order of addition of the ingredients in the system can play an important role in determining the ease of preparing a surfactant-based delivery system, as well as its structure and physicochemical properties (for liposomes). Surfactants are often suspended in the phase that they are most soluble in; that is, water-soluble surfactants are dispersed in an aqueous phase whereas oil-soluble surfactants are dispersed in an oil phase. A variety of processing treatments may be required to ensure that the surfactant is adequately dispersed, such as high shear mixing or thermal treatments (Rao and McClements 2011a,b).

## 5.2.2 Physicochemical Properties

This section provides a brief overview of some of the most important physicochemical properties of surfactants that are relevant to their ability to form delivery systems.

### 5.2.2.1 Molecular Organization in Solution

At relatively low concentrations, surfactants exist as individual molecules in solution because the entropy of mixing overweighs the attractive forces operating between the surfactant molecules (Israelachvili 2011; Rosen and Kunjappu 2012). However, as their concentration is increased, they spontaneously aggregate into thermodynamically stable structures known as association colloids. The type of association colloid formed depends on surfactant concentration, surfactant structure (head and tail groups), solution composition (e.g., pH, ionic strength, and solvent type), and environmental conditions (e.g., temperature). The primary driving force for the formation of these association colloids is the hydrophobic effect, which favors a molecular organization that minimizes the thermodynamically unfavorable contact between nonpolar surfactant tails and water (Evans and Wennerstrom 1999; Israelachvili 2011). Some of the structures that may be formed when surfactants are dispersed into solvents at surfactant concentrations that are not too high are shown in Figure 5.1. Micelles, bilayers, and vesicles may be formed within aqueous

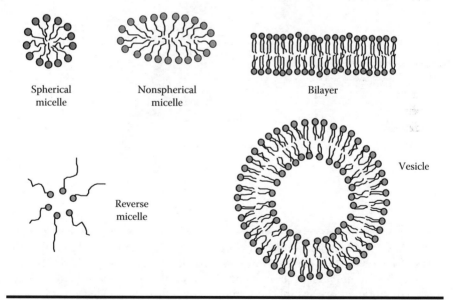

Spherical
micelle

Nonspherical
micelle

Bilayer

Reverse
micelle

Vesicle

**Figure 5.1 A variety of association colloids can be formed from surfactants when they are dispersed into a water or oil phase at relatively low concentrations. At higher concentrations, liquid crystals may be formed.**

solutions, whereas reverse micelles may be formed within organic solutions. At higher surfactant concentrations, the surfactant molecules may assemble into a variety of liquid crystalline structures, such as hexagonal, lamellar, and reversed hexagonal phases (Evans and Wennerstrom 1999). In addition, a surfactant solution may separate into a number of different phases, with each phase having a different composition and structural organization.

The molecular organization of surfactants in solutions depends mainly on the geometry and interactions of the surfactant molecules, the nature of the solvent, the solution composition, and the temperature. The influence of solution and environmental conditions on the molecular organization of a particular surfactant is conveniently described by a *phase diagram* (Figure 5.2). Phase diagrams are usually determined empirically for a given surfactant system and allow one to determine the type, number, and composition of each of the phases formed under a given set of experimental conditions. The surfactant concentrations present in final food and beverage products are normally too small to lead to the formation of liquid crystalline structures. However, some of these structures may be present in the intermediate steps during the formation of colloidal delivery systems.

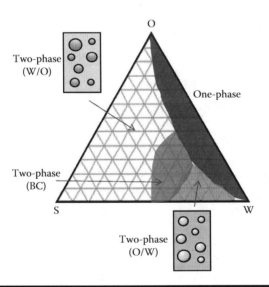

**Figure 5.2 Ternary phase diagrams are particularly useful for describing the kinds of structures that can be formed by mixtures of surfactant (S), oil (O), and water (W). The number of phases and the type of structures formed for each SOW combination are established empirically and then plotted on the phase diagram. For example, SOR combinations that lead to O/W, W/O, or bicontinuous (BC) phases can be established.**

## 5.2.2.2 Critical Micelle Concentration

When a surfactant is added to water, it forms micelles when its concentration exceeds a critical level known as the *critical micelle concentration* or CMC (Rosen and Kunjappu 2012). At surfactant concentrations lower than the CMC, the surfactant molecules are dispersed in the water predominantly as isolated individual molecules (monomers). However, at surfactant concentrations above the CMC, any additional surfactant added to the system will form micelles, and the monomer concentration remains relatively constant (Hiemenz and Rajagopalan 1997). Micelles are highly dynamic structures, with surfactant molecules continually exchanging with the surrounding liquid, because they are only held together by relatively weak physical interactions (Israelachvili 2011). Nevertheless, they do have well-defined average sizes and shapes under a given set of solution and environmental conditions. When additional surfactant is incorporated into an aqueous solution above the CMC, the *number* of micelles increases, rather than the size or shape of the individual micelles (although this may not be true at higher surfactant concentrations). There is an abrupt change in the physico-chemical properties of a surfactant solution when the CMC is exceeded, for example, surface tension, turbidity, electrical conductivity, and osmotic pressure (Hiemenz and Rajagopalan 1997; Rosen and Kunjappu 2012). This is because the physicochemical properties of surfactant monomers are different from those of surfactant micelles; for example, monomers are amphiphilic and so are surface active, whereas micelles are hydrophilic and therefore have little surface activity. Consequently, the surface tension of a solution decreases steeply with increasing surfactant concentration below the CMC but remains fairly constant above it (Figure 5.3).

The CMC of a surfactant solution depends on the chemical structure of the surfactant molecules, as well as on solution composition (e.g., pH, ionic strength, and solvent type) and environmental conditions (e.g., temperature) (Israelachvili 2011; Rosen and Kunjappu 2012). The CMC tends to decrease as the hydrophobicity of surfactant molecules increases (e.g., by increasing the hydrocarbon tail length) or their hydrophilicity decreases (e.g., by decreasing the length of a nonionic head group or by exchanging an ionic head group for a nonionic one). For ionic surfactants, the CMC decreases appreciably with increasing ionic strength, since counterions screen the electrostatic repulsion between charged head groups, thereby reducing the magnitude of this unfavorable contribution to micelle formation. For these systems, CMCs are not usually strongly temperature dependent over the temperature ranges normally found in foods (e.g., 0°C to 100°C). For many commercial food-grade surfactants, the CMC does not occur at a well-defined concentration, but over a range of concentrations, because the surfactant ingredient contains a mixture of components with different chain lengths, degrees of unsaturation, and head group properties.

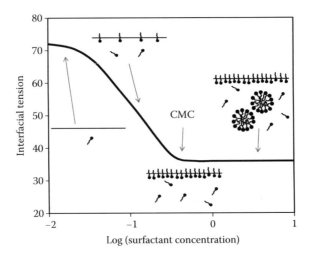

**Figure 5.3 Surfactants are surface-active molecules that exist as monomers in aqueous solutions below the CMC, but monomers + micelles above the CMC. The CMC can be determined by measuring the interfacial tension versus surfactant concentration profile.**

### 5.2.2.3 Krafft Point

The proper functional performance of a surfactant, including its ability to form colloidal delivery systems, usually requires that it be adequately dispersed within a suitable solvent prior to use. At relatively low temperatures, the water solubility of some surfactants ($C_S^*$) is below their CMC and so they tend to form crystals rather than micelles when added at high concentrations (Rosen and Kunjappu 2012). These surfactants can be characterized by a critical temperature, known as the *Krafft point*, which is the temperature where the surfactant solubility is equal to the CMC. At temperatures below the Krafft point, the amount of surfactant that can be dispersed into an aqueous solution is limited by its solubility and is therefore relatively small. On the other hand, at temperatures above the Krafft point, the amount of surfactant that can be dispersed is much higher because micelles can be formed.

### 5.2.2.4 Cloud Point

A nonionic surfactant solution will become turbid when it is heated above a certain temperature, which is known as the *cloud point* (Rosen and Kunjappu 2012). This phenomenon occurs because the hydrophilic head groups of the surfactant molecules become progressively dehydrated as the temperature is raised, which alters their molecular geometry and decreases the hydration repulsion between them (Israelachvili 2011). At the *cloud point*, the surfactant molecules associate with each other and form new structures that are large enough to scatter light and therefore make the solution appear turbid. As the temperature is increased further, these structures may grow so large that they sediment under the influence of gravity and form a separate phase at the bottom of the container. The cloud point of nonionic surfactants tends to increase as the size of their hydrophilic head group increases and depends on the type and concentration of electrolytes, cosurfactants, and cosolvents present. Knowledge of the cloud point may be an important factor to consider when selecting a surfactant for preparing a colloidal delivery system for a particular food application. For example, a delivery system that is stored at elevated temperatures or subjected to thermal processing may turn turbid and break down because of this effect. Typically, the interfacial tension decreases appreciably as the temperature is raised toward the cloud point, which provides a convenient means of determining it. Alternatively, the cloud point can simply be determined by measuring the turbidity versus temperature profile for the surfactant.

### 5.2.2.5 Solubilization Capacity

Nonpolar active ingredients that cannot normally be dispersed in water can be solubilized within the hydrophobic cores of micelles or microemulsions (Israelachvili 2011; McClements 2005). There are a number of important factors that influence the solubilization properties of micelle and microemulsion delivery systems: (i) the location of the active ingredient within the micelles, (ii) the maximum amount of active ingredient that can be solubilized per unit mass of surfactant, and (iii) the solubilization rate. Consider the changes that occur when increasing amounts of oil droplets are added into an aqueous solution that initially consists of a fixed amount of surfactant dispersed as micelles in water (Figure 5.4). Initially, all of the added oil droplets dissolve because the oil molecules are solubilized by the micelles, and so the solution appears transparent. However, once the solubilization capacity ($C_{Sat}$) of the micelles has been exceeded, they cannot incorporate oil molecules anymore, and so any additional oil droplets added do not dissolve, and so the system appears turbid or opaque. If this system were allowed to come to equilibrium, it would consist of a layer of oil on top of an aqueous solution containing saturated swollen micelles [an oil-in-water (O/W) microemulsion]. If this mixture is then homogenized, a metastable system is formed that consists of surfactant-coated oil droplets dispersed within an aqueous phase containing saturated swollen micelles (Figure 5.4). This approach offers a simple and convenient method for measuring the factors that influence the solubilization capacity of different systems (Rao and McClements 2011a,b, 2012).

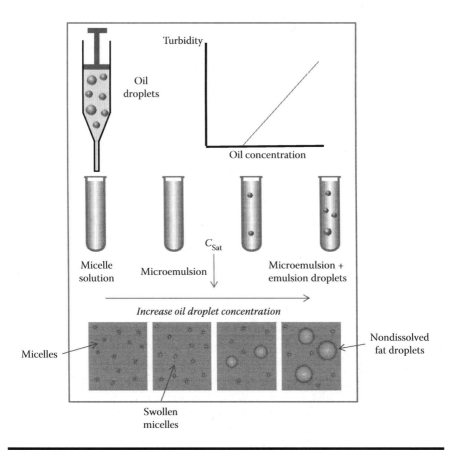

**Figure 5.4** **Schematic representation of the emulsion dilution method. Increasing amounts of emulsion droplets are titrated into an aqueous surfactant solution. Initially, the oil is solubilized within surfactant micelles to form microemulsion droplets. Above a critical oil concentration, the micelles become saturated with oil and any further droplets added remain.**

## 5.2.3 Surfactant Classification Schemes

A large number of different surfactants are available for commercial utilization, and a manufacturer must select the most appropriate one for fabricating a colloidal delivery system with the required functional performance. For this reason, various classification schemes have been proposed to facilitate the selection of surfactants for particular applications (McClements 2005; McClements and Rao 2011). In this section, some of the most widely used classification schemes for characterizing surfactants are reviewed.

## 5.2.3.1 Bancroft's Rule

One of the earliest empirical rules used to classify surfactants according to their functional properties was proposed by Bancroft based on their preferential solubility in oil and water (Ruckenstein 1996). A water-soluble surfactant mainly partitions into the water phase, whereas an oil-soluble surfactant mainly partitions into the oil phase (Figure 5.5). This rule was initially proposed to facilitate the selection of surfactants for forming emulsions. According to Bancroft's rule, the phase in which a surfactant is most soluble will form the continuous phase of an emulsion. Hence, a water-soluble surfactant should stabilize O/W emulsions, whereas an oil-soluble surfactant should stabilize water-in-oil (W/O) emulsions. In reality, the solubility of a surfactant in a particular phase should be defined as the total concentration (monomers + micelles) in a phase, rather than just the monomers (Binks 1993). The water solubility of surfactant monomers may be relatively low (determined by the CMC), but the overall amount of surfactant that can be dispersed in water can be quite large because of the high water solubility of micelles. The Bancroft rule provides a good starting point for formulating colloidal delivery systems using surfactants (in particular O/W and W/O emulsions), but it does have a number of important limitations. For example, two surfactants could both be predominately water soluble, but they may function very differently in forming and stabilizing colloidal delivery systems.

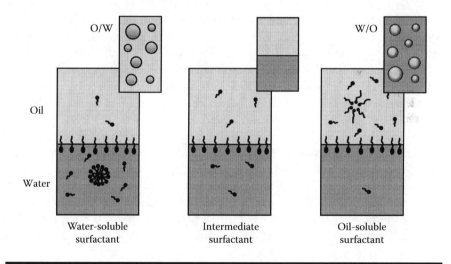

**Figure 5.5**  **Bancroft's rule states that a surfactant that is more soluble in a water phase will form an O/W emulsion, whereas one that is more soluble in an oil phase will form a W/O emulsion.**

## 5.2.3.2 Hydrophile–Lipophile Balance (HLB)

The HLB classification scheme has been used for many years to aid in the selection of surfactants for particular applications (Becher 1966). It is a more sophisticated method than Bancroft's rule because it takes into account the molecular structure of the surfactant molecules involved (Friberg et al. 2004; McClements 2005). The HLB of a surfactant is described by a number that depends on the balance of hydrophilic and lipophilic groups on the surfactant molecule. The HLB number therefore provides an indication of the relative affinity of a surfactant molecule for the oil and water phases, and its tendency to form different kinds of association colloids and emulsions. Each surfactant is assigned an HLB number according to its chemical structure. A molecule with a high HLB number has a high ratio of hydrophilic to lipophilic groups, and vice versa. The HLB number of a surfactant can be estimated from knowledge of the number and type of hydrophilic and lipophilic groups it contains using the following semiempirical equation:

$$HLB = 7 + \Sigma(HGN) - \Sigma(LGN) \tag{5.1}$$

Here, HGN and LGN are hydrophilic and lipophilic group numbers, respectively. The magnitudes of specific group numbers are determined by the polarity of the functional groups on the surfactant molecule. The sum of all the HGN and LGN values associated with the different chemical groups on a surfactant molecule is inserted into Equation 5.1 and the HLB number is calculated. Alternatively, the HLB number of a surfactant can often be estimated from experimental measurements, such as cloud point determinations. The HLB numbers of many food-grade surfactants have been calculated or determined experimentally (Table 5.1).

The HLB number of a surfactant gives a useful indication of its solubility characteristics, the type of association colloids it tends to form, and the types of emulsions it can stabilize (Israelachvili 2011; McClements 2005). A surfactant with a low HLB number (3–6) is predominantly hydrophobic, dissolves preferentially in oil, stabilizes W/O emulsions and microemulsions, and forms reverse micelles in oil. A surfactant with an intermediate HLB number (7–9) has no particular preference for oil or water and tends to form bilayers or vesicles in water. A surfactant with a high HLB number (10–18) is predominantly hydrophilic, dissolves preferentially in water, stabilizes O/W emulsions and microemulsions, and forms micelles in water.

Knowledge of the HLB number is often important when forming emulsion-based delivery systems such as emulsions or nanoemulsions (Chapter 6). Emulsion droplets are highly susceptible to coalescence when they are coated by surfactants that have extreme or intermediate HLB numbers. At very high or low HLB numbers, surfactants have such low surface activity that they do not accumulate appreciably at the droplet surfaces and therefore do not provide good protection against coalescence. At intermediate HLB numbers (7–9), emulsions are unstable to coalescence because the interfacial tension is so low that very little free energy is required to disrupt the interface and cause the droplets to merge together (Kabalnov 1998; Kabalnov and Wennerstrom 1996). Empirical observations suggest that maximum emulsion stability is obtained for O/W emulsions using surfactants with an HLB number around 8–18, and for W/O emulsions, it is around 3–6 (Brooks et al. 1998). This is because the surfactants are surface active but do not lower the interfacial tension so much that the droplets are easily disrupted. Under certain circumstances, it is possible to adjust the "effective" HLB number by using a combination of two or more surfactants with different HLB numbers (Brooks et al. 1998). Surfactant blends are often used in the food industry to improve the overall functionality of surfactant systems in commercial products.

A major drawback of the HLB concept is that it does not take into account the fact that the functional performance of surfactant molecules may be altered significantly when the solution composition or temperature is changed. Hence, a surfactant may be capable of stabilizing O/W emulsions at one temperature, but W/O emulsions at another temperature, even though it has exactly the same chemical structure. In addition, the ability of surfactants to encapsulate and stabilize different kinds of oil phases depends on oil type. The hydrophilic–lipophilic deviation (HLD) concept discussed in Section 5.2.3.4 has been developed to overcome some of the limitations of the HLB classification system.

### 5.2.3.3 Molecular Geometry

The HLB number scheme takes into account the type of chemical groups on surfactant molecules, but it does not take into account their molecular geometry, which is important because it influences the packing of surfactant molecules at interfaces in colloidal systems. The molecular geometry of a surfactant molecule is conveniently described by a packing parameter, $p$ (Israelachvili 2011):

$$p = \frac{v}{la_H} = \frac{a_T}{a_H} \qquad (5.2)$$

where $a_T$, $v$ and $l$ are the cross-sectional area volume and length of the hydrophobic tail and $a_H$ is the cross-sectional area of the hydrophilic head group (Figure 5.6). When surfactant molecules associate with each other, they tend to form monolayers that have a curvature that allows the most efficient packing of the head and tail groups. At this *optimum curvature*, the monolayer has its lowest free energy, and any deviation from this curvature requires an increase in free energy. The optimum curvature of a monolayer depends on the packing parameter of the surfactant: for $p = 1$, monolayers with zero curvature are preferred; for $p < 1$, the optimum curvature is convex; and for $p > 1$, the optimum curvature is concave (Figure 5.6). Geometrical considerations indicate that spherical micelles are formed when $p$ is less than 1/3, nonspherical micelles are formed when $p$ is between 1/3 and 1/2, and bilayers are formed when $p$ is between 1/2 and 1 (Israelachvili 2011). Bilayers tend to join and form vesicles at relatively high

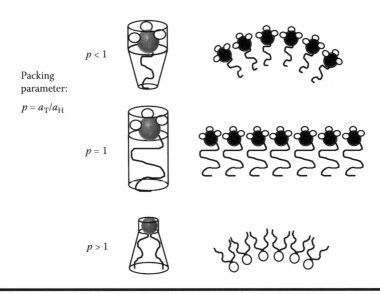

Packing parameter:

$p = a_T/a_H$

$p < 1$

$p = 1$

$p > 1$

**Figure 5.6 Surfactants can be classified according to their molecular geometry and optimum curvature.**

surfactant concentrations to minimize unfavorable end effects. At $p > 1$, reverse micelles are formed, in which the hydrophilic head groups are located in the interior (away from the oil) and the hydrophobic tail groups are located at the exterior (in contact with the oil) (Figure 5.1). The packing parameter therefore gives a useful indication of the type of association colloid that a surfactant molecule tends to form in solution.

The packing parameter is also useful because it provides a physical framework for understanding the temperature dependence of the physicochemical properties of surfactant solutions, for example, the cloud point or phase inversion temperature (PIT) (Kabalnov and Wennerstrom 1996). The PIT is the temperature at which an O/W emulsion stabilized by a nonionic surfactant changes to a W/O emulsion (Shinoda and Arai 1964). The utilization of the PIT method to form emulsion- and nanoemulsion-based delivery systems is considered in more detail in Chapter 6.

### 5.2.3.4 Hydrophilic–Lipophilic Deviation

Recently, a more comprehensive approach has been developed to characterize surfactant performance that takes into account the molecular characteristics of the surfactant, as well as the environment it operates in (Queste et al. 2007; Salager et al. 2004). This approach involves describing the behavior of a surfactant–oil–water (SOW) system in terms of a *formulation–composition map* (Figure 5.7). The *x*-axis of one of these maps represents changes in the "composition" of the SOW system, which is usually expressed as the water-to-oil ratio: WOR = mass of water/mass of oil. The *y*-axis represents changes in the "formulation" of the SOW system,

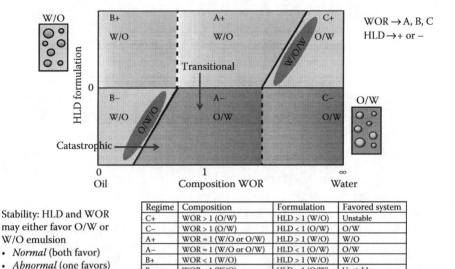

| Regime | Composition | Formulation | Favored system |
|---|---|---|---|
| C+ | WOR > 1 (O/W) | HLD > 1 (W/O) | Unstable |
| C− | WOR > 1 (O/W) | HLD < 1 (O/W) | O/W |
| A+ | WOR ≈ 1 (W/O or O/W) | HLD > 1 (W/O) | W/O |
| A− | WOR ≈ 1 (W/O or O/W) | HLD < 1 (O/W) | O/W |
| B+ | WOR < 1 (W/O) | HLD > 1 (W/O) | W/O |
| B− | WOR < 1 (W/O) | HLD < 1 (O/W) | Unstable |

Stability: HLD and WOR may either favor O/W or W/O emulsion
- *Normal* (both favor)
- *Abnormal* (one favors)

**Figure 5.7  HLD: A formulation–composition map can be used to describe the behavior of a surfactant–oil–water system for different formulation variables (HLD) and compositions (WOR).**

which is usually expressed as the HLD. The HLD is a measure of the relative affinity of the surfactant for the hydrophilic (water) phase and the lipophilic (oil) phase (Salager et al. 2005; Witthayapanyanon et al. 2008). In some publications, the *surfactant affinity difference* (SAD) is used to represent the formulation variable, which is closely related to the HLD number: HLD = SAD/$RT$, where $R$ is the gas constant and $T$ is the absolute temperature. A formulation–composition map is a convenient means of specifying the types of colloidal systems that can be formed by a particular SOW mixture when either the surfactant properties are altered (e.g., by altering surfactant type, pH, ionic strength, or temperature) or the system composition is changed (e.g., by varying the relative amounts of oil and water present). It is also a useful means of predicting the behavior of a colloidal delivery system if it is prepared in a concentrated form and then diluted prior to use.

Empirical equations have been developed to calculate the HLD numbers of various classes of surfactants from knowledge of their molecular structure, oil phase characteristics (such as equivalent alkane carbon number), aqueous phase characteristics (such as salt and alcohol content), and environmental conditions (such as temperature) (Bouton et al. 2009, 2010; Phan et al. 2010; Queste et al. 2007). These relationships can be used to predict how a surfactant behaves under different conditions and to establish the type of delivery systems that it may be able to form and stabilize.

The relationship between the HLD number of a surfactant and its ability to form colloidal delivery systems is highlighted below:

- **HLD < 0:** The surfactant has (i) a higher affinity for the water phase than the oil phase; (ii) the surfactant tends to form micelles or O/W microemulsions in the water phase; and (iii) the surfactant tends to stabilize O/W (rather than W/O) emulsions. The more negative the HLD number, the greater the affinity for the water phase.
- **HLD = 0:** The surfactant has (i) an equal affinity for the water and oil phases; (ii) the surfactant tends to form bilayers, vesicles, bicontinuous microemulsions, or liquid crystalline phases; and (iii) the surfactant tends to stabilize neither O/W nor W/O emulsions.
- **HLD > 0:** The surfactant has (i) a higher affinity for the oil phase than for the water phase; (ii) the surfactant tends to form reverse micelles or W/O microemulsions in the oil phase; and (iii) the surfactant tends to stabilize W/O (rather than O/W) emulsions. The more positive the HLD number, the greater the affinity for the oil phase.

Knowledge of the HLD number, WOR, and formulation–composition map for a particular system can be used to rationalize its behavior under different conditions (Figure 5.7). A formulation–composition map can be conveniently divided into a number of regimes designated by a *letter* (A, B, C) and a *sign* (+/–) (Leal-Calderon et al. 2007). The sign determines the influence of the formulation parameter (HLD number) on the type of colloidal dispersion that tends to be stable in that regime. In regimes (A–, B–, and C–) where the formulation parameter is negative (HLD < 0), the surfactant prefers to form micelles, O/W microemulsions, or O/W emulsions, whereas in regimes (A+, B+, and C+) where the formulation parameter is positive (HLD > 0), the surfactant prefers to form reverse micelles, W/O microemulsions, and W/O emulsions (Salager et al. 2005; Witthayapanyanon et al. 2008). The overall composition of a SOW system also has a major influence on the formation of different kinds of colloidal dispersions. The letter (A, B, C) determines the influence of composition (WOR) on the type of colloidal dispersion formed, particularly for emulsions. *A* refers to a system where the oil and water phases have fairly similar amounts (WOR ≈ 1) and so formation of both O/W and W/O emulsions is equally favorable; *B* refers to a system where the oil phase is in excess and so the formation of W/O emulsions is favored; and *C* refers to a system where the water phase is in excess and so the formation of O/W emulsions is favored. If both the formulation variable (HLD number, sign) and composition variable (WOR, letter) favor the formation of a particular emulsion type (e.g., O/W), then this emulsion is said to be "normal" and will tend to be stable (Mira et al. 2003; Rondón-Gonzaléz et al. 2006). On the other hand, if the formulation variable favors one emulsion type (e.g., W/O) while the composition variable favors the other type (e.g., O/W), then this system is said to be "abnormal" and will tend to be unstable. In the formulation–composition map shown in Figure 5.7, there are four regimes where the emulsions should be stable (A–, A+, B+, and C–) and two regimes where they should be unstable (B– and C+). The emulsions formed in the abnormal regimes are usually highly unstable to droplet coalescence and phase separation, but multiple emulsions may be formed near certain phase boundaries. For example, O/W/O emulsions may be formed near the B– to A– boundary, whereas W/O/W emulsions may be formed near the C+ to A+ boundary (Figure 5.7).

Formulation–composition maps are also useful for describing *transitional* and *catastrophic* phase inversions involving surfactant-stabilized colloidal particles (Leal-Calderon et al. 2007). For example, if one moves in a horizontal direction from regime B– to regime A– by increasing the amount of water present (increasing WOR), the system undergoes a catastrophic phase inversion from a W/O to an O/W emulsion, passing through an intermediate regime where a multiple emulsion (O/W/O) is formed. Conversely, if one moves in a vertical direction from regime A+ to regime A– by decreasing the HLD value from positive to negative (e.g., decreasing temperature from above to below the PIT), the system undergoes a transitional phase inversion from a W/O to an O/W emulsion, passing through an intermediate regime where a bicontinuous system or liquid crystalline phase was formed (Figure 5.7). The temperature at which this transition occurs is the PIT. Knowledge of these potential transitions is often important when trying to formulate surfactant-based or emulsion-based delivery systems. For example, nanoemulsions can be formed using a PIT method that involves heating a SOW mixture to a temperature where a microemulsion is formed (near the PIT) and then rapidly cooling it with continuous stirring (Chapter 6).

### 5.2.3.5 Additional Factors

The classification schemes mentioned above provide some important information about the type of micelle, microemulsion, or emulsion that a surfactant tends to form and stabilize (i.e., O/W or W/O), but they do not provide much insight into the size of the particles formed, the amount of surfactant required to form a stable system, or the stability of the particles formed. These factors must often be considered when selecting a surfactant to produce a colloidal delivery system suitable for a particular application.

## 5.2.4 Food-Grade Surfactants

The properties of a number of food-grade small-molecule surfactants that can be used to form colloidal delivery systems are briefly discussed below and summarized in Table 5.1. Water-soluble surfactants with relatively high HLB numbers (8–18) are normally used to form micelles, O/W microemulsions, and O/W emulsions that can be incorporated into aqueous-based products (Myers 2006). Oil-soluble surfactants with relatively low HLB numbers (2–6) are used to form reverse micelles, W/O microemulsions, and W/O emulsions. Surfactants with intermediate HLB numbers (7–9), such as some lecithin-based emulsifiers, can be used to form liposomes. As mentioned earlier, surfactants with different characteristics are often used in combination to facilitate the formation and stability of colloidal delivery systems (Myers 2006; Schick 1987). In addition, to their direct application in forming colloidal delivery systems such as micelles, microemulsions, and emulsions, surfactants can also be used indirectly as processing aids during the formation of colloidal delivery systems. For example, surfactants can be used to alter the functional properties of biopolymers by binding with them, to displace biopolymers from oil–water interfaces through competitive adsorption, or to modulate the crystallization of oil or water phases.

As mentioned earlier, most surfactants do not consist of an individual molecular species but consist of a complex mixture of different types of molecular species. Some of the impurities in surfactant mixtures may have a large impact on their functional performance. Hence, it is often necessary to ensure that a surfactant is of a reliable high purity and quality before it is used to prepare a product. Detailed descriptions of various surfactants have been given elsewhere (Friberg et al. 2004; Hasenhuettl and Hartel 2010; Kralova and Sjoblom 2009; McClements 2005; Whitehurst 2004), and so only a brief outline is given here. It should also be noted that many ingredient manufacturers provided detailed information about the composition, properties, and functional performances of the surfactants they produce.

### 5.2.4.1 Mono- and Diglycerides

The terms *monoglyceride* and *diglyceride* are commonly used to describe a series of surfactants produced by interesterification of fats or oils with glycerol (Moonen and Bas 2004). This procedure produces a complex mixture of monoacylglycerides, diacylglycerides, triacylglycerides, glycerol, and free fatty acids. The monoacylglyceride fraction can be separated (>90% purity) from the other fractions by molecular distillation to produce a more pure surface-active ingredient, referred to as *distilled monoglycerides*. Distilled monoglycerides are available with hydrocarbon chains of differing lengths and degrees of unsaturation. Generally, monoglycerides are nonionic oil-soluble surfactants with relatively low HLB numbers (~2 to 5) and are therefore most suitable for forming reverse micelles, W/O microemulsions, and W/O emulsions.

### 5.2.4.2 Organic Acid Esters of Mono- and Diglycerides

Mono- and diglycerides can be esterified with a variety of different organic acids (e.g., acetic, citric, diacetyl tartaric, and lactic acids) to form small-molecule surfactants with different functional attributes (Gaupp and Adams 2004). The most common examples of this type of surfactant are acetylated monoglycerides (ACETEM), lactylated monoglycerides (LACTEM), diacetyl tartaric acid monoglycerides (DATEM), and citric acid esters of monoglycerides (CITREM). Each of these surfactants is available with hydrocarbon chains of differing lengths and degrees of unsaturation. ACETEM and LACTEM are nonionic oil-soluble surfactants with low HLB numbers, whereas DATEM and CITREM are anionic water-dispersible surfactants with intermediate or high HLB numbers.

## 5.2.4.3 Polyol Esters of Fatty Acids

Small-molecule surfactants with different functional attributes can be produced by esterification of different types of polyols with different kinds of fatty acids (Cottrell and van Peij 2004; Nelen and Cooper 2004; Norn 2004; Sparso and Krog 2004). The polyols form the hydrophilic head group, whereas the fatty acid chains form the hydrophobic tail group. The functional characteristics of a particular surfactant produced using this approach depend on the type of polyols and fatty acids used. The most commonly used polyols are polyglycerol, propylene glycol, sorbitan, polyoxyethylene sorbitan, and sucrose. The fatty acids used to prepare these kinds of surfactants may vary in their chain length and degree of unsaturation. The solubility and functional properties of polyol esters of fatty acids depend on the relative sizes of the hydrophilic and lipophilic parts of the molecules. Surfactants with relatively large polyol head groups tend to be water dispersible and have high HLB numbers (e.g., sucrose, polyglycerol, and polyoxyethylene sorbitan esters), whereas those with small polyol head groups tend to be oil soluble and have low HLB numbers (e.g., propylene glycol esters). The ratio of hydrophilic to lipophilic groups can also be varied appreciably for some of these surfactants by changing the number of fatty acids attached to the polyol group, which leads to both oil-soluble and water-dispersible surfactants in the same class, for example, sucrose, sorbitan, or polyoxyethylene sorbitan esters (Table 5.2). Sorbitan esters of fatty acids are one of the most commonly used oil-soluble nonionic surfactants, which are often sold under the trade name "Span." On the other hand, polyoxyethylene sorbitan esters are one of the most commonly used water-soluble nonionic surfactants, which are often

**Table 5.2   Selected HLB Group Numbers**

| Hydrophilic Group | Group Number | Lipophilic Group | Group Number |
|---|---|---|---|
| $-SO_3Na^+$ | 20.7 | $-CH-$ | 0.475 |
| $-COO^-H^+$ | 21.2 | $-CH_2-$ | 0.475 |
| Tertiary amine | 9.4 | $-CH_3$ | 0.475 |
| Sorbitan ester | 6.8 | $-CH=$ | 0.475 |
| Glyceryl ester | 5.25 | | |
| $-COOH$ | 2.1 | | |
| $-OH$ | 1.9 | | |
| $-O-$ | 1.3 | | |
| $-(CH_2-CH_2-O)-$ | 0.33 | | |
| $-CH_2-CH_2-OH$ | 0.95 | | |

*Source:* Adapted from McClements, D. J., *Food Emulsions: Principles, Practice, and Techniques.* Boca Raton, CRC Press, 2005.

sold under the trade names "Polysorbate" or "Tween." Oil-soluble and water-soluble surfactants are often used in combination to facilitate the formation and stability of colloidal delivery systems (Myers 2006).

### 5.2.4.4 Stearoyl Lactylate Salts

Surfactants can be produced by esterification of lactic acid with fatty acids in the presence of either sodium or calcium hydroxide (Boutte and Skogerson 2004). Sodium stearoyl lactylate is an anionic water-dispersible surfactant with an intermediate HLB number, whereas calcium stearoyl lactylate is an anionic oil-soluble surfactant with a low HLB number.

### 5.2.4.5 Lecithins

Lecithins are naturally occurring surface-active molecules that can be extracted from a variety of sources, including soybeans, milk, rapeseed, and egg (Bueschelberger 2004). In nature, they are present in the cell and organelle walls of plants, animals, and microorganisms, where they form a natural barrier with important functions in protection, separation, and transport of components. Lecithins isolated from natural sources contain a complex mixture of different types of phospholipids and other lipids, although they can be fractionated to form more pure ingredients that are enriched with specific fractions. The most common phospholipids in lecithin are phosphatidylcholine (PC), phosphatidylethanolamine (PE), and phosphatidylinositol (PI). The hydrophilic head groups of these molecules are either anionic (PI) or zwitterionic (PC and PE), while the lipophilic tail groups consist of two fatty acids. Natural lecithins tend to have low to intermediate HLB numbers (2–8) and are therefore most suitable for stabilizing W/O systems (low HLB) or forming bilayers or liposomes in aqueous solutions (intermediate HLB). However, lecithins can also be used in combination with other types of surfactants to improve stability and to form different structures. In addition, lecithin can be chemically or enzymatically hydrolyzed to break off one of the hydrocarbon tails to produce more hydrophilic surfactants called lysolecithins that are capable of forming micelles, microemulsions, or O/W emulsions.

## 5.3 Micelle and Microemulsion Delivery Systems

In this section, we primarily focus on those colloidal delivery systems that can be used to encapsulate lipophilic active ingredients into aqueous-based systems, such as swollen micelles and O/W microemulsions. Nevertheless, much of the information presented is also relevant to colloidal delivery systems that are suitable for encapsulating hydrophilic active ingredients into oily systems, such as reverse micelles or W/O microemulsions.

## 5.3.1 Composition and Structure

The colloidal particles in delivery systems containing swollen micelles and microemulsions are thermodynamically stable structures formed primarily from surfactants, but which may also contain other structural components, such as cosurfactants, cosolvents, and carrier oils (Brodskaya 2012; Flanagan and Singh 2006; Garti and Aserin 2012; Torchilin 2007) (Figure 5.1). For the sake of clarity, we use the term *micelle* to refer to structures that are only formed from surfactant (and possibly cosurfactants and cosolvents), whereas we use the term *microemulsion* to refer to structures formed from surfactant and carrier oil (as well as possibly cosurfactants and cosolvents). When a lipophilic active ingredient is incorporated into a micelle, we refer to it as a "swollen micelle." Under certain circumstances, the terms *swollen micelle* and *microemulsion* can be used interchangeably, that is, if the lipophilic active agent is actually oil (such as flavor oil, essential oil, or $\omega$-3 oil).

Micelles and microemulsions tend to be characterized by surfactants that have high water solubility/dispersibility, high HLB numbers, negative HLD numbers, and packing parameters less than unity ($p < 1$) (Israelachvili 2011; Leser et al. 2006). The colloidal particles formed are often roughly spherical in shape, although they may be nonspherical under certain circumstances. The nonpolar tails form a hydrophobic core within the particle interior, whereas the polar head groups form a hydrophilic shell in contact with the surrounding aqueous phase (Flanagan and Singh 2006; Spernath and Aserin 2006; Spernath et al. 2002). Lipophilic active ingredients and carrier oils may also be present within the hydrophobic core (Huang et al. 2010; Marze 2013). Micelles and microemulsions may be formed using a single surfactant type or they may contain a mixture of surfactants (Myers 2006). Typically, the dimensions of micelles are in the 2 to 10 nm range and microemulsions are in the 10 to 100 nm range depending on their composition and environmental conditions, such as pH, ionic composition, and temperature. As mentioned previously, micelles and microemulsions have highly dynamic structures with surfactants and other molecules continually entering and leaving the individual colloidal particles, so that there may be molecular exchange between different particles and with the surrounding liquid. Solubilized lipophilic active ingredients are mainly present within the hydrophobic core, although any polar groups they may have can protrude into the hydrophilic shell and water phase. The physical location of the active ingredients within a micelle or microemulsion influences their availability for chemical reactions with other components in the system (e.g., in the oil, water, or interfacial phases), which may have an important impact on the chemical stability of the delivery system.

The particles in micelles and microemulsions may have electrical charges ranging from highly negative to highly positive depending on the type of surfactants used, which may be important in some applications, such as particle stability, interactions, or targeting. Nonionic surfactants tend to form neutral particles, anionic surfactants form negative particles, and cationic surfactants form positive particles.

The electrical charge on the particles can also be modulated by using mixtures of surfactants with different charge characteristics, such as a nonionic and an ionic surfactant (Ziani et al. 2011a,b).

## 5.3.2 Formation

In principle, swollen micelles and microemulsions form spontaneously when suitable ingredients (surfactant, oil, water, and possibly cosolvent/cosurfactant) are mixed together under appropriate environmental conditions (Flanagan and Singh 2006; Spernath and Aserin 2006; Spernath et al. 2002). In practice, it is often necessary to apply some kind of processing treatment to facilitate the formation of these thermodynamically stable structures, for example, high shear mixing, sonication, or heating (Rao and McClements 2011a,b). The reason for this can be attributed to the existence of kinetic energy barriers that separate the individual components from the final system (Figure 5.8).

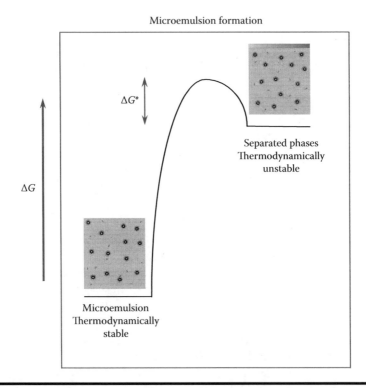

Microemulsion formation

$\Delta G^*$

$\Delta G$

Separated phases
Thermodynamically
unstable

Microemulsion
Thermodynamically
stable

**Figure 5.8** **Schematic diagram of the free energy of microemulsion systems compared to the phase-separated state. Microemulsions have a lower free energy than the phase-separated state and therefore should form spontaneously. However, there may be an activation energy ($\Delta G^*$) that must be exceeded first.**

## 5.3.3 *Properties*

### 5.3.3.1 *Optical Properties*

The size of the colloidal particles in micelles and microemulsions is typically much smaller than the wavelength of light ($d \ll \lambda$), and therefore they do not scatter light strongly and appear optically transparent. Typically, colloidal particles with diameters less than approximately 40 nm appear clear and are therefore suitable for applications where optical clarity is necessary (McClements 2011; Wooster et al. 2008). Some microemulsions may have droplet sizes that approach the wavelength of light and may therefore have a more turbid or cloudy appearance.

### 5.3.3.2 *Rheological Properties*

At the relatively low concentrations present in final food and beverage applications, micelles and microemulsions should not have a large impact on the bulk rheological properties of a product. Nevertheless, knowledge of the rheological properties may be important for optimizing the formation, transport, storage, and utilization of this type of colloidal dispersion. The rheology of SOW mixtures may vary from very low to very high depending on system composition and environmental conditions (such as temperature). For example, the change in viscosity when water is titrated into a surfactant–oil mixture with a constant surfactant-to-oil ratio (SOR = 0.5) is shown in Figure 5.9 (Mayer et al. 2013). A large increase in viscosity (and gel-like behavior) is observed at intermediate water contents,

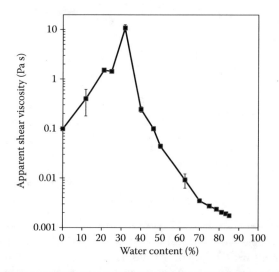

**Figure 5.9   Change in the apparent shear viscosity of a surfactant–oil–water system with changing water content at a constant surfactant-to-oil ratio (SOR = 0.5).**

which may have important implications for the practical application of this type of system.

### 5.3.3.3 Stability

Micelles and microemulsions are thermodynamically stable systems (Figure 5.8), and therefore once they have been formed, they should remain stable indefinitely provided (i) there is no chemical degradation of the components during storage (e.g., surfactant hydrolysis), (ii) there is no change in the overall chemical composition of the system (e.g., owing to dilution, salt addition, alcohol addition, pH adjustment, etc.), and (iii) there is no appreciable change in environmental conditions (such as temperature or pressure). In other words, they should be stable to gravitational separation, particle aggregation, Ostwald ripening, and phase separation (unlike many other kinds of colloidal delivery systems) as long as the initial conditions are not changed appreciably. Nevertheless, micelle- and microemulsion-based delivery systems are often prepared in a concentrated form (consisting of mainly surfactant and oil) and then diluted with water prior to application, and so it is important to ensure that the particles remain stable after the dilution process (Garti et al. 2004; Spernath et al. 2002). This can be established empirically or by using phase diagrams that describe the phase changes that occur when the surfactant, oil, and water concentrations are altered (Figure 5.2).

Many micelle and microemulsion systems become unstable when they are heated because the surfactant head groups become dehydrated, which changes the

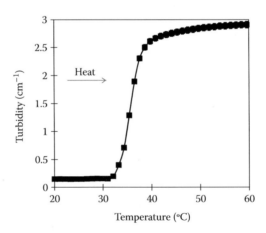

**Figure 5.10 Change in the turbidity of a surfactant–oil–water system with increasing temperature. A system existed as a transparent nanoemulsion at low temperatures but underwent coalescence as the nonionic head groups were dehydrated, leading to an increase in turbidity.**

optimum curvature and solubility of the surfactant (Israelachvili 2011; Salager et al. 2003). Consequently, one may observe an increase in turbidity when the system is heated above a particular temperature, which may be undesirable for some commercial applications. This phenomenon should be reversible for a true micelle or microemulsion system but may be irreversible if the system is trapped in a metastable state after heating. The susceptibility of a colloidal delivery system to this kind of thermal change can be established using temperature-scanning experiments (Figure 5.10).

### 5.3.3.4 Flavor

There are often practical limits associated with the use of high levels of surfactants in food and beverages related to their undesirable flavor profiles. High levels of surfactants may lead to undesirable off-flavor, astringency, or bitterness in a commercial product. Consequently, micelles and microemulsions are most suitable for applications where only relatively low levels of a lipophilic active ingredient need to be incorporated into a food or beverage product. In addition, micelles and microemulsions may indirectly alter the flavor profile of a product owing to partitioning of nonpolar flavor molecules into their hydrophobic interiors, thereby altering the distribution of flavor molecules between the headspace and product (Lloyd et al. 2011; McClements 2005).

## 5.3.4 Applications

In this section, we briefly highlight some of the potential applications of micelles and microemulsions as colloidal delivery systems for active ingredients. These delivery systems are most suitable for encapsulating and delivering lipophilic active ingredients that cannot normally be incorporated into aqueous-based food and beverage products, for example, hydrophobic nutraceuticals, vitamins, colors, flavors, and antimicrobials.

A major advantage of micelle and microemulsion systems is their ease of preparation. They can usually be prepared by simple mixing or heating without the need for any expensive or specialized equipment (such as colloid mills, high-pressure homogenizers, or microfluidizers). Commercially, it is often more convenient to transport materials in a concentrated form, so as to reduce transport costs and the need for preparation equipment at all production facilities. For this reason, this type of surfactant-based delivery system is often prepared in a concentrated form that contains the active ingredient and is then diluted with water after it reaches the final production facility. This not only greatly simplifies the ease of handling and utilization of the ingredient but also reduces transport and storage costs. This approach is widely used for the production and transport of nonpolar flavor ingredients that are going to be incorporated into soft drinks and other beverages.

Another important advantage of micelles and microemulsions is that they can be formulated to be optically transparent and so they can be used in applications where optical clarity is essential, such as fortified waters or soft drinks. The incorporation of lipophilic active ingredients into surfactant-based delivery systems may also be beneficial because it may increase their oral bioavailability after ingestion (McClements and Xiao 2012). The small colloidal particles present in these systems may interact with the biological surfactants in the gastrointestinal tract (GIT) (such as bile salts and phospholipids) and form mixed micelles that can easily be absorbed (Rozner et al. 2010). There are a number of ingredient suppliers who utilize micelles or microemulsions to encapsulate and deliver lipophilic ingredients for use in the food, cosmetic, personal care, and other industries for these reasons. For example, AquaNova is a German company that uses this kind of colloidal delivery system to encapsulate a wide variety of lipophilic active agents, including oil-soluble vitamins, nutraceuticals, antioxidants, colors, and flavors (www.aquanova.de).

## 5.4 Liposome Delivery Systems

The potential applications of liposomes as delivery systems for various kinds of active food ingredients have been reviewed in detail elsewhere (Mozafari et al. 2008; Singh et al. 2012; Taylor et al. 2005), and so only a brief overview of their formation and properties is given here. Although there are many potential applications of liposomes as delivery systems in the food industry, there are a number of challenges that currently limit their widespread utilization. First, it is difficult to produce them reliably and economically on a large scale. Second, they tend to have poor stability under the complex environments found in many food products. Third, they usually have a poor loading capacity for hydrophilic bioactive components. Nevertheless, advances in ingredient technology, processing operations, and stabilization mechanisms may lead to their more widespread utilization in the future.

### 5.4.1 Composition and Structure

Liposomes are typically fabricated from surface-active substances that have intermediate HLB numbers and optimum curvatures close to zero, such as phospholipids (Israelachvili 2011). The most common sources of phospholipids in foods are those made from egg, soy, sunflower, or milk lecithin (Mozafari et al. 2008; Singh et al. 2012; Taylor et al. 2005). Commercial lecithins contain a mixture of different kinds of phospholipids as well as various other components, such as triglycerides, sterols, and fatty acids (Bueschelberger 2004). The composition of a lecithin ingredient plays a large role in determining the formation, stability, and properties of liposome-based delivery systems. For example, the nature of the phospholipid head groups will determine the electrical characteristics of the liposomes, whereas the nature of the tail groups (chain length and unsaturation) will determine the fluidity, permeability, and solubilization capacity of the lipid bilayers. The type and amount of sterols present will also influence the fluidity and permeability of the lipid bilayers; for example, addition of cholesterol can cause a large increase in membrane rigidity.

Structurally, liposomes consist of single or multiple bilayers, with each bilayer consisting of two layers of surfactant molecules with their nonpolar tail groups facing toward each other (Sharma and Sharma 1997; Torchilin and Weissig 2003). Liposomes are frequently classified according to their structures (Figure 5.11). A unilamellar liposome has a balloon-like structure consisting of a single bilayer, whereas a multilamellar liposome has an onion-like structure consisting of multiple concentric bilayer shells (Reineccius 1995; Taylor et al. 2005). Unilamellar liposomes may be further classified as small unilamellar vesicles (SUV) ($r < 100$ nm) or large unilamellar vesicles ($r > 100$ nm). A multivesicular vesicle consists of a number of smaller vesicles trapped within a larger vesicle.

The electrical characteristics of liposomes depend on the type and amount of phospholipids they contain (Singh et al. 2012). For example, the head group of phosphatidic acid is entirely negative, whereas the head groups of PE, PC, and phosphatidylserine have both positive and negative charges. The ionization of the

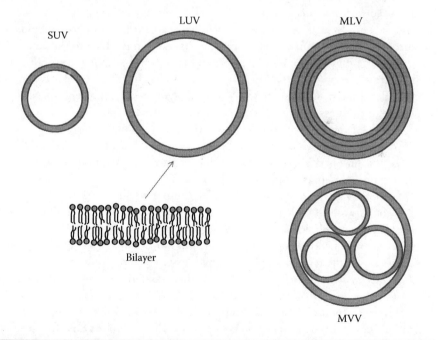

**Figure 5.11 A variety of liposome structures can be formed, including small unilamellar vesicles (SUV), large unilamellar vesicles (LUV), multilamellar vesicles (MLV), and multivesicular vesicles (MVV).**

charged groups also changes with pH, with the charge on the zwitterionic phospholipids going from negative at high pH to positive at low pH. The fact that the head groups are charged means that delivery systems containing them are often sensitive to changes in ionic composition. In addition, their charge can be used to create more complex delivery systems, for example, liposomes coated with a layer of oppositely charged biopolymer (Laye et al. 2008).

## 5.4.2 Formation

### 5.4.2.1 Vesicle Fabrication

It is not usually possible to form liposomes by simply dispersing phospholipids into aqueous solutions using simple mixing procedures, even though bilayers should form spontaneously. Instead, specialized preparation procedures are usually needed to form liposomes with the desired characteristics, for example, morphology, size, loading capacity, and encapsulation efficiency (Maherani et al. 2011a,b; Torchilin and Weissig 2003). In general, these preparation procedures can be classified into a number of different groups. Some of the most important groups of preparation procedures are outlined below.

*Solvent Evaporation/Rehydration*: In this group of methods, the phospholipids are dissolved within an organic solvent first. This mixture is then placed into a container and the solvent is evaporated, which leaves a thin film of phospholipids on the surface of the container. Finally, an aqueous solution is added to the container that causes the phospholipid layers to peel off and form liposomes. Typically, this method creates relatively large multilamellar vesicles, but they can be converted into smaller unilamellar vesicles (SUVs) by applying mechanical energy (such as sonication, high-pressure homogenization, or microfluidization).

*Solvent Displacement*: In this method, the phospholipids are dissolved in an amphiphilic organic solvent (such as ethanol), and then this mixture is injected into an aqueous solution. The organic solvent moves into the aqueous solution and the remaining phospholipids assemble into liposomes.

*Surfactant Displacement*: This method is similar to the solvent displacement method, but a water-soluble surfactant is used rather than an amphiphilic organic solvent. The phospholipids are mixed with the surfactant, and then this mixture is injected into an aqueous solution. The surfactant moves into the surrounding aqueous solution and the remaining phospholipids assemble into liposomes.

*Homogenization*: In these methods, the phospholipids and aqueous phase are mixed together to form a crude suspension, which is then homogenized using a high-energy method, such as sonication, high-pressure valve homogenization, microfluidization, or membrane homogenization. These methods can often be used to form SUVs. The size of the vesicles formed can be reduced by increasing the homogenization intensity or duration, for example, by increasing homogenization pressure and number of passes in a microfluidizer.

## 5.4.2.2 Loading

There are a number of possibilities for loading active ingredients into liposomes, which depend on the nature of the active ingredients and preparation methods used (Maherani et al. 2011a,b; Torchilin and Weissig 2003). In general, liposome loading approaches can be classified as either passive or active:

*Passive Loading*: In passive loading, the active ingredients are simply mixed with the liposomes (either before or after they have been formed). For hydrophilic active ingredients, the encapsulation efficiency for passive loading is simply determined by the volume of the internal aqueous phase relative to the total volume of the aqueous phase in the overall system: EE% = 100 × $V_{Internal}/V_{Total}$. This equation assumes that no hydrophilic active ingredient is incorporated into the lipid membranes. Typically, the loading efficiency for hydrophilic active ingredients is considerably less than 50% because of challenges associated with creating high internal aqueous phase volumes. For lipophilic active ingredients, the encapsulation efficiency depends on the solubilization capacity of the lipid membranes, which depends on the total amount and type of phospholipids present. The encapsulation efficiency of lipophilic active ingredients is therefore often relatively high (>90%) since most of them are incorporated into the lipid membranes. The incorporation of active ingredients into liposomes is sometimes achieved by adding them to a liposome solution and then freeze-drying the entire system. The lyophilized system is then rehydrated, which leads to the incorporation of the bioactive ingredients into the liposomes.

*Active Loading*: In the case of active loading, the system is designed so that a specific physicochemical mechanism drives the active ingredient into the liposome structure, such as electrostatic attraction or directed mass transfer. An electrically charged active ingredient may be loaded into an oppositely charged liposome through electrostatic attraction; for example, a positive active ingredient may be incorporated into an anionic liposome. The directed mass transfer method can only be used for weak acids and bases. The lipid membrane has a low permeability for the charged form of these molecules but has a high permeability for the uncharged form. A liposome solution can be prepared using a particular buffer so that the active ingredient is initially in its charged form. The pH of the external aqueous phase can then be adjusted so that the active ingredient is in its uncharged form and diffuses through the lipid bilayer and into the internal aqueous phase where it converts back into the charged form and is therefore trapped. Relatively high encapsulation efficiencies (>90%) can be achieved using active loading mechanisms, but the types of ingredient that can be encapsulated by this mechanism are relatively limited (e.g., weak acids and bases).

## 5.4.3 Properties

### 5.4.3.1 Optical Properties

The size of the colloidal particles in liposome suspensions may be appreciably smaller ($d \ll \lambda$) or comparable ($d \approx \lambda$) to the wavelength of light. Suspensions containing relatively small liposomes do not scatter light strongly and appear optically transparent, whereas those containing larger ones appear turbid or cloudy. The optical properties of liposomes also depend on the thickness and number of bilayers present (Khlebtsov et al. 2001). A unilamellar liposome should scatter light more weakly than a multilamellar one.

### 5.4.3.2 Rheological Properties

Liposome suspensions have rheological properties similar to other types of colloidal dispersions. The viscosity initially increases slowly with increasing particle concentration but then increases more steeply as the liposomes become more closely packed together. The rheological properties of liposome suspensions can be described by similar equations as those used to describe other colloidal dispersions (McClements 2005; Pal 2001):

$$\frac{\eta}{\eta_1} = \left( 1 - \frac{\phi_{\text{eff}}}{\phi_c} \right)^{-2} \tag{3.12}$$

Here, $\eta_1$ is the viscosity of the continuous phase, $\phi_{\text{eff}}$ is the effective volume fraction of the liposome particles, and $\phi_c$ is the critical packing parameter ($\approx 0.65$). In this case, the effective volume fraction is given by the expression $\phi_{\text{eff}} = \phi R_V$, where $\phi$ is the volume fraction of the phospholipid molecules themselves and $R_V$ is the volume ratio (= effective particle volume/volume of phospholipid molecules). For liposomes, the volume ratio may be considerably greater than unity because the effective particle volume is much larger than the volume of phospholipid molecules caused by the water trapped inside the vesicles. Thus, a liposome suspension containing a certain amount of phospholipids may be much more viscous than an O/W emulsion containing the same amount of oil. A liposome suspension may also be more prone to shear thinning behavior because the structure of the vesicles may be disrupted at high shear rates (Danker et al. 2008).

### 5.4.3.3 Stability

One of the major factors limiting the practical application of liposomes in commercial food and beverage products is their physical and chemical instability (Maherani et al. 2011a,b; Mozafari et al. 2008; Singh et al. 2012; Taylor et al. 2005). Although the formation of lipid bilayers in aqueous solutions is thermodynamically favorable, the formation of colloidal particles with particular characteristics is not. As a result, the composition, dimensions, and structure of liposomes may change considerably during storage. Liposomes are sensitive to flocculation or fusion when solution or environmental conditions are altered, such as pH, ionic strength, temperature, or solvent quality. Liposomes are partly prevented from aggregating because of the electrostatic repulsion acting between them, and so any factors that reduce this repulsive interaction can promote aggregation. For example, altering the pH so the liposomes lose their charge or adding salts to increase the ionic strength may lead to liposome instability. Studies have shown that the physical stability of liposomes to aggregation can often be improved by coating them with biopolymers, such as cationic chitosan (Laye et al. 2008).

Heating a liposome solution may also promote aggregation or lead to leakage of any encapsulated components owing to changes in membrane permeability and structure (Taylor et al. 2005). The bilayers in liposomes typically have a characteristic thermal transition temperature ($T_C$), below which they are more solid-like and above which they are more fluid-like. The permeability of the membranes is often highest around this transition temperature because of the more disorganized packing of the phospholipids, which may lead to leakage of any encapsulated components. This may be an advantage for triggered release applications but a disadvantage for long-term storage at elevated temperatures. The thermal transition temperature is typically higher for saturated than unsaturated phospholipid tails and increases when cholesterol or other sterols are incorporated into the bilayers.

Liposomes have a different density than the surrounding aqueous phase and are therefore susceptible to gravitational separation, with the rate of separation increasing as their size increases.

Liposomes may also break down during long-term storage because of various chemical degradation reactions (Maherani et al. 2011a,b). Chemical or enzymatic reactions may lead to hydrolysis of the phospholipid molecules leading to the formation of lysolecithin and fatty acids. As a result, the optimum curvature of the surfactant monolayers changes, which will change the properties and stability of the liposomes. The lecithin ingredients used to form liposomes may contain unsaturated fatty acids as part of the phospholipid tails, which are susceptible to lipid oxidation. Lipid oxidation involves a complex series of degradation reactions that eventually lead to the formation of undesirable volatile reaction products that are perceived as "rancidity." Other components within liposomes may also be susceptible to chemical degradation reactions, such as sterols and encapsulated active ingredients. The chemical degradation of liposomes can be inhibited in a number of ways, including the following: avoiding elevated temperatures, high oxygen levels, and exposure to light; controlling solution pH; removing or deactivating prooxidants; and adding antioxidants (McClements and Decker 2000; Waraho et al. 2011).

One of the most important factors determining the shelf life of liposome delivery systems is their ability to retain any encapsulated ingredients inside. Some active ingredients have relatively high solubilities in both oil and water, and can therefore diffuse out of the internal water phase, through the lipid membrane, and into the external aqueous phase. This process can be retarded by controlling the permeability of the lipid membrane to molecular transport. Membrane permeability depends strongly on the composition of the lipid bilayers and the temperature of the system (Singh et al. 2012). Phospholipids with longer and more saturated tails tend to be more rigid and therefore less permeable than those with shorter and more unsaturated tails. The presence of cholesterol within the lipid bilayers also makes them more rigid and less permeable. The utilization of cholesterol to form delivery systems suitable for food and beverage applications is usually undesirable because of the potential adverse health effects associated with this lipid. However, other sterols (such as plant sterols) can be used instead of cholesterol to decrease lipid permeability (Singh et al. 2012).

### 5.4.3.4 Flavor

Liposomes are made from large molecules (phospholipids) that are already commonly found in many foods and do not have adverse flavor profiles. Nevertheless, the chemical degradation of unsaturated phospholipids owing to lipid oxidation may lead to the formation of rancid off-flavors. In addition, liposomes may indirectly influence the flavor profile of a food or beverage by altering the partitioning of volatile and nonvolatile flavor molecules in the system. For example, nonpolar flavor molecules may partition into the lipid membranes, whereas polar ones may partition into the internal aqueous core. Indeed, studies have shown that liposomes may be useful for controlling flavor release, or for masking undesirable flavors by trapping them within their internal structures (see below).

### 5.4.3.5 Release Characteristics

Any encapsulated components within a liposome may be released through a number of different physicochemical mechanisms depending on the nature of the system (Maherani et al. 2011a,b; Taylor et al. 2005; Torchilin and Weissig 2003). Encapsulated active ingredients may be released from the aqueous core or from the lipid membrane into the surrounding aqueous phase by simple molecular diffusion driven by a concentration gradient. Simple diffusion may occur when a liposome solution is diluted with the aqueous phase (e.g., saliva) or when an active ingredient is absorbed from the external aqueous phase by some mechanism. Alternatively, an active ingredient may be released by a specific trigger associated with changes in environmental conditions, such as pH, ionic strength, solvent quality, surface-active substances, enzyme activity, or temperature. A liposome may dissociate when it is mixed with bile salts in the small intestine, thereby releasing any encapsulated components. The membrane permeability of liposome particles increases when the temperature is raised to the thermal transition temperature, leading to release of encapsulated components.

## 5.4.4 Applications

In this section, some of the potential applications of liposome-based delivery systems are highlighted below. More detailed discussions have been given in various review articles (Maherani et al. 2011a,b; Mozafari et al. 2008; Singh et al. 2012; Taylor et al. 2005).

### 5.4.4.1 Enzymes

Enzymes are often used in the food industry as processing aids to alter the flavor profile, texture, stability, or nutritional properties of foods (Whitaker et al. 2002). However, enzymes are often susceptible to chemical or biochemical degradation within a food or after ingestion. Most enzymes are hydrophilic or amphiphilic molecules that are primarily water soluble. These enzymes can therefore be trapped within the internal aqueous core of liposomes, where they can be protected from the external aqueous phase by the lipid membrane (Figure 5.12). Enzyme encapsulation may have a number of potential advantages for commercial applications, including enhancing their handling, increasing their activity, improving their stability, and controlling their release (Liu et al. 2013). Studies have shown that liposomes can be used to improve enzyme functionality in cheese ripening and lactose removal from dairy products (Elsoda and Pandian 1991; Gibbs et al. 1999;

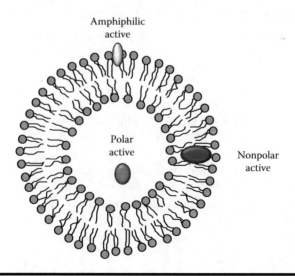

Amphiphilic active

Polar active

Nonpolar active

**Figure 5.12 An active component may be present in various regions within a liposome structure depending on its polarity.**

Kheadr et al. 2003; Kim et al. 1999; Singh et al. 2012). They have also been shown to reduce the loss of activity of enzymes within simulated gastric conditions (Hsieh et al. 2002).

### 5.4.4.2 Nutraceuticals and Minerals

Nutraceuticals are bioactive components present in foods that have specific health benefits over and above their normal nutritional role (i.e., as an energy source or a building block). In particular, many nutraceuticals have been claimed to be capable of inhibiting chronic diseases (such as obesity, heart disease, diabetes, and cancer) or improving human performance (such as attention, energy levels, and stamina). Nutraceuticals may be polar, nonpolar, or amphiphilic molecules (Chapter 2). One of the potential major advantages of liposomes is that they can be used to encapsulate all of these different types of active ingredients within their structures (Hashida et al. 2005; Maherani et al. 2011a,b; Singh et al. 2012). Polar nutraceuticals can be incorporated into the internal aqueous core, nonpolar nutraceuticals can be incorporated within the nonpolar regions of the bilayers, and amphiphilic nutraceuticals can be encapsulated between bilayers (Figure 5.12). Studies have shown that a variety of nonpolar (e.g., vitamins A, D, and E; carotenoids; and Coenzyme Q10) and polar (e.g., vitamin C, iron, and calcium) nutraceuticals can be incorporated into liposomes (Singh et al. 2012; Tan et al. 2013; Xia et al. 2006). Many of these studies have shown that encapsulation may be used to improve their handling, dispersion, stability, and bioavailability.

### 5.4.4.3 Controlling Biological Fate

A number of bioactive components may be degraded during their passage through the GIT and therefore need to be protected. For example, many peptides or proteins may be denatured or hydrolyzed within the stomach because of the highly acidic conditions and protease activity (Iwanaga et al. 1999). These water-soluble components can be trapped within the internal core of a liposome and therefore may be protected during their passage through the stomach (Fricker et al. 2010; Hsieh et al. 2002; Niu et al. 2011). In this case, the system will have to be designed so that the liposome remains intact in the food, mouth, and stomach but falls apart in the small intestine and releases the active ingredient so that it can be absorbed. Liposomes must be carefully formulated to ensure that they are resistant to dissociation within the stomach, for example, by increasing bilayer rigidity or coating them with biopolymer layers.

### 5.4.4.4 Taste Masking

Some active ingredients may give an undesirable flavor profile in the mouth, such as some water-soluble peptides, proteins, and minerals. These water-soluble components can be trapped within the internal aqueous phase of liposomes so they are prevented from being released into the mouth (Fricker et al. 2010). In this case, the liposome should be designed so that it remains intact in the food product and mouth but breaks down after swallowing so that the active agent is released within the GIT.

### 5.4.4.5 Inhibiting Physicochemical Reactions

A food or beverage product may contain a number of different functional ingredients that can adversely interact with each other. For example, multivalent mineral ions (such as calcium or iron) can promote precipitation of some proteins (Pitkowski et al. 2009), whereas transition metals (such as iron or copper) can promote oxidation of some proteins and lipids (Elias et al. 2008; Waraho et al. 2011). In this case, it may be necessary to isolate the different reactive components from each other to avoid undesirable chemical changes in the system. This may be achieved by having one of the components distributed within the internal aqueous core, whereas the other component is distributed within the external aqueous phase. In this case, the different types of molecules should not diffuse across the lipid membrane during storage, which may require control of its permeability.

## 5.5 Summary

Surfactants are the key building blocks for constructing micelle-, microemulsion-, and liposome-based delivery systems suitable for use in the food industry. They are also important elements of many of the emulsion-based systems discussed in the following chapter. Micelles and microemulsions are relatively easy to prepare using simple and inexpensive processing methods. They are primarily suitable for encapsulating nonpolar or amphiphilic active ingredients in their internal hydrophobic cores. Micelles and microemulsions are both thermodynamically favorable systems that should remain stable indefinitely provided that solution and environmental conditions are not changed appreciably or chemical degradation of the structural components does not occur. A major advantage of this type of delivery system is that it can be formulated to be optically transparent and therefore incorporated into food and beverage products that should be clear, such as many soft drinks and beverages. The major disadvantages of these systems are that they usually have low loading capacities and cannot be used at high levels in foods because of concerns with the flavor profile, cost, toxicity, and legal status of many of the surfactants used to fabricate them. Colloidal delivery systems based on micelles and microemulsions are already widely used in the food industry, particularly to incorporate lipophilic flavors, colors, and nutraceuticals into soft drinks and beverages. These products are particularly suitable for this kind of delivery system because they only require very low levels of surfactant to encapsulate the active ingredients.

Liposomes are more difficult to prepare than micelles and microemulsions and usually require the use of specialized processing operations (such as solvent or surfactant displacement) or equipment (such as homogenizers). The formation of bilayers in solution is thermodynamically favorable, but the formation of liposomes with a specific size and structure is not. Consequently, liposomes may break down over time owing to mechanisms such as flocculation, fusion, gravitational separation, and release of active ingredients if they are not formulated appropriately. Liposomes may also be fabricated from components that are susceptible to chemical degradation, such as oxidation and hydrolysis of phospholipids. Liposome delivery systems must therefore be carefully designed to avoid these physical and chemical instability mechanisms. Liposomes are prepared from natural surfactants (phospholipids) that can be incorporated into food products at higher levels than synthetic surfactants. In addition, there is considerable scope in designing novel functional performance into these systems: encapsulating both lipophilic and hydrophilic active agents in the same system, masking off-flavors, controlled or targeted release, and separation of chemically reactive components. Nevertheless, their widespread use in the food industry is currently limited because of the difficulties associated with economically manufacturing them and their poor physical stability under typical food conditions. Research is ongoing into developing cost-effective liposome-based delivery systems suitable for utilization in commercial food and beverage products.

# References

Becher, P. (1966). *Emulsions: Theory and Practice*. New York, Reinhold.

Binks, B. P. (1993). "Relationship between microemulsion phase-behavior and macroemulsion type in systems containing nonionic surfactant." *Langmuir* **9**(1): 25–28.

Bouton, F., M. Durand, V. Nardello-Rataj, A. P. Borosy, C. Quellet and J. M. Aubry (2010). "A QSPR model for the prediction of the 'fish-tail' temperature of CiE4/water/polar hydrocarbon oil systems." *Langmuir* **26**(11): 7962–7970.

Bouton, F., M. Durand, V. Nardello-Rataj, M. Serry and J. M. Aubry (2009). "Classification of terpene oils using the fish diagrams and the Equivalent Alkane Carbon (EACN) scale." *Colloids and Surfaces A-Physicochemical and Engineering Aspects* **338**(1–3): 142–147.

Boutte, T. and L. Skogerson (2004). Stearoyl-2-lactylates and oleoyl lactylates. In *Emulsifiers in Food Technology*, R. J. Whitehurst (ed.). Oxford, UK, Blackwell Publishing: 206–225.

Brodskaya, E. N. (2012). "Computer simulations of micellar systems." *Colloid Journal* **74**(2): 154–171.

Brooks, B. W., H. N. Richmond and M. Zerfa (1998). Phase inversion and drop formation in agitated liquid-liquid dispersions in the presence of non-ionic surfactants. In *Modern Aspects of Emulsion Science*, B. P. Binks (ed.). Cambridge, UK, Royal Society of Chemistry: 173–204.

Bueschelberger, H. G. (2004). Lecithins. In *Emulsifiers in Food Technology*, R. J. Whitehurst (ed.). Oxford, UK, Blackwell Publishing: 1–39.

Cottrell, T. and J. van Peij (2004). Sorbitan esters and polysorbates. In *Emulsifiers in Food Technology*, R. J. Whitehurst (ed.). Oxford, UK, Blackwell Publishing: 162–165.

Danker, G., C. Verdier and C. Misbah (2008). "Rheology and dynamics of vesicle suspension in comparison with droplet emulsion." *Journal of Non-Newtonian Fluid Mechanics* **152**(1–3): 156–167.

Elias, R. J., S. S. Kellerby and E. A. Decker (2008). "Antioxidant activity of proteins and peptides." *Critical Reviews in Food Science and Nutrition* **48**(5): 430–441.

Elsoda, M. and S. Pandian (1991). "Recent developments in accelerated cheese ripening." *Journal of Dairy Science* **74**(7): 2317–2335.

Evans, D. F. and H. Wennerstrom (1999). *The Colloidal Domain: Where Physics, Chemistry, Biolology and Technology Meet.* New York, VCH Publishers, Inc.

Flanagan, J. and H. Singh (2006). "Microemulsions: A potential delivery system for bioactives in food." *Critical Reviews in Food Science and Nutrition* **46**(3): 221–237.

Friberg, S., K. Larsson and J. Sjoblom (2004). *Food Emulsions.* New York, Marcel Dekker.

Fricker, G., T. Kromp, A. Wendel, A. Blume, J. Zirkel, H. Rebmann, C. Setzer, R. O. Quinkert, F. Martin and C. Muller-Goymann (2010). "Phospholipids and lipid-based formulations in oral drug delivery." *Pharmaceutical Research* **27**(8): 1469–1486.

Garti, N. and A. Aserin (2012). Micelles and microemulsions as food ingredient and nutraceutical delivery systems. In *Encapsulation Technologies and Delivery Systems for Food Ingredients and Nutraceuticals*, N. Garti and D. J. McClements (eds.). Oxford, UK, Woodhead Publishing: 211–251.

Garti, N., I. Zakharia, A. Spernath, A. Yaghmur, A. Aserin, R. E. Hoffman and L. Jacobs (2004). Solubilization of water-insoluble nutraceuticals in nonionic microemulsions for water-based use. In *Trends in Colloid and Interface Science XVII*, V. Cabuil, P. Levitz and C. Treiner (eds.) New York, Springer: **126**: 184–189.

Gaupp, R. and W. Adams (2004). Acid esters of mono- and diglycerides. In *Emulsifiers in Food Technology*, R. J. Whitehurst (ed.). Oxford, UK, Blackwell Publishing: 59–85.

Gibbs, B. F., S. Kermasha, I. Alli and C. N. Mulligan (1999). "Encapsulation in the food industry: A review." *International Journal of Food Sciences and Nutrition* **50**(3): 213–224.

Hasenhuettl, G. L. and R. W. Hartel (2010). *Food Emulsifiers and Their Applications.* New York, Springer Science.

Hashida, M., S. Kawakami and F. Yamashita (2005). "Lipid carrier systems for targeted drug and gene delivery." *Chemical and Pharmaceutical Bulletin* **53**(8): 871–880.

Hiemenz, P. C. and R. Rajagopalan (1997). *Principles of Colloid and Surface Chemistry.* New York, Marcel Dekker.

Hsieh, Y. F., T. L. Chen, Y. T. Wang, J. H. Chang and H. M. Chang (2002). "Properties of liposomes prepared with various lipids." *Journal of Food Science* **67**(8): 2808–2813.

Huang, Q. R., H. L. Yu and Q. M. Ru (2010). "Bioavailability and delivery of nutraceuticals using nanotechnology." *Journal of Food Science* **75**(1): R50–R57.

Israelachvili, J. (2011). *Intermolecular and Surface Forces*, 3rd Edition. London, Academic Press.

Iwanaga, K., S. Ono, K. Narioka, M. Kakemi, K. Morimoto, S. Yamashita, Y. Namba and N. Oku (1999). "Application of surface coated liposomes for oral delivery of peptide: Effects of coating the liposome's surface on the GI transit of insulin." *Journal of Pharmaceutical Sciences* **88**(2): 248–252.

Kabalnov, A. (1998). "Thermodynamic and theoretical aspects of emulsions and their stability." *Current Opinion in Colloid and Interface Science* **3**(3): 270–275.

Kabalnov, A. and H. Wennerstrom (1996). "Macroemulsion stability: The oriented wedge theory revisited." *Langmuir* **12**(2): 276–292.

Kheadr, E. E., J. C. Vuillemard and S. A. El-Deeb (2003). "Impact of liposome-encapsulated enzyme cocktails on cheddar cheese ripening." *Food Research International* **36**(3): 241–252.

Khlebtsov, N. G., L. A. Kovler, S. V. Zagirova, B. N. Khlebtsov and V. A. Bogatyrev (2001). "Spectroturbidimetry of liposome suspensions." *Colloid Journal* **63**(4): 491–498.

Kim, C. K., H. S. Chung, M. K. Lee, L. N. Choi and M. H. Kim (1999). "Development of dried liposomes containing beta-galactosidase for the digestion of lactose in milk." *International Journal of Pharmaceutics* **183**(2): 185–193.

Kralova, I. and J. Sjoblom (2009). "Surfactants used in food industry: A review." *Journal of Dispersion Science and Technology* **30**(9): 1363–1383.

Laye, C., D. J. McClements and J. Weiss (2008). "Formation of biopolymer-coated liposomes by electrostatic deposition of chitosan." *Journal of Food Science* **73**(5): N7–N15.

Leal-Calderon, F., V. Schmitt and J. Bibette (2007). *Emulsion Science: Basic Principles*. Berlin, Springer Verlag.

Leser, M. E., L. Sagalowicz, M. Michel and H. J. Watzke (2006). "Self-assembly of polar food lipids." *Advances in Colloid and Interface Science* **123**: 125–136.

Liu, W. L., A. Q. Ye, W. Liu, C. M. Liu and H. Singh (2013). "Liposomes as food ingredients and nutraceutical delivery systems." *Agro Food Industry Hi-Tech* **24**(2): 68–71.

Lloyd, N. W., E. Kardaras, S. E. Ebeler and S. R. Dungan (2011). "Measuring local equilibrium flavor distributions in SDS solution using headspace solid-phase microextraction." *Journal of Physical Chemistry B* **115**(49): 14484–14492.

Maherani, B., E. Arab-Tehrany and M. Linder (2011a). "Mechanism of bioactive transfer through liposomal bilayers." *Current Drug Targets* **12**(4): 531–545.

Maherani, B., E. Arab-Tehrany, M. R. Mozafari, C. Gaiani and M. Linder (2011b). "Liposomes: A review of manufacturing techniques and targeting strategies." *Current Nanoscience* **7**(3): 436–452.

Marze, S. (2013). "Bioaccessibility of nutrients and micronutrients from dispersed food systems: Impact of the multiscale bulk and interfacial structures." *Critical Reviews in Food Science and Nutrition* **53**(1): 76–108.

Mayer, S., J. Weiss and D. J. McClements (2013). "Vitamin E-enriched nanoemulsions formed by emulsion phase inversion: Factors influencing droplet size and stability." *Journal of Colloid and Interface Science* **402**: 122–130.

McClements, D. J. (2005). *Food Emulsions: Principles, Practice, and Techniques*. Boca Raton, FL, CRC Press.

McClements, D. J. (2011). "Edible nanoemulsions: Fabrication, properties, and functional performance." *Soft Matter* **7**(6): 2297–2316.

McClements, D. J. and E. A. Decker (2000). "Lipid oxidation in oil-in-water emulsions: Impact of molecular environment on chemical reactions in heterogeneous food systems." *Journal of Food Science* **65**(8): 1270–1282.

McClements, D. J. and J. Rao (2011). "Food-grade nanoemulsions: Formulation, fabrication, properties, performance, biological fate, and potential toxicity." *Critical Reviews in Food Science and Nutrition* **51**(4): 285–330.

McClements, D. J. and H. Xiao (2012). "Potential biological fate of ingested nanoemulsions: Influence of particle characteristics." *Food and Function* **3**(3): 202–220.

Mira, I., N. Zambrano, E. Tyrode, L. Márquez, A. Peña, A. Pizzino and J. Salager (2003). "Emulsion catastrophic inversion from abnormal to normal morphology. 2. Effect of the stirring intensity on the dynamic inversion frontier." *Industrial and Engineering Chemistry Research* **42**(1): 57–61.

Moonen, H. and H. Bas (2004). Mono- and Di-glycerides. In *Emulsifiers in Food Technology*, R. J. Whitehurst (ed.). Oxford, UK, Blackwell Publishing: 40–58.

Mozafari, M. R., C. Johnson, S. Hatziantoniou and C. Demetzos (2008). "Nanoliposomes and their applications in food nanotechnology." *Journal of Liposome Research* **18**(4): 309–327.

Myers, D. (2006). *Surfactant Science and Technology*. Hoboken, NJ, John Wiley and Sons.

Nelen, B. A. P. and J. M. Cooper (2004). Sucrose esters. In *Emulsifiers in Food Technology*, R. J. Whitehurst (ed.). Oxford, UK, Blackwell Publishing: 131–161.

Niu, M. M., Y. Lu, L. Hovgaard and W. Wu (2011). "Liposomes containing glycocholate as potential oral insulin delivery systems: Preparation, in vitro characterization, and improved protection against enzymatic degradation." *International Journal of Nanomedicine* **6**: 1155–1166.

Norn, V. (2004). Polyglycerol esters. In *Emulsifiers in Food Technology*, R. J. Whitehurst (ed.). Oxford, UK, Blackwell Publishing: 110–130.

Pal, R. (2001). "Novel viscosity equations for emulsions of two immiscible liquids." *Journal of Rheology* **45**(2): 509–520.

Phan, T. T., J. H. Harwell and D. A. Sabatini (2010). "Effects of triglyceride molecular structure on optimum formulation of surfactant-oil-water systems." *Journal of Surfactants and Detergents* **13**(2): 189–194.

Pitkowski, A., T. Nicolai and D. Durand (2009). "Stability of caseinate solutions in the presence of calcium." *Food Hydrocolloids* **23**(4): 1164–1168.

Queste, S., J. L. Salager, R. Strey and J. M. Aubry (2007). "The EACN scale for oil classification revisited thanks to fish diagrams." *Journal of Colloid and Interface Science* **312**(1): 98–107.

Rao, J. J. and D. J. McClements (2011a). "Food-grade microemulsions, nanoemulsions and emulsions: Fabrication from sucrose monopalmitate and lemon oil." *Food Hydrocolloids* **25**(6): 1413–1423.

Rao, J. J. and D. J. McClements (2011b). "Formation of flavor oil microemulsions, nanoemulsions and emulsions: Influence of composition and preparation method." *Journal of Agricultural and Food Chemistry* **59**(9): 5026–5035.

Rao, J. J. and D. J. McClements (2012). "Lemon oil solubilization in mixed surfactant solutions: Rationalizing microemulsion and nanoemulsion formation." *Food Hydrocolloids* **26**(1): 268–276.

Reineccius, G. A. (1995). Liposomes for controlled-release in the food-industry. In *Encapsulation and Controlled Release of Food Ingredients*, S. J. Risch and G. A. Reineccius (eds.) American Chemical Society, Washington. **590**: 113–131.

Rondón-Gonzaléz, M., V. Sadtler, L. Choplin and J. Salager (2006). "Emulsion catastrophic inversion from abnormal to normal morphology. 5. Effect of the water-to-oil ratio and surfactant concentration on the inversion produced by continuous stirring." *Industrial and Engineering Chemistry Research* **45**(9): 3074–3080.

Rosen, M. J. and J. T. Kunjappu (2012). *Surfactants and Interfacial Phenomena*. Hoboken, NJ, John Wliey and Sons.

Rozner, S., D. E. Shalev, A. I. Shames, M. F. Ottaviani, A. Aserin and N. Garti (2010). "Do food microemulsions and dietary mixed micelles interact?" *Colloids and Surfaces B-Biointerfaces* **77**(1): 22–30.

Ruckenstein, E. (1996). "Microemulsions, macroemulsions, and the Bancroft rule." *Langmuir* **12**(26): 6351–6353.

Salager, J., R. Antón, D. Sabatini, J. Harwell, E. Acosta and L. Tolosa (2005). "Enhancing solubilization in microemulsions—State of the art and current trends." *Journal of Surfactants and Detergents* **8**(1): 3–21.

Salager, J. L., R. E. Anton, M. I. Briceno, L. Choplin, L. Marquez, A. Pizzino and M. P. Rodriguez (2003). "The emergence of formulation engineering in emulsion making—Transferring know-how from research laboratory to plant." *Polymer International* **52**(4): 471–478.

Salager, J. L., A. Forgiarini, L. Marquez, A. Pena, A. Pizzino, M. P. Rodriguez and M. Rondo-Gonzalez (2004). "Using emulsion inversion in industrial processes." *Advances in Colloid and Interface Science* **108**: 259–272.

Schick, J. J. (1987). *Non-ionic Surfactants: Physical Chemistry*. Boca Raton, FL, CRC Press.

Sharma, A. and U. S. Sharma (1997). "Liposomes in drug delivery: Progress and limitations." *International Journal of Pharmaceutics* **154**(2): 123–140.

Shinoda, K. and H. Arai (1964). "Correlation between phase inversion temperature in emulsion + cloud point in solution of nonionic emulsifier." *Journal of Physical Chemistry* **68**(12): 3485–3490.

Singh, H., A. Thompson, W. Liu and M. Corredig (2012). Liposomes as food ingredients and nutraceutical delivery systems. In *Encapsulation Technologies and Delivery Systems for Food Ingredients and Nutraceuticals*, N. Garti and D. J. McClements (eds.). Oxford, UK, Woodhead Publishing: 287–318.

Sparso, F. V. and N. Krog (2004). Propylene glycol fatty acid esters. In *Emulsifiers in Food Technology*, R. J. Whitehurst (ed.). Oxford, UK, Blackwell Publishing: 186–205.

Spernath, A. and A. Aserin (2006). "Microemulsions as carriers for drugs and nutraceuticals." *Advances in Colloid and Interface Science* **128**: 47–64.

Spernath, A., A. Yaghmur, A. Aserin, R. E. Hoffman and N. Garti (2002). "Food-grade microemulsions based on nonionic emulsifiers: Media to enhance lycopene solubilization." *Journal of Agricultural and Food Chemistry* **50**(23): 6917–6922.

Stauffer, S. E. (1999). *Emulsifiers*. St Paul, MN, Eagen Press.

Tan, C., S. Q. Xia, J. Xue, J. H. Xie, B. A. Feng and X. M. Zhang (2013). "Liposomes as vehicles for lutein: Preparation, stability, liposomal membrane dynamics, and structure." *Journal of Agricultural and Food Chemistry* **61**(34): 8175–8184.

Taylor, T. M., P. M. Davidson, B. D. Bruce and J. Weiss (2005). "Liposomal nanocapsules in food science and agriculture." *Critical Reviews in Food Science and Nutrition* **45**(7–8): 587–605.

Torchilin, V. P. (2007). "Micellar nanocarriers: Pharmaceutical perspectives." *Pharmaceutical Research* **24**(1): 1–16.

Torchilin, V. P. and W. Weissig (2003). *Liposomes: A Practical Approach*. Oxford, UK, Oxford University Press.

Waraho, T., D. J. McClements and E. A. Decker (2011). "Mechanisms of lipid oxidation in food dispersions." *Trends in Food Science and Technology* **22**(1): 3–13.

Whitaker, J. R., A. G. J. Voragen and D. W. S. Wong (2002). *Handbook of Food Enzymology*. Boca Raton, FL, CRC Press.

Whitehurst, R. J. (2004). *Emulsifiers in Food Technology*. Oxford, UK, Blackwell Publishing.

Witthayapanyanon, A., J. Harwell and D. Sabatini (2008). "Hydrophilic-lipophilic deviation (HLD) method for characterizing conventional and extended surfactants." *Journal of Colloid and Interface Science* **325**(1): 259–266.

Wooster, T., M. Golding and P. Sanguansri (2008). "Impact of oil type on nanoemulsion formation and Ostwald ripening stability." *Langmuir* **24**(22): 12758–12765.

Xia, S. Q., S. Y. Xu and X. M. Zhang (2006). "Optimization in the preparation of coenzyme Q(10) nanoliposomes." *Journal of Agricultural and Food Chemistry* **54**(17): 6358–6366.

Ziani, K., J. A. Barish, D. J. McClements and J. M. Goddard (2011a). "Manipulating interactions between functional colloidal particles and polyethylene surfaces using interfacial engineering." *Journal of Colloid and Interface Science* **360**(1): 31–38.

Ziani, K., Y. H. Chang, L. McLandsborough and D. J. McClements (2011b). "Influence of surfactant charge on antimicrobial efficacy of surfactant-stabilized thyme oil nanoemulsions." *Journal of Agricultural and Food Chemistry* **59**(11): 6247–6255.

# Chapter 6

# Emulsion-Based Delivery Systems

## 6.1 Introduction

Emulsions are one of the most common delivery systems currently used in the food and beverage industry. Conventional emulsions, such as oil-in-water (O/W) and water-in-oil (W/O) emulsions, have been utilized for many years in the preparation of a wide range of food and beverage products (Dickinson 1992; Friberg et al. 2004; McClements 2005a). More recently, there have been considerable advances in the design of more complex structured emulsions that have improved on novel performances (McClements 2010b, 2012a; McClements and Li 2010; McClements et al. 2009). A number of different kinds of these emulsion-based delivery systems are shown schematically in Figure 6.1. In this chapter, we mainly focus on emulsion-based delivery systems suitable for application in aqueous-based products, such as O/W emulsions, although much of the discussion will also be relevant to delivery systems based on W/O emulsions. We begin by discussing the physicochemical characteristics of the major building blocks in emulsion-based delivery systems (i.e., fat droplets) and then discuss the major types of conventional and structured delivery systems that can be assembled from them. Special emphasis is given to the types of ingredients and preparation methods that can be used in their fabrication, the structural basis of their physicochemical properties and stability, and their potential applications within the food industry.

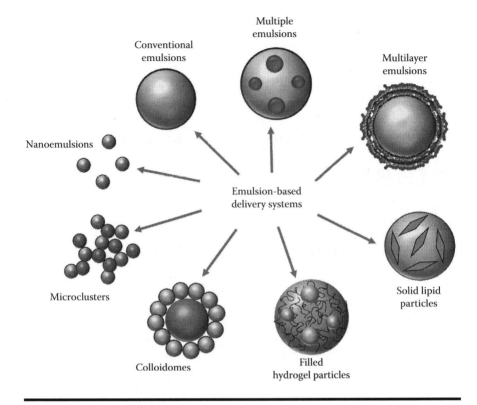

**Figure 6.1** **Examples of emulsion-based delivery systems that can be prepared using fat droplets as a building block.**

## 6.2 Building Blocks: Fat Droplets

In general, an emulsion consists of at least two immiscible liquids, with one of the liquids being dispersed as small spherical droplets in the other. In food applications, these two liquids are usually oil and water, but other types of liquids can also be used (e.g., glycerol instead of water). Emulsion-based systems can conveniently be classified according to the spatial arrangement of the oil and water phases. A system that consists of oil droplets dispersed in water is called an O/W emulsion, whereas a system that consists of water droplets dispersed in oil is called a W/O emulsion. The substance that forms the droplets is usually referred to as the *dispersed phase*, whereas the substance that forms the surrounding liquid is referred to as the *continuous phase*. In addition to conventional O/W and W/O emulsions, it is also possible to prepare various types of *multiple emulsions*, for example, oil-in-water-in-oil (O/W/O) or water-in-oil-in-water (W/O/W) emulsions (Benichou et al. 2004; Garti and Bisperink 1998; van der Graaf et al. 2005). It is even possible to form "emulsions" from two aqueous phases with different compositions, such as water-in-water (W/W) or oil-in-water-in-water (O/W/W) emulsions (Kim et al. 2006; Norton and Frith

2001; Norton et al. 2006). Emulsions are thermodynamically unstable systems that will always revert back to the separate phases given sufficient time. Emulsions may physically break down through numerous processes, including gravitational separation, flocculation, coalescence, and Ostwald ripening (Dickinson 1992; Friberg et al. 2004; McClements 2005a). Consequently, they are different from micelles and microemulsions, which are thermodynamically stable systems. Emulsions that are kinetically stable (metastable) can be formed using substances known as *stabilizers*, for example, emulsifiers, texture modifiers, weighting agents, and ripening inhibitors (McClements and Rao 2011). Emulsifiers are surface-active molecules that adsorb to the surface of the droplets formed during homogenization, where they form a coating that protects the droplets against aggregating. Texture modifiers thicken or gel the continuous phase, which improves emulsion stability by retarding or preventing droplet movement. Weighting agents are dense hydrophobic substances that are added to the oil phase so that its density more closely matches that of the aqueous phase, thereby reducing the driving force for gravitational separation. Ripening inhibitors are hydrophobic substances with very low water solubility that are added to the oil phase to prevent droplet growth through Ostwald ripening. Selection of the most appropriate stabilizer(s) is one of the most important factors determining the shelf life, physicochemical properties, and functional performance of emulsion-based delivery systems. As mentioned earlier, in this chapter, we mainly focus on the properties of emulsions whose continuous phase is aqueous (i.e., O/W, W/O/W, or O/W/W), since these are the most commonly used for the encapsulation and delivery of active ingredients. In addition, we will primarily focus on the formation of the initial liquid systems, rather than any subsequent drying steps. However, it should be noted that many emulsion-based delivery systems are converted into powdered form prior to utilization because this eases their storage, transport, handling, and shelf life (Desai and Park 2005; Drusch and Mannino 2009; Kuang et al. 2010; Madene et al. 2006; Murugesan and Orsat 2012; Vega and Roos 2006). Some of the most common methods used to convert fluid delivery systems into powdered form are discussed in Chapter 4.

The stability, physicochemical properties, and functional performance of emulsion-based delivery systems are highly dependent on the properties of the particles they contain (McClements 2005a, 2012a). Some of the most important particle characteristics that a food or ingredient manufacturer can control during the preparation of emulsion-based delivery systems are highlighted below. A more detailed discussion of particle properties in general is given in Chapter 3.

## 6.2.1 Droplet Concentration

The droplet concentration is usually expressed as the number, mass, or volume of droplets per unit volume or mass of emulsion (McClements 2005a). For example, the disperse phase volume fraction ($\varphi$) is the volume of droplets per unit volume of emulsion. The droplet concentration of an emulsion can usually be controlled by varying the proportions of the two immiscible liquids used to prepare it. Alternatively, an emulsion may be prepared with a particular droplet concentration and then be either diluted (e.g., by adding continuous phase) or concentrated (e.g., by gravitational separation, centrifugation, filtration, or solvent evaporation). A manufacturer therefore normally has good control over the droplet concentration in an emulsion-based delivery system. Typically, the droplet concentration in food products varies from less than 0.1% (e.g., some soft drinks and beverages) to more than 50% (e.g., some dressings and mayonnaise). The droplet concentration has a pronounced influence on the appearance, texture, stability, flavor, and release characteristics of emulsions and should therefore be carefully controlled.

## 6.2.2 Particle Size

The particle size distribution (PSD) of an emulsion represents the fraction of particles in different size classes (McClements 2005a). It is typically represented as either a table or a plot of particle concentration (e.g., volume or number percent) versus droplet dimensions (e.g., diameter or radius). It is often convenient to represent the full PSD by a measure of the central tendency of the distribution (e.g., mean, median, or mode) and a measure of the width of the distribution (e.g., standard deviation or polydispersity index). The PSD of an emulsion can usually be controlled by varying homogenizer type (e.g., high-shear mixer, colloid mill, sonicator, high-pressure valve homogenizer, microfluidizer, or membrane homogenizer), homogenization conditions (e.g., intensity or duration of energy input), or system composition (e.g., the types and concentrations of emulsifier, oil, and water used). Smaller droplets can usually be produced by increasing the intensity or duration of homogenization or by increasing the emulsifier concentration (Jafari et al. 2007, 2008; Qian and McClements 2011).

Typically, the mean diameter of the droplets within food systems is somewhere between 100 nm and 100 μm, but in some systems, it may be either smaller or larger than these values. For example, there have been considerable advances recently in the development of food-grade nanoemulsions using both low- and high-energy methods, and these systems have droplet sizes that range from around 20 to 100 nm (McClements and Rao 2011). The size of the particles in an emulsion-based delivery system has a major impact on its optical properties, rheology, stability, and functional performance, and so it is important for food manufacturers to determine the optimum PSD required for a particular application and to ensure that this is achieved during the manufacturing process and maintained throughout storage.

### 6.2.3 Particle Charge

The electrical characteristics of emulsion droplets are primarily determined by the type, concentration, and organization of any ionized molecular species adsorbed to the droplet surfaces during or after homogenization (McClements 2005a). The electrical properties of particles in colloidal suspensions can be conveniently characterized in terms of their $\zeta$-potential ($\zeta$) versus pH profiles (McClements and Rao 2011). The $\zeta$-potential is the electrical potential at the "shear plane," which is defined as the distance away from the particle surface below which any counterions remain attached to the particle surface as it moves within an applied electrical field (Hunter 1986). The ionized molecular species present at the droplet surfaces may be charged emulsifiers, biopolymers, fatty acids, or mineral ions originally located in either the oil or water phases. The $\zeta$-potential can be conveniently measured using commercially available analytical instruments, such as particle electrophoresis and electroacoustic methods (Hunter 1986; McClements 2005a). The electrical characteristics of emulsion droplets can be controlled by careful selection of emulsifier types and other charged species used to fabricate the system. Droplets stabilized by nonionic surfactants tend to have a small droplet charge (e.g., Tweens and Spans), those stabilized by anionic surfactants have a negative charge (e.g., lecithin, DATEM, CITREM, fatty acids), those stabilized by polysaccharide emulsifiers tend to have a negative charge (e.g., gum arabic, modified starch, and beet pectin), and those stabilized by proteins have a positive charge below the isoelectric point and negative charge above it (e.g., whey protein, casein, soy proteins).

The electrical charge on the droplets in emulsions plays an important role in determining their interactions with other charged materials. Consequently, it is important for ensuring their stability (since electrostatic repulsion is often used to prevent aggregation), their interactions with nonbiological and biological surfaces (such as processing equipment, packaging materials, microbes, and the gastrointestinal tract), and their ability to form complex structures on the basis of electrostatic interactions (such as multilayer emulsions, microclusters, or filled hydrogel particles). It is therefore important for food manufacturers to determine the optimum charge characteristics required for a particular application and to design the system so that it meets these requirements.

## 6.2.4 Interfacial Characteristics

The droplets in emulsions are usually coated by a layer of adsorbed species to protect them from aggregation, for example, emulsifiers (Dickinson 2003; McClements 2005a). The characteristics of the interfacial region are determined by the type, concentration, and interactions of any surface-active species present during homogenization, as well as by the events that occur before, during, and after emulsion formation, for example, competitive adsorption, coadsorption, and complexation (Dickinson 2003). The properties of the interfacial region can be controlled by altering system composition or processing variables. Some of the most important properties that can be controlled are the charge, thickness, composition, density, and permeability of the interfacial region. Controlling these interfacial characteristics is one of the most powerful methods of designing delivery systems with specific functional performances, for example, improved stability, targeted delivery, and controlled release. The interfacial characteristics of emulsion droplets can be controlled by selection of specific emulsifier types. For example, small-molecule surfactants form thin fluid-like layers, globular proteins form thin gel-like layers, and polysaccharides form thick gel-like layers (Dickinson 1992, 2003). In addition, the interfacial properties can often be changed after emulsion formation, for example, by emulsifier exchange, coadsorption, cross-linking, or multilayer formation (Littoz and McClements 2008; McClements and Rao 2011).

## 6.2.5 Physical State

The droplets that make up the dispersed phase of conventional O/W emulsions are usually liquid, but there are a number of emulsion-based delivery systems that contain either partially or fully solidified droplets, for example, solid lipid nanoparticles (SLNs), nanostructured lipid carriers (NLCs), or partly crystalline microparticles (McClements 2005a; Muller and Keck 2004; Walstra 2003; Wissing et al. 2004). The droplets in an O/W emulsion can be made to crystallize by reducing the temperature sufficiently below the melting point of an oil phase (McClements 2012). However, the crystallization temperature may be appreciably less in emulsified oil than bulk oil because of supercooling effects. Alternatively, the droplets in an O/W emulsion can be made to crystallize by removing an organic solvent from the disperse phase that initially contained dissolved crystals. The nature of the crystals formed by an emulsified oil phase may be different from those formed by a bulk oil phase because of curvature effects and the limited volume present in individual emulsion droplets, for example, crystal structure, dimensions, and melting behavior (Muller and Keck 2004; Wissing et al. 2004). The concentration, nature, and location of the fat crystals within the oil droplets in an O/W emulsion can be controlled by careful selection of oil type (e.g., solid fat content vs. temperature profile), thermal history (e.g., temperature vs. time), emulsifier type, and droplet size (McClements 2012b; Muller and Keck 2004; Muller et al. 2000; Walstra 2003).

## 6.2.6 Colloidal Interactions

The stability of fat droplets to aggregation depends on the nature of the attractive and repulsive colloidal interactions that operate between them (Israelachvili 2011; McClements 2005a, 2011). In general, any colloidal interaction can be described by its sign (attractive to repulsive), its magnitude (weak to strong), and its range (short to long). The types of colloidal interactions that are important in a particular emulsion-based delivery system depend on the nature of the particles present (such as their composition, size, charge, and interfacial structure), the properties of the surrounding aqueous solution (such as pH, ionic strength, dielectric constant, and osmolarity), and environmental conditions (such as temperature and pressure). The most common attractive forces are van der Waals, hydrophobic, depletion, and bridging attraction, while the most common repulsive forces are electrostatic and steric (McClements 2005a). Van der Waals forces are long-range attractive interactions that act between all types of particles and will cause a colloidal system to aggregate unless there is a sufficiently large opposing repulsive force. Hydrophobic forces are long-range attractive interactions that act between nonpolar groups on particle surfaces and therefore depend on the surface hydrophobicity of droplets. Depletion attraction occurs because of the presence of nonadsorbed polymers or colloidal particles in the aqueous phase surrounding the oil droplets owing to a steric exclusion effect that generates an osmotic pressure and depends on the dimensions and number concentration of the nonadsorbing substances. Bridging attraction occurs because of the presence of ions, polymers, or colloidal particles that are simultaneously attracted to the surfaces of two or more oil droplets to promote aggregation. Electrostatic repulsion occurs between electrically charged oil droplets and is highly dependent on pH and ionic strength—decreasing as the droplet charge decreases or ionic strength increases. Steric repulsion occurs owing to overlap of polymer groups on the surfaces of different oil droplets and mainly arises because of a reduction in the entropy that occurs when the polymer layers overlap. In general, the stability of an emulsion-based system to aggregation can be predicted by calculating the sum of the various attractive and repulsive interactions as a function of the distance between them (Figure 6.2). The main factors that determine the sign, magnitude, and range of colloidal interactions, as well as their influence on the stability and properties of emulsions, are discussed in more detail in Chapter 10 and elsewhere (Israelachvili 2011; McClements 2005a).

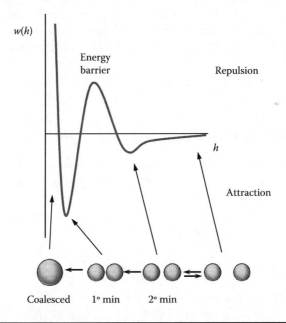

**Figure 6.2  Schematic representation of the interaction potential between two emulsion droplets. The droplets may be stable, weakly flocculated, strongly flocculated, or coalesced depending on the attractive and repulsive forces between them.**

## 6.2.7 Controlling Droplet Characteristics for Improved Performance

Manipulation of the droplet characteristics described in this section can be carried out to create emulsion-based delivery systems with specific functional performances suitable for encapsulating, protecting, and releasing different types of active ingredients into different types of food matrices. For example, some bioactive lipids are susceptible to chemical degradation through oxidative reactions, for example, ω-3 fatty acids, conjugated linoleic acids (CLAs), and carotenoids. In these systems, it is important to protect the bioactive lipids from coming into contact with prooxidants, such as transition metals. Previous studies have shown that this can be achieved by engineering the properties of the interfacial layer surrounding the lipid droplets, for example, by making it cationic so that it repels cationic transition metal ions or by increasing its thickness to form a steric boundary (McClements and Decker 2000). Another problem that has to be overcome for certain bioactive lipids is that they are crystalline at ambient temperatures, for example, phytosterols and carotenoids. Crystalline lipids cause instability problems in conventional emulsions because they promote partial coalescence, that is, a crystal from one droplet penetrates into the liquid region of a another droplet leading to aggregation (McClements 2012b). It may be possible to restrict the degree of partial coalescence in an emulsion-based delivery system containing a crystalline bioactive lipid by increasing the strength of the repulsive interactions operating between the droplets, by increasing the thickness of the adsorbed layer to inhibit crystal penetration, or by crystallizing the carrier oil phase (Coupland 2002; McClements 2012b; Muller and Keck 2004; Thanasukarn et al. 2004; Vanapalli et al. 2002). In general, a wide variety of different strategies are available to tailor emulsion droplet properties for particular applications.

# 6.3 Physicochemical Properties of Emulsions

Any emulsion-based delivery system should be compatible with the specific food matrix that it is going to be incorporated into; that is, it should not adversely affect the appearance, texture, stability, or taste profile of the product. The potential influence of fat droplet characteristics on the physicochemical properties of food products is therefore briefly highlighted below.

## 6.3.1 Appearance

The overall appearance of a food or beverage product is one of the most important factors influencing consumer perception and product acceptability (McClements 2005a). The optical properties of commercial products are primarily determined by their opacity and color, which can be quantitatively described by color coordinates, such as the $L^*a^*b^*$ system (McClements 2002a,b). In this color system, $L^*$ represents the lightness, and $a^*$ and $b^*$ are color coordinates: where $+a^*$ is the red direction, $-a^*$ is the green direction, $+b^*$ is the yellow direction, $-b^*$ is the blue direction, low $L^*$ is dark, and high $L^*$ is light. The opacity of a food product can therefore by characterized by its lightness ($L^*$), while its overall color intensity can be characterized by its chroma: $C = (a^{*2} + b^{*2})^{1/2}$. The addition of fat droplets to a food or beverage product alters its appearance by an amount that depends on their refractive index contrast, concentration, and size (McClements 2002a,b, 2005a). The lightness of a product tends to increase with increasing refractive index contrast and increasing droplet concentration and has a maximum value at an intermediate droplet size (close to the wavelength of visible light). The influence of adding increasing amounts of oil droplets to an aqueous phase on the overall optical properties of the system is shown in Figure 6.3: the lightness increases steeply as the oil content is increased from 0% to 5% but then increases more gradually at higher droplet concentrations. The influence of droplet size on the optical properties of colloidal dispersions is shown in Figure 6.4:

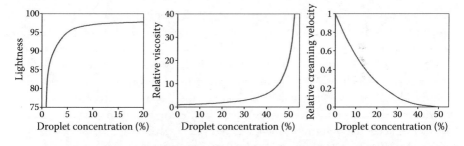

**Figure 6.3    Dependence of the lightness, relative viscosity, and relative creaming stability of emulsions on droplet concentration.**

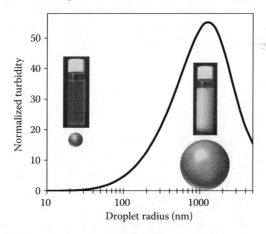

**Figure 6.4    Dependence of the turbidity of an emulsion on the droplet size: very small droplets lead to transparent systems.**

the turbidity is relatively low when the droplets are much smaller than the wavelength of light, increases appreciably as the droplet size increases toward the wavelength of light, and then decreases when the droplet size is much larger than the wavelength of light. The impact of the droplet characteristics on the overall appearance of an emulsion may be an important consideration when designing a delivery system for a specific food or beverage application. Some food products are transparent or only slightly turbid (e.g., soft drinks and fruit beverages) and so the delivery system should not cause a large increase in opacity. Other food products are optically opaque (e.g., dressings, sauces, and mayonnaise) and so the light scattering characteristics of the delivery system are less important. For products where optical clarity is required, it is important to use emulsion-based delivery systems with small droplet sizes, low droplet concentrations, or low refractive index contrast.

## 6.3.2 Rheology

Emulsion-based delivery systems exhibit various rheological characteristics depending on their compositions, structures, and interactions: liquids, viscoelastic liquids, viscoelastic solids, plastics, or elastic solids (Genovese et al. 2007; McClements 2005a; Pal 2011; Walstra 2003). In addition, the rheological properties of the food matrices into which delivery systems are going to be incorporated may also vary widely. The rheology of colloidal dispersions was discussed in Chapter 3, and so only a brief overview is given here. For relatively dilute emulsions, the rheology is normally characterized in terms of their apparent shear viscosity. The shear viscosity of an emulsion is mainly determined by the continuous phase viscosity ($\eta_1$), the droplet concentration ($\varphi$), and the nature of any droplet–droplet interactions ($w$): $\eta = \eta_1 \times f(\varphi,w)$ (Genovese et al. 2007; McClements 2005a). For dilute systems containing noninteracting droplets, the shear viscosity is given by $\eta = \eta_1 \times (1 + 2.5\varphi)$. For concentrated systems, the following semiempirical equation can be used to describe the dependence of viscosity on droplet concentration (Chapter 3):

$$\frac{\eta}{\eta_1} = \left( 1 - \frac{\varphi_{eff}}{\varphi_c} \right)^{-2} \qquad (6.1)$$

Here, $\eta_1$ is the viscosity of the liquid surrounding the droplets, $\varphi_{eff}$ is the effective dispersed phase volume fraction, and $\varphi_c$ is the critical packing parameter ($\approx 0.65$). The critical packing parameter represents the concentration where the droplets become jammed together and the system gains solid-like characteristics. For a conventional O/W emulsion, $\varphi_{eff}$ is approximately equal to the disperse phase volume fraction, $\varphi$, that is, the oil content. However, $\varphi_{eff}$ may be considerably higher than $\varphi$ for nonconventional systems, such as nanoemulsions, multilayer emulsions, multiple emulsions, filled hydrogel particles, and microclusters (see later sections).

Normally, the viscosity of an emulsion increases with increasing droplet concentration, gradually at first and then steeply as the droplets become more closely packed

(Figure 6.3). Around and above the droplet concentration where close packing occurs (typically around 50%–60% for a conventional O/W emulsion), the emulsion exhibits solid-like characteristics, such as viscoelasticity and plasticity. The droplet concentration where the steep increase in emulsion viscosity is observed depends on the nature of the droplet interactions in the system, decreasing for either strong attractive or strong repulsive interactions (McClements 2005a). The viscosity of an emulsion tends to increase when the droplets are flocculated because the effective particle concentration is increased because of the continuous phase trapped within the floc structure. In addition, shear thinning behavior is observed in flocculated emulsions because of deformation and breakdown of the floc structure as shear stresses increase. The viscosity of an emulsion may also depend appreciably on droplet size if there are either strong attractive or repulsive interactions operating between the droplets (McClements and Rao 2011). For example, if the attractive interactions between the droplets are strong enough to promote flocculation, then there may be a large increase in viscosity when the particle size decreases (Mao and McClements 2012a–e). Conversely, if there is a large long-range repulsive interaction between the droplets (e.g., strong electrostatic repulsion), then there may also be a large increase in viscosity with decreasing droplet size (Weiss and McClements 2000).

The impact of the droplet characteristics on the overall rheology of an emulsion may be an important consideration when designing a delivery system for a particular food application. Some food systems have a relatively low viscosity (such as beverages) and therefore the delivery system itself should not significantly increase the viscosity. Other food systems are highly viscous or gel-like (e.g., dressings, dips, deserts), and in these cases, the delivery system should not decrease the viscosity or disrupt the gel network.

## 6.3.3 Stability

Emulsion-based delivery systems are thermodynamically unfavorable systems that will therefore tend to break down over time through a variety of different mechanisms (Figure 6.5), including gravitational separation, flocculation, coalescence, partial coalescence, Ostwald ripening, and phase separation (Dickinson 1992; Friberg et al. 2004; McClements 2005a). The physicochemical basis of these different mechanisms is described in some detail in Chapter 3, and so only a brief overview of their importance in emulsions is given here. It should be noted that many of these instability mechanisms are linked to each other: flocculation can promote coalescence; coalescence or flocculation may promote gravitational separation; gravitational separation may promote coalescence or flocculation; coalescence promotes phase separation through oiling-off.

*Gravitational separation*: Gravitational separation is one of the most common forms of instability in emulsion-based delivery systems and may take the form of either *creaming* or *sedimentation* depending on the relative densities of the dispersed and continuous phases (McClements 2005a). Particles tend to move upward when they have a lower density than the surrounding liquid, which is referred to as

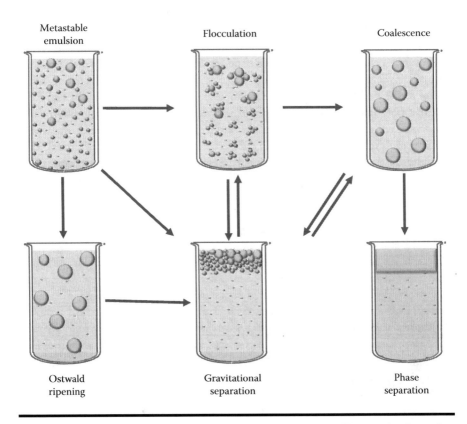

**Figure 6.5** **Schematic diagram of the most common instability mechanisms that occur in food emulsions: flocculation, coalescence, creaming, sedimentation, Ostwald ripening, and phase separation. (Prepared with the help of Cheryl Chung, University of Massachusetts.)**

creaming. On the other hand, they tend to move downward when they have a higher density than the surrounding liquid, which is referred to as sedimentation. Liquid oils normally have lower densities than liquid water and so creaming is more prevalent in O/W emulsions, whereas sedimentation is more prevalent in W/O emulsions. Nevertheless, the particles in emulsion-based delivery systems can sediment under certain circumstances, for example, if the liquid oil is dense, if the oil phase crystallizes, if they contain high concentrations of weighting agent, if they are coated by thick dense polymer layers, or if they are trapped within hydrogel particles.

The creaming rate of noninteracting rigid spherical particles in a dilute Newtonian liquid is given by the following expression:

$$U = -\frac{2gr^2(\rho_2 - \rho_1)}{9\eta_1} \tag{6.2}$$

Here, $U$ is the creaming velocity (+ for creaming; – for sedimentation), $g$ is the acceleration owing to gravity, $r$ is particle radius, $\rho$ is density, $\eta$ is shear viscosity, and the subscripts 1 and 2 refer to the continuous and disperse phases, respectively. More sophisticated mathematical models that take into account polydispersity, nonspherical particles, particle fluidity, particle–particle interactions, and non-Newtonian fluids are available (McClements 2005a). This equation shows that gravitational separation can be controlled in emulsion-based delivery systems by reducing the density contrast between the dispersed and continuous phases, decreasing the particle size, or increasing the continuous phase viscosity.

*Droplet aggregation*: The aggregation of the droplets in an emulsion-based delivery system may be desirable or undesirable depending on the application. Controlled aggregation of emulsion droplets can be used to create products with desirable textural properties, such as high viscosity or paste-like characteristics (Mao and McClements 2012a–e; Wu et al. 2013). On the other hand, droplet aggregation can promote undesirable changes in the stability, texture, and appearance of some products. Droplet aggregation occurs when the attractive colloidal forces (such as van der Waals, hydrophobic, depletion, or bridging interactions) outweigh the repulsive colloidal forces (such as steric and electrostatic interactions) acting between droplets (McClements 2005a). In general, the three most important forms of droplet aggregation are flocculation, coalescence, and partial coalescence. Flocculation is the process whereby two or more droplets associate with each other but maintain their individual integrities. Coalescence is the process whereby two or more droplets merge together and form a larger droplet. Extensive droplet coalescence may lead to the formation of a layer of oil on top of an emulsion ("oiling off") and eventually to complete phase separation. Partial coalescence is the process where two or more partially crystalline droplets form clumps held together by fat crystals (Fredrick et al. 2010). The rate and extent of these aggregation processes depend on the nature of the colloidal interactions, as well as particle size, concentration, physical state, and interfacial properties (McClements 2005a).

*Ostwald ripening*: Ostwald ripening is the process whereby large droplets grow at the expense of small ones owing to molecular diffusion of the dispersed phase through the continuous phase. It is important in emulsion-based delivery systems containing oil phases that have an appreciable solubility in the aqueous phase, for example, essential oils, flavor oils, and short-chain triacylglycerols. It is not usually important in systems that contain highly water-insoluble oils, such as long-chain triacylglycerols or waxes. The thermodynamic driving force for Ostwald ripening is the higher water solubility of the disperse phase around small droplets than around large droplets. As a consequence, disperse phase molecules move from the small droplets through the intervening continuous phase and into the large droplets, thereby leading to a net increase in particle size. In O/W emulsions, this process can be inhibited by adding highly water-insoluble hydrophobic materials ("ripening inhibitors") into the oil phase prior to homogenization (Li et al. 2009; Wooster et al. 2008) or by creating strong elastic interfacial coatings around the oil droplets (Mun and McClements 2006; Zeeb et al. 2012).

## 6.3.4 Molecular Distribution and Release Characteristics

When an emulsion-based delivery system is incorporated into a food or beverage product, there may be a redistribution of the various types of molecules present among the different phases (e.g., oil, water, and interfacial phases), which is governed by their equilibrium partition coefficients and the kinetics of molecular motion (McClements 2005a). For example, if an emulsion-based delivery system is incorporated into a fatty food, then some of the lipophilic active ingredients may move from the oil droplets in the delivery system into the fat phase of the food. This molecular transfer process may have an adverse, neutral, or beneficial impact on the stability and functionality of an active ingredient in the system and should be considered when designing delivery systems for specific food and beverage matrices. Another important physicochemical property of emulsion-based delivery systems is their ability to release encapsulated materials. In particular, it is important to establish any potential trigger mechanisms for release (e.g., pH, ionic strength, temperature, enzymes, etc.), as well as the rate and extent of release. In an emulsion, release is usually characterized in terms of the increase in concentration of the active ingredient at some specified location as a function of time (Chapter 3). For a volatile ingredient, this may be the headspace above a food or the human nose, whereas for a nonvolatile ingredient, this may be the aqueous phase surrounding the droplets or some target tissue, such as the mouth, stomach, or small intestine. A number of parameters can be derived from intensity–time curves, such as the area under the curve (AUC), the maximal concentration released ($C_{max}$), and the time to reach the maximum concentration ($t_{max}$) (Aguilera 2006). The release rate of encapsulated components from within emulsions depends on many factors, including their equilibrium partition coefficients, their original location, the mass transfer coefficients of the components in the different phases, mechanical agitation, and the microstructure of the system, for example, particle size and layer thickness (McClements 2005a). Consequently, it is possible to structurally design emulsion-based systems that are capable of controlling the release of encapsulated components by selecting appropriate ingredients and microstructures (Lian et al. 2004).

## 6.3.5 Implications for Design of Delivery Systems

The particle properties of emulsion-based delivery systems (such as droplet size, concentration, charge, interactions, and physical state) determine their physicochemical properties (such as optical properties, rheology, stability, and molecular partitioning), which in turn influence the properties of any food or beverage products they are incorporated into (e.g., appearance, texture, mouthfeel, flavor, and shelf life) (McClements 2005a). Consequently, it is important when designing an emulsion-based delivery system to ensure that it does not adversely affect the quality attributes of the food or beverage product that it is going to be utilized in. For example, some food products are required to have a low overall viscosity yet contain particulate matter that is stable to gravitational separation (e.g., beverage emulsions), so that an emulsion-based delivery system used in these products should contain small droplets that do not cream or sediment during the life time of the product and that do not appreciably increase product viscosity. On the other hand, other food products are highly viscous or gelled (e.g., dressings, dips, sauces, and deserts) so that the size of the droplets in the delivery system may be less crucial since gravitational separation is not a major problem. An important factor to be aware of when designing a delivery system is that changing the droplet characteristics to achieve one goal (e.g., reducing gravitational separation) may have an impact on all the other quality attributes of the final product (e.g., appearance, rheology, and flavor release) (McClements and Demetriades 1998). Consequently, one should take an integrated approach when designing an emulsion-based delivery system to meet all of the required quality specifications for a particular application.

## 6.4 Emulsion-Based Delivery Systems

In the remainder of this chapter, an overview of the major kinds of conventional and structured emulsion-based delivery systems suitable for use in the food industry is given. The primary focus will be on those delivery systems that are dispersible in aqueous solutions, that is, those in which water is the continuous phase. Methods to covert liquid emulsion-based delivery systems into powders are discussed in Chapter 4.

### 6.4.1 Emulsions and Nanoemulsions

#### 6.4.1.1 Composition and Structure

In general, O/W emulsions consist of oil droplets dispersed in an aqueous continuous phase, with each droplet being surrounded by a thin interfacial layer consisting of emulsifier molecules (Figure 6.6) (Dickinson 1992; Friberg et al. 2004; McClements 2005a). The concentration and PSD of the oil droplets can be controlled, as can the nature of the emulsifier used to stabilize them. The oil phase used to prepare emulsions may contain a variety of different nonpolar substances, including triacylglycerol oils, flavor oils, essential oils, oil-soluble vitamins, nutraceuticals, weighting agents, and ripening inhibitors. The aqueous phase may also contain a range of different polar substances, such as water, salts, acids, bases, carbohydrates, proteins, and cosolvents. A range of different emulsifiers may be used to stabilize emulsions, which can be categorized as small-molecule surfactants, phospholipids, proteins, polysaccharides, and fine particles. The molecular and physical properties

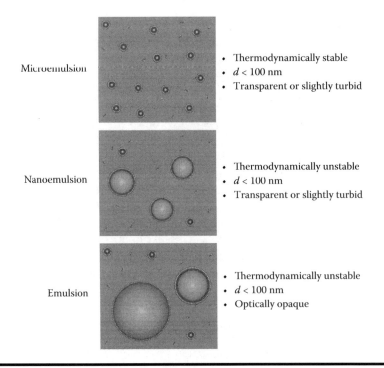

**Figure 6.6 Schematic representation to highlight the difference between emulsions, nanoemulsions, and microemulsions.**

of the different ingredients used to assemble emulsions influence their formation, stability, and physicochemical properties. There is therefore considerable scope in designing emulsions with different functional properties by varying the types and concentrations of ingredients used.

Emulsions with a wide range of different droplet sizes can be produced depending on the ingredients and processing methods used. Emulsions containing different droplet sizes have different physicochemical and functional properties, and therefore it is often convenient to classify them according to their particle dimensions (McClements 2011; McClements and Rao 2011). One possible classification scheme is based on differences in particle radius ($r$): *nanoemulsions*: $r < 100$ nm; *miniemulsions:* $100 < r < 1000$ nm; and *macroemulsions*: $r > 1000$ nm. Alternatively, emulsions may simply be divided into two categories: nanoemulsions ($r < 100$ nm) and conventional emulsions ($r > 100$ nm) (McClements and Rao 2011). At present, there appears to be no consensus about the classification of different emulsion systems. It should also be noted that there is no distinct change in the physicochemical or biological properties of oil droplets at particular particle sizes (McClements 2012c). Instead, there tends to be more gradual changes in different emulsion properties as the particle size changes. Overall, differences in the size of the droplets in an emulsion may have a major impact on its stability, physicochemical properties, and functional performance (see later). The thickness of the interfacial coating around the oil droplets may also vary considerably depending on the nature of the surface-active molecules present but is typically in the range of around 2 to 20 nm. Typically, the interfacial thickness decreases in the following order for different classes of emulsifier: polysaccharides > proteins > surfactants (Dickinson 1992). As mentioned in Section 6.2, the electrical characteristics of the droplets can be controlled by selecting an appropriately charged emulsifier, which may be positive, neutral, or negative.

It is important to clearly distinguish emulsion-based delivery systems from microemulsion-based ones (McClements 2012a). These two types of colloidal systems have similar structural characteristics since they both contain small particles with a hydrophobic core and a hydrophilic shell (Figure 6.6). However, microemulsions are thermodynamically stable systems since the free energy of the colloidal system is less than that of the separated phases, whereas emulsions are thermodynamically unstable systems since the opposite is true (Figure 6.7). This has important consequences for the preparation and stabilization of colloidal delivery systems since different mechanisms are involved for these two types of systems.

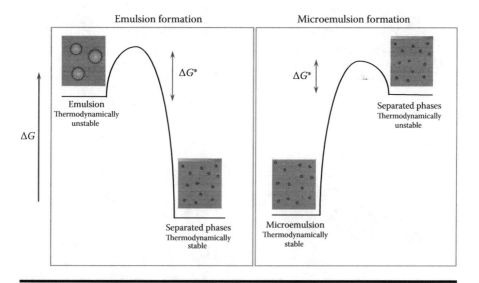

**Figure 6.7** **Schematic diagram of the free energy of microemulsion and emulsion systems compared to the phase-separated state. Microemulsions have a lower free energy than the phase-separated state, whereas emulsions have a higher free energy. The two states are separated by an activation energy $\Delta G^*$.**

## 6.4.1.2 Formation

A large number of different methods have been developed to form emulsions and nanoemulsions, which can be classified broadly as either high-energy or low-energy approaches.

*High-energy approaches*: High-energy approaches use mechanical devices ("homogenizers") capable of generating intense disruptive forces that break down and mix the oil and water phases, leading to the formation of small droplets of one phase dispersed in the other phase. The phase that eventually forms the droplets depends on the nature of the emulsifier and the water-to-oil ratio (Chapter 5). O/W emulsions are usually prepared by homogenizing an oil phase and an aqueous phase together in the presence of a water-soluble emulsifier. A variety of homogenizers are available for creating emulsions, including high-shear mixers, high-pressure homogenizers, colloid mills, ultrasonic homogenizers, and membrane homogenizers (McClements 2005a; Walstra 1993, 2003). Some of the most commonly used homogenizer types in the food industry are shown schematically in Figure 6.8. The principles behind these methods are discussed in more detail in Chapter 4. The choice of a particular type of homogenizer and its operating conditions depend on the characteristics of the materials being homogenized (e.g., viscosity, interfacial tension, shear sensitivity) and the required final properties of the emulsion (e.g., droplet concentration, droplet size, viscosity). For example, the size of the droplets in an O/W emulsion produced by a

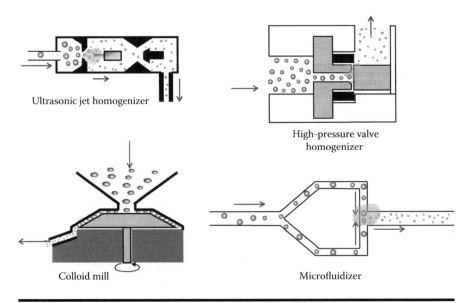

**Figure 6.8  Schematic representation of three mechanical devices that can be used to continuously produce food-grade nanoemulsions using a high-energy approach: high-pressure valve homogenizer, colloid mill, microfluidizer, and ultrasonic homogenizer.**

high-pressure homogenizer can usually be decreased by increasing the homogenizer pressure or number of passes (Qian and McClements 2011). Alternatively, the droplet size can be decreased by homogenizing an oil and organic solvent together to form an O/W emulsion and then evaporating the organic solvent to cause the droplets to shrink (Lee and McClements 2010; Lee et al. 2011). A lipophilic active ingredient would normally be dispersed in the oil phase prior to homogenization with the water phase. If the active ingredient was crystalline (e.g., phytosterols or carotenoids), then it may be necessary to ensure that it is used at a level below its saturation concentration in the carrier oil or to warm the lipid phase prior to homogenization to melt any crystals present (since fat crystals can cause fouling of homogenizers) (McClements 2012b). If the active ingredient was susceptible to chemical degradation (e.g., ω-3 fatty acids, CLAs, and carotenoids), then it may be necessary to carefully control homogenization conditions to avoid exposure to factors that increase the degradation rate, for example, high temperatures, oxygen, light, or transition metals (McClements and Decker 2000; Waraho et al. 2011).

*Low-energy approaches*: There have been considerable advances in recent years in utilizing low-energy methods to produce emulsions and nanoemulsions because of their potential advantages for certain applications, for example, low equipment costs, simplicity, and small particle sizes. These methods rely on the spontaneous formation of oil droplets in surfactant–oil–water (SOW) mixtures when either their composition or their environment is altered in a specific way (Anton and Vandamme 2009; McClements and Rao 2011; Solans and Sole 2012). A number of these low-energy methods can be used to produce food-grade emulsions or nanoemulsions (Figure 6.9):

■ *Spontaneous emulsification (SE) methods*: SE methods rely on titrating a mixture of oil- and water-soluble surfactant into a water phase with continuous stirring (Saberi et al. 2013). Small oil droplets are formed at the oil–water boundary as the surfactant molecules move from the oil phase to the water phase. In this case, lipophilic active ingredients are added to the surfactant–oil mixture prior to titration.

■ *Emulsion inversion point (EIP) methods*: EIP methods rely on titrating water into a mixture of oil- and water-soluble surfactant with continuous stirring. As increasing amounts of water are added to the system, a W/O emulsion is formed, then an O/W/O emulsion, and then an O/W emulsion. It has been proposed that the small internal oil droplets in the O/W/O (that later become the oil droplets in the O/W emulsion) are formed by spontaneous emulsification at the oil–water boundary (Ostertag et al. 2012). Again, a lipophilic active ingredient would normally be added to the surfactant–oil mixture prior to titration.

■ *Phase inversion temperature (PIT) methods*: PIT methods rely on heating a SOW mixture around or slightly above its phase inversion temperature and then quench cooling with continuous stirring (Anton and Vandamme 2009). This method relies on changes in the optimum curvature and solubility of

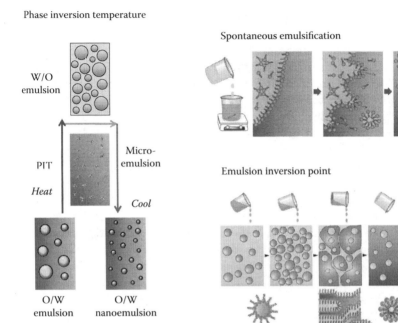

**Figure 6.9** **Examples of some low-energy methods for producing nanoemulsions and emulsions: phase inversion temperature (PIT), spontaneous emulsification (SE), and emulsion inversion point (EIP).**

nonionic surfactant molecules when they are heated. At temperatures above the PIT, the surfactant is more soluble in the oil phase and has a curvature that favors W/O emulsions. When the emulsion passes through the PIT, the optimum curvature tends toward unity, thereby leading to an ultralow interfacial tension and a highly dynamic interface. In addition, the surfactant molecules become more hydrophilic and water soluble as the head groups become hydrated at lower temperatures. It has been proposed that ultrafine oil droplets are formed when a SOW mixture is rapidly cooled from around the PIT because of movement of surfactant molecules from the oil to the aqueous phase (in a process similar to spontaneous emulsification) (Anton and Vandamme 2009). In this case, a lipophilic active ingredient would be mixed with the oil phase prior to heating the SOW mixture.

## 6.4.1.3 Properties

The general physicochemical properties of any emulsion-based system, such as appearance, rheology, stability, and molecular distribution, were covered in Section 6.2.

This section therefore focuses on the specific physicochemical attributes of emulsions and nanoemulsions that make them suitable as colloidal delivery systems for active ingredients (McClements 2012a; McClements and Li 2010).

Conventional emulsions should be one of the first systems considered when selecting an emulsion-based delivery system for an active ingredient because of their relative ease of preparation, low cost, and suitability for scale-up, especially when compared to some of the more sophisticated delivery systems discussed later in this chapter. Despite the relative simplicity of conventional emulsions, there is still considerable scope in designing delivery systems with a range of physicochemical properties and functional attributes. Emulsions contain a nonpolar region (the oil phase), a polar region (the aqueous phase), and an amphiphilic region (the interfacial layer), and so it is possible to incorporate active ingredients that are polar, nonpolar, or amphiphilic within the same delivery system. For example, a lipophilic ingredient could be dispersed in the oil phase prior to homogenization, while a hydrophilic ingredient could be incorporated in the aqueous phase either before or after homogenization (McClements 2005a). The heterogeneous structure of emulsions means that it is possible to develop novel strategies for controlling the chemical stability of any encapsulated active ingredients, for example, by engineering the interface or controlling the physical location of different reactants (Coupland and McClements 1996; McClements and Decker 2000). Emulsions can be prepared with various rheological characteristics (ranging from viscous liquids, to plastic pastes, to elastic solids) by controlling their composition and microstructure, so that they can be made in a form that is most convenient for each specific application (Genovese et al. 2007). Emulsions can be used directly in their "wet" state or they can be dehydrated to form powdered ingredients, which may facilitate their transport, reduce storage costs, and improve their utilization (Desai and Park 2005; Klinkesorn et al. 2005a–c; Soottitantawat et al. 2003; Vega and Roos 2006). Finally, emulsions can be created entirely from food-grade ingredients (such as water, oil, surfactants, phospholipids, proteins, and polysaccharides) using fairly simple processing operations (mixing and homogenization). Indeed, it is often possible to form them from all-natural ingredients, which is desirable for some applications in the food and beverage industries.

Nevertheless, there are limitations to the use of conventional emulsions as delivery systems for certain specialist applications, which means that more sophisticated systems are needed. Conventional emulsions are often prone to physical instability when exposed to environmental stresses, such as heating, chilling, freezing, drying, pH extremes, and high mineral concentrations. In addition, one often has limited control over the ability to protect and control the release of functional components because the small size of the droplets (approximately in micrometers) and the interfacial layers (approximately in nanometers) means that the time scales for molecular diffusion of substances out of emulsion droplets are extremely short (McClements 2005a). Finally, there are a limited number of emulsifiers that can be used to form the interfacial layers that surround the oil droplets, which limits one's ability to create delivery systems exhibiting a wide range of protection and release characteristics. Consequently, there are continuing attempts to identify novel emulsifiers or for developing new approaches to utilizing existing emulsifiers more effectively to facilitate emulsion formation and improve emulsion stability (Benichou et al. 2002; Dickinson 2003, 2011b; Drusch 2007; Drusch and Schwarz 2006).

Nanoemulsions have many different physicochemical properties compared with conventional emulsions owing to their small droplet sizes, which makes them more suitable for certain applications in the food industry (McClements 2011; McClements and Rao 2011). Nanoemulsions often have a greater stability to droplet aggregation because the strength of the attractive colloidal forces falls off more rapidly than the repulsive forces with decreasing droplet size. Nanoemulsions also have better stability to creaming and sedimentation because Brownian motion dominates gravitational forces for very small particles and because the density contrast between the oil and water phases may be reduced owing to the presence of the interfacial coating around the droplets (see Section 6.4.3.3). This coating makes an appreciable contribution to particle density for the small droplets in nanoemulsions but not for the large droplets in conventional emulsions. The very small size of the droplets in nanoemulsions also means that they only scatter light weakly, and so they may be transparent or slightly turbid, whereas conventional emulsions are usually highly turbid or opaque (Wooster et al. 2008). Nanoemulsions may also have a much higher viscosity than conventional emulsions at the same droplet content because of the fact that the particles have a core–shell structure, and the contribution of the shell becomes more important for smaller particles (McClements 2011). In addition, the biological fate of nanoemulsions in the gastrointestinal tract is often different from that of conventional emulsions, which leads to a greater bioavailability of encapsulated lipophilic active ingredients (McClements and Xiao 2012). Many of the equations presented in the section on multilayer emulsions, which deal with the influence of a thick interfacial coating on the physicochemical properties of colloidal delivery systems, are also relevant to nanoemulsions (Section 6.4.3.3).

### 6.4.1.4 Applications

Emulsions and nanoemulsions have been used widely used to encapsulate and deliver lipophilic active ingredients, which are normally incorporated into the oil phase prior to homogenization. This type of emulsion has been used as a delivery system for encapsulating ω-3 fatty acids and incorporating them into food products, such as milk, yogurt, ice cream, and meat patties (Chee et al. 2005, 2007; Lee et al. 2005, 2006a,b; McClements and Decker 2000; Sharma 2005). These delivery systems had to be carefully designed to prevent oxidation of the polyunsaturated fats within the lipid droplets (McClements and Decker 2000; Sun et al.

2011; Waraho et al. 2011). Reduction in the rate and extent of lipid oxidation in emulsions can be achieved using a number of different strategies. For example, fat droplets can be coated by cationic emulsifiers that repel cationic transition metal ions ($Fe^{2+}$) that normally catalyze lipid oxidation from the droplets' surfaces. Alternatively, they can be coated by thick dense interfacial layers that prevent the transition metals from reaching the encapsulated lipids through steric hindrance effects. Chelating agents (such as EDTA) can be added to the aqueous phase of emulsions to sequester transition metal ions, thereby preventing them from coming into proximity to the lipid phase. Antioxidants that partition into regions where lipid oxidation reactions occur (such as the droplet surface and interior) can also be added to inhibit oxidation.

Emulsions and nanoemulsions have also been used to encapsulate various other types of bioactive lipids, such as lycopene (Ribeiro et al. 2006; Tyssandier et al. 2001), astaxanthin (Ribeiro et al. 2005, 2006), lutein (Losso et al. 2005; Santipanichwong and Suphantharika 2007), β-carotene (Santipanichwong and Suphantharika 2007), plant sterols (Sharma 2005), CLAs (Jimenez et al. 2004), and oil-soluble vitamins (Saberi et al. 2013; Yang and McClements 2013; Ziani et al. 2012). A number of these lipophilic active ingredients are crystalline at ambient temperature (e.g., carotenoids, flavonoids, and phytosterols), which can cause problems during emulsion formation or for their long-term stability (Li et al. 2012b,c). For this reason, these types of active ingredients have to be dissolved in carrier oils or heated above their melting points prior to homogenization (McClements 2012b). As an illustration of this approach, we consider the encapsulation of lycopene (melting point = 173°C) in O/W emulsions (Ribeiro et al. 2003). Emulsions were prepared by dispersing lycopene crystals into carrier oil (medium-chain triglycerides) and then heating the mixture to a temperature where the crystals melted. The hot oil phase was then homogenized with hot aqueous phase (containing water-soluble emulsifier) to form an O/W emulsion. Lycopene is highly susceptible to chemical degradation and therefore it was necessary to minimize the time spent at elevated temperatures and to reduce the oxygen levels in the system during homogenization. The chemical stability of the emulsified lycopene was shown to depend on the nature of the food matrix it was incorporated into: milk, orange juice, or water (Ribeiro et al. 2003). Lycopene was more stable in orange juice than in milk or water, and its stability improved considerably when antioxidants (α-tocopherol) were added. A range of factors have also been shown to influence the chemical stability of another carotenoid (β-carotene) in emulsions, for example, emulsifier type, pH, temperature, and antioxidant addition (Qian et al. 2012a,b). As mentioned earlier, O/W emulsions can usually be converted into a powdered form by spray drying, which increases their long-term stability, facilitates their transport, and improves their ease of utilization (Beristain et al. 2001; Soottitantawat et al. 2003, 2005).

A number of in vitro and in vivo studies have shown that the bioavailability or bioactivity of encapsulated lipophilic components may be increased when they are incorporated into emulsions or nanoemulsions, with the extent of the effect depending on particle size, concentration, composition, and interfacial characteristics (Pinheiro et al. 2013; Rao et al. 2013; Salvia-Trujillo et al. 2013; Speranza et al. 2013). These types of delivery systems may therefore be suitable for increasing the bioactivity of lipophilic nutraceutical agents that normally have low oral bioavailability.

## 6.4.2 Multiple Emulsions

### 6.4.2.1 Composition and Structure

W/O/W emulsions consist of small water droplets contained within larger oil droplets that are dispersed within a water continuous phase (Dickinson 2011a; Garti 1997a,b; Garti and Benichou 2004; Garti and Bisperink 1998; Jimenez-Colmenero 2013; Muschiolik 2007) (Figure 6.10). This type of multiple emulsion is therefore more accurately described as a $W_1/O/W_2$ emulsion, where $W_1$ is the internal water phase and $W_2$ is the external water phase, which may have different compositions and physicochemical properties. There are also two different interfaces in this type of emulsion: the $W_1$–O interface at the boundary between the internal water droplets and the oil phase and the O–$W_2$ interface at the boundary between the oil droplets and the external water phase. Consequently, two different types of

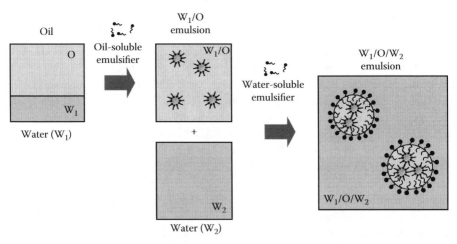

**Figure 6.10  Multiple emulsions (W/O/W) are usually prepared using a two-step procedure: (i) An oil and aqueous phase are homogenized together in the presence of an oil-soluble emulsifier to form a W/O emulsion. (ii) The W/O emulsion is homogenized with a water phase in the presence of a water-soluble emulsifier to form a W/O/W emulsion.**

emulsifiers are usually needed to stabilize W/O/W emulsions: an oil-soluble emulsifier for the internal water droplets and a water-soluble emulsifier for the oil droplets. The structural organization of this kind of system can be altered by controlling the PSD and concentration of both the internal water droplets and the oil droplets, as well as the properties of the $W_1$–O and O–$W_2$ interfaces surrounding the droplets (e.g., thickness, charge, permeability, and environmental responsiveness). The dimensions of the oil droplets in W/O/W emulsions can also be controlled after emulsion formation by osmotic swelling, for example, by changing the solute concentration gradient in the internal or external aqueous phases (Mezzenga et al. 2004). Typically, the sizes of the oil droplets in W/O/W emulsions are considerably larger than those in conventional O/W emulsions because of the need to encapsulate water droplets inside them. Thus, the mean droplet diameter of the internal aqueous phase may be around 100 to 1000 nm, whereas the mean droplet diameter of the oil droplets may be from 1 to 100 μm.

## 6.4.2.2 Formation

$W_1/O/W_2$ emulsions are normally produced using a two-step procedure (Aserin 2008). First, a $W_1/O$ emulsion is produced by homogenizing a water phase ($W_1$) with an oil phase (O), containing oil and an oil-soluble emulsifier (Figure 6.10). Second, a $W_1/O/W_2$ emulsion is produced by homogenizing the $W_1/O$ emulsion with a different water phase ($W_2$), containing water and a water-soluble emulsifier. Similar kinds of homogenizers can be used to produce $W_1/O/W_2$ emulsions as to produce O/W emulsions, for example, high-shear mixers, high-pressure homogenizers, colloid mills, ultrasonic homogenizers, and membrane homogenizers (McClements 2005a). Nevertheless, the homogenization conditions used in the second stage should usually be less intense than those used in the first stage, so as to avoid disruption or expulsion of the $W_1$ droplets within the oil phase. Thus, a high-pressure homogenizer may be used to form the $W_1/O$ emulsion, and then a blender or membrane homogenizer may be used to form the $W_1/O/W_2$ emulsion. The size of the water droplets in the $W_1/O$ emulsion can be controlled by varying emulsifier type, emulsifier concentration, and homogenization conditions (e.g., energy intensity and duration) in the first homogenization stage. Similarly, the size of the oil droplets in the final $W_1/O/W_2$ emulsion can be controlled by varying emulsifier type, emulsifier concentration, and homogenization conditions in the second homogenization stage. The concentration of water droplets in the $W_1/O$ emulsion can be controlled by using different ratios of $W_1$ to O phases in the first homogenization step, whereas the concentration of oil droplets in the final $W_1/O/W_2$ emulsion can be controlled by using different ratios of $W_1/O$ emulsion to $W_2$ phase in the second homogenization step. The droplet dimensions can also be altered after emulsion formation using osmotic swelling methods (Leal-Calderon et al. 2012; Mezzenga et al. 2004). If the solute concentration in the external aqueous phase is higher than that in the internal aqueous phase, then water diffuses out of the oil droplets causing them to shrink. Conversely, if the solute concentration in the internal aqueous phase is greater than that in the external aqueous phase, then water diffuses into the oil droplets and they swell. In addition, it is possible to alter the physical state of the different phases, for example, the internal and external aqueous phases can be gelled using appropriate gelling agents (Iancu et al. 2009; Surh et al. 2007), whereas the oil phase can be fully or partially crystallized by controlling the nature of the fat phase and the temperature (Vladisavljevic and Williams 2005). Consequently, one has great scope for designing multiple emulsions with different compositions, structures, and functional properties using this approach.

Active ingredients can potentially be located in a number of different environments within a $W_1/O/W_2$ emulsion (Aserin 2008; Jimenez-Colmenero 2013). Hydrophilic ingredients can be incorporated within the internal water phase by dispersing them in the $W_1$ phase prior to the first homogenization step or in the external water phase by dispersing them in the $W_2$ phase either before or after the second homogenization step. Lipophilic ingredients can be incorporated into the oil droplets by dispersing them in the oil phase either before or after the first homogenization step. Surface-active ingredients could be located at either the $W_1/O$ or the $O/W_2$ interface depending on when they were incorporated into the system during the homogenization procedures.

### 6.4.2.3 Properties

The properties of W/O/W emulsions may differ appreciably from those of the simple O/W emulsions discussed in Section 6.2. The light scattering behavior of complex particle structures, such as those in multiple emulsions, is more complex than that from spheres because light waves may be scattered from both the small internal water droplets and the larger oil droplets containing them (Bohren and Huffman 1983; Wriedt 1998). This can cause complications when analyzing the PSDs of multiple emulsions using light scattering methods. The analytical instruments based on this principle assume that the scattering entities are spherical homogeneous particles with a specific refractive index, which the particles within multiple emulsions are not. In this case, it may be better to use microscopy methods to determine the dimensions of the different kinds of particles present within the system. The rheology of W/O/W emulsions may also be appreciably different from that of O/W emulsions at the same fat content because of the higher effective volume fraction of the particles in multiple emulsions (Pal 2011). In an O/W emulsion, the particle volume fraction is approximately equal to the volume fraction of the oil phase ($\varphi_O$), but in a $W_1/O/W_2$ emulsion, it is approximately equal to the sum of the volume fractions of the oil and internal water phases ($\varphi_O + \varphi_{W1}$). This phenomenon can be used in the development of reduced-fat food products, since highly viscous products can be made at lower fat contents (Pal 2011). The predicted dependence of the shear viscosity on fat content for conventional

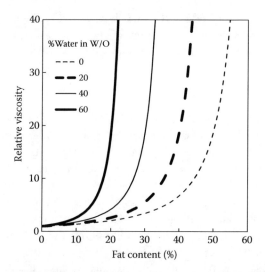

**Figure 6.11** **The viscosity of a W/O/W emulsion may be considerably higher than that of a conventional emulsion with the same fat content depending on how much water is in the W/O phase.**

emulsions and multiple emulsions with different internal water contents is shown in Figure 6.11.

The introduction of a multiple emulsion delivery system into a food or beverage product may influence its flavor profile by altering the distribution of polar and nonpolar molecules within the system. For example, nonpolar volatile molecules within a food may move into the oil phase of the W/O/W emulsion, which would alter their overall flavor profile. Conversely, any polar or nonpolar flavor molecules originally in the W/O/W emulsion may move into the aqueous, oil, or gas phases of a food as a new equilibrium distribution is established. Indeed, multiple emulsions may be used as flavor delivery systems. Polar flavors, such as salts, sugars, acids, or bases, could be trapped within the internal or external aqueous phases. Those in the external aqueous phase will be released and perceived immediately after ingestion, whereas those in the internal aqueous phase may have a specific release profile. Some of them may not be released, some of them may be released slowly owing to diffusion, and some of them may be released rapidly if the multiple emulsion is disrupted and the internal water droplets are released from the oil droplets. W/O/W emulsions may therefore be specifically designed by altering their composition and structure to get different flavor release profiles within the mouth. However, it is also important to design them so that any encapsulated flavors are not released within the product during storage. This may be achieved by controlling the physical state or binding properties of the oil or aqueous phases.

The fact that the oil droplets in W/O/W emulsions contain some water droplets may be used to modulate their stability to gravitational separation (Muschiolik

2007). The overall density of a W/O particle in a W/O/W emulsion depends on the densities (ρ) and volume fractions (φ) of the oil and water phases within the particle. To a first approximation, the overall density of a W/O particle is given by

$$\rho_{Particle} = \varphi_O \rho_O + (1 - \varphi_O) \rho_{W1} \qquad (6.3)$$

Here, the subscripts O and $W_1$ refer to the oil and internal water phases, respectively. Typically, the density of the internal water phase is higher than that of the oil phase. Consequently, the overall $W_1$/O particle density can be increased (and therefore the density contrast with the $W_2$ phase decreased) by incorporating more water droplets into the oil droplets, particularly if the aqueous phase contains dense components such as proteins or polysaccharides (that do not promote osmotic swelling).

One of the potential advantages of W/O/W emulsions as delivery systems is that they can be used to encapsulate both lipophilic and hydrophilic active ingredients within the same system (Jimenez-Colmenero 2013; Muschiolik 2007). Hydrophilic active ingredients (such as minerals, sugars, vitamins, enzymes, proteins, peptides, and dietary fibers) could be trapped within the internal water phase, which may have benefits in a number of applications. First, hydrophilic ingredients can be trapped within the internal water droplets and released at a controlled rate or in response to specific environmental triggers, for example, in the mouth, stomach, or small intestine. This phenomenon may be useful for flavor release applications where a controlled or burst release of a hydrophilic flavor is required in the mouth. Conversely, it may be useful for flavor-masking purposes, where a bioactive ingredient with an unpleasant taste (such as a bitter peptide) remains trapped in the internal aqueous phase in the mouth but is released in the stomach and small intestine after swallowing. Furthermore, the system could be designed to release an encapsulated bioactive ingredient in different regions of the gastrointestinal tract. Second, hydrophilic ingredients can be protected from chemical degradation by isolating them from other hydrophilic ingredients that they might normally react with by trapping one in the internal aqueous phase and the other in the external aqueous phase. Third, the overall fat content of food products that normally exist as O/W emulsions (e.g., dressings, mayonnaise, dips, sauces, desserts) could be reduced by loading the oil phase with water droplets (Jimenez-Colmenero 2013; Muschiolik 2007). Thus, a W/O/W emulsion could be produced with the same particle concentration and dimensions as a conventional O/W emulsion, but with a reduced fat content. W/O/W emulsions may therefore be particularly useful for developing delivery systems for health-promoting bioactive lipids.

The considerable potential of multiple emulsions for utilization as delivery systems in the food industry has been realized for many years (Aserin 2008; Benichou et al. 2004; Garti 1997; Garti and Bisperink 1998). Nevertheless, there are still few examples of multiple emulsions being successfully used in commercial food and beverage products. One of the primary reasons for this phenomenon is that multiple emulsions are highly susceptible to physical instability during storage or when exposed to environmental conditions commonly experienced by commercial food products, such as mechanical stresses, chilling, freezing, heat treatments, and dehydration. A variety of instability mechanisms are responsible for the breakdown of W/O/W emulsions, with some of these being similar to those in conventional O/W emulsions and others being unique to multiple emulsions. The oil droplets in W/O/W emulsions are susceptible to creaming, flocculation, coalescence, and Ostwald ripening just as they are in O/W emulsions. The internal water droplets in W/O/W emulsions are also susceptible to flocculation, coalescence, and Ostwald ripening processes; however, they may also become unstable owing to diffusion of water molecules between the internal and external aqueous phases (leading to swelling or shrinkage) or to expulsion of entire water droplets from inside the oil droplets. A variety of different strategies have been developed in an attempt to overcome these problems (Benichou et al. 2004; Garti 1997a,b; Garti and Benichou 2004; Garti and Bisperink 1998), including identification of appropriate combinations of oil- and water-soluble emulsifiers, gelation of the $W_1$ phase, solidification of the oil phase, covering the oil droplets with a protective coating, and osmotic balancing of the inner and outer water phases to prevent water diffusion (Aserin 2008; Benichou et al. 2004; Garti 1997a,b; Garti and Benichou 2004; Garti and Bisperink 1998; Grigoriev and Miller 2009; Mezzenga et al. 2004; Surh et al. 2007).

It is clear that multiple emulsions do have great potential as delivery systems for use in the food industry but that there are still a number of important challenges that need to be addressed before this technology can be successfully employed commercially. In particular, it is important that they can be produced from all food-grade ingredients using simple and economic processing operations and that they have the physical stability to resist the stresses typically encountered in foods, for example, mechanical stresses, heating, chilling, freezing, drying, and so on.

### 6.4.2.4 Applications

In this section, potential applications of W/O/W emulsions for encapsulating and delivering lipophilic or hydrophilic agents in the food industry are highlighted. W/O/W emulsions have occasionally been used to encapsulate lipophilic active ingredients, such as β-carotene (Rodriguez-Huezo et al. 2004) and ω-3 fatty acids (Cournarie et al. 2004; Onuki et al. 2003). However, conventional O/W emulsions are probably more suitable for this purpose because of their simpler preparation and better storage stability. W/O/W emulsions are more widely used to encapsulate hydrophilic bioactive components, such as water-soluble minerals (Jimenez-Colmenero 2013), vitamins (Fechner et al. 2007; Giroux et al. 2013; Kukizaki and Goto 2007; Owusu et al. 1992), pigments (Frank et al. 2011, 2012), nutraceuticals (Hemar et al. 2010), immunoglobulins (Chen et al. 1999; Lee et al. 2004), insulin (Cournarie et al. 2004), proteins (Balcao et al. 2013; Su et al. 2006), and amino acids (Owusu et al. 1992; Weiss et al. 2005). Multiple emulsions may be designed to control the release of active ingredients within different locations of the gastrointestinal tract by controlling their breakdown characteristics in the mouth, stomach, or small intestine (Frank et al. 2012). As mentioned earlier, they may also be used to encapsulate hydrophilic active ingredients that have off-flavors (such as bitter or astringent compounds) in the internal aqueous phase so that they are not perceived within the mouth (Jimenez-Colmenero 2013). Multiple emulsions may also be suitable for improving the stability of products to gravitational separation by creating density-matched particles, for example, by loading the fat droplets with dense internal water droplets.

The main challenge with the commercial utilization of multiple emulsions has been to make products that have a sufficiently long shelf life for utilization within the food industry and which are capable of withstanding the fairly harsh processing operations mentioned above. In conclusion, W/O/W emulsions are most suitable for encapsulation of hydrophilic bioactive components, but they may be useful if one wants to prepare a delivery system that contains both lipophilic and hydrophilic bioactive components in the same system (Cournarie et al. 2004).

### 6.4.3 Multilayer Emulsions

#### 6.4.3.1 Composition and Structure

There has been considerable interest in using interfacial engineering methods to create multilayer emulsions with improved functional performances over conventional emulsions (Dickinson 2011b; Grigoriev and Miller 2009; Guzey and McClements 2006; McClements 2012a; Shchukina and Shchukin 2012). Multilayer O/W emulsions consist of oil droplets dispersed in an aqueous medium, with each oil droplet being coated by a nanolaminated interfacial structure, which usually consists of emulsifier and biopolymer molecules (McClements 2010a,b) (Figure 6.1). The PSD and concentration of the oil droplets in multilayer emulsions can be controlled, as can the characteristics of their nanolaminated coatings, for example, composition, thickness, charge, permeability, environmental responsiveness, and chemical reactivity. This flexibility gives one great scope in designing colloidal delivery systems with improved functional properties, such as greater protection against physical and chemical instability, and controlled or targeted release.

#### 6.4.3.2 Formation

The physicochemical principle behind the fabrication of multilayer emulsions is shown in Figure 6.12 (McClements 2005a). Initially, an O/W emulsion is prepared by homogenizing an oil phase and an aqueous phase together in the presence

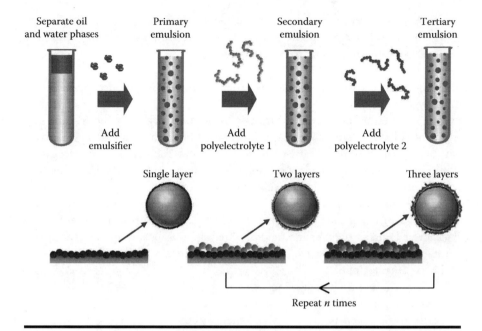

**Figure 6.12 Multilayer emulsions are produced by a multistep procedure: (i)** *Primary emulsion*: **an oil and aqueous phase are homogenized together in the presence of a charged water-soluble emulsifier. (ii)** *Secondary emulsion*: **an oppositely charged polyelectrolyte is added to coat the droplets. (iii)** *Multilayer emulsions*: **sequential polyelectrolyte adsorption steps can be carried out.**

of a charged water-soluble emulsifier. The resulting "primary" emulsion consists of small charged oil droplets dispersed within an aqueous continuous phase. An oppositely charged polyelectrolyte is then added to the system so that it adsorbs to the droplet surfaces and forms a "secondary" emulsion containing oil droplets coated by a two-layer (emulsifier–polyelectrolyte) interface. This procedure can be repeated to form oil droplets coated by nanolaminated interfaces containing multiple polyelectrolyte layers by mixing the particles formed at a particular stage with further oppositely charged polyelectrolyte solutions. Each polyelectrolyte layer can be deposited onto the droplet surfaces using either a one or two-step mixing procedure (Guzey and McClements 2006):

i. *One-Step Mixing.* An O/W emulsion containing electrically charged droplets is prepared, and it is then mixed with an oppositely charged polyelectrolyte that directly adsorbs to the droplet surfaces through electrostatic attraction.
ii. *Two-Step Mixing.* An O/W emulsion is prepared and then mixed with a polyelectrolyte solution under conditions where there is not a strong electrostatic attraction between the droplets and polyelectrolyte. The solution conditions

are then altered so that the electrical charge on the droplets and that on the polyelectrolyte are opposite so that the polyelectrolyte adsorb to the droplet surfaces through electrostatic attraction.

Some studies have shown that two-step mixing often produces more stable multilayer emulsions than one-step mixing (Guzey et al. 2004). A *washing step* may be required between each electrostatic deposition step to remove excess non-adsorbed polyelectrolyte remaining in the continuous phase, as this may interfere with subsequent multilayer formation. Washing can be achieved by centrifugation or filtration to remove nonadsorbed polyelectrolytes prior to any subsequent electrostatic deposition step. Alternatively, solution conditions can be optimized during the electrostatic deposition step so that there is little or no free polyelectrolyte remaining in the aqueous phase. In some cases, it is possible to break down any flocs formed during the preparation of multilayer emulsions by applying mechanical agitation to the system, such as sonication, blending, or homogenization (Guzey and McClements 2006).

One of the main challenges to the successful implementation of the multilayer emulsion method is the formation of irreversible flocs during the electrostatic deposition step (Cho and McClements 2009; McClements 2005b). This may occur if there are insufficient polyelectrolytes present to completely cover the oil droplet surfaces or if the adsorption rate of the polyelectrolyte molecules is too slow to cover the droplets before they collide with each other. Flocculation may also occur during the formation of multilayer emulsions if the polyelectrolyte concentration is too high because of an osmotic (depletion flocculation) mechanism. Theoretical stability maps have been developed to predict the influence of polyelectrolyte characteristics (concentration and radius of hydration) and droplet characteristics (concentration and size) on the formation of stable multilayer emulsions, for example, Figure 6.13 (Cho and McClements 2009; McClements 2005b). These stability maps can be generated using software that is freely available online (Normand et al. 2010). They show regions where stable multilayer emulsions can be produced by controlling biopolymer and droplet characteristics.

One of the major advantages of using multilayer emulsions as delivery systems is that the properties of the interfacial layer surrounding the oil droplets can be carefully controlled, for example, composition, structure, charge, thickness, permeability, rheology, and environmental responsiveness (Decher and Schlenoff 2003). This can be achieved by careful control of system composition and preparation conditions during the fabrication of the multilayer emulsions, for example, emulsifier type and concentration, polyelectrolyte type and concentration, pH, ionic strength, order of ingredient addition, and mixing conditions (Guzey and McClements 2006).

A range of different food-grade emulsifiers and biopolymers can be used to assemble nanolaminated coatings around lipid droplets, including electrically

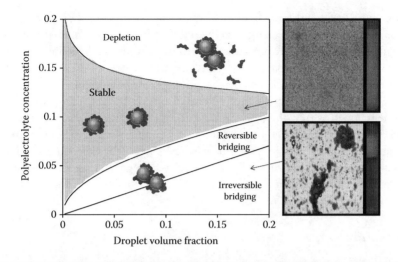

**Figure 6.13** **Theoretical stability map showing the effect of droplet and polymer concentration on the formation of multilayer emulsions. Droplet flocculation may occur because of bridging or depletion mechanisms under certain conditions.**

charged surfactants, phospholipids, proteins, and polysaccharides. Food-grade emulsifiers and biopolymers with different molecular characteristics (e.g., electrical charge, molecular weight, conformation, hydrophobicity, thermal stability) and functional properties (e.g., solubility, viscosity, gelation, surface activity) can be used, which gives one a great deal of scope in designing the functional performances of nanolaminated coatings. For example, the functional performance of coatings can be controlled by the following:

■ Changing the nature of the *emulsifier* used to prepare the initial lipid droplets
■ Changing the nature of the *polyelectrolytes* used to form the individual layers within the nanolaminated coatings
■ Changing the total *number* of electrostatic deposition steps (and therefore layers) used to prepare the nanolaminated coatings
■ Changing the *order* that the various polyelectrolytes (P) are deposited onto the lipid droplet surfaces (e.g., $E - P_1/P_2/P_3$ vs. $E - P_3/P_2/P_1$)
■ Controlling the properties of the *solutions* used during the preparation of the nanolaminated coatings (e.g., pH, ionic strength, and solvent quality)
■ Cross-linking one or more of the polyelectrolyte layers in the nanolaminated coatings, for example, physically, chemically, or enzymatically

The ability to systematically control interfacial properties in a rational manner enables one to design droplets with improved stability or novel functional performance.

In multilayer emulsions, lipophilic active ingredients can be trapped within the oil droplets, whereas charged active ingredients can be trapped within the nanolaminated coatings surrounding the droplets. For example, a lipophilic active ingredient could be dissolved in the oil phase prior to homogenization, whereas a charged hydrophilic active ingredient could be incorporated into one or more of the polyelectrolyte layers surrounding the oil droplets. The active ingredients could then be retained within the delivery system until they were released at the site of action in response to a specific environmental stimulus, such as pH, ionic strength, temperature, or enzyme activity. This can be achieved by designing the interfacial coating so that its permeability or integrity changes in a well-defined way in response to an environmental trigger (Gu et al. 2006, 2007; Guzey and McClements 2006; Ogawa et al. 2003a,b).

### 6.4.3.3 Properties

Multilayer emulsions can be tailored to have novel physicochemical and functional properties because of the ability to manipulate the characteristics of the polyelectrolyte coatings surrounding the oil droplets, for example, their composition, thickness, charge, and packing. When the coatings are relatively thin compared to the droplet radius ($\delta_S \ll r$), then the bulk physicochemical properties of multilayer emulsions are fairly similar to those of conventional emulsions, for example, optical properties, rheology, gravitational separation, and molecular partitioning and release. On the other hand, if the interfacial coating is relatively thick ($\delta_S \approx r$), then these physicochemical properties may be changed appreciably. In this case, the composite particles can be considered to have a core–shell structure with a lipid core of radius $r$ and a polyelectrolyte shell of thickness $\delta_S$. The core and shell have different refractive indices, which may influence the overall optical properties of the system because of changes in the light scattering pattern. In addition, the effective volume fraction of the composite particles ($\varphi_{eff}$) may be appreciably larger than that of the oil phase ($\varphi$): $\varphi_{eff} = \varphi \times (1 + \Phi_S \times [1 + \delta_S/r]^3)$, where $\Phi_S$ is the fraction of the composite particle that comprises the shell:

$$\Phi_S = \frac{(r+\delta_S)^3 - r^3}{(r+\delta_S)^3} \tag{6.4}$$

Here, $\Phi_S$ ($=V_S/V_{CP}$) is the volume of the shell divided by the volume of the composite particle (= core + shell). In conventional emulsions, the thickness of the emulsifier shell is much smaller than the radius of the lipid core, and so the composite particles can be thought to consist predominantly of the oil phase. On the other hand, in multilayer emulsions (and nanoemulsions), the shell may make up an appreciable contribution to the overall particle composition. This increase in effective particle concentration may have a major impact on the viscosity of multilayer emulsions (Equation 6.1), particularly as the thickness of the shell increases relative to the core (Equation 6.4).

The presence of a thick coating around the oil droplets in multilayer emulsions may also have an influence on the direction and rate of gravitational separation. The overall density of a particle consisting of a core (oil) surrounded by a shell (multilayer) is given by

$$\rho_{\text{particle}} = \Phi_S \rho_S + (1 - \Phi_S)\rho_C. \tag{6.5}$$

A polyelectrolyte layer usually has a higher density than the oil or aqueous phases, and so an increase in shell thickness leads to an increase in the overall density of the composite particle (McClements 2010a,b). This phenomenon has important implications for inhibiting gravitational separation in emulsions since it leads to a reduction in the density contrast between the composite particles and surrounding fluid. Indeed, very small oil droplets may actually sediment rather than cream if they contain sufficiently thick and dense polyelectrolyte layers (Cho et al. 2010). It may therefore be possible to produce delivery systems that are stable to gravitational separation by ensuring that the density of the composite particles is the same as that of the surrounding aqueous phase.

The properties of polyelectrolyte coatings can also be engineered to alter other physicochemical attributes of multilayer emulsions (Decher and Schlenoff 2003; Guzey and McClements 2006; McClements 2010a,b):

■ Improved physical stability to environmental stresses by control of the composition and properties of the interfacial layer, for example, pH, salt, thermal processing, chilling, freezing, dehydration, and mechanical agitation. Enhanced physical stability can often be achieved by manipulating the thickness and charge on the polyelectrolyte coating so as to modulate the repulsive steric and electrostatic interactions acting between particles.
■ Improved chemical stability of encapsulated components, for example, the oxidative stability of ω-3 fatty acids can be improved by minimizing the interaction between the lipids and transition metal ions in the aqueous phase by controlling interfacial charge and thickness.
■ Greater control over the release rate of functional agents owing to the ability to manipulate the thickness and selective permeability of the interfacial layer.
■ Ability to trigger release of functional ingredients in response to specific changes in environmental conditions, such as dilution, pH, ionic strength, or temperature. For example, it is possible to cause interfacial coatings to detach from fat droplet surfaces when pH is altered, since this causes either the emulsifier or polyelectrolyte to change its charge. Alternatively, the permeability of an interfacial coating may be tuned by altering pH or ionic strength.

Multilayer emulsions can be fabricated entirely from food-grade ingredients (such as surfactants, proteins, polysaccharides, and phospholipids) using relatively simple processing operations that are already widely used in the food industry (such as homogenization and mixing). Nevertheless, the formation of stable multilayer emulsions requires careful control over system composition and preparation procedures to avoid droplet flocculation (Aoki et al. 2005; Guzey et al. 2004; McClements 2005a; Ogawa et al. 2006a,b). The main limitations of this method are that additional ingredients and processing steps are required over conventional emulsion formation. In addition, it is often only possible to fabricate relatively dilute emulsions (<5 wt%) using the electrostatic deposition approach because of the tendency for droplet flocculation to occur during their preparation (Cho and McClements 2009; McClements 2005a).

It should be stressed that the electrostatic deposition method has more general applications within the food industry than simply coating lipid droplets with oppositely charged polyelectrolytes. In principle, it can be used to coat any type of charged object with another charged object. For example, it has been used to coat large, electrically charged fat droplets with smaller, oppositely charged fat droplets (Gu et al. 2006) or protein fibrils (Humblet-Hua et al. 2012). It can also be used to coat charged particles other than fat droplets, such as air bubbles, oil bodies, liposomes, biological cells, or hydrogel beads (Chen et al. 2010; Chun et al. 2013; Laye et al. 2008; Rossier-Miranda et al. 2012). Electrostatic deposition is therefore a particularly versatile method for tailoring the properties of colloidal delivery systems for different applications.

### 6.4.3.4 Applications

Multilayer emulsions are most suitable for encapsulation of lipophilic active ingredients since they can be trapped within the hydrophobic core and protected by the polyelectrolyte shell. Studies have shown that multilayer emulsions can be used to encapsulate oil-soluble flavors (Benjamin et al. 2012; Djordjevic et al. 2007; Yang et al. 2012), ω-3 fatty acids (Klinkesorn et al. 2005a–c; Lesmes et al. 2010; Lomova et al. 2010), vitamin A (Szczepanowicz et al. 2010), and carotenoids (Hou et al. 2010; Mao et al. 2013; Szczepanowicz et al. 2010). In these studies, either the lipophilic ingredients are used as the oil phase themselves (e.g., ω-3 oils) or they are dissolved within carrier oil (e.g., β-carotene in corn oil). Primary emulsions are then prepared by homogenizing the oil and aqueous phases together in the presence of a charged water-soluble emulsifier. Secondary emulsions are then prepared by mixing the primary emulsions with oppositely charged polyelectrolyte solutions. In some cases, additional coatings are added by mixing the coated droplets with additional oppositely charged polyelectrolyte solutions. The preparation conditions have to be carefully controlled to avoid flocculation in the multilayer emulsions, for example, by controlling droplet concentration and size, polyelectrolyte type and concentration, order of addition, and stirring conditions (Guzey and McClements 2006). In many of these studies, the chemical stability of the encapsulated lipids was shown to be improved by being encapsulated within the nanolaminated coatings, which was attributed to a number of reasons: (i) *charge repulsion*—positively charged coatings may repel cationic prooxidants (such as iron or copper ions) from the droplet surfaces; (ii) *steric effects*—the polyelectrolyte coatings may be impermeable to the diffusion of prooxidants because of the tight packing of the molecules; (iii) *antioxidant activity*—some polyelectrolytes within the coatings have antioxidant activity (such as proteins).

Numerous studies have shown that coating oil droplets with nanolaminated polyelectrolyte coatings can also improve the physical stability of emulsions to environmental stresses, such as pH changes, high ionic strengths, thermal processing, freezing, and dehydration (Guzey and McClements 2006, 2007; Harnsilawat et al. 2006). This improvement in stability has been attributed to the ability of the thick charged coatings to increase the repulsive interactions between the droplets. Multilayer emulsions may be converted into a powdered form by spray drying, which would facilitate their utilization in many commercial products (Klinkesorn et al. 2005a–c, 2006). An additional advantage of multilayer emulsions in terms of drying is that the amount of wall material (hydrophilic polymer) required to form stable powders is considerably reduced when the oil droplets have a polyelectrolyte coating around them (Shaw et al. 2007).

Multilayer emulsions may also be useful for overcoming other challenges associated with the encapsulation and protection of bioactive lipids in foods. For example, the tendency for partially crystalline lipid droplets to aggregate because of partial coalescence could be prevented by building a thick biopolymer coating around them (Gu et al. 2007a; Walstra 2003). Consequently, the electrostatic deposition technology may prove particularly useful for improving the stability of emulsions in which the bioactive lipids or carrier oils have a tendency to crystallize, for example, carotenoids, phytosterols, or high-melting-point fats.

A particularly useful attribute of multilayer emulsions for their utilization as delivery systems is their ability to control the biological fate of encapsulated lipophilic components. Nanolaminated biopolymer coatings have been shown to modulate the digestibility of emulsified lipids (Hu et al. 2011; Klinkesorn and McClements 2010; Li et al. 2010; Mun et al. 2006; Tokle et al. 2012) and the release of encapsulated lipophilic ingredients (Trojer et al. 2013a,b). The ability to modulate the biological fate of ingested lipids is associated with changes in the integrity and permeability of the nanolaminated coatings around the fat droplets within the gastrointestinal tract. For encapsulated triacylglycerols to become bioavailable, it is necessary for lipases to adsorb to the fat droplet surfaces and catalyze their conversion into monoacylglycerols and free fatty acids (Tso 2000). For encapsulated lipophilic nutraceuticals (such as vitamins, carotenoids, or phytosterols) to become bioavailable, it is usually necessary for the triacylglycerol carrier oil surrounding them to be digested first, since this leads to the formation of mixed micelles that can solubilize and transport them (Tso 2000). Hence, the access of lipases to encapsulated lipids is a critical step in determining their bioactivity. The most important properties of nanol-aminated coatings that could potentially influence these processes are as follows (McClements 2010a):

- *Coating Integrity.* If a fat droplet is surrounded by a nanolaminated coating that prevents digestive enzymes from reaching the encapsulated lipids, then the lipid phase may not be digested and the lipophilic nutraceuticals will not be released. The *integrity* of a nanolaminated coating may be altered during the passage of a multilayer emulsion through the gastrointestinal tract owing to changes in pH, ionic strength, levels of surface-active components, and enzyme activities. Surface-active components may displace polyelectrolyte layers through a competitive adsorption process. Changes in pH and ionic strength may cause polyelectrolyte layers to become detached by weakening electrostatic interactions. Enzymes such as proteases or amylases may degrade certain types of polyelectrolytes (such as proteins or starches). The bioavail-ability of encapsulated lipids may therefore be controlled by controlling the responsiveness of a coating's integrity to different biological fluids (e.g., oral, gastric, or intestinal fluids).
- *Coating Permeability.* Even if a nanolaminated coating remains intact around the fat droplets within the gastrointestinal tract, the digestibility of the encapsulated lipids will still depend on the ability of the digestive enzymes to penetrate through the layers and access the lipid phase (Angelatos et al. 2007). The *permeability* of a nanolaminated coating depends on its pore size, as well as specific interactions between the enzyme molecules and polyelec-trolytes in successive layers. Studies have shown that the permeability of nanolaminated coatings depends strongly on ionic strength and pH because they are held together by electrostatic interactions. For example, when the solution pH or ionic strength is changed so that the electrostatic interac-tions between the polyelectrolyte layers are weakened, a coating may swell considerably, thereby increasing its permeability to small molecules (Rubner 2003; Schlenoff 2003). It is therefore possible to "tune" the permeability of coatings by careful control of polyelectrolyte type and assembly condi-tions (Angelatos et al. 2007). Consequently, it may be possible to control the bioavailability of an encapsulated lipid by controlling the responsiveness of a coating's permeability to different biological fluids (e.g., oral, gastric, or intestinal fluids).

In summary, the major advantage of the multilayer technology is that inter-facial coatings with specific physicochemical properties (e.g., thickness, charge, permeability, composition, and digestibility) can be designed to achieve particular functional performances (e.g., antioxidant properties, stability to environmental stresses, controlled digestibility, and targeted release profiles). Nevertheless, they are more expensive to fabricate than conventional emulsions and have to be prepared carefully to avoid droplet flocculation.

## 6.4.4 Solid Lipid Particles

### 6.4.4.1 Composition and Structure

Solid lipid particle (SLP) suspensions have similar compositions and structures to nanoemulsions or emulsions, since they consist of emulsifier-coated lipid particles dispersed within an aqueous continuous phase (Figure 6.1). However, the lipid phase inside the particles is either fully or partially solidified, and the particles may have a nonspherical shape. Delivery systems with different functional attributes can be created by modulating SLP composition, size, shape, and concentration, as well as the characteristics of the interfacial layer coating them. In addition, the morphology and packing of the crystals within the lipid phase can be controlled (Saupe et al. 2005; Souto et al. 2004; Uner et al. 2004; Wissing and Muller 2002; Wissing et al. 2004). In this section, the term *SLP* is used to refer generally to fully or partially solidified lipid particles that may have different sizes, shapes, and internal structures. In the literature, a number of other expressions are typically used to refer to this type of particle. The term *solid lipid nanoparticles* (SLNs) is commonly used to refer to systems in which the lipid phase is fully crystalline, whereas the term *nanostructured lipid carriers* (NLCs) is often used to refer to systems where the lipid phase is partially crystalline or amorphous. The physical state of the lipid phase has a major impact on the stability, encapsulation, and release properties of lipid particles. SLNs are often susceptible to particle aggregation and to expulsion of encapsulated lipophilic molecules, whereas NLCs are specifically designed to be more stable to these instability mechanisms. The terms SLN and NLC are often used to refer to suspensions that contain relatively small particles ($d < 1000$ nm), but they are also sometimes used more loosely to include larger particles.

### 6.4.4.2 Formation

SLP suspensions are usually prepared using a two-step procedure (Figure 6.14) (Saupe et al. 2005; Schubert and Muller-Goymann 2005; Souto et al. 2004; Uner et al. 2004; Wissing and Muller 2002; Wissing et al. 2004). First, an O/W emulsion (or nanoemulsion) is formed by homogenizing an oil and water phase together in the presence of a water-soluble emulsifier at a temperature above the melting point of the lipid phase. The initial hot emulsion may be fabricated using either high-energy or low-energy approaches used to prepare conventional emulsions or nanoemulsions (Section 6.4.1). The size of the SLPs produced depends on the initial

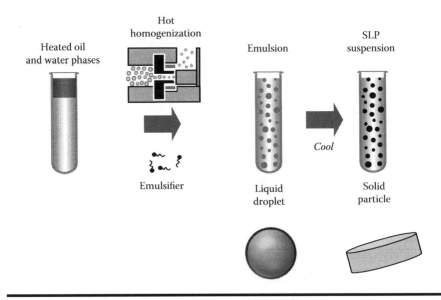

**Figure 6.14** **Solid lipid particles are typically produced by homogenizing an oil phase and aqueous phase together in the presence of a water-soluble emulsifier at a temperature above the melting point of the lipid phase. The emulsion is then cooled to promote fat crystallization. In some cases, crystallization may cause a change in particle morphology.**

size of the droplets in this hot emulsion and therefore can be controlled in a similar manner as used for emulsions. Second, the hot emulsion is cooled so that some or all of the lipid phase within the droplets crystallizes. It is important to maintain the temperature of the emulsion appreciably above the crystallization temperature of the highest melting lipid throughout homogenization to prevent fat solidification. In high-energy methods, fat crystallization during homogenization can cause blocking of homogenizer chambers, which could damage the equipment or mean extensive cleaning is required. The morphology and stability of the SLPs produced, as well as the spatial organization of any crystals within them, can be controlled by careful selection of the type of lipids present, the nature of the emulsifier(s) used to stabilize the droplets, the initial oil droplet size and concentration, and cooling and storage conditions (Muller and Keck 2004; Muller et al. 2000). In principle, a variety of different colloidal delivery systems can be fabricated from crystalline active ingredients or crystalline carrier lipids (McClements 2012b). For example, a crystalline lipophilic ingredient can be dissolved or dispersed within a liquid or solid carrier lipid (Figure 6.15). Alternatively, top–down (such as milling) or bottom–up (such as antisolvent precipitation) methods could be used to form a nanocrystal suspension.

The structure and spatial organization of fat crystals within lipid particles have important consequences for designing SLP suspensions for encapsulating,

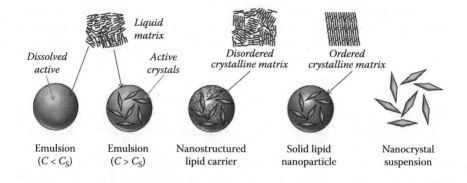

**Figure 6.15    Examples of different kinds of emulsion-based delivery systems containing crystalline bioactive components. Either the bioactive component or the surrounding carrier lipid may be crystalline. (Adapted from D. J. McClements,** *Advances in Colloid and Interface Science,* **174, 1–30, 2012b.)**

protecting, and releasing lipophilic active ingredients. For example, in some applications, it may be beneficial to locate the lipophilic component within the core (away from any hydrophilic reactive species), whereas in other situations, it may be more important to locate it close to the droplet surface (to enhance its release). Typically, two or more lipids with different melting profiles are used to create specific microstructures within SLP, for example, mixtures of purified triglycerides, complex triglyceride mixtures, or waxes (Dubes et al. 2003; Saupe et al. 2005; Souto et al. 2004; Uner et al. 2004; Wissing and Muller 2002; Wissing et al. 2004). Utilization of a number of lipids, rather than an individual lipid, usually increases the loading capacity and retention of encapsulated lipophilic ingredients because they can fit better into a more imperfect crystalline or amorphous structure than a highly ordered one (McClements 2012b).

The nature of the emulsifier used to stabilize the lipid droplets may also be important in creating specific internal structures. For example, the tail groups of certain surfactants act as templates that promote nucleation within lipids located at the oil–water interface, that is, heterogeneous surface nucleation (Sonoda et al. 2006). This principle can be used to form core–shell particles with a solid shell and a liquid core. A lipophilic active ingredient is usually dissolved or dispersed in the carrier lipid phase at ~10°C above the melting temperature of the highest melting lipid. The hot lipid phase is then homogenized with hot aqueous phase in the presence of a hydrophilic emulsifier to produce an O/W emulsion, which is then cooled in a controlled manner to promote lipid crystallization.

### 6.4.4.3 Properties

The physicochemical properties and stability of SLP suspensions are somewhat similar to those of conventional emulsions (Section 6.4.1), but there are also some notable differences. The refractive index of an oil phase typically increases when it crystallizes, which increases the refractive index contrast between the oil and water phases, thereby increasing the light scattering efficiency of the particles. Consequently, a suspension of SLPs will tend to have a higher turbidity and lightness than a suspension of liquid lipid droplets of similar size and concentration. In principle, the viscosity of a colloidal dispersion should increase somewhat when the particles go from liquid to solid because of the greater resistance to fluid flow around a solid particle (McClements 2005a). In practice, the difference in rheology between liquid and solid particles is usually small because the layer of emulsifier molecules around liquid droplets gives them some elastic-like surface properties anyway. Nevertheless, the rheology of SLP suspensions may be considerably different from that of an emulsion if the lipid particles undergo a shape change or aggregate after crystallization occurs, since nonspherical and aggregated particles give higher viscosities than individual ones. The crystallization of the lipid phase may also alter the flavor distribution within an SLP suspension. Lipophilic flavors often have a much lower solubility in a solid fat than in a liquid fat, and therefore they may be expelled from inside the lipid particles after crystallization occurs (Ghosh et al. 2007; Mei et al. 2010; Yucel et al. 2013).

The stability of SLP suspensions is influenced by many of the same phenomenon that influences the stability of emulsions, for example, gravitational separation, aggregation, and Ostwald ripening. SLPs that are partially crystalline may aggregate through partial coalescence, where a crystal in one particle sticks into a liquid region in another particle. SLPs may also be susceptible to flocculation if the attractive forces acting between them exceed the repulsive forces. The stability of SLPs to particle aggregation can be improved by increasing the repulsive forces (such as electrostatic or steric) acting between them or by ensuring they are coated by a thick interfacial coating that is resistant to rupture. The liquid oil droplets in emulsions will normally cream because they are less dense than the surrounding aqueous phase. On the other hand, the SLPs in SLP suspensions tend to sediment because they are denser. In principle, gravitational separation can be completely inhibited by having a mixture of solid and liquid lipids that gives a density similar to that of the aqueous phase. Ostwald ripening may also occur in SLPs owing to diffusion of lipid molecules through the continuous phase, but the rate of this process is usually less than that in emulsions, because of the additional energy required to remove the lipids from the crystalline state. Aggregation may also occur after lipid crystallization occurs because of an increase in particle surface area associated with a change in particle morphology. If there is insufficient surfactant present to cover the newly formed oil–water interfaces, then particle aggregation may occur because of hydrophobic attraction between exposed nonpolar patches (Helgason et al. 2008).

SLPs have a number of physicochemical and structural characteristics that may give them advantages over conventional emulsions or nanoemulsions for certain applications (McClements 2012; Muller and Keck 2004; Muller et al. 2000; Weiss et al. 2008; Wissing et al. 2004):

■ The stability of chemically labile lipophilic ingredients can be improved by trapping them within a solid lipid matrix (Tikekar and Nitin 2012). The molecular mobility of the lipophilic ingredient or of any reactive chemical species (such as oxygen) can be altered by controlling the physical state and structure of the solid lipid matrix. For example, a lipophilic ingredient that normally reacts with hydrophilic components in the aqueous phase could be encapsulated within the interior of the lipid core and surrounded by a protective shell of inert lipid. In this case, it is important that the lipophilic ingredient remains within the interior of the lipid particles after their solidification, which is not always true (Mei et al. 2010; Okuda et al. 2005).

■ The release of encapsulated lipophilic ingredients may be controlled by trapping them within a solid lipid matrix. For example, the digestion rate of solidified lipids is slower than that of liquid lipids within gastrointestinal tract conditions, and therefore it may be possible to delay the release of an encapsulated lipophilic ingredient (Bonnaire et al. 2008). Alternatively, a solid lipid phase could be designed to melt at a particular temperature thereby releasing an encapsulated lipophilic ingredient. In principle, this phenomenon could be used to design delivery systems that release flavor molecules within the mouth.

■ Under some circumstances, the physical stability of colloidal delivery systems is improved when the lipid phase is solidified. For example, the tendency for coalescence or Ostwald ripening would be reduced if the lipid phase was fully crystalline. In addition, the encapsulation of crystalline active ingredients within conventional emulsions or nanoemulsions may cause problems owing to partial coalescence, that is, the crystals from one droplet forming a bridge with another droplet (McClements 2005a). This problem may be prevented by trapping a solid lipophilic active ingredient within a solid carrier lipid matrix.

A major limitation of SLP suspensions is that they must be prepared at elevated temperatures to avoid crystallization of the lipid phase during the homogenization process. If crystallization occurs using a high-energy homogenization method, then the homogenizer may be blocked and potentially damaged or require extensive cleaning. In addition, the use of high temperatures may cause chemical degradation of heat-sensitive lipophilic components (such as carotenoids or ω-3 oils) and should therefore be limited as much as possible (Ribeiro et al. 2003). Finally, a carrier lipid phase usually has to be highly saturated so that it has a sufficiently high melting point to form SLP emulsions, which may be undesirable for the development of functional foods designed to improve health and wellness. There are also potential problems associated with particle aggregation and uncontrolled release of encapsulated substances. Certain kinds of lipid phase produce particles that undergo a pronounced change in morphology when they are transformed from the liquid to solid state, which can lead to extensive particle aggregation (Bunjes 2010, 2011; Bunjes et al. 2007). This shape change is associated with the transformation of the crystalline lipid from a less to a more stable polymorphic form (e.g., α to β). In addition, the change in crystal packing leads to expulsion of lipophilic active ingredients originally trapped inside the lipid particles, which can decrease their chemical stability owing to greater interactions with aqueous phase components (Mei et al. 2010; Okuda et al. 2005). The lipid phase composition and SLP preparation conditions must therefore be carefully controlled to avoid these problems.

### 6.4.4.4 Applications

Research in the pharmaceutical industry has shown that delivery systems based on SLP suspensions (such as SLN and NLC) can be used to increase the stability and bioavailability of a variety of highly lipophilic drugs (Muchow et al. 2008; Muller et al. 2000, 2011; Williams et al. 2013). The potential of using SLPs as delivery systems for a variety of lipophilic food ingredients is currently being explored. Studies have shown that SLP emulsions can be used to encapsulate, protect, and deliver fat-soluble vitamins (Iscan et al. 2005; Jee et al. 2006; Pople and Singh 2006; Souto and Muller 2005; Uner et al. 2005), carotenoids (Cornacchia and Roos 2011; Helgason et al. 2009; Hentschel et al. 2008), ω-3 oils (Awad et al. 2009), and fat-soluble flavors (Ghosh et al. 2007). Most of these SLP suspensions were formed using the high-energy hot homogenization method described above. It is particularly important to select an appropriate lipid carrier and processing conditions to avoid expulsion of the encapsulated lipophilic component during storage. A number of studies have shown that the chemical stability of lipophilic ingredients is actually less in SLP suspensions than in emulsions because they are expelled to the particle surface after droplet crystallization, where they are then exposed to aqueous phase reactants (Okuda et al. 2005; Qian et al. 2013). Additional applications of SLPs within the food industry have been reviewed elsewhere (Fathi et al. 2012; McClements and Rao 2011; Tamjidi et al. 2013; Weiss et al. 2008).

### 6.4.5 Filled Hydrogel Particles

#### 6.4.5.1 Composition and Structure

Suspensions of filled hydrogel particles consist of oil droplets contained within hydrogel particles that are dispersed within an aqueous continuous phase (Figure 6.1) (Matalanis et al. 2011). They can therefore be thought of as a type of oil-in-water-in-water ($O/W_1/W_2$) emulsion, with two different aqueous phases ($W_1$ and $W_2$).

The composition and structure of filled hydrogel particles can be varied in a number of ways to create delivery systems with different functional performances. The size, concentration, physical state, and location of the oil droplets within the hydrogel particles can be changed. The size, concentration, charge, and interfacial properties of the hydrogel particles themselves can be varied. The structural and physicochemical properties of the hydrogel matrix can be controlled, for example, composition, packing, interactions, permeability, stability, and environmental responsiveness. There is therefore considerable scope for creating emulsion-based delivery systems with different physicochemical properties and functional attributes based on filled hydrogel particles.

## 6.4.5.2 Formation

Numerous methods are available to fabricate hydrogel particles and these are discussed in more detail in Chapter 7. In this section, a brief overview of some of the most suitable methods for fabricating filled hydrogel particles that can be utilized as delivery systems in the food industry is given. As a first step, most of these methods involve preparing an O/W emulsion (or nanoemulsion) by homogenizing an oil phase with an aqueous phase containing a water-soluble emulsifier. The size, concentration, and charge of the droplets in these emulsions can be controlled by selecting an appropriate emulsifier (type and concentration) and appropriate homogenization procedure (homogenizer type and operating conditions). A filled hydrogel particle can then be created by combining this O/W emulsion with an appropriate biopolymer solution and then adjusting the solution or environmental conditions to promote hydrogel particle formation. A number of methods that can be used to form filled hydrogel particles are summarized below:

> *Antisolvent Precipitation*: Biopolymer molecules dissolved within an appropriate solvent can be made to associate with each other and form colloidal particles by adjusting solvent quality (Chen et al. 2006). Initially, the biopolymer is dissolved in a good solvent, and then solution conditions are altered so that it is dispersed within an antisolvent. This could be achieved by dissolving a hydrophobic biopolymer (such as zein) in an alcohol solution and then titrating it into water (Patel et al. 2010; Zhong et al. 2008). Conversely, it could be achieved by dissolving a hydrophilic biopolymer (such as β-lactoglobulin) in water and then titrating it into an ethanol solution (Gulseren et al. 2012). The biopolymer molecules aggregate when they are dispersed in the antisolvent and form nanoparticles or microparticles. This process is often referred to as *simple coacervation* (since it only requires a single biopolymer type) and can be used to form filled hydrogel particles. For example, the lipid droplets could be mixed with the biopolymer solution prior to adjusting the solvent quality. The solvent quality could then be adjusted to promote the formation

of small hydrogel particles that trapped the lipid droplets inside them. If necessary, the solution conditions could then be further adjusted to promote internal gelation of the biopolymer molecules within the hydrogel particles to improve their stability or change their permeability. In this case, it is important to ensure that the biopolymer concentration is not too high, or else gelation of the whole system may occur.

*Injection Methods*: Many biopolymer solutions are capable of forming gels when the solution or environmental conditions are altered in a specific manner, for example, mineral addition, pH adjustment, temperature change, or enzyme treatment (BeMiller and Huber 2008; Cui 2005). This phenomenon can be utilized to form small hydrogel particles. A biopolymer solution that is fluid is introduced into another liquid that promotes biopolymer gelation, for example, by injection, spraying, atomizing, or extruding (Senuma et al. 2000; Zhang et al. 2007). The size of the particles produced can be controlled by varying the injection conditions, for example, injection flow rates, pore sizes, or stirring conditions (Zhang et al. 2007). This process can be adapted to form filled hydrogel particles (Zhang et al. 2006). The lipid droplets are mixed with the biopolymer solution prior to gelation, and the resulting mixture is then injected into another liquid that promotes rapid biopolymer gelation.

*Gel Disruption Methods*: Filled hydrogel particles can be prepared by first of all forming a macroscopic filled gel and then applying mechanical forces to break it down (Guo et al. 2013). The size and structure of the resulting hydrogel particles depend on the nature of the forces holding the biopolymer molecules together in the gel network, as well as the mechanical method used to carry out gel disruption.

*Emulsion Template Methods*: In this approach, an O/W/O emulsion is formed by homogenizing an O/W emulsion with an (outer) oil phase containing a lipophilic emulsifier (Cho et al. 2003; Ribeiro et al. 1999). The (inner) oil phase of the O/W emulsion contains the lipophilic component that is to be encapsulated, while the water phase contains a biopolymer that is capable of forming a gel. After the O/W/O emulsion has been produced, the gelation of the biopolymer is induced, for example, enzymatically, chemically, or thermally (Cho et al. 2003; Hwang et al. 2005; Ribeiro et al. 1999). The outer oil phase is then removed from the O/W/O emulsion by centrifuging or filtering to collect the W/O droplets, followed by washing with an organic solvent and drying to remove any excess external oil. Alternatively, the O/W/O emulsion can be formed by homogenizing an O/W emulsion with an organic solvent that acts as the outer oil phase, which can then be removed by evaporation (Freitas et al. 2005). After separation, the filled hydrogel particles are collected and can be used directly or dispersed in water. The size and properties of the internal oil droplets can be controlled by varying the composition (e.g., emulsifier type and concentration) and homogenization conditions (e.g., duration and intensity) used to produce the initial O/W emulsion, whereas the size and properties of the hydrogel particles can be controlled by varying the composition and homogenization conditions used to produce the O/W/O emulsion.

*Aggregative Phase Separation Methods*: If an aqueous solution contains two biopolymers that have a sufficiently strong attractive force between them (Chapter 7), then it will separate into two phases: a biopolymer-enriched phase and a biopolymer-depleted phase (Burgess 1990; Cooper et al. 2005; de Kruif et al. 2004; Renard et al. 2002; Weinbreck et al. 2003). In most food applications, the main driving force for this type of aggregative phase separation is electrostatic attraction between oppositely charged biopolymers, for example, an anionic polysaccharide and a cationic protein. The biopolymer-enriched phase may be a *coacervate* or a *precipitate* depending on the strength of the electrostatic attraction and the charge densities of the two biopolymers (Cooper et al. 2005). Coacervates tend to have fairly loose open structures that hold a lot of water, whereas precipitates have dense structures that contain less water. Coacervates are usually more suitable for use as delivery systems because they tend to form particles with more well-defined properties (size and charge) and they have better stability to aggregation and sedimentation. A W/W-type emulsion consisting of the coacervate phase dispersed as small droplets within the surrounding aqueous phase can be formed by agitating this type of phase-separated system (Cooper et al. 2005; Norton and Frith 2001, 2006). However, the coacervate "droplets" are usually susceptible to coalescence and may dissociate when either the pH or the ionic strength of the solution is adjusted since this changes the electrostatic forces holding them together. Consequently, it is usually necessary to stabilize the coacervate droplets so that one or both of the biopolymers form a gel, for example, by heating, cooling, mineral addition, or enzyme treatment. In this case, filled hydrogel particles could be created by mixing an O/W emulsion with a mixed biopolymer solution before inducing complex coacervation. After coacervation is induced, the coacervate phase forms around the oil droplets and traps them within the coacervate particles. Alternatively, the O/W emulsion could be mixed with a preformed coacervate phase and then the filled coacervate phase could be injected into or blended with an aqueous solution that promotes particle gelation.

*Segregative Phase Separation Methods*: If an aqueous solution contains two biopolymers that have a sufficiently strong repulsive force between them (Chapter 7), then it may separate into two aqueous phases (Benichou et al. 2002; Norton and Frith 2001; Schmitt et al. 1998; Tolstoguzov 2002, 2003). One of the aqueous phases is rich in one type of biopolymer and depleted in the other type of biopolymer, whereas the opposite is true for the other aqueous phase. Typically, the driving force for this kind of phase separation is steric exclusion or electrostatic repulsion. When the solution conditions are adjusted so that segregative separation is promoted in a mixed biopolymer solution, the system often initially forms droplets of one phase dispersed in a continuous medium of the other phase. This kind of system is often referred to as a W/W emulsion. As with coacervates, the water droplets formed because of segregative separation are often unstable to coalescence and gravitational separation and so it may be necessary to stabilize them by adjusting the solution or environmental conditions so that one or both of the water phases gel, for example, by heating, cooling, mineral addition, or enzyme treatment. Filled hydrogel particles could be formed by mixing an O/W emulsion with the mixed biopolymer solution prior to inducing segregative phase separation or by mixing it with the biopolymer phase that will become the dispersed phase of the W/W emulsion and then incorporating this into the biopolymer phase that will become the continuous phase (e.g., by mixing, injecting, or spraying).

Filled hydrogel particles with different microstructures and properties can be created by varying the nature of the biopolymers involved (e.g., charge, size, branching, flexibility, and hydrophobicity), the solution composition (e.g., pH, ionic strength, cosolvent addition), the mixing conditions (e.g., order of addition, nature, intensity, and duration of applied forces), and environmental conditions (e.g., temperature). Thus, it is possible to prepare hydrogel particles that have different sizes, electrical charges, permeability, loading capacities, release properties, and environmental responsiveness. Lipophilic active ingredients can be incorporated into the oil phase prior to the formation of the initial O/W emulsion, whereas hydrophilic active ingredients can be encapsulated within the hydrogel particles or the surrounding aqueous phase, for example, peptides, minerals, and chelating agents.

### 6.4.5.3 Properties

The physicochemical properties and stability of a suspension of filled hydrogel particles may differ considerably from those of conventional emulsions (Section 6.2). The light scattering behavior of filled hydrogel particles is complex because light waves are scattered from the oil droplets and from the hydrogel particles. In general, the lightness of a filled hydrogel particle suspension will increase as the particle concentration increases, but by an amount that depends on the size and properties of the oil droplets and the hydrogel particles. The complex nature of the light scattering by filled hydrogel particles also causes complications when analyzing particle size measurements made using light scattering methods. Light scattering instruments normally use mathematical models that assume that the scatterers are homogeneous spheres with a well-defined refractive index, in which filled hydrogel particles clearly are not. In this case, optical or electron microscopy methods are often more appropriate to determine the structure and dimensions of this kind of system.

The rheology of filled hydrogel particle suspensions ($[O/W_1]/W_2$) is different from that of conventional emulsions with the same fat content because of the higher effective volume fraction of the particles present (Matalanis et al. 2011). The particle volume fraction in an emulsion is approximately equal to that of the oil phase ($\varphi_O$), but in a filled hydrogel particle emulsion, it is approximately equal to the sum of the volume fractions of the oil and hydrogel phases ($\varphi_O + \varphi_{W1}$). This effect can be used in the development of reduced-fat food products, since highly viscous products can be made at lower fat contents.

Filled hydrogel particles are susceptible to some of the same instability mechanisms as conventional emulsions, but they also have some additional ones. The hydrogel particles are susceptible to coalescence (merging together), flocculation (associating with each other), and gravitational separation (upward or downward movement). Coalescence can usually be avoided by ensuring that there are strong physical or covalent cross-links between the biopolymer molecules within the hydrogel matrix. Flocculation can be avoided by ensuring that the sum of the repulsive interactions (e.g., electrostatic and steric) is stronger than the sum of the attractive interactions (e.g., van der Waals, hydrophobic, depletion, or bridging). The tendency for gravitational separation to occur depends on the size and density of the filled hydrogel particles, as well as the rheology of the surrounding continuous phase (Matalanis et al. 2011). The creaming rate of noninteracting rigid spherical particles in a dilute Newtonian liquid is given in Equation 6.2. This equation indicates that the rate of gravitational separation can be decreased by decreasing the hydrogel particle size, decreasing the density contrast, or increasing the viscosity of the aqueous phase.

The overall density of a filled hydrogel particle depends on the densities ($\rho$) and volume fractions ($\varphi$) of the different components within the particle. To a first approximation, the overall density of a filled hydrogel particle containing biopolymer, water, and oil droplets in different ratios is given by (Matalanis et al. 2011)

$$\rho_{Particle} = \varphi_B \rho_B + \varphi_O \rho_O + (1 - \varphi_B - \varphi_O)\rho_W. \tag{6.6}$$

Here, the subscripts B, W, and O refer to the biopolymer, water, and oil phase, respectively. In many applications, biopolymer particles may contain other components (such as minerals or solids) and so the above equation must be extended. Typically, the densities of biopolymers are higher than that of water, whereas the densities of liquid oils are lower. Consequently, the overall particle density will be higher than that of water at high biopolymer concentrations, thereby favoring sedimentation, but lower than that of water at low biopolymer concentrations, thereby favoring creaming. At a particular particle composition, the density of the particle will equal the density of the surrounding aqueous phase, which will stop gravitational separation. This effect can be seen in Figure 6.16, which shows calculations of the rate of gravitational separation for filled hydrogel particles with different compositions (Matalanis et al. 2011).

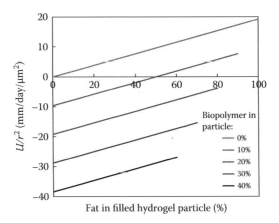

**Figure 6.16**  **The stability of filled hydrogel particles to gravitational separation depends on their composition, that is, the amount of oil, biopolymer, and water they contain. Here, $U$ is the creaming velocity and $r$ is the particle radius.**

## 6.4.5.4 Applications

Studies have shown that filled hydrogel particles can be formed from food-grade ingredients using a variety of different approaches, including injection methods (Chen et al. 2006; Li et al. 2012a; Malone et al. 2003), antisolvent precipitation (Patel et al. 2010; Zhong et al. 2008), aggregative phase separation (Benichou et al. 2002; Desai and Park 2005; Li et al. 2012a; Madene et al. 2006; Renken and Hunkeler 1998; Schmitt et al. 1998; Weinbreck et al. 2004), and segregative phase separation (Kim et al. 2006; Lian et al. 2004; Malone and Appelqvist 2003; Matalanis et al. 2010; Matalanis and McClements 2013; Norton and Frith 2001). Filled hydrogel particles based on aggregative phase separation have been used to encapsulate and protect ω-3 fatty acids (Lamprecht et al. 2001; Wu et al. 2005) and flavor oils (Weinbreck et al. 2004), while those based on segregative phase separation have also been used to encapsulate fish oils (Kim et al. 2006). As a specific example of this approach, we will briefly review the formation of filled hydrogel particles based on complex coacervation to stabilize ω-3 fatty acids (Lamprecht et al. 2001). In this study, an O/W emulsion stabilized by gelatin was initially formed, and then it was mixed with a solution containing gum acacia. The pH was then adjusted to promote complex coacervation of the gelatin and gum acacia, which led to the formation of a thick biopolymer shell around the lipid droplets. Finally, different methods were used to cross-link or harden the biopolymer shells so as to increase the long-term stability of the filled hydrogel particles (Lamprecht et al. 2001). The authors reported that filled hydrogel particles hardened by ethanol were the most stable to lipid oxidation. The encapsulation of lipids within hydrogel particles may be beneficial for the protection and delivery of a variety of lipophilic bioactive components in foods. The biopolymer shell could be constructed so that it provides chemical or physical protection of the encapsulated lipids. For example, the composition, thickness, permeability, and environmental responsiveness of the biopolymer shell could be designed to prevent lipid oxidation or to prevent instability owing to partial coalescence (which may occur in conventional emulsions). The shell can also be designed so that it releases the bioactive ingredients at the appropriate site within the human digestive system, for example, the mouth, stomach, or small intestine (Matalanis et al. 2010; McClements and Li 2010). This kind of system is widely used in the pharmaceutical industry for the delivery of drugs and is likely to gain increasing utilization within the food industry once suitable formulations and preparation conditions have been identified.

## 6.4.6 Microclusters

### 6.4.6.1 Composition and Structure

Microclusters are formed by controlled aggregation of oil droplets (Dickinson 2012b; Mao and McClements 2013). This process can be carried out by aggregation of similar oil droplets ("homoaggregation") or of dissimilar oil droplets ("heteroaggregation"). A suspension of microclusters consists of flocculated oil droplets dispersed within an aqueous continuous phase (Figure 6.17). The physicochemical properties of microcluster suspensions can be manipulated by controlling the properties of the oil droplets they are fabricated from (such as particle size, concentration, physical state, and interfacial properties), solution conditions (such as pH and ionic strength), and preparation procedures (such as particle ratio, order

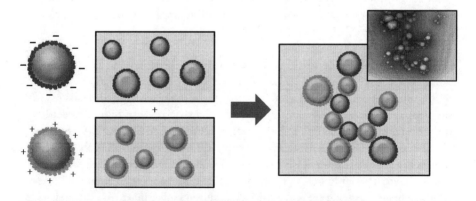

**Figure 6.17** **Microclusters can be formed by mixing together two emulsions containing oppositely charged droplets owing to heteroaggregation. The image is an electron microscopy image of a system formed from positively and negatively charged protein-coated droplets.**

of addition, and mixing conditions). The functional performance of a microcluster suspension depends on the characteristics of the microclusters formed, for example, the total number of flocs present, the strength of the attractive forces between the droplets, the number of droplets per floc, and the dimensions and shape of the flocs (McClements 2005a).

### 6.4.6.2 Formation

Microcluster dispersions may be fabricated in a variety of different ways depending on the main driving force for droplet flocculation. In general, flocculation tends to occur when the attractive forces acting between the oil droplets dominate the repulsive forces (McClements 2005a). Droplet flocculation may therefore be induced by increasing the strength of the attractive forces or reducing the strength of the repulsive forces between the oil droplets. The approach used is highly system dependent and is largely governed by the nature of the emulsifier(s) stabilizing the droplets. As mentioned earlier, flocculation may involve the association of similar oil droplets (homoaggregation) or dissimilar oil droplets (heteroaggregation). Dissimilar oil droplets may differ in terms of their charges, compositions, physical states, or sizes, although droplets with different charges are most commonly used. In an emulsion containing uncharged droplets, microcluster formation can be induced by adding nonadsorbing biopolymers or surfactant micelles to the continuous phase to induce depletion flocculation (McClements 2005a). In emulsions containing similarly charged droplets, flocculation can be induced by altering the pH to reduce the charge on the droplets, adding electrolyte to screen the electrostatic interactions between the droplets, or mixing oppositely charged polyelectrolytes with the droplets to promote bridging (McClements 2005a). Flocculation can also be induced through manipulation of electrostatic interactions by mixing an emulsion containing positively charged droplets with one containing negatively charged droplets, as shown in Figure 6.17 (Mao and McClements 2011, 2012a–e).

To control the physicochemical and functional properties of microcluster dispersions, it is important to control the nature of the microclusters formed, for example, the strength of the attractive forces holding the droplets together, and the number and structural organization of the droplets within the clusters. The way this can be achieved depends on the mechanism used to induce flocculation in the system. Experimental and theoretical studies have shown that a variety of structures can be formed using two or more different types of colloidal particles as building blocks (Lopez-Lopez et al. 2006, 2009; Yates et al. 2005). Recently, it was shown that microclusters with different morphologies could be created through heteroaggregation of positively charged (lactoferrin) and negatively charged (β-lactoglobulin) protein-coated oil droplets (Mao and McClements 2011). When one of the particle types is in excess, the microclusters formed consist of a central particle surrounded by oppositely charged particles (Figure 6.18). On the other hand, large microclusters containing many droplets are formed when the

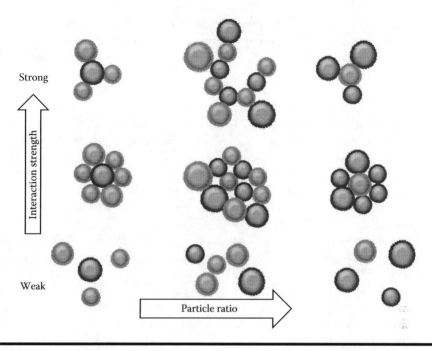

**Strong**

Interaction strength

**Weak**

Particle ratio

**Figure 6.18** **The microstructure of microclusters formed by heteroaggregation depends on the particle ratio and strength of the colloidal interactions between the oil droplets.**

two types of particles are present in approximately equal amounts. The packing of the droplets within the microclusters can be altered by modulating the strength of the colloidal interactions between them. When the attractive interactions are very weak, then the particles will not aggregate. When the attractive interactions are moderate, then the particles will form dense microclusters since the droplets are free to move and reposition themselves after aggregating. When the attractive interactions are strong, then the particles will form open microclusters because the droplets tend to remain in the initial location where they first encountered each other. Experimental studies have shown that the physicochemical properties of microcluster suspensions can be manipulated by controlling the total droplet concentration, the ratio of positive-to-negative droplets, mixing procedures, and solution conditions, for example, pH, ionic strength, and temperature (Mao and McClements 2011, 2012a–e).

### 6.4.6.3 Properties

Microcluster suspensions have properties very different from those of suspensions of nonflocculated oil droplets. The flocculation of oil droplets changes their light scattering efficiency and therefore alters the overall lightness of emulsions.

Experimental studies have shown that droplet flocculation causes a slight decrease in the lightness of O/W emulsions (Chantrapornchai et al. 2001; Chung et al. 2013). Microcluster formation may also have an indirect influence on the overall appearance of emulsion-based delivery systems by altering the spatial location of the oil droplets within the system owing to gravitational separation (Moschakis et al. 2010). Dilute microcluster suspensions are highly susceptible to creaming because of the increase in particle size after flocculation (Chanamai and McClements 2000). Consequently, a white cream layer typically forms at the top of an emulsion, while a clear serum layer forms at the bottom, which may be undesirable for many commercial products. On the other hand, concentrated microcluster suspensions may be more stable to gravitational separation than conventional emulsions because the droplets form a three-dimensional network that inhibits their movement (Mao and McClements 2012a–e). The creaming stability of microcluster suspensions is less important in products that have highly viscous or gelled aqueous phases, because then the droplet movement would be inhibited (Equation 6.2).

The viscosity of a microcluster suspension may be considerably higher than that of a nonflocculated emulsion of similar oil content because of the presence of the water trapped within the floc structure (Quemada and Berli 2002). The effective volume fraction of the particles in a flocculated system is greater than that of the oil phase ($\varphi$) because of this trapped water: $\varphi_{eff} = \varphi \times (1 + \Phi_W)$, where $\Phi_W$ is the fraction of water phase trapped within the microcluster particles. The value of $\Phi_W$ increases as the packing of the oil droplets within the flocs becomes more open, which usually occurs when the attractive interactions between the droplets becomes stronger (Figure 6.18). Thus, the viscosity of a microcluster suspension will be higher than that of a conventional emulsion by an amount that depends on the packing of the droplets within the flocs. The viscosity of microcluster suspensions usually decreases steeply with increasing applied shear stress ("shear thinning"), which can be attributed to deformation and disruption of the flocs reducing the effective volume fraction of the microclusters (Berli et al. 2002; Quemada and Berli 2002). The critical shear stress where shear thinning occurs depends on the strength of the attractive interactions between the droplets. One of the major potential advantages of microcluster suspensions is therefore that their rheological properties can be controlled by modulating droplet–droplet interactions. Recent experiments have shown that the rheological properties of suspensions of microclusters can be controlled by altering the size, concentration, and interactions of the oil droplets (Mao and McClements 2011, 2012a–e).

## 6.4.6.4 Applications

One of the most promising potential applications of microclusters is controlling the rheological properties of emulsion-based delivery systems. As mentioned earlier, a considerable amount of water is trapped between the oil droplets in microclusters. This trapped water greatly increases the effective particle volume fraction of the system, thereby leading to an increase in shear viscosity (Berli et al. 2002; Quemada and Berli 2002). In addition, at sufficiently high total particle concentrations, they can be used to form three-dimensional networks that extend throughout the volume of the material and thus provide gel-like or paste-like properties (Dickinson 2012b). Indeed, recent experiments have shown that mixtures of positively and negatively charged oil droplets can have viscosities that are over three orders of magnitude higher than the individual emulsions from which they were prepared (Mao and McClements 2011). These systems were fabricated by mixing an emulsion containing cationic droplets (lactoferrin coated) with one containing anionic droplets (β-lactoglobulin coated), leading to heteroaggregation (Figure 6.19). The controlled aggregation of oil droplets may also have an appreciable influence upon their behavior within the human gastrointestinal tract. Recent experiments have shown that flocculated droplets may be digested at a slower rate by the enzyme lipase than nonflocculated droplets

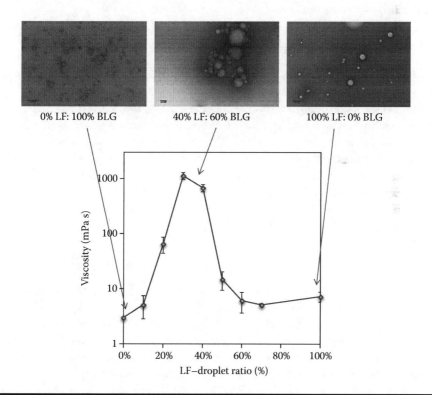

**Figure 6.19**  **Microclusters formed by heteroaggregation may have much greater viscosities than either of the two emulsions they are prepared from.**

(Golding et al. 2011), which may have important consequences for controlled or targeted release applications. However, we recently found that the lipid phase was digested at a similar rate and extent in conventional emulsions and microcluster suspensions, which was attributed to dissociation of the microclusters studied under gastrointestinal conditions (Tokle et al. 2013). Microclusters could also be used as a method of encapsulating two or more different lipophilic components in a single delivery system. For example, one functional lipophilic component could be trapped within negatively charged droplets and a different functional lipophilic component could be trapped within positively charged droplets, and then the two systems could be mixed together to form microclusters. The rate at which the two components are released could then be varied by using droplets with different core compositions, interfacial layers, or dimensions.

## 6.4.7 Miscellaneous Systems

There are a number of emulsion-based delivery systems or closely related systems that do not fit into the various categories discussed in previous sections. In this section, a brief overview of some of these systems is given.

### 6.4.7.1 Particle-Stabilized Emulsions

Particle-stabilized emulsions (sometimes called Pickering emulsions or colloidosomes) consist of oil droplets coated by a layer of solid particles (Aveyard et al. 2003; Dickinson 2012a,b, 2013; Yi et al. 2011; Zeng et al. 2006). To prepare an O/W emulsion, it is necessary to use solid particles that are better wetted by the water phase than the oil phase so that they protrude into the aqueous phase after adsorption to the fat droplet surfaces (McClements 2012). There has been considerable interest in these systems because of their high stability to droplet aggregation and Ostwald ripening (Aveyard et al. 2003). The major challenge to developing these systems is finding suitable food-grade solid particles to adsorb to the oil–water interfaces (Dickinson 2012b).

### 6.4.7.2 Emulsified Microemulsions and Cubosomes

Emulsified microemulsions have a similar structure to W/O/W emulsions but the oil droplets contain W/O microemulsions (or reverse micelles), rather than water droplets (Lutz et al. 2007). In this case, a W/O microemulsion solution is formed by dispersing water- and oil-soluble surfactant in an oil phase. The W/O microemulsion solution is then homogenized with an aqueous phase containing a water-soluble emulsifier to form an emulsified microemulsion. Similarly, emulsified cubosomes can be formed by homogenizing an oily cubosome phase with an aqueous phase containing a water-soluble emulsifier (Kulkarni et al. 2010; Muller et al. 2010; Siekmann et al. 2002). Hydrophilic active ingredients can be incorporated into the polar interior of the internal microemulsion or cubosome phase inside the oil droplets.

### 6.4.7.3 Nanocrystal Suspensions

Nanocrystal suspensions have some characteristics similar to those of emulsions, as they consist of small lipid particles (usually coated by emulsifier) dispersed in an aqueous solution. Nanocrystal suspensions can be formed by bottom–up methods (e.g., controlled crystallization or antisolvent precipitation) or by top–down methods (e.g., breaking larger crystals into smaller ones using mechanical devices) (Junghanns and Muller 2008; Muller et al. 2011; Shegokar and Muller 2010). Nanocrystals may be useful for delivering active ingredients that cannot be dissolved in oil or water phases and therefore have to be delivered in a solid form.

## 6.5 Summary

As shown in this chapter, a wide variety of emulsion-based delivery systems is currently available for the encapsulation of active ingredients, each with its own advantages and disadvantages. Most of these systems are suitable for encapsulating lipophilic ingredients that can be incorporated into an oil phase, while some of them are also suitable for encapsulating hydrophilic ingredients (e.g., multiple emulsions and filled hydrogel particles). Conventional O/W emulsions are currently the most widely used method of encapsulating lipophilic ingredients, but they are often susceptible to breakdown over time or when they are exposed to certain environmental stresses during production, transport, storage, or utilization. In addition, they have limited ability to protect and control the release of certain types of lipophilic ingredients. Consequently, there has been increasing interest in alternative types of emulsion-based delivery systems, such as nanoemulsions, multiple emulsions, multilayer emulsions, SLPs, filled hydrogel particles, and microclusters. These more complex structured emulsions often have certain advantages over conventional emulsions, but they are often more complicated to prepare and are sometimes more unstable, and therefore there needs to be a strong reason for utilizing them.

## References

Aguilera, J. M. (2006). "Food microstructure affects the bioavailability of several nutrients." *Journal of Food Science* **72**(2): R21–R32.

Angelatos, A. S., A. P. R. Johnston, Y. J. Wang and F. Caruso (2007). "Probing the permeability of polyelectrolyte multilayer capsules via a molecular beacon approach." *Langmuir* **23**(8): 4554–4562.

Anton, N. and T. F. Vandamme (2009). "The universality of low-energy nano-emulsification." *International Journal of Pharmaceutics* **377**(1–2): 142–147.

Aoki, T., E. A. Decker and D. J. McClements (2005). "Influence of environmental stresses on stability of O/W emulsions containing droplets stabilized by multilayered membranes produced by a layer-by-layer electrostatic deposition technique." *Food Hydrocolloids* **19**(2): 209–220.

Aserin, A. (2008). *Multiple Emulsions: Technology and Applications.* Hoboken, NJ, John Wiley and Sons.

Aveyard, R., B. P. Binks and J. H. Clint (2003). "Emulsions stabilised solely by colloidal particles." *Advances in Colloid and Interface Science* **100**: 503–546.

Awad, T. S., T. Helgason, J. Weiss, E. A. Decker and D. J. McClements (2009). "Effect of omega-3 fatty acids on crystallization, polymorphic transformation and stability of tripalmitin solid lipid nanoparticle suspensions." *Crystal Growth and Design* **9**(8): 3405–3411.

Balcao, V. M., C. I. Costa, C. M. Matos, C. G. Moutinho, M. Amorim, M. E. Pintado, A. P. Gomes, M. M. Vila and J. A. Teixeira (2013). "Nanoencapsulation of bovine lactoferrin for food and biopharmaceutical applications." *Food Hydrocolloids* **32**(2): 425–431.

BeMiller, J. N. and K. C. Huber (2008). Carbohydrates. In *Food Chemistry*, S. Damodaran, K. L. Parkin and O. R. Fennema (eds.). Boca Raton, FL, CRC Press: 83–154.

Benichou, A., A. Aserin and N. Garti (2002). "Protein-polysaccharide interactions for stabilization of food emulsions." *Journal of Dispersion Science and Technology* **23**(1–3): 93–123.

Benichou, A., A. Aserin and N. Garti (2004). "Double emulsions stabilized with hybrids of natural polymers for entrapment and slow release of active matters." *Advances in Colloid and Interface Science* **108–109**: 29–41.

Benjamin, O., P. Silcock, M. Leus and D. W. Everett (2012). "Multilayer emulsions as delivery systems for controlled release of volatile compounds using pH and salt triggers." *Food Hydrocolloids* **27**(1): 109–118.

Beristain, C. I., H. S. Garcia and E. J. Vernon-Carter (2001). "Spray-dried encapsulation of cardamom (Elettaria cardamomum) essential oil with mesquite (Prosopis juliflora) gum." *Lebensmittel-Wissenschaft Und-Technologie-Food Science and Technology* **34**(6): 398–401.

Berli, C. L. A., D. Quemada and A. Parker (2002). "Modelling the viscosity of depletion flocculated emulsions." *Colloids and Surfaces A-Physicochemical and Engineering Aspects* **203**(1–3): 11–20.

Bohren, C. F. and D. R. Huffman (1983). *Absorption and Scattering of Light by Small Particles.* New York, John Wiley and Sons, Inc.

Bonnaire, L., S. Sandra, T. Helgason, E. A. Decker, J. Weiss and D. J. McClements (2008). "Influence of lipid physical state on the in vitro digestibility of emulsified lipids." *Journal of Agricultural and Food Chemistry* **56**(10): 3791–3797.

Bunjes, H. (2010). "Lipid nanoparticles for the delivery of poorly water-soluble drugs." *Journal of Pharmacy and Pharmacology* **62**(11): 1637–1645.

Bunjes, H. (2011). "Structural properties of solid lipid based colloidal drug delivery systems." *Current Opinion in Colloid and Interface Science* **16**(5): 405–411.

Bunjes, H., F. Steiniger and W. Richter (2007). "Visualizing the structure of triglyceride nanoparticles in different crystal modifications." *Langmuir* **23**(7): 4005–4011.

Burgess, D. J. (1990). "Practical analysis of complex coacervate systems." *Journal of Colloid and Interface Science* **140**(1): 227–238.

Chanamai, R. and D. J. McClements (2000). "Creaming stability of flocculated mono-disperse oil-in-water emulsions." *Journal of Colloid and Interface Science* **225**(1): 214–218.

Chantrapornchai, W., F. M. Clydesdale and D. J. McClements (2001). "Influence of flocculation on optical properties of emulsions." *Journal of Food Science* **66**(3): 464–469.

Chee, C. P., D. Djordjevic, H. Faraji, E. A. Decker, R. Hollender, D. J. McClements, D. G. Peterson, R. F. Roberts and J. N. Coupland (2007). "Sensory properties of vanilla and strawberry flavored ice cream supplemented with omega-3 fatty acids." *Milchwissenschaft-Milk Science International* **62**(1): 66–69.

Chee, C. P., J. J. Gallaher, D. Djordjevic, H. Faraji, D. J. McClements, E. A. Decker, R. Hollender, D. G. Peterson, R. F. Roberts and J. N. Coupland (2005). "Chemical and sensory analysis of strawberry flavoured yogurt supplemented with an algae oil emulsion." *Journal of Dairy Research* **72**(3): 311–316.

Chen, B. C., D. J. McClements, D. A. Gray and E. A. Decker (2010). "Stabilization of soybean oil bodies by enzyme (laccase) cross-linking of adsorbed beet pectin coatings." *Journal of Agricultural and Food Chemistry* **58**(16): 9259–9265.

Chen, L. Y., G. E. Remondetto and M. Subirade (2006). "Food protein-based materials as nutraceutical delivery systems." *Trends in Food Science and Technology* **17**(5): 272–283.

Chen, C. C., Y. Y. Tu and H. M. Chang (1999). "Efficiency and protective effect of encapsulation of milk immunoglobulin G in multiple emulsion." *Journal of Agricultural and Food Chemistry* **47**(2): 407–410.

Cho, Y. H., E. A. Decker and D. J. McClements (2010). "Formation of protein-rich coatings around lipid droplets using the electrostatic deposition method." *Langmuir* **26**(11): 7937–7945.

Cho, Y. H. and D. J. McClements (2009). "Theoretical stability maps for guiding preparation of emulsions stabilized by protein-polysaccharide interfacial complexes." *Langmuir* **25**(12): 6649–6657.

Cho, Y. H., H. K. Shim and J. Park (2003). "Encapsulation of fish oil by an enzymatic gelation process using transglutaminase cross-linked proteins." *Journal of Food Science* **68**(9): 2717–2723.

Chun, J. Y., M. J. Choi, S. G. Min and J. Weiss (2013). "Formation and stability of multiple-layered liposomes by layer-by-layer electrostatic deposition of biopolymers." *Food Hydrocolloids* **30**(1): 249–257.

Chung, C., B. Degner and D. J. McClements (2013). "Designing reduced-fat food emulsions: Locust bean gum-fat droplet interactions." *Food Hydrocolloids* **32**(2): 263–270.

Cooper, C. L., P. L. Dubin, A. B. Kayitmazer and S. Turksen (2005). "Polyelectrolyte-protein complexes." *Current Opinion in Colloid and Interface Science* **10**(1–2): 52–78.

Cornacchia, L. and Y. H. Roos (2011). "State of dispersed lipid carrier and interface composition as determinants of beta-carotene stability in oil-in-water emulsions." *Journal of Food Science* **76**(8): C1211–C1218.

Coupland, J. N. (2002). "Crystallization in emulsions." *Current Opinion in Colloid and Interface Science* **7**(5–6): 445–450.

Coupland, J. N. and D. J. McClements (1996). "Lipid oxidation in food emulsions." *Trends in Food Science and Technology* **7**(3): 83–91.

Cournarie, F., M. P. Savelli, W. Rosilio, F. Bretez, C. Vauthier, J. L. Grossiord and M. Seiller (2004). "Insulin-loaded W/O/W multiple emulsions: Comparison of the performances of systems prepared with medium-chain-triglycerides and fish oil." *European Journal of Pharmaceutics and Biopharmaceutics* **58**(3): 477–482.

Cui, S. W. (2005). *Food Carbohydrates: Chemistry, Physical Properties and Applications*. Boca Raton, FL, Taylor and Francis.

de Kruif, C. G., F. Weinbreck and R. de Vries (2004). "Complex coacervation of proteins and anionic polysaccharides." *Current Opinion in Colloid and Interface Science* **9**(5): 340–349.

Decher, G. and J. B. Schlenoff (2003). *Multilayer Thin Films: Sequential Assembly of Nanocomposite Materials*. Weinheim, Wiley-VCH.

Desai, K. G. H. and H. J. Park (2005). "Recent developments in microencapsulation of food ingredients." *Drying Technology* **23**(7): 1361–1394.

Dickinson, E. (1992). *Introduction to Food Colloids*. Cambridge, UK, Royal Society of Chemistry.

Dickinson, E. (2003). "Hydrocolloids at interfaces and the influence on the properties of dispersed systems." *Food Hydrocolloids* **17**(1): 25–39.

Dickinson, E. (2011a). "Double emulsions stabilized by food biopolymers." *Food Biophysics* **6**(1): 1–11.

Dickinson, E. (2011b). "Mixed biopolymers at interfaces: Competitive adsorption and multilayer structures." *Food Hydrocolloids* **25**(8): 1966–1983.

Dickinson, E. (2012a). "Emulsion gels: The structuring of soft solids with protein-stabilized oil droplets." *Food Hydrocolloids* **28**(1): 224–241.

Dickinson, E. (2012b). "Use of nanoparticles and microparticles in the formation and stabilization of food emulsions." *Trends in Food Science and Technology* **24**(1): 4–12.

Dickinson, E. (2013). "Stabilising emulsion-based colloidal structures with mixed food ingredients." *Journal of the Science of Food and Agriculture* **93**(4): 710–721.

Djordjevic, D., L. Cercaci, J. Alamed, D. J. McClements and E. A. Decker (2007). "Chemical and physical stability of citral and limonene in sodium dodecyl sulfate-chitosan and gum arabic-stabilized oil-in-water emulsions." *Journal of Agricultural and Food Chemistry* **55**(9): 3585–3591.

Drusch, S. (2007). "Sugar beet pectin: A novel emulsifying wall component for microencapsulation of lipophilic food ingredients by spray-drying." *Food Hydrocolloids* **21**(7): 1223–1228.

Drusch, S. and S. Mannino (2009). "Patent-based review on industrial approaches for the microencapsulation of oils rich in polyunsaturated fatty acids." *Trends in Food Science and Technology* **20**(6–7): 237–244.

Drusch, S. and K. Schwarz (2006). "Microencapsulation properties of two different types of n-octenylsuccinate-derivatised starch." *European Food Research and Technology* **222**(1–2): 155–164.

Dubes, A., H. Parrot-Lopez, W. Abdelwahed, G. Degobert, H. Fessi, P. Shahgaldian and A. W. Coleman (2003). "Scanning electron microscopy and atomic force microscopy imaging of solid lipid nanoparticles derived from amphiphilic cyclodextrins." *European Journal of Pharmaceutics and Biopharmaceutics* **55**(3): 279–282.

Fathi, M., M. R. Mozafari and M. Mohebbi (2012). "Nanoencapsulation of food ingredients using lipid based delivery systems." *Trends in Food Science and Technology* **23**(1): 13–27.

Fechner, A., A. Knoth, I. Scherze and G. Muschiolik (2007). "Stability and release properties of double-emulsions stabilised by caseinate-dextran conjugates." *Food Hydrocolloids* **21**(5–6): 943–952.

Frank, K., K. Kohler and H. P. Schuchmann (2011). "Formulation of labile hydrophilic ingredients in multiple emulsions: Influence of the formulation's composition on the emulsion's stability and on the stability of entrapped bioactives." *Journal of Dispersion Science and Technology* **32**(12): 1753–1758.

Frank, K., E. Walz, V. Graf, R. Greiner, K. Kohler and H. P. Schuchmann (2012). "Stability of anthocyanin-rich W/O/W-emulsions designed for intestinal release in gastrointestinal environment." *Journal of Food Science* **77**(12): N50–N57.

Fredrick, E., P. Walstra and K. Dewettinck (2010). "Factors governing partial coalescence in oil-in-water emulsions." *Advances in Colloid and Interface Science* **153**(1–2): 30–42.

Freitas, S., H. P. Merkle and B. Gander (2005). "Microencapsulation by solvent extraction/ evaporation: Reviewing the state of the art of microsphere preparation process technology." *Journal of Controlled Release* **102**(2): 313–332.

Friberg, S., K. Larsson and J. Sjoblom (2004). *Food Emulsions*. New York, Marcel Dekker.

Garti, N. (1997a). "Double emulsions—Scope, limitations and new achievements." *Colloids and Surfaces A-Physicochemical and Engineering Aspects* **123**: 233–246.

Garti, N. (1997b). "Progress in stabilization and transport phenomena of double emulsions in food applications." *Food Science and Technology-Lebensmittel-Wissenschaft and Technologie* **30**(3): 222–235.

Garti, N. and A. Benichou (2004). Recent developments in double emulsions for food applications. In *Food Emulsions*, S. E. Friberg, K. Larsson and J. Sjoblom (eds.). New York, Marcel Dekker.

Garti, N. and C. Bisperink (1998). "Double emulsions: Progress and applications." *Current Opinion in Colloid and Interface Science* **3**(6): 657–667.

Genovese, D. B., J. E. Lozano and M. A. Rao (2007). "The rheology of colloidal and noncolloidal food dispersions." *Journal of Food Science* **72**(2): R11–R20.

Ghosh, S., D. G. Peterson and J. N. Coupland (2007). "Aroma release from solid droplet emulsions: Effect of lipid type." *Journal of the American Oil Chemists Society* **84**(11): 1001–1014.

Giroux, H. J., S. Constantineau, P. Fustier, C. P. Champagne, D. St-Gelais, M. Lacroix and M. Britten (2013). "Cheese fortification using water-in-oil-in-water double emulsions as carrier for water soluble nutrients." *International Dairy Journal* **29**(2): 107–114.

Golding, M., T. J. Wooster, L. Day, M. Xu, L. Lundin, J. Keogh and P. Clifton (2011). "Impact of gastric structuring on the lipolysis of emulsified lipids." *Soft Matter* **7**(7): 3513–3523.

Grigoriev, D. O. and R. Miller (2009). "Mono- and multilayer covered drops as carriers." *Current Opinion in Colloid and Interface Science* **14**(1): 48–59.

Gu, Y., E. Decker and D. McClements (2006). "Irreversible thermal denaturation of beta-lactoglobulin retards adsorption of carrageenan onto beta-lactoglobulin-coated droplets " *Langmuir* **22**: 7480–7486.

Gu, Y. S., E. A. Decker and D. J. McClements (2007a). "Application of multi-component biopolymer layers to improve the freeze-thaw stability of oil-in-water emulsions: Beta-lactoglobulin-iota-carrageenan-gelatin." *Journal of Food Engineering* **80**(4): 1246–1254.

Gu, Y. S., E. A. Decker and D. J. McClements (2007b). "Formation of colloidosomes by adsorption of small charged oil droplets onto the surface of large oppositely charged oil droplets." *Food Hydrocolloids* **21**(4): 516–526.

Gulseren, I., Y. Fang and M. Corredig (2012). "Whey protein nanoparticles prepared with desolvation with ethanol: Characterization, thermal stability and interfacial behavior." *Food Hydrocolloids* **29**(2): 258–264.

Guo, Q., A. Q. Ye, M. Lad, D. Dalgleish and H. Singh (2013). "The breakdown properties of heat-set whey protein emulsion gels in the human mouth." *Food Hydrocolloids* **33**(2): 215–224.

Guzey, D., H. J. Kim and D. J. McClements (2004). "Factors influencing the production of O/W emulsions stabilized by beta-lactoglobulin-pectin membranes." *Food Hydrocolloids* **18**(6): 967–975.

Guzey, D. and D. J. McClements (2006). "Formation, stability and properties of multi-layer emulsions for application in the food industry." *Advances in Colloid and Interface Science* **128**: 227–248.

Guzey, D. and D. J. McClements (2007). "Impact of electrostatic interactions on formation and stability of emulsions containing oil droplets coated by beta-lactoglobulin-pectin complexes." *Journal of Agricultural and Food Chemistry* **55**(2): 475–485.

Harnsilawat, T., R. Pongsawatmanit and D. J. McClements (2006). "Characterization of beta-lactoglobulin-sodium alginate interactions in aqueous solutions: A calorimetry, light scattering, electrophoretic mobility and solubility study." *Food Hydrocolloids* **20**(5): 577–585.

Helgason, T., T. S. Awad, K. Kristbergsson, E. A. Decker, D. J. McClements and J. Weiss (2009). "Impact of surfactant properties on oxidative stability of beta-carotene encapsulated within solid lipid nanoparticles." *Journal of Agricultural and Food Chemistry* **57**(17): 8033–8040.

Helgason, T., T. S. Awad, K. Kristbergsson, D. J. McClements and J. Weiss (2008). "Influence of polymorphic transformations on gelation of tripalmitin solid lipid nanoparticle suspensions." *Journal of the American Oil Chemists Society* **85**(6): 501–511.

Hemar, Y., L. J. Cheng, C. M. Oliver, L. Sanguansri and M. Augustin (2010). "Encapsulation of resveratrol using water-in-oil-in-water double emulsions." *Food Biophysics* **5**(2): 120–127.

Hentschel, A., S. Gramdorf, R. H. Muller and T. Kurz (2008). "Beta-carotene-loaded nano-structured lipid carriers." *Journal of Food Science* **73**(2): N1–N6.

Hou, Z. Q., Y. X. Gao, F. Yuan, Y. W. Liu, C. L. Li and D. X. Xu (2010). "Investigation into the physicochemical stability and rheological properties of beta-carotene emulsion stabilized by soybean soluble polysaccharides and chitosan." *Journal of Agricultural and Food Chemistry* **58**(15): 8604–8611.

Hu, M., Y. Li, E. A. Decker, H. Xiao and D. J. McClements (2011). "Impact of layer structure on physical stability and lipase digestibility of lipid droplets coated by biopolymer nanolaminated coatings." *Food Biophysics* **6**(1): 37–48.

Humblet-Hua, N. P. K., E. van der Linden and L. M. C. Sagis (2012). "Microcapsules with protein fibril reinforced shells: Effect of fibril properties on mechanical strength of the shell." *Journal of Agricultural and Food Chemistry* **60**(37): 9502–9511.

Hunter, R. J. (1986). *Foundations of Colloid Science*. Oxford, UK, Oxford University Press.

Hwang, Y. J., C. Oh and S. G. Oh (2005). "Controlled release of retinol from silica particles prepared in O/W/O emulsion: The effects of surfactants and polymers." *Journal of Controlled Release* **106**(3): 339–349.

Iancu, M. N., Y. Chevalie, M. Popa and T. Hamaide (2009). "Internally gelled W/O and W/O/W double emulsions." *E-Polymers* 99, 2009.

Iscan, Y., S. A. Wissing, S. Hekimoglu and R. H. Muller (2005). "Solid Lipid Nanoparticles (SLN (TM)) for topical drug delivery: Incorporation of the lipophilic drugs N,N-diethyl-m-toluamide and vitamin K." *Pharmazie* **60**(12): 905–909.

Israelachvili, J. (2011). *Intermolecular and Surface Forces*, 3rd Edition. London, Academic Press.

Jafari, S. M., E. Assadpoor, Y. H. He and B. Bhandari (2008). "Re-coalescence of emulsion droplets during high-energy emulsification." *Food Hydrocolloids* **22**(7): 1191–1202.

Jafari, S. M., Y. He and B. Bhandari (2007). "Optimization of nano-emulsions production by microfluidization." *European Food Research and Technology* **225**(5–6): 733–741.

Jee, J. P., S. J. Lim, J. S. Park and C. K. Kim (2006). "Stabilization of all-trans retinol by loading lipophilic antioxidants in solid lipid nanoparticles." *European Journal of Pharmaceutics and Biopharmaceutics* **63**(2): 134–139.

Jimenez, M., H. S. Garcia and C. I. Beristain (2004). "Spray-drying microencapsulation and oxidative stability of conjugated linoleic acid." *European Food Research and Technology* **219**(6): 588–592.

Jimenez-Colmenero, F. (2013). "Potential applications of multiple emulsions in the development of healthy and functional foods." *Food Research International* **52**(1): 64–74.

Junghanns, J. and R. H. Muller (2008). "Nanocrystal technology, drug delivery and clinical applications." *International Journal of Nanomedicine* **3**(3): 295–309.

Kim, H. J., E. A. Decker and D. J. McClements (2006). "Preparation of multiple emulsions based on thermodynamic incompatibility of heat-denatured whey protein and pectin solutions." *Food Hydrocolloids* **20**(5): 586–595.

Klinkesorn, U. and D. J. McClements (2010). "Impact of lipase, bile salts, and polysaccharides on properties and digestibility of tuna oil multilayer emulsions stabilized by Lecithin-Chitosan." *Food Biophysics* **5**(2): 73–81.

Klinkesorn, U., P. Sophanodora, P. Chinachoti, E. A. Decker and D. J. McClements (2005a). "Encapsulation of emulsified tuna oil in two-layered interfacial membranes prepared using electrostatic layer-by-layer deposition." *Food Hydrocolloids* **19**(6): 1044–1053.

Klinkesorn, U., P. Sophanodora, P. Chinachoti, E. A. Decker and D. J. McClements (2006). "Characterization of spray-dried tuna oil emulsified in two-layered interfacial membranes prepared using electrostatic layer-by-layer deposition." *Food Research International* **39**(4): 449–457.

Klinkesorn, U., P. Sophanodora, P. Chinachoti, D. J. McClements and E. A. Decker (2005b). "Increasing the oxidative stability of liquid and dried tuna oil-in-water emulsions with electrostatic layer-by-layer deposition technology." *Journal of Agricultural and Food Chemistry* **53**(11): 4561–4566.

Klinkesorn, U., P. Sophanodora, P. Chinachoti, D. J. McClements and E. A. Decker (2005c). "Stability of spray-dried tuna oil emulsions encapsulated with two-layered interfacial membranes." *Journal of Agricultural and Food Chemistry* **53**(21): 8365–8371.

Kuang, S. S., J. C. Oliveira and A. M. Crean (2010). "Microencapsulation as a tool for incorporating bioactive ingredients into food." *Critical Reviews in Food Science and Nutrition* **50**(10): 951–968.

Kukizaki, M. and M. Goto (2007). "Preparation and evaluation of uniformly sized solid lipid microcapsules using membrane emulsification." *Colloids and Surfaces A-Physicochemical and Engineering Aspects* **293**(1–3): 87–94.

Kulkarni, C. V., R. Mezzenga and O. Glatter (2010). "Water-in-oil nanostructured emulsions: Towards the structural hierarchy of liquid crystalline materials." *Soft Matter* **6**(21): 5615–5624.

Lamprecht, A., U. Schafer and C. M. Lehr (2001). "Influences of process parameters on preparation of microparticle used as a carrier system for omega-3 unsaturated fatty acid ethyl esters used in supplementary nutrition." *Journal of Microencapsulation* **18**(3): 347–357.

Laye, C., D. J. McClements and J. Weiss (2008). "Formation of biopolymer-coated liposomes by electrostatic deposition of chitosan." *Journal of Food Science* **73**(5): N7–N15.

Leal-Calderon, F., S. Homer, A. Goh and L. Lundin (2012). "W/O/W emulsions with high internal droplet volume fraction." *Food Hydrocolloids* **27**(1): 30–41.

Lee, J. J., I. B. Park, Y. H. Cho, C. S. Huh, Y. J. Baek and J. Park (2004). "Whey protein-based IgY microcapsules prepared by multiple emulsification and heat gelation." *Food Science and Biotechnology* **13**(4): 494–497.

Lee, S., E. A. Decker, C. Faustman and R. A. Mancini (2005). "The effects of antioxidant combinations on color and lipid oxidation in n-3 oil fortified ground beef patties." *Meat Science* **70**(4): 683–689.

Lee, S., C. Faustman, D. Djordjevic, H. Faraji and E. A. Decker (2006a). "Effect of antioxidants on stabilization of meat products fortified with n-3 fatty acids." *Meat Science* **72**(1): 18–24.

Lee, S., P. Hernandez, D. Djordjevic, H. Faraji, R. Hollender, C. Faustman and E. A. Decker (2006b). "Effect of antioxidants and cooking on stability of n-3 fatty acids in fortified meat products." *Journal of Food Science* **71**(3): C233–C238.

Lee, S. J., S. J. Choi, Y. Li, E. A. Decker and D. J. McClements (2011). "Protein-stabilized nanoemulsions and emulsions: Comparison of physicochemical stability, lipid oxidation, and lipase digestibility." *Journal of Agricultural and Food Chemistry* **59**(1): 415–427.

Lee, S. J. and D. J. McClements (2010). "Fabrication of protein-stabilized nanoemulsions using a combined homogenization and amphiphilic solvent dissolution/evaporation approach." *Food Hydrocolloids* **24**(6–7): 560–569.

Lesmes, U., S. Sandra, E. A. Decker and D. J. McClements (2010). "Impact of surface deposition of lactoferrin on physical and chemical stability of omega-3 rich lipid droplets stabilised by caseinate." *Food Chemistry* **123**(1): 99–106.

Li, Y., M. Hu, H. Xiao, Y. M. Du, E. A. Decker and D. J. McClements (2010). "Controlling the functional performance of emulsion-based delivery systems using multi-component biopolymer coatings." *European Journal of Pharmaceutics and Biopharmaceutics* **76**(1): 38–47.

Li, Y., J. Kim, Y. Park and D. J. McClements (2012a). "Modulation of lipid digestibility using structured emulsion-based delivery systems: Comparison of in vivo and in vitro measurements." *Food and Function* **3**(5): 528–536.

Li, Y., S. Le Maux, H. Xiao and D. J. McClements (2009). "Emulsion-based delivery systems for tributyrin, a potential colon cancer preventative agent." *Journal of Agricultural and Food Chemistry* **57**(19): 9243–9249.

Li, Y., H. Xiao and D. J. McClements (2012b). "Encapsulation and delivery of crystalline hydrophobic nutraceuticals using nanoemulsions: Factors affecting polymethoxyflavone solubility." *Food Biophysics* **7**(4): 341–353.

Li, Y., J. K. Zheng, H. Xiao and D. J. McClements (2012c). "Nanoemulsion-based delivery systems for poorly water-soluble bioactive compounds: Influence of formulation parameters on polymethoxyflavone crystallization." *Food Hydrocolloids* **27**(2): 517–528.

Lian, G. P., M. E. Malone, J. E. Homan and I. T. Norton (2004). "A mathematical model of volatile release in mouth from the dispersion of gelled emulsion particles." *Journal of Controlled Release* **98**(1): 139–155.

Littoz, F. and D. J. McClements (2008). "Bio-mimetic approach to improving emulsion stability: Cross-linking adsorbed beet pectin layers using laccase." *Food Hydrocolloids* **22**(7): 1203–1211.

Lomova, M. V., G. B. Sukhorukov and M. N. Antipina (2010). "Antioxidant coating of microsize droplets for prevention of lipid peroxidation in oil-in-water emulsion." *ACS Applied Materials and Interfaces* **2**(12): 3669–3676.

Lopez-Lopez, J. M., A. Schmitt, A. Moncho-Jorda and R. Hidalgo-Alvarez (2006). "Stability of binary colloids: Kinetic and structural aspects of heteroaggregation processes." *Soft Matter* **2**(12): 1025–1042.

Lopez-Lopez, J. M., A. Schmitt, A. Moncho-Jorda and R. Hidalgo-Alvarez (2009). "Electrostatic heteroaggregation regimes in colloidal suspensions." *Advances in Colloid and Interface Science* **147–148**: 186–204.

Losso, J. N., A. Khachatryan, M. Ogawa, J. S. Godber and F. Shih (2005). "Random centroid optimization of phosphatidylglycerol stabilized lutein-enriched oil-in-water emulsions at acidic pH." *Food Chemistry* **92**(4): 737–744.

Lutz, R., A. Aserin, E. J. Wachtel, E. Ben-Shoshan, D. Danino and N. Garti (2007). "A study of the emulsified microemulsion by SAXS, Cryo-TEM, SD-NMR, and electrical conductivity." *Journal of Dispersion Science and Technology* **28**(8): 1149–1157.

Madene, A., M. Jacquot, J. Scher and S. Desobry (2006). "Flavour encapsulation and controlled release—A review." *International Journal of Food Science and Technology* **41**(1): 1–21.

Malone, M. E. and I. A. M. Appelqvist (2003). "Gelled emulsion particles for the controlled release of lipophilic volatiles during eating." *Journal of Controlled Release* **90**(2): 227–241.

Malone, M. E., I. A. M. Appelqvist and I. T. Norton (2003). "Oral behaviour of food hydrocolloids and emulsions. Part 2. Taste and aroma release." *Food Hydrocolloids* **17**(6): 775–784.

Mao, Y. Y., M. Dubot, H. Xiao and D. J. McClements (2013). "Interfacial engineering using mixed protein systems: Emulsion-based delivery systems for encapsulation and stabilization of beta-carotene." *Journal of Agricultural and Food Chemistry* **61**(21): 5163–5169.

Mao, Y. Y. and D. J. McClements (2011). "Modulation of bulk physicochemical properties of emulsions by hetero-aggregation of oppositely charged protein-coated lipid droplets." *Food Hydrocolloids* **25**(5): 1201–1209.

Mao, Y. Y. and D. J. McClements (2012a). "Fabrication of functional micro-clusters by heteroaggregation of oppositely charged protein-coated lipid droplets." *Food Hydrocolloids* **27**(1): 80–90.

Mao, Y. Y. and D. J. McClements (2012b). "Fabrication of reduced fat products by controlled heteroaggregation of oppositely charged lipid droplets." *Journal of Food Science* **77**(5): E144–E152.

Mao, Y. Y. and D. J. McClements (2012c). "Fabrication of viscous and paste-like materials by controlled heteroaggregation of oppositely charged lipid droplets." *Food Chemistry* **134**(2): 872–879.

Mao, Y. Y. and D. J. McClements (2012d). "Influence of electrostatic heteroaggregation of lipid droplets on their stability and digestibility under simulated gastrointestinal conditions." *Food and Function* **3**(10): 1025–1034.

Mao, Y. Y. and D. J. McClements (2012e). "Modulation of emulsion rheology through electrostatic heteroaggregation of oppositely charged lipid droplets: Influence of particle size and emulsifier content." *Journal of Colloid and Interface Science* **380**: 60–66.

Mao, Y. Y. and D. J. McClements (2013). "Modification of emulsion properties by heteroaggregation of oppositely charged starch-coated and protein-coated fat droplets." *Food Hydrocolloids* **33**(2): 320–326.

Matalanis, A., O. G. Jones and D. J. McClements (2011). "Structured biopolymer-based delivery systems for encapsulation, protection, and release of lipophilic compounds." *Food Hydrocolloids* **25**(8): 1865–1880.

Matalanis, A., U. Lesmes, E. A. Decker and D. J. McClements (2010). "Fabrication and characterization of filled hydrogel particles based on sequential segregative and aggregative biopolymer phase separation." *Food Hydrocolloids* **24**(8): 689–701.

Matalanis, A. and D. J. McClements (2013). "Hydrogel microspheres for encapsulation of lipophilic components: Optimization of fabrication and performance." *Food Hydrocolloids* **31**(1): 15–25.

McClements, D. J. (2002a). "Colloidal basis of emulsion color." *Current Opinion in Colloid and Interface Science* **7**(5–6): 451–455.

McClements, D. J. (2002b). "Theoretical prediction of emulsion color." *Advances in Colloid and Interface Science* **97**(1–3): 63–89.

McClements, D. J. (2005a). *Food Emulsions: Principles, Practice, and Techniques.* Boca Raton, FL, CRC Press.

McClements, D. J. (2005b). "Theoretical analysis of factors affecting the formation and stability of multilayered colloidal dispersions." *Langmuir* **21**(21): 9777–9785.

McClements, D. J. (2010a). "Design of nano-laminated coatings to control bioavailability of lipophilic food components." *Journal of Food Science* **75**(1): R30–R42.

McClements, D. J. (2010b). "Emulsion design to improve the delivery of functional lipophilic components." *Annual Review of Food Science and Technology* **1**(1): 241–269.

McClements, D. J. (2011). "Edible nanoemulsions: Fabrication, properties, and functional performance." *Soft Matter* **7**(6): 2297–2316.

McClements, D. J. (2012a). "Advances in fabrication of emulsions with enhanced functionality using structural design principles." *Current Opinion in Colloid and Interface Science* **17**(5): 235–245.

McClements, D. J. (2012b). "Crystals and crystallization in oil-in-water emulsions: Implications for emulsion-based delivery systems." *Advances in Colloid and Interface Science* **174**(0): 1–30.

McClements, D. J. (2012c). "Nanoemulsions versus microemulsions: Terminology, differences, and similarities." *Soft Matter* **8**(6): 1719–1729.

McClements, D. J. and E. A. Decker (2000). "Lipid oxidation in oil-in-water emulsions: Impact of molecular environment on chemical reactions in heterogeneous food systems." *Journal of Food Science* **65**(8): 1270–1282.

McClements, D. J., E. A. Decker, Y. Park and J. Weiss (2009). "Structural design principles for delivery of bioactive components in nutraceuticals and functional foods." *Critical Reviews in Food Science and Nutrition* **49**(6): 577–606.

McClements, D. J. and K. Demetriades (1998). "An integrated approach to the development of reduced-fat food emulsions." *Critical Reviews in Food Science and Nutrition* **38**(6): 511–536.

McClements, D. J. and Y. Li (2010). "Structured emulsion-based delivery systems: Controlling the digestion and release of lipophilic food components." *Advances in Colloid and Interface Science* **159**(2): 213–228.

McClements, D. J. and J. Rao (2011). "Food-grade nanoemulsions: Formulation, fabrication, properties, performance, biological fate, and potential toxicity." *Critical Reviews in Food Science and Nutrition* **51**(4): 285–330.

McClements, D. J. and H. Xiao (2012). "Potential biological fate of ingested nanoemulsions: Influence of particle characteristics." *Food and Function* **3**(3): 202–220.

Mei, L. Y., S. J. Choi, J. Alamed, L. Henson, M. Popplewell, D. J. McClements and E. A. Decker (2010). "Citral stability in oil-in-water emulsions with solid or liquid octadecane." *Journal of Agricultural and Food Chemistry* **58**(1): 533–536.

Mezzenga, R., B. M. Folmer and E. Hughes (2004). "Design of double emulsions by osmotic pressure tailoring." *Langmuir* **20**(9): 3574–3582.

Moschakis, T., B. S. Murray and C. G. Biliaderis (2010). "Modifications in stability and structure of whey protein-coated o/w emulsions by interacting chitosan and gum arabic mixed dispersions." *Food Hydrocolloids* **24**(1): 8–17.

Muchow, M., P. Maincent and R. H. Muller (2008). "Lipid nanoparticles with a solid matrix (SLN, NLC, LDC) for oral drug delivery." *Drug Development and Industrial Pharmacy* **34**(12): 1394–1405.

Muller, F., A. Salonen and O. Glatter (2010). "Monoglyceride-based cubosomes stabilized by Laponite: Separating the effects of stabilizer, pH and temperature." *Colloids and Surfaces A-Physicochemical and Engineering Aspects* **358**(1–3): 50–56.

Muller, R. H., S. Gohla and C. M. Keck (2011). "State of the art of nanocrystals—Special features, production, nanotoxicology aspects and intracellular delivery." *European Journal of Pharmaceutics and Biopharmaceutics* **78**(1): 1–9.

Muller, R. H. and C. M. Keck (2004). "Challenges and solutions for the delivery of biotech drugs—A review of drug nanocrystal technology and lipid nanoparticles." *Journal of Biotechnology* **113**(1–3): 151–170.

Muller, R. H., K. Mader and S. Gohla (2000). "Solid lipid nanoparticles (SLN) for controlled drug delivery—A review of the state of the art." *European Journal of Pharmaceutics and Biopharmaceutics* **50**(1): 161–177.

Mun, S., E. A. Decker, Y. Park, J. Weiss and D. J. McClements (2006). "Influence of interfacial composition on in vitro digestibility of emulsified lipids: Potential mechanism for chitosan's ability to inhibit fat digestion." *Food Biophysics* **1**(1): 21–29.

Mun, S. H. and D. J. McClements (2006). "Influence of interfacial characteristics on Ostwald ripening in hydrocarbon oil-in-water emulsions." *Langmuir* **22**(4): 1551–1554.

Murugesan, R. and V. Orsat (2012). "Spray drying for the production of nutraceutical ingredients—A review." *Food and Bioprocess Technology* **5**(1): 3–14.

Muschiolik, G. (2007). "Multiple emulsions for food use." *Current Opinion in Colloid and Interface Science* **12**(4–5): 213–220.

Normand, M. D., U. Lesmes, M. G. Corradini and M. Peleg (2010). "Wolfram demonstrations: Free interactive software for food engineering education and practice." *Food Engineering Reviews* **2**(3): 157–167.

Norton, I. T. and W. J. Frith (2001). "Microstructure design in mixed biopolymer composites." *Food Hydrocolloids* **15**(4–6): 543–553.

Norton, I. T., W. J. Frith and S. Ablett (2006). "Fluid gels, mixed fluid gels and satiety." *Food Hydrocolloids* **20**(2–3): 229–239.

Ogawa, S., E. A. Decker and D. J. McClements (2003a). "Influence of environmental conditions on the stability of oil in water emulsions containing droplets stabilized by lecithin-chitosan membranes." *Journal of Agricultural and Food Chemistry* **51**(18): 5522–5527.

Ogawa, S., E. A. Decker and D. J. McClements (2003b). "Production and characterization of O/W emulsions containing cationic droplets stabilized by lecithin-chitosan membranes." *Journal of Agricultural and Food Chemistry* **51**(9): 2806–2812.

Okuda, S., D. J. McClements and E. A. Decker (2005). "Impact of lipid physical state on the oxidation of methyl linolenate in oil-in-water emulsions." *Journal of Agricultural and Food Chemistry* **53**(24): 9624–9628.

Onuki, Y., M. Morishita, H. Watanabe, Y. Chiba, S. Tokiwa, K. Takayama and T. Nagai (2003). "Improved insulin enteral delivery using water-in-oil-in-water multiple emulsion incorporating highly purified docosahexaenoic acid." *STP Pharma Sciences* **13**(4): 231–235.

Ostertag, F., J. Weiss and D. J. McClements (2012). "Low-energy formation of edible nano-emulsions: Factors influencing droplet size produced by emulsion phase inversion." *Journal of Colloid and Interface Science* **388**: 95–102.

Owusu, R. K., Q. H. Zhu and E. Dickinson (1992). "Controlled release of L-Tryptophan and vitamin-B2 from model water oil-water multiple emulsions." *Food Hydrocolloids* **6**(5): 443–453.

Pal, R. (2011). "Rheology of simple and multiple emulsions." *Current Opinion in Colloid and Interface Science* **16**(1): 41–60.

Patel, A. R., E. C. M. Bouwens and K. P. Velikov (2010). "Sodium caseinate stabilized zein colloidal particles." *Journal of Agricultural and Food Chemistry* **58**(23): 12497–12503.

Pinheiro, A. C., M. Lad, H. D. Silva, M. A. Coimbra, M. Boland and A. A. Vicente (2013). "Unravelling the behaviour of curcumin nanoemulsions during in vitro digestion: Effect of the surface charge." *Soft Matter* **9**(11): 3147–3154.

Pople, P. V. and K. K. Singh (2006). "Development and evaluation of topical formulation containing solid lipid nanoparticles of vitamin A." *AAPS Pharmscitech* **7**(4), 91, 2006.

Qian, C., E. A. Decker, H. Xiao and D. J. McClements (2012a). "Inhibition of beta-carotene degradation in oil-in-water nanoemulsions: Influence of oil-soluble and water-soluble antioxidants." *Food Chemistry* **135**(3): 1036–1043.

Qian, C., E. A. Decker, H. Xiao and D. J. McClements (2012b). "Physical and chemical stability of beta-carotene-enriched nanoemulsions: Influence of pH, ionic strength, temperature, and emulsifier type." *Food Chemistry* **132**(3): 1221–1229.

Qian, C., E. A. Decker, H. Xiao and D. J. McClements (2013). "Impact of lipid nanoparticle physical state on particle aggregation and beta-carotene degradation: Potential limitations of solid lipid nanoparticles." *Food Research International* **52**(1): 342–349.

Qian, C. and D. J. McClements (2011). "Formation of nanoemulsions stabilized by model food-grade emulsifiers using high-pressure homogenization: Factors affecting particle size." *Food Hydrocolloids* **25**(5): 1000–1008.

Quemada, D. and C. Berli (2002). "Energy of interaction in colloids and its implications in rheological modeling." *Advances in Colloid and Interface Science* **98**(1): 51–85.

Rao, J. J., E. A. Decker, H. Xiao and D. J. McClements (2013). "Nutraceutical nanoemulsions: Influence of carrier oil composition (digestible versus indigestible oil) on -carotene bioavailability." *Journal of the Science of Food and Agriculture* **93**(13): 3175–3183.

Renard, D., P. Robert, L. Lavenant, D. Melcion, Y. Popineau, J. Gueguen, C. Duclairoir, E. Nakache, C. Sanchez and C. Schmitt (2002). "Biopolymeric colloidal carriers for encapsulation or controlled release applications." *International Journal of Pharmaceutics* **242**(1–2): 163–166.

Renken, A. and D. Hunkeler (1998). "Microencapsulation: A review of polymers and technologies with a focus on bioartificial organs." *Polimery* **43**(9): 530–539.

Ribeiro, A. J., R. J. Neufeld, P. Arnaud and J. C. Chaumeil (1999). "Microencapsulation of lipophilic drugs in chitosan-coated alginate microspheres." *International Journal of Pharmaceutics* **187**(1): 115–123.

Ribeiro, H. S., K. Ax and H. Schubert (2003). "Stability of lycopene emulsions in food systems." *Journal of Food Science* **68**(9): 2730–2734.

Ribeiro, H. S., J. M. M. Guerrero, K. Briviba, G. Rechkemmer, H. P. Schuchmann and H. Schubert (2006). "Cellular uptake of carotenoid-loaded oil-in-water emulsions in colon carcinoma cells in vitro." *Journal of Agricultural and Food Chemistry* **54**(25): 9366–9369.

Ribeiro, H. S., L. G. Rico, G. G. Badolato and H. Schubert (2005). "Production of O/W emulsions containing astaxanthin by repeated Premix membrane emulsification." *Journal of Food Science* **70**(2): E117–E123.

Rodriguez-Huezo, M. E., R. Pedroza-Islas, L. A. Prado-Barragan, C. I. Beristain and E. J. Vernon-Carter (2004). "Microencapsulation by spray drying of multiple emulsions containing carotenoids." *Journal of Food Science* **69**(7): E351–E359.

Rossier-Miranda, F. J., K. Schroen and R. Boom (2012). "Microcapsule production by an hybrid colloidosome-layer-by-layer technique." *Food Hydrocolloids* **27**(1): 119–125.

Rubner, M. F. (2003). pH-controlled fabrication of polyelectrolyte multilayers: Assembly and applications. In *Multilayer Thin Films: Sequential Assembly of Nanocomposite Materials*, G. Decher and J. B. Schlenoff (eds.). Weinheim, Wiley-VCH: 133–154.

Saberi, A. H., Y. Fang and D. J. McClements (2013). "Fabrication of vitamin E-enriched nanoemulsions: Factors affecting particle size using spontaneous emulsification." *Journal of Colloid and Interface Science* **391**: 95–102.

Salvia-Trujillo, L., C. Qian, O. Martin-Belloso and D. J. McClements (2013). "Influence of particle size on lipid digestion and beta-carotene bioaccessibility in emulsions and nanoemulsions." *Food Chemistry* **141**(2): 1472–1480.

Santipanichwong, R. and M. Suphantharika (2007). "Carotenoids as colorants in reduced-fat mayonnaise containing spent brewer's yeast beta-glucan as a fat replacer." *Food Hydrocolloids* **21**(4): 565–574.

Saupe, A., S. A. Wissing, A. Lenk, C. Schmidt and R. H. Muller (2005). "Solid lipid nanoparticles (SLN) and nanostructured lipid carriers (NLC)—Structural investigations on two different carrier systems." *Bio-Medical Materials and Engineering* **15**(5): 393–402.

Schlenoff, J. B. (2003). Charge balance and transport in polyelectrolyte multilayers. In *Multilayer Thin Films: Sequential Assembly of Nanocomposite Materials*, G. Decher and J. B. Schlenoff (eds.). Weinheim, Wiley-VCH: 99–132.

Schmitt, C., C. Sanchez, S. Desobry-Banon and J. Hardy (1998). "Structure and techno-functional properties of protein-polysaccharide complexes: A review." *Critical Reviews in Food Science and Nutrition* **38**(8): 689–753.

Schubert, M. A. and C. C. Muller-Goymann (2005). "Characterisation of surface-modified solid lipid nanoparticles (SLN): Influence of lecithin and nonionic emulsifier." *European Journal of Pharmaceutics and Biopharmaceutics* **61**(1–2): 77–86.

Senuma, Y., C. Lowe, Y. Zweifel, J. G. Hilborn and I. Marison (2000). "Alginate hydrogel microspheres and microcapsules prepared by spinning disk atomization." *Biotechnology and Bioengineering* **67**(5): 616–622.

Sharma, R. (2005). "Market trends and opportunities for functional dairy beverages." *Australian Journal of Dairy Technology* **60**(2): 195–198.

Shaw, L. A., D. J. McClements and E. A. Decker (2007). "Spray-dried multilayered emulsions as a delivery method for omega-3 fatty acids into food systems." *Journal of Agricultural and Food Chemistry* **55**(8): 3112–3119.

Shchukina, E. M. and D. G. Shchukin (2012). "Layer-by-layer coated emulsion microparticles as storage and delivery tool." *Current Opinion in Colloid and Interface Science* **17**(5): 281–289.

Shegokar, R. and R. H. Muller (2010). "Nanocrystals: Industrially feasible multifunctional formulation technology for poorly soluble actives." *International Journal of Pharmaceutics* **399**(1–2): 129–139.

Siekmann, B., H. Bunjes, M. M. H. Koch and K. Westesen (2002). "Preparation and structural investigations of colloidal dispersions prepared from cubic monoglyceride-water phases." *International Journal of Pharmaceutics* **244**(1–2): 33–43.

Solans, C. and I. Sole (2012). "Nano-emulsions: Formation by low-energy methods." *Current Opinion in Colloid and Interface Science* **17**(5): 246–254.

Sonoda, T., Y. Takata, S. Ueno and K. Sato (2006). "Effects of emulsifiers on crystallization behavior of lipid crystals in nanometer-size oil-in-water emulsion droplets." *Crystal Growth and Design* **6**(1): 306–312.

Soottitantawat, A., F. Bigeard, H. Yoshii, T. Furuta, M. Ohkawara and P. Linko (2005). "Influence of emulsion and powder size on the stability of encapsulated D-limonene by spray drying." *Innovative Food Science and Emerging Technologies* **6**(1): 107–114.

Soottitantawat, A., H. Yoshii, T. Furuta, M. Ohkawara and P. Linko (2003). "Microencapsulation by spray drying: Influence of emulsion size on the retention of volatile compounds." *Journal of Food Science* **68**(7): 2256–2262.

Souto, E. B. and R. H. Muller (2005). "SLN and NLC for topical delivery of ketoconazole." *Journal of Microencapsulation* **22**(5): 501–510.

Souto, E. B., S. A. Wissing, C. M. Barbosa and R. H. Muller (2004). "Development of a controlled release formulation based on SLN and NLC for topical clotrimazole delivery." *International Journal of Pharmaceutics* **278**(1): 71–77.

Speranza, A., M. G. Corradini, T. G. Hartman, D. Ribnicky, A. Oren and M. A. Rogers (2013). "Influence of emulsifier structure on lipid bioaccessibility in oil-water nanoemulsions." *Journal of Agricultural and Food Chemistry* **61**(26): 6505–6515.

Su, J. H., J. Flanagan, Y. Hemar and H. Singh (2006). "Synergistic effects of polyglycerol ester of polyricinoleic acid and sodium caseinate on the stabilisation of water-oil-water emulsions." *Food Hydrocolloids* **20**(2–3): 261–268.

Sun, Y. E., W. D. Wang, H. W. Chen and C. Li (2011). "Autoxidation of unsaturated lipids in food emulsion." *Critical Reviews in Food Science and Nutrition* **51**(5): 453–466.

Surh, J., G. T. Vladisavljevic, S. Mun and D. J. McClements (2007). "Preparation and characterization of water/oil and water/oil/water emulsions containing biopolymer-gelled water droplets." *Journal of Agricultural and Food Chemistry* **55**(1): 175–184.

Szczepanowicz, K., H. J. Hoel, L. Szyk-Warszynska, E. Bielanska, A. M. Bouzga, G. Gaudernack, C. Simon and P. Warszynski (2010). "Formation of biocompatible nanocapsules with emulsion core and pegylated shell by polyelectrolyte multi layer adsorption." *Langmuir* **26**(15): 12592–12597.

Tamjidi, F., M. Shahedi, J. Varshosaz and A. Nasirpour (2013). "Nanostructured lipid carriers (NLC): A potential delivery system for bioactive food molecules." *Innovative Food Science and Emerging Technologies* **19**: 29–43.

Thanasukarn, P., R. Pongsawatmanit and D. J. McClements (2004). "Influence of emulsifier type on freeze–thaw stability of hydrogenated palm oil-in-water emulsions." *Food Hydrocolloids* **18**(6): 1033–1043.

Tikekar, R. V. and N. Nitin (2012). "Distribution of encapsulated materials in colloidal particles and its impact on oxidative stability of encapsulated materials." *Langmuir* **28**(25): 9233–9243.

Tokle, T., U. Lesmes, E. A. Decker and D. J. McClements (2012). "Impact of dietary fiber coatings on behavior of protein-stabilized lipid droplets under simulated gastrointestinal conditions." *Food and Function* **3**(1): 58–66.

Tokle, T., Y. Mao, and D. J. McClements (2013). "Potential biological fate of emulsion-based delivery systems: Lipid particles nanolaminated with lactoferrin and beta-lactoglobulin coatings." *Pharmaceutical Research* **30**(12): 3200–3213.

Tolstoguzov, V. (2002). "Thermodynamic aspects of biopolymer functionality in biological systems, foods, and beverages." *Critical Reviews in Biotechnology* **22**(2): 89–174.

Tolstoguzov, V. (2003). "Some thermodynamic considerations in food formulation." *Food Hydrocolloids* **17**(1): 1–23.

Trojer, M. A., Y. Li, C. Abrahamsson, A. Mohamed, J. Eastoe, K. Holmberg and M. Nyden (2013a). "Charged microcapsules for controlled release of hydrophobic actives. Part I: Encapsulation methodology and interfacial properties." *Soft Matter* **9**(5): 1468–1477.

Trojer, M. A., Y. Li, M. Wallin, K. Holmberg and M. Nyden (2013b). "Charged microcapsules for controlled release of hydrophobic actives Part II: Surface modification by LbL adsorption and lipid bilayer formation on properly anchored dispersant layers." *Journal of Colloid and Interface Science* **409**: 8–17.

Tso, P. (2000). Overview of digestion and absorption. In *Biochemical and Physiological Aspects of Human Nutrition*, K. D. Crissinger and M. H. Stipanuk (eds.). Philadelphia, PA, W.B. Saunders Company: 75–90.

Tyssandier, V., B. Lyan and P. Borel (2001). "Main factors governing the transfer of carotenoids from emulsion lipid droplets to micelles." *Biochimica Et Biophysica Acta-Molecular and Cell Biology of Lipids* **1533**(3): 285–292.

Uner, M., S. A. Wissing, G. Yener and R. H. Muller (2004). "Influence of surfactants on the physical stability of Solid Lipid Nanoparticle (SLN) formulations." *Pharmazie* **59**(4): 331–332.

Uner, M., S. A. Wissing, G. Yener and R. H. Muller (2005). "Skin moisturizing effect and skin penetration of ascorbyl palmitate entrapped in Solid Lipid Nanoparticles (SLN) and Nanostructured Lipid Carriers (NLC) incorporated into hydrogel." *Pharmazie* **60**(10): 751–755.

van der Graaf, S., C. G. P. H. Schroen and R. M. Boom (2005). "Preparation of double emulsions by membrane emulsification—A review." *Journal of Membrane Science* **251**(1–2): 7–15.

Vanapalli, S. A., J. Palanuwech and J. N. Coupland (2002). "Stability of emulsions to dispersed phase crystallization: Effect of oil type, dispersed phase volume fraction, and cooling rate." *Colloids and Surfaces A-Physicochemical and Engineering Aspects* **204**(1–3): 227–237.

Vega, C. and Y. H. Roos (2006). "Invited review: Spray-dried dairy and dairy-like—Emulsions compositional considerations." *Journal of Dairy Science* **89**(2): 383–401.

Vladisavljevic, G. T. and R. A. Williams (2005). "Recent developments in manufacturing emulsions and particulate products using membranes." *Advances in Colloid and Interface Science* **113**(1): 1–20.

Walstra, P. (1993). "Principles of emulsion formation." *Chemical Engineering Science* **48**: 333.

Walstra, P. (2003). *Physical Chemistry of Foods*. New York, Marcel Decker.

Waraho, T., D. J. McClements and E. A. Decker (2011). "Mechanisms of lipid oxidation in food dispersions." *Trends in Food Science and Technology* **22**(1): 3–13.

Weinbreck, F., R. de Vries, P. Schrooyen and C. G. de Kruif (2003). "Complex coacervation of whey proteins and gum arabic." *Biomacromolecules* **4**(2): 293–303.

Weinbreck, F., M. Minor and C. G. De Kruif (2004). "Microencapsulation of oils using whey protein/gum arabic coacervates." *Journal of Microencapsulation* **21**(6): 667–679.

Weiss, J., E. A. Decker, D. J. McClements, K. Kristbergsson, T. Helgason and T. Awad (2008). "Solid lipid nanoparticles as delivery systems for bioactive food components." *Food Biophysics* **3**(2): 146–154.

Weiss, J. and D. J. McClements (2000). "Influence of Ostwald ripening on rheology of oil-in-water emulsions containing electrostatically stabilized droplets." *Langmuir* **16**(5): 2145–2150.

Weiss, J., I. Scherze and G. Muschiolik (2005). "Polysaccharide gel with multiple emulsion." *Food Hydrocolloids* **19**(3): 605–615.

Williams, H. D., N. L. Trevaskis, S. A. Charman, R. M. Shanker, W. N. Charman, C. W. Pouton and C. J. H. Porter (2013). "Strategies to address low drug solubility in discovery and development." *Pharmacological Reviews* **65**(1): 315–499.

Wissing, S. A., O. Kayser and R. H. Muller (2004). "Solid lipid nanoparticles for parenteral drug delivery." *Advanced Drug Delivery Reviews* **56**(9): 1257–1272.

Wissing, S. A. and R. G. Muller (2002). "The influence of the crystallinity of lipid nanoparticles on their occlusive properties." *International Journal of Pharmaceutics* **242**(1–2): 377–379.

Wooster, T., M. Golding and P. Sanguansri (2008). "Impact of oil type on nanoemulsion formation and Ostwald ripening stability." *Langmuir* **24**(22): 12758–12765.

Wriedt, T. (1998). "A review of elastic light scattering theories." *Particle and Particle Systems Characterization* **15**(2): 67–74.

Wu, B. C., B. Degner and D. J. McClements (2013). "Microstructure and rheology of mixed colloidal dispersions: Influence of pH-induced droplet aggregation on starch granule-fat droplet mixtures." *Journal of Food Engineering* **116**(2): 462–471.

Wu, K. G., X. H. Chai and Y. Chen (2005). "Microencapsulation of fish oil by simple coacervation of hydroxypropyl methylcellulose." *Chinese Journal of Chemistry* **23**(11): 1569–1572.

Yang, X. Q., H. X. Tian, C. T. Ho and Q. R. Huang (2012). "Stability of citral in emulsions coated with cationic biopolymer layers." *Journal of Agricultural and Food Chemistry* **60**(1): 402–409.

Yang, Y. and D. J. McClements (2013). "Encapsulation of vitamin E in edible emulsions fabricated using a natural surfactant." *Food Hydrocolloids* **30**(2): 712–720.

Yates, P. D., G. V. Franks, S. Biggs and G. J. Jameson (2005). "Heteroaggregation with nanoparticles: Effect of particle size ratio on optimum particle dose." *Colloids and Surfaces A-Physicochemical and Engineering Aspects* **255**(1–3): 85–90.

Yi, C. L., Y. Q. Yang, J. Q. Jiang, X. Y. Liu and M. Jiang (2011). "Research and application of particle emulsifiers." *Progress in Chemistry* **23**(1): 65–79.

Yucel, U., R. J. Elias and J. N. Coupland (2013). "Effect of liquid oil on the distribution and reactivity of a hydrophobic solute in solid lipid nanoparticles." *Journal of the American Oil Chemists Society* **90**(6): 819–824.

Zeeb, B., M. Gibis, L. Fischer and J. Weiss (2012). "Influence of interfacial properties on Ostwald ripening in crosslinked multilayered oil-in-water emulsions." *Journal of Colloid and Interface Science* **387**: 65–73.

Zeng, C., H. Bissig and A. D. Dinsmore (2006). "Particles on droplets: From fundamental physics to novel materials." *Solid State Communications* **139**(11–12): 547–556.

Zhang, F. J., G. X. Cheng, Z. Gao and C. P. Li (2006). "Preparation of porous calcium alginate membranes/microspheres via an emulsion templating method." *Macromolecular Materials and Engineering* **291**(5): 485–492.

Zhang, H., E. Tumarkin, R. M. A. Sullan, G. C. Walker and E. Kumacheva (2007). "Exploring microfluidic routes to microgels of biological polymers." *Macromolecular Rapid Communications* **28**(5): 527–538.

Zhong, Q. X., M. F. Jin, D. Xiao, H. L. Tian and W. N. Zhang (2008). "Application of supercritical anti-solvent technologies for the synthesis of delivery systems of bioactive food components." *Food Biophysics* **3**(2): 186–190.

Ziani, K., Y. Fang and D. J. McClements (2012). "Encapsulation of functional lipophilic components in surfactant-based colloidal delivery systems: Vitamin E, vitamin D, and lemon oil." *Food Chemistry* **134**(2): 1106–1112.

# Chapter 7

# Biopolymer-Based Delivery Systems

## 7.1 Introduction

There are many examples of biopolymers being used to create food-grade colloidal dispersions suitable for encapsulation and delivery of active ingredients (Burey et al. 2008; Jones and McClements 2010; Matalanis et al. 2011; Norton and Frith 2001). Biopolymer colloidal particles can be assembled from proteins or polysaccharides using various bottom–up and top–down methods (Aguilera 2000; Matalanis et al. 2011; Norton and Frith 2001; Tolstoguvoz 2007). These particles must be carefully designed so that they exhibit the required functional attributes within the final product. In this chapter, we discuss the various types of building blocks (proteins or polysaccharides) and processing methods (physicochemical or mechanical) that can be used to fabricate biopolymer particles. We then discuss the influence of biopolymer-based colloidal delivery systems on the physicochemical properties of foods, and then discuss their potential applications within the food industry for the encapsulation, protection, and release of active ingredients.

## 7.2 Building Blocks: Biopolymers

A variety of food-grade proteins and polysaccharides can be used to fabricate biopolymer particles, including whey proteins, casein, soy proteins, gelatin, zein, starch, cellulose, and various other hydrocolloids (Table 7.1). As with any polymer, the type, number, and distribution of monomers along the polymer chain as well as the nature of the bonds between the monomers determine their overall molecular characteristics, such as molecular weight, conformation, flexibility, hydrophobicity, electrical charge, and reactivity. In turn, these molecular characteristics affect the ability of a biopolymer to form colloidal particles, as well as the physicochemical properties and functional performance of any colloidal delivery systems created from it. It should be noted that most food-grade biopolymer ingredients are actually complex mixtures of different polymers and other ingredients, which may vary from batch-to-batch owing to the inherent variability associated with their biological origin, isolation, and processing. Consequently, it is often necessary to carefully characterize the molecular characteristics of an ingredient that is going to be used for this purpose.

The selection of particular proteins or polysaccharides to form biopolymer particles depends on a number of factors: (i) the ability of the biopolymers to be assembled into colloidal particles, (ii) the functional requirements for the

**Table 7.1  Summary of Important Molecular Characteristics of Some Common Food-Grade Proteins Used to Assemble Biopolymer Particles**

| Name | Main Source | Main Structural Type | pI | ~$T_m$ (°C) | Solubility |
|---|---|---|---|---|---|
| β-Lactoglobulin | Milk | Globular | ~5.0 | ~75 | Water |
| Caseins | Milk | Flexible | ~4.6 | ~125–140 | Water |
| Bovine serum albumin | Milk/ blood | Globular | ~4.7 | ~80 | Water |
| Lactoferrin | Milk | Globular | ~8.0 | ~60 and 90 | Water |
| Ovalbumin | Egg white | Globular | ~4.6 | ~74; 82[S] | Water |
| Lysozyme | Egg white | Globular | ~11.0 | ~74 | Water |
| Phosvitin | Egg yolk | Globular | ~4.0 | ~80 | Water |
| Gelatin | Animal collagen | Flexible | ~8[A] ~5[B] | ~5 (fish) ~40 (animal) | Water |
| Soy glycinin | Soybean | Globular | ~5.0 | ~67[7S]; 87[11S] | Water |
| Zein | Corn | Globular | ~6 | ~90 | Organic solvent |

*Source:* Here, pI is the isoelectric point and $T_m$ is the thermal transition temperature.

*Note:* A, Type A gelatin; B, Type B gelatin; S, S-type ovalbumin; 7S and 11S, soy glycinin fractions.

particles (e.g., size, charge, permeability, and stability to environmental conditions), and (iii) the legal status, cost, ease of use, and consistency of the ingredients and processing operations. For commercial applications, biopolymer particles should be designed so that they have the specific physicochemical and functional characteristics required for a particular application. For example, a biopolymer particle meant to deliver a flavor component should be designed to break down in the mouth and release the encapsulated flavor molecules there. Conversely, a biopolymer particle meant to deliver an anticancer component to the colon should be designed to resist disruption within the mouth, stomach, and small intestine but break down when it reaches the colon. The design of biopolymer particles with specific functional attributes requires knowledge of the building blocks used to assemble them (proteins and polysaccharides), the forces holding these building blocks together (physical and covalent), and the physicochemical principles used to assemble the building blocks (top–down and bottom–up methods). In this section, we provide a brief overview of the molecular and physicochemical attributes of some of the most widely used proteins (Table 7.1) and polysaccharides (Table 7.2).

## 7.2.1 Proteins

Proteins are primarily formed from polypeptide chains consisting of amino acids linked together through peptide bonds (Damodaran 2007). There are 20 main amino acids from which proteins are typically constructed in nature, which are distinguished from each other by their side groups. Side groups differ in their dimensions, charge, polarity, and chemical reactivity. The type, number, and sequence of amino acids along the polypeptide chain determine the molecular weight, conformation, electrical charge, hydrophobicity, interactions, and chemical reactivity of proteins.

### 7.2.1.1 Molecular Conformation

Food proteins come in a variety of conformations with the three major structural features being random coil, fibrous (rigid rod), and globular (Figure 7.1) (Belitz et al. 2009). The molecular structure adopted by a particular protein depends on its amino acid sequence, prevailing environmental conditions, and history, for example, exposure to different temperatures, pH values, ionic compositions, and solvents (Damodaran 2007). Proteins tend to adopt a structure that minimizes the overall free energy of the system, provided there are no kinetic constraints (energy barriers) that prevent them from reaching the lowest energy state. In many food applications, proteins are trapped in a metastable state.

In globular proteins, the polypeptide chain is folded into a compact spheroid conformation with most of the nonpolar amino acids located in the interior and the polar amino acids located at the exterior (Seno and Trovato 2007). The major

**Table 7.2  Summary of Important Molecular Characteristics of Some Common Food-Grade Polysaccharides Used for Assembling Biopolymer Particles**

| Name | Source | Main Structure Type | Major Monomer | Gelation |
|---|---|---|---|---|
| Carrageenan | Algal | Linear/helical | Sulfated galactan | Cold set ($K^+$ or $Ca^{2+}$) |
| Xanthan gum | *Xanthomonas campestris* exudate | Linear/helical (high MW) | β-D-Glucose (backbone) | Forms gels at high concentration |
| Methyl cellulose | Wood pulp | Linear | Methylated glucose | Heat set (reversible) |
| Pectin | Plant cell walls | Highly branched coil | Glucuronate (backbone) | Sugar/heat (HM); $Ca^{2+}$ (LM) |
| Beet pectin | Sugar beet pulp | Branched coil with protein | Glucuronate (backbone) | Sugar/heat (HM); $Ca^{2+}$ (LM) Laccase |
| Gum arabic | Acacia sap | Branched coil domains on protein scaffold | Galactose | Forms gels at high concentration |
| Inulin | Plants or bacteria | Linear with occasional branches | β-D-Fructose | Forms gels at high concentration |
| Chitosan | Crustaceans, invertebrates | Linear | 2-Amino-2-deoxy-β-D-glucose | Polyphosphate cross-linking |
| Alginate | Algal | Linear | β-D-Mannuronic Acid | $Ca^{2+}$ cross-linking |

*Note:* Commercially available polysaccharide ingredients typically contain appreciably different molecular and functional properties; the listed information describes general characteristics for industrial usage.

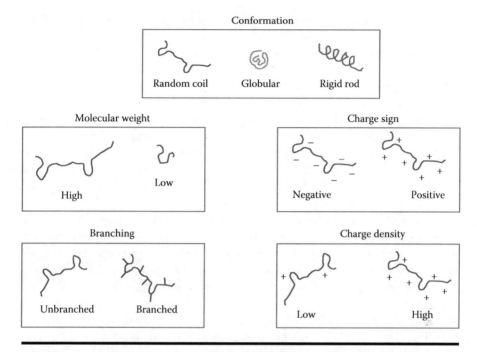

**Figure 7.1  Schematic representation of different structures that protein and polysaccharide molecules may adopt in aqueous solutions. Biopolymers vary in molecular characteristics depending on biological, functional extraction and purification procedures, and any physical, chemical, or enzyme modifications made.**

driving force for the formation of this type of compact structure is the *hydrophobic effect*, which is an attempt to minimize the unfavorable contact area between nonpolar groups and water. Nevertheless, the globular structure is also partly stabilized by other types of physical interaction, such as van der Waals, hydrogen bonding, and electrostatic forces. Thus, an appreciable fraction of the amino acids in a globular protein may also be involved in secondary-structure formation involving extensive hydrogen bonding, such as α-helices or β-sheets. Covalent bonds (disulfides) also play a key role in maintaining the internal structure of many globular proteins (Fox 2003; Hoffman and van Mill 1997; Wolf 1993). In fibrous proteins, the polypeptide chain tends to form a helical structure that is held together by relatively strong hydrogen bonds between polar amino acid side groups. In these systems, the free energy gain from the formation of hydrogen bonds is large enough to offset the free energy loss associated with the low entropy of the helical structure (Damodaran 2007). Conversely, in random coil proteins, the configuration entropy dominates the various attractive interactions operating between the amino acids, leading to a highly disorganized and flexible structure. In practice, many proteins have a combination of globular, fibrous, and random coil structures and may undergo transitions from one form to another in response to changes in their environment.

In nature, proteins have a specific biological function (e.g., enzyme activity, signaling, transport, molecular recognition, or structural), which requires that they have a specific three-dimensional conformation usually referred to as the *native* state. In food systems, proteins typically exist in various nonnative (denatured) conformations, which differ from the native state by an amount that depends on the history of the system. For example, the structure of a protein may be altered during the extraction, isolation, and purification of a commercial protein ingredient (such as milk, egg, or soy protein isolates) or after incorporation into a food product owing to changes in their environment, such as pH, ionic strength, solvent type, temperature, adsorption to interfaces, high pressure, dehydration, or chemical treatments (Damodaran 2007). The temperature at which a protein undergoes a substantial conformation change upon heating is referred to as the thermal denaturation temperature ($T_m$). Denaturation may be reversible or irreversible depending on protein type and the conditions used to induce the conformation changes. When a protein is denatured, its physical and chemical interactions change appreciably, which leads to alterations in its functional performance. An understanding of the factors that influence protein denaturation is therefore helpful in the rational formation of biopolymer particles with specific functional attributes. Previous studies have shown that a variety of structures can be formed by carefully controlling the denaturation and aggregation of proteins, such as spheroids, filaments, and tubes (Akkermans et al. 2007; Graveland-Bikker et al. 2009; Jung et al. 2008).

### 7.2.1.2 Electrical Characteristics

An understanding of the electrical properties of proteins is particularly important for assembling biopolymer particles using electrostatic interactions. The electrical characteristics of a protein depend on the number, type, and distribution of ionizable amino acids along the polypeptide chain, as well as solution conditions such as pH and ionic composition. Amino side groups contribute a positive charge at relatively low pH values (pH < 9) because they are protonated ($-NH_3^+$) whereas they are neutral at higher pH values because they are nonprotonated ($-NH_2$). Conversely, carboxylate side groups contribute a negative charge at relatively high pH values (>3) because they are nonprotonated ($-CO_2^-$) but are neutral at lower pH values because they are protonated ($-CO_2H$). Proteins therefore possess a net positive charge at low pH, a net negative charge at high pH, and a point of zero charge at an intermediate pH (Figure 7.2).

The pH at the point of zero net charge is usually referred to as the *isoelectric point* (pI), and it is one of the most important parameters needed to select proteins for particle fabrication purposes. It is important to note that even though the net charge is zero at the pI, there are localized regions of positive and negative charge on a protein's

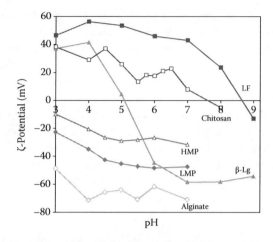

**Figure 7.2 Biopolymers, such as proteins and ionic polysaccharides, exhibit a range of different electrical characteristics, which can be characterized by their ζ-potential versus pH profiles. Key: LF = lactoferrin; HMP = high methoxy pectin; LMP = low methoxy pectin; β-Lg = β-lactoglobulin.**

surface (Cooper et al. 2005). Thus, even though the net charge on a protein may be negative (or positive), it may still bind negative (or positive) components (such as minerals, polysaccharides, or surfactants). Finally, very high (>9) and low (<2) pH values often lead to physical and chemical instability of the protein structure through a combination of denaturation and chemical degradation processes. The isoelectric points of some common food-grade proteins are listed in Table 7.1. Most commonly used proteins have pI values near 5 (e.g., whey protein, casein, bovine serum albumin [BSA], soy protein, and ovalbumin), but some have appreciably higher values (e.g., lactoferrin and phosvitin) (Williams 2007). Knowledge of the pH dependence of the net charge and surface charge distribution of different proteins is extremely useful for the rational design of biopolymer particles. This can often be categorized by measuring the electrophoretic mobility (or ζ-potential) versus pH profile using commercially available electrophoresis instruments (Figure 7.2).

## 7.2.1.3 Hydrophobic Characteristics

The functional attributes of many proteins are determined by the number of nonpolar groups exposed to the surrounding aqueous phase, rather than by the total number of nonpolar groups along the polypeptide chain (Damodaran 2007). The nonpolar properties of proteins are therefore often characterized by their *surface hydrophobicity*. The surface hydrophobicity is an important parameter to consider when fabricating biopolymer particles for a number of reasons: (i) it affects their ability to bind small nonpolar molecules (such as fatty acids, vitamins, flavors, or surfactants), (ii) it affects their ability to adsorb to nonpolar surfaces (such as oil or air) and therefore stabilize fat droplets or air bubbles, and (iii) it affects their ability to associate with other protein molecules to form aggregates or gels. The surface hydrophobicity of globular proteins tends to increase after they become denatured, which affects their binding properties, surface activity, and aggregation behavior. Many proteins have a relatively low surface hydrophobicity and can therefore be dispersed in aqueous solutions (e.g., whey proteins and caseins), whereas others are so hydrophobic that they are effectively insoluble in water but soluble in some organic solvents (such as zein). Differences in the solubility of proteins in different solvents can be used to prepare biopolymer-based delivery systems. For example, zein can be dissolved in ethanol and then titrated into water, which leads to the formation of zein nanoparticles or microparticles (Patel et al. 2010, 2012).

### 7.2.1.4 Physical Interactions

Proteins may interact with a variety of other molecules in their neighborhood, such as similar proteins, different proteins, polysaccharides, solvent molecules, cosolvent molecules, surfactants, phospholipids, sugars, and minerals. The most important physical forces involved in these interactions are van der Waals, hydrogen bonding, hydrophobic bonding, and electrostatic interactions (Section 7.3). Proteins can often be made to form different kinds of structures by modulating these physical interactions, for example, by altering pH, ionic strength, solvent, or temperature. It is therefore important to be aware of the various types of physical interactions that a particular protein molecule can be involved in, as well as the factors that affect them, when selecting a protein to create a particular type of biopolymer particle.

### 7.2.1.5 Chemical Reactivity

Proteins may participate in various types of chemical reactions that can be useful in the design of biopolymer-based delivery systems, such as sulfhydryl, hydrolysis, dehydration, oxidation, Maillard, and protein–protein cross-linking reactions (Gerrard 2002).

Disulfide interchange occurs naturally during thermal denaturation of many globular proteins. High temperatures expose the cysteine and cystine residues and provide thermal energy for the transitions (Sava et al. 2005). Partial deprotonation of the free cysteine group is a prerequisite for disulfide interchange, so it tends to occur more rapidly at high pH values (Hoffman and van Mill 1997, 1999), but some disulfide interactions also occur at low pH (Hoffmann and van Mill 1999; Monahan et al. 1995). Disulfide interchange may occur within a single protein molecule or between different protein molecules.

Dehydration reactions can be used to form cross-links between protein molecules, for example, by heating them under alkali conditions (Gerrard 2002).

Hydrolysis reactions can be used to break down a protein into smaller fragments, which can be achieved using strong acids, strong bases, or enzyme treatments.

Maillard reactions involve the formation of a covalent link between amine and aldehyde groups and therefore usually involve a protein and a carbohydrate (Oliver et al. 2006). This reaction usually occurs faster under alkaline conditions and can be carried out at high temperatures in dried powders containing mixtures of proteins and carbohydrates. These reactions have been used to fabricate glycoproteins, such as conjugates of β-lactoglobulin and dextran that have different functional properties to the parent protein molecule (Akhtar and Dickinson 2003; Zhu et al. 2008). These glycoproteins may therefore be used as alternatives to conventional proteins to fabricate biopolymer-based delivery systems with different functional attributes.

### 7.2.1.6 Commonly Used Proteins

In this section, a brief overview of some of the proteins that are most commonly used as building blocks for assembling biopolymer-based delivery systems is given. More comprehensive reviews are given elsewhere (Jones and McClements 2010; Matalanis et al. 2011; McClements 2005).

*Milk Proteins*: Milk proteins are divided into two major categories according to the method used in their isolation (selective precipitation): caseins (~80 wt%) and whey proteins (~20 wt%). Both casein and whey ingredients contain a complex mix of different proteins, but it is possible to fractionate them into individual purified protein fractions (e.g., β-lactoglobulin, α-lactalbumin, κ-casein, or lactoferrin). These purified fractions are often too expensive for widespread application as ingredients in the food industry, but they are frequently used in research studies because they facilitate a more fundamental understanding of protein functionality. In addition, they may be suitable for the formation of biopolymer-based delivery systems for high-value applications.

There are four major protein types in casein with different molecular and functional attributes: $\alpha_{S1}$ (~44%), $\alpha_{S2}$ (~11%), β (~32%), and κ (~11%). Despite their differences, all caseins have relatively disordered and flexible structures in aqueous solutions, but they do have some secondary and tertiary structure. Caseins also have regions along their backbones that are either highly nonpolar or highly charged, which plays a major role in determining their functional performance in foods. In their natural state in milk, caseins mainly exist as molecular clusters referred to as "casein micelles" that are typically between 50 and 250 nm in diameter and are partly held together by mineral ions (such as calcium phosphate). These micelles may be useful as naturally occurring biopolymer-based delivery systems owing to their ability to solubilize hydrophobic ingredients in their interiors, such as curcumin, carotenoids, and oil-soluble vitamins (Benzaria et al. 2013; Chevalier-Lucia et al. 2011; Saiz-Abajo et al. 2013).

Caseinates have an isoelectric point around pH 4.6, and casein molecules or casein-coated oil droplets will tend to aggregate strongly around this pH value because of the reduction in electrostatic repulsion between polymers or particles (Dickinson 2010). Because they have fairly disordered flexible structures, caseins are much less sensitive to thermal treatments than globular proteins (such as whey protein), which may be an important consideration when selecting a protein for a particular application.

Whey protein is a mixture of mainly globular proteins with different molecular characteristics, with the most dominant being β-lactoglobulin (~55%), α-lactalbumin (~24%), serum albumin (~5%), and immunoglobulins (~15%). Native whey protein molecules tend to have a high water solubility across a wide range of pH, although a limited amount of aggregation may occur near their isoelectric point (pI ~5) (Majhi et al. 2006; Yan et al. 2013). On the other hand, whey protein-coated oil droplets are highly susceptible to aggregation near the isoelectric point, because of the reduction in electrostatic repulsion (McClements 2004). The whey proteins have a globular structure that tends to unfold when they are heated above a critical temperature ($T_m$), which is around 70°C for β-lactoglobulin. When the proteins unfold, they expose reactive amino acid side groups normally located in their interiors, such as nonpolar and sulfhydryl groups. Exposure of these groups can promote aggregation and cross-linking of whey proteins, which may be detrimental in some applications since it leads to instability but is beneficial in other applications because specific structures can be built or stabilized.

*Meat and Fish Proteins*: Meat and fish contain numerous proteins that may be used to fabricate biopolymer-based delivery systems, with gelatin being one of the most widely used proteins (Tarte 2011). Gelatin is a relatively high molecular weight protein derived from animal collagen, for example, pig, cow, or fish (Abd Elgadir et al. 2013; Djagny et al. 2001; Gomez-Guillen et al. 2011). It is usually prepared commercially by boiling collagen in the presence of either strong acid (Type A gelatin) or strong alkaline (Type B gelatin) to denature and hydrolyze the original structure. The isoelectric point of Type A gelatin (pI ~7 to 9) is higher than that of Type B gelatin (pI ~4.7 to 5.4) as a result of the different preparation methods. Gelatin exists as a flexible disordered (random coil) molecule at relatively high temperatures but undergoes a coil-to-helix transition when it is cooled below a critical temperature, which is approximately 10°C to 25°C for pig and cow gelatin and approximately 0°C to 5°C for fish gelatin. Gelatin has some surface activity and can therefore be used to coat oil droplets in emulsions, but it is mainly used in foods to thicken and gel aqueous solutions. It may also be used as a structuring agent to form novel structures in biopolymer-based delivery systems, for example, because of its ability to form gels, create coatings, and induce phase separation.

*Egg Proteins*: Both egg yolk and egg white contain globular proteins that can be used to construct biopolymer-based delivery systems (Anton 2013; Anton et al. 2009; Mine 2002). For example, egg white contains ovalbumin and lysozyme, whereas egg yolk contains phosvitin. These proteins are surface-active molecules that are capable of adsorbing to oil droplets and stabilizing oil-in-water (O/W) emulsions. In addition, they can be used to thicken solutions or form gels by heating them above their thermal denaturation temperature as this promotes protein unfolding and aggregation. Finally, they can also be used as structuring agents to assemble novel structures in delivery systems because of their ability to interact with other molecules through hydrophobic and electrostatic interactions. The isoelectric point of egg proteins varies considerably, for example, pI ≈ 4.0 for phosvitin, pI ≈ 4.6 for ovalbumin, and pI ≈ 11.0 for lysozyme. This gives considerable scope for selecting proteins with different charge characteristics for building delivery systems using electrostatic interactions. Phosvitin is a highly phosphorylated protein that can bind cationic ions (such as calcium and iron), which gives it good antioxidant activity.

*Plant Proteins*: A number of different plant species contain proteins that can be used as building blocks, including legumes and cereals (Day 2013; Moure et al. 2006). One of the most widely used plant proteins is soy protein. Soy protein ingredients are actually a complex mixture of globular proteins with different molecular and functional characteristics, for example, 2S, 7S, 11S, and 15S fractions. Like other sources of globular proteins, soy proteins are surface active and can be used to build structures through hydrophobic and electrostatic interactions. Soy proteins may therefore be used as emulsifiers, thickening agents, gelling agents, and structure formers. Another widely used protein from a plant source is zein from corn. Zein is a highly hydrophobic protein that is insoluble in pure water but soluble in concentrated ethanol solutions. This characteristic can be utilized to create colloidal particles capable of encapsulating lipophilic active ingredients using antisolvent precipitation methods (Patel et al. 2010).

## 7.2.1.7 Protein Selection

A number of factors must be considered when selecting a suitable protein or combination of proteins to fabricate biopolymer-based delivery systems. First, it is important to establish the conditions where the protein molecules are able to associate with other protein or nonprotein structure-forming components, for example, environmental and solution conditions. This usually requires knowledge of specific physicochemical characteristics of the proteins involved, such as thermal denaturation temperatures (for globular proteins), helix–coil transition temperatures (for gelatin or collagen), isoelectric points (pI), sensitivities to specific monovalent or multivalent ions, or susceptibility to specific enzyme, chemical cross-linking, or degradation reactions. Second, it is often important to establish the electrical characteristics of the protein molecules involved since electrostatic interactions are often utilized in structure formation, which can be conveniently described by the ζ-potential versus pH profile (Figure 7.2). Proteins can vary widely in their isoelectric points depending on their biological origin and the processing procedures used to extract them (Table 7.1). Third, it is usually important to have knowledge of the nature of the structures that can be formed after protein association, such as their morphology (e.g., particulate, fibrous, or tubular), physical properties (density, refractive index), size, charge, and stability (e.g., to pH, ionic strength, temperature, and enzymes). These factors will determine how the particles may affect the optical, rheological, stability, and functional characteristics of the products into which they are incorporated (LaClair and Etzel 2010).

## 7.2.2 Polysaccharides

Polysaccharides are biopolymers assembled by covalent linking of monosaccharides (BeMiller and Huber 2007; Cui 2005; Rinaudo 2008; Stephen et al. 2006; Williams 2007). Numerous monosaccharides are available to form polysaccharides, which vary in their electrical charge (neutral, negative, or positive), polarity (polar or nonpolar), interactions, and chemical reactivity. The type, number, sequence, and bonding of these monosaccharides along the biopolymer chain determine the unique molecular properties and functional attributes of different polysaccharides. Food-grade polysaccharides are usually isolated from plants (e.g., cellulose, starch, pectin, gum arabic, locust bean gum), algae (e.g., carrageenan, alginate), bacteria (e.g., xanthan), or animals (e.g., chitosan). Polysaccharides vary in their molecular weights, conformations, flexibility, branching, electrical characteristics, hydrophobicities, and chemical reactivities, which leads to differences in their physicochemical and functional properties, for example, solubilities, thickening and gelation characteristics, binding properties, and surface activities (Table 7.2).

The ability of polysaccharides to self-associate and form gels is widely used in the formation of biopolymer-based delivery systems (Jones and McClements 2010). Comprehensive reviews of the major categories of gelling polysaccharides have been published in the literature (Cui 2005; Imeson 2010; Williams 2007). In general, polysaccharides can be conveniently divided into different categories depending on the gelation mechanism involved: swelling (e.g., starch granules), cooling (e.g., agar and carrageenan), heating (e.g., methyl cellulose [MC]), ion addition (e.g., alginate and low methoxyl [LM] pectin), and retrogradation (e.g., amylopectin). Knowledge of the gelation mechanism and the factors that influence it is important when selecting a particular polysaccharide to form a biopolymer-based delivery system. A brief overview of the most important molecular and physicochemical properties of polysaccharides is given in the remainder of this section.

### 7.2.2.1 Molecular Conformation

The molecular characteristics of specific polysaccharides are often not as clearly defined as those of specific proteins, which is mainly a result of their different functional roles in nature. Consequently, the chemical composition, molecular weight, and electrical characteristics of the polysaccharides within a food ingredient are often highly variable and may change considerably from supplier to supplier and from batch to batch. There may also be considerable variations in the type and quantity of impurities present within polysaccharide ingredients depending on their origin and the processing methods used to isolate, purify, and possibly modify them, such as mineral ions, sugars, acids, bases, and proteins. For these reasons, it is usually important to have a good understanding of the potential variability in the composition and functional performance of polysaccharide ingredients used to prepare biopolymer-based delivery systems (Autio 2006). Consequently, a range of analytical methods are often needed to establish the composition and structural characteristics of polysaccharide ingredients (Eliasson 2006; Muralikrishna and Rao 2007; Stephen et al. 2006). One must often decide whether to use more costly, highly purified and well-defined ingredients, or to use cheaper, but more variable and inconsistent food-grade ingredients.

Despite the inherent variability in the properties of specific polysaccharide ingredients, they do have general features that can be used to distinguish them, such as typical monosaccharide compositions, molecular weight ranges, degrees of branching, electrical characteristics, and functional properties. Some of the molecular properties of common food-grade polysaccharides are shown in Table 7.2. The upper molecular weight of a particular polysaccharide is usually determined by its biological origin, but the molecular size can often be reduced using processing operations, such as chemical (e.g., strong acids or bases), enzyme (e.g., amylases, pectinases), or mechanical treatments (e.g., ultrasonic processing) (Coffey et al. 2006; Gidley and Reid 2006; Wurzburg 2006). In some cases, the molecular weight of a polysaccharide can be increased by cross-linking reactions.

Many polysaccharides have either random coil or helical structures when dispersed in aqueous solutions and may be transformed from one form to another by changing temperature (Cui 2005). Rotation of the glycosidic bonds linking monosaccharides in carbohydrates is much less flexible than rotation of the amide bonds linking peptides in proteins, thus making linear structuring more common in polysaccharides than in proteins. The electrical charge of ionic polysaccharides also limits chain flexibility through internal electrostatic repulsion. Many polysaccharides have regions that consist of fairly rigid helical structures that optimize the formation of intrachain hydrogen bonds between the monosaccharide subunits. This type of polysaccharide can be characterized by a thermal transition temperature ($T_m$), which is the temperature where the molecule undergoes a helix-to-coil transition upon heating. The presence of helical structures is often a prerequisite for the formation of the cross-links responsible for polysaccharide aggregation and gelation (Cui 2005; Imeson 2010; Williams 2007). These junction zones may consist of two or more helices from different molecules held together by strong hydrogen bonds (as in agar), or they may consist of two similarly charged helices held together by oppositely charged counterions (as in potassium-carrageenan, calcium-pectin, or calcium-alginate gels).

Polysaccharides may have either linear or branched backbones depending on their origin and processing. The frequency, location, length, and composition of the branches have a major impact on the functional attributes of branched polysaccharides, such as water solubility, viscosity enhancement, gelation characteristics, and water holding capacity. Pectin is a commonly used branched ionic polysaccharide, which has an anionic linear backbone with neutral "hairy regions" attached (Williams 2007). Knowledge of polysaccharide branching is often important for the fabrication of biopolymer-based delivery systems since it influences the ability of the molecules to form complexes with other molecules through steric hindrance effects.

## 7.2.2.2 Electrical Characteristics

Polysaccharides may be anionic, cationic, or nonionic depending on the nature of their functional groups (Table 7.2). The pH dependence of the electrical charge on ionic polysaccharides depends on the $pK_a$ values of their ionizable side groups (Figure 7.2). The most common anionic side groups on polysaccharides are carboxylate groups ($-CO_2^-$, $pK_a \approx 2.5$–4.5) and sulfate groups ($-SO_4^-$, $pK_a < 0$), while the most common cationic groups are amino groups ($-NH_3^+$, $pK_a \approx 6.5$–9.5). For example, pectins and alginates derive their negative charge from carboxylate groups (Ridley et al. 2001), whereas carrageenans and agars derive their negative charges from sulfate groups (Piculell 1995; Whistler and BeMiller 1997). Chitosan is one of the few positively charged polysaccharides available for forming biopolymer-based delivery systems, and therefore it has been widely used in research on the development of these systems. However, it is not generally accepted as safe in all countries, which currently limits its widespread commercial application as a food ingredient. The magnitude of the electrical charge on ionic polysaccharides varies depending on their biological origin and processing conditions used in their isolation and preparation. For example, the magnitude of the negative charge of pectin molecules depends on their degree of esterification: the greater the number of carboxyl groups that are esterified (methoxylated), the lower the charge density. Hence, LM pectin has a higher anionic charge density than high methoxyl (HM) pectin. The degree of methoxylation can be controlled using chemical (strong acids or bases) or enzyme treatments, and it is possible to control whether the charge distribution is random or blocky (Einhorn-Stoll and Kunzek 2009). The negative charge density on carrageenan molecules depends on their biological origin: $\kappa < \iota < \lambda$ (Piculell 1995). The positive charge on chitosan molecules depends on the degree of deacetylation of chitin, which varies with chitin source and the degree of alkali treatment (Kurita 2006; Rinaudo 2006). The overall charge on polysaccharide molecules also depends on the pH of the surrounding solution relative to the $pK_a$ values of their ionizable groups. Knowledge of the charge density of polysaccharides is particularly important when fabricating biopolymer-based delivery systems using electrostatic interactions.

## 7.2.2.3 Hydrophobic Characteristics

Most of the polysaccharide ingredients that are widely used in the food industry have relatively low hydrophobicities because they are primarily composed of polar monosaccharides. Nevertheless, some polysaccharides do have an appreciable nonpolar character, which may arise because of the unique molecular conformation or because they have some nonpolar groups attached to their backbone (Cui 2005). Maltodextrin and starch molecules are capable of binding nonpolar fatty acids and surfactants because they form helical structures that have a hydrophobic pocket capable of accommodating nonpolar groups (Siswoyo and Morita 2003). Some polysaccharides have nonpolar chains covalently attached to their polar backbones (e.g., OSA-starch or HPMC), whereas others have amphiphilic proteins attached (e.g., gum arabic) (Dickinson 2003). These polysaccharides tend to be surface-active molecules that can bind to nonpolar groups and can therefore be used to stabilize the oil droplets in emulsions or to form specific structures based on hydrophobic interactions.

## 7.2.2.4 Physical Interactions

Knowledge of the physical interactions that polysaccharides can form with other molecules is often important when designing biopolymer particles (Jones and McClements 2010; Matalanis et al. 2011). For uncharged polysaccharides, such as starch, the primary physical interactions are hydrogen bonding and van der Waals interactions, with the former being the most important in most situations. These polysaccharides form hydrogen bonds with the surrounding water molecules when they are dispersed in solution, but they may also form intra- or intermolecular hydrogen bonds with polar functional groups on polysaccharide molecules, especially when helical structures are formed. The formation of junction zones between helices on different molecules is important for aggregation and gelation of many polysaccharides, for instance, amylase, amylopectin, and carrageenan. For ionic polysaccharides, such as alginate, agar, carrageenan, pectin, and xanthan, the most important physical interactions are electrostatic, hydrogen bonding, and van der Waals interactions. Electrostatic interactions are particularly important for the assembly of biopolymer particles from two oppositely charged polysaccharides or from proteins and ionic polysaccharides. Hydrophobic interactions are important for polysaccharides that have some nonpolar character, such as hydrophobically modified starches and celluloses; for example, MC tends to self-associate when heated.

## 7.2.2.5 Chemical Reactivity

Polysaccharides may be involved in a number of different types of chemical reactions, which can be useful for modifying their molecular properties and functional performance (BeMiller and Huber 2007). Polysaccharides can be hydrolyzed or cross-linked to decrease or increase their molecular weight, or they can be conjugated with polar, nonpolar, and ionic molecules to form new molecules with different functional properties (Cui 2005). Glycoproteins with novel functional attributes can be formed by covalently linking proteins and carbohydrates together using the Maillard reaction (Evans et al. 2013). Celluloses and starches can be conjugated with various forms of side groups (e.g., acetate, pyruvate, or carboxylic acids) to alter their solubility characteristics, structure-forming ability, and gelation properties (Cui 2005; Whistler and BeMiller 1997).

### 7.2.2.6 Commonly Used Polysaccharides

In this section, a brief overview of some of the polysaccharides most commonly used as building blocks for biopolymer-based delivery systems is given. More comprehensive reviews can be found elsewhere (Cui 2005; Eliasson 2006; Piculell 1995; Stephen et al. 2006; Williams 2007). One of the most important biological characteristics of polysaccharides is their susceptibility to enzymatic digestion in the upper gastrointestinal tract (GIT), since this has important implications for the development of delivery systems for releasing active ingredients in different regions of the GIT. Some polysaccharides are susceptible to digestion in the GIT (such as many forms of starch), whereas others are indigestible and can be considered to act as dietary fibers.

*Starch*: Starch can be isolated from a number of different sources, such as corn, potato, wheat, tapioca, and rice (BeMiller and Huber 2007). There are two main molecular types in starch: amylose and amylopectin. Amylose is primarily a linear molecule consisting of α-D-(1–4)-linked glucose units, whereas amylopectin is a highly branched molecule consisting of a main backbone of α-D-(1–4)-linked glucose units with a limited number of α-D-(1–6)-linked branches. Starches from different natural sources vary in the ratio of amylose to amylopectin, as well as the molecular weights of these different components, which accounts for some of the differences in their physicochemical properties and functional performances. They also vary in the nature of the starch granules in which the starch molecules are normally found.

In nature, amylose and amylopectin are packed into starch granules that consist of crystalline regions separated by amorphous regions (Perez and Bertoft 2010). When aqueous solutions of starch granules are heated above a critical temperature, they incorporate water and the crystalline regions are disrupted. The resultant swelling of the starch granule leads to an appreciable increase in solution viscosity and may lead to gelation if the starch concentration is sufficiently high ("gelatinization"). Upon further heating, (nonmodified) starch granules disintegrate and amylose and amylopectin molecules leach out of the granules, which results in a decrease in viscosity. When a gelatinized starch solution is cooled, linear regions in starch molecules associate with each other through hydrogen bonding, which leads to an increase in viscosity or even gelation ("retrogradation"). The characteristics of a particular native starch depend on the structural organization of the molecules within the starch granule, the ratio of amylose to amylopectin, the precise molecular characteristics of each of these fractions, the solution composition (e.g., pH, ionic strength, sugar content), and environmental factors (e.g., shearing, temperature, pressure).

Nonmodified native starches often have limited commercial applications because of their low water solubility, sensitivity to environmental conditions, restricted functional properties, and instability during processing or storage. For this reason, native starches are often physically, chemically, or enzymatically modified to improve their functional performance, for example, pregelatinization, hydrolysis, addition of side groups, or cross-linking (Belitz et al. 2009; BeMiller and Huber 2007). Starch ingredients that are soluble in cold or hot water, that thicken or gel with or without heating, that exhibit a wide range of gelation characteristics (e.g., opacity, gel strength, water holding capacity), and that have different stabilities to environmental conditions (e.g., heating, freezing, pH, ionic strength, shearing) are available. Modified starches have also been developed that are effective emulsifiers because they have nonpolar groups attached to their backbones, for example, OSA-starch (Sweedman et al. 2013). There are also appreciable differences in the digestibility of different kinds of starch, so that they can be classified as rapidly digestible, slowly digestible, or resistant starches (Singh et al. 2010; Taggart and Mitchell 2009). These differences may be useful for designing biopolymer-based delivery systems that release bioactive components in different regions of the GIT. In summary, there is considerable scope for using different forms of starch (such as natural and modified granules and molecular forms) as building blocks to create colloidal delivery systems with different functional attributes.

*Carrageenans*: Carrageenans are natural hydrocolloids extracted from certain species of red seaweed (Campo et al. 2009; Cui 2005; Necas and Bartosikova 2013; Williams 2007). They are linear sulfated polysaccharides consisting of alternating $\beta(1-3)$- and $\alpha(1-4)$-linked galactose residues. There are three major types of carrageenan that primarily differ in the number and position of sulfate ester groups on the galactose residues: κ, ι, and λ. These differences in structure have a major impact on the functional characteristics of different carrageenans, for example, solubility, thickening, gelation, environmental sensitivity, ingredient compatibility, and structure formation. For example, λ-carrageenan is commonly used as a thickening agent, whereas κ- and ι-carrageenans are widely used as cold-setting reversible gelling agents. Commercially, carrageenan ingredients come in numerous forms with different attributes, for example, molecular weights, counterion type, and blends. Typically, they are sold as salts (Na, K, Ca) and have number-average molecular weights between approximately 200 and 400 kDa.

Carrageenans usually have a random coil conformation at high temperatures but undergo a helix-to-coil transition when they are cooled below their thermal transition temperature (~40°C to 70°C). The transition temperature depends on carrageenan structure, salt type and concentration, and the presence of sugars. In the presence of sufficiently high quantities of salt, helical regions of gelling carrageenans (κ and ι) associate with each other to form hydrogen-bonded junction zones that promote gelation. Knowledge of the transition temperature is important when utilizing carrageenans to construct delivery systems since it determines the temperature above which they must be heated to adequately disperse and solubilize them in water and the temperature they must be cooled below in order to form gels. Carrageenans have a number of characteristics that make them useful for building blocks in biopolymer-based delivery systems, for example, their ability to form gels, inhibit ice crystallization, and be used to create novel structures through electrostatic interactions.

*Agars*: Agars are another group of natural hydrocolloids extracted from certain species of red seaweed (Belitz et al. 2009; Cui 2005; Imeson 2010; Williams 2007). They are linear polysaccharides consisting primarily of alternating β(1–3)- and α(1–4)-linked galactose units. Different kinds of agar vary in the number and type of substituents (e.g., sulfate, pyruvate, urinate, or methoxyl) on the hydroxyl groups of the sugar residues and in the fraction of α(1–4)-linked galactose units present in the 3–6 anhydride form. Agar can be divided into two main fractions: *agarose*, a nonionic gelling polysaccharide, and *agaropectin*, a slightly anionic nongelling polysaccharide. The agaropectin fraction contains anionic substituents (such as sulfates) along its backbone. Commercial agars vary in the relative proportions of the nonionic and ionic fractions present. Agars usually require heating in aqueous solutions to dissolve them prior to use. When the system is cooled, it forms a viscous solution that gels over time without the need for specific additives. Gels can be formed at very low polysaccharide concentrations (<0.1%). The gelation temperature upon cooling (30°C–40°C) is usually much lower than the melting temperature upon heating (85°C–95°C), which may be an important factor when using agar to form specific structures. The gelation mechanism has been attributed to the transition of an appreciable part of the agar molecules from a random coil to a helical structure upon cooling and subsequent aggregation of the helical structures to form junction zones separated by irregular flexible chain regions. The gels formed by agars are thermo-reversible cold-set gels.

*Alginates*: Alginates are typically extracted from certain species of brown seaweed (Cui 2005; Imeson 2010; Williams 2007). They are linear copolymers of D-manuronic acid (M) and L-guluronic acids (G), which are distributed as blocks of M, blocks of G, or blocks of alternating M and G residues. The M-blocks have flexible conformations, the G-blocks have inflexible conformations, and the MG-blocks tend to have an intermediate flexibility. Alginates vary in the proportions and distributions of M and G groups along the chain, which leads to differences in their functional attributes. The alginic acid extracted from brown seaweed is usually reacted with bases to produce sodium, potassium, calcium, or ammonium alginate salts that are used as commercial ingredients. Alternatively, alginic acid can be reacted with propylene oxide to produce propylene glycol alginate, in which partial esterification of the carboxylic acid groups on the uronic acid residues occurs. The monovalent salts of alginate have good water solubility, whereas alginic acid and multivalent salts of alginate have relatively poor water solubility and form paste-like materials. Monovalent salts of alginate can be made to gel by adding multivalent cations (such as calcium) that form ion bridges between the relatively stiff G-block regions on different alginate molecules. The gels formed by alginates tend to be cold-set thermo-irreversible gels. Different gelation characteristics can therefore be obtained by selecting alginates with different numbers and lengths of the G-blocks or by using different salt forms.

*Pectin*: Most pectin used in the food industry is extracted from either citrus or apple pomace (Cui 2005; Imeson 2010; Williams 2007). The term *pectin* actually refers to a group of molecules that have some common features. In particular, pectins tend to have a structure that consists of "smooth" linear regions of $\alpha(1-4)$-linked D-galacturonic acids separated by "hairy" branched regions consisting of various neutral sugars. A fraction of the galacturonic acid groups may be esterified with methyl groups, either naturally or as a result of postisolation modifications. The electrical charge on pectin molecules therefore depends on the ratio of esterified to nonesterified galacturonic groups, as well as the pH relative to the $pK_a$ value of the acid groups ($pK_a \approx 3.5$). Pectins therefore tend to be negatively charged at high pH values, but neutral at lower ones. The fraction of esterified galacturonic groups is one of the main factors influencing the functional characteristics of pectins. Pectins may be classified as HM or LM pectins depending on whether the degree of methylation is greater or less than 50%, respectively. HM pectin gels under acidic conditions in the presence of high sugar levels, which is attributed to the reduction of electrostatic repulsion between the chains at low pH and the increased osmotic attraction at high sugar contents. Gels formed by HM pectins are thermo-irreversible cold-setting gels held together by hydrogen bonding and hydrophobic cross-links between helical regions in the smooth regions of the pectin backbone. LM pectin forms gels in the presence of calcium, which is attributed to the ability of the divalent cationic calcium ions ($Ca^{2+}$) to form electrostatic bridges between the anionic smooth regions of the pectin backbone. Gels formed by LM pectins are thermo-reversible cold-setting gels. Certain natural sources of pectins can be made to gel using alternative mechanisms. For example, sugar beet pectin has phenolic (ferulic acid) side groups that can be covalently cross-linked by enzymes, such as laccase (Saulnier and Thibault 1999).

The precise gelation characteristics of a particular pectin ingredient depend on its molecular characteristics (e.g., linear charge density, molecular weight, branching, and side groups), as well as the prevailing environmental conditions (e.g., pH, ionic strength, sugar content, and temperature). Pectins are water soluble, but usually have to be dispersed in warm water prior to use to ensure proper dissolution. Pectins are relatively stable to heating at certain pH values (pH 3 to 5) but may degrade at higher or lower pH values owing to hydrolysis, which affects their utilization in certain types of delivery systems.

*Seed Gums (Galactomannans)*: Polysaccharide ingredients can be isolated from the seeds of various bushes, trees, and plants, for example, locust bean, guar, and tara gums (Cui 2005; Imeson 2010; Williams 2007). These polysaccharides are primarily nonionic linear polysaccharides known as galactomannans, which consist of β(1–4)-linked D-mannose residues with single α-D-galactose residues linked to the main chain. One of the main differences between galactomannans from different sources is the galactose-to-mannose ratio ≈1:4.5 for LBG, ≈1:3 for tara gum, and ≈1:2 for guar gum. The galactose side chains tend to inhibit close molecular associations, and hence these variations in galactose content lead to pronounced differences in the functional properties of different galactomannans, for example, solubility, thickening, and gelation. For example, guar gum can be dissolved in cold water, whereas LBG and tara gum require hot water for dissolution.

At ambient temperatures, galactomannans tend to exist as individual molecules in aqueous solutions because close intermolecular associations are inhibited by the presence of the galactose side chains. For this reason, seed gums are primarily used as thickening agents, rather than as gelling agents. Galactomannan solutions tend to be highly viscous and shear thinning, and their rheological characteristics are not strongly influenced by pH or ionic strength because they are nonionic biopolymers. Galactomannans are sensitive to thermal degradation in acidic solutions (pH < 4.5), which limits their application in certain types of delivery application.

*Tree Gum Exudates*: A variety of polysaccharides can be extracted from the exudates of certain trees (Cui 2005; Imeson 2010; Williams 2007). Gum arabic is the most widely used type of exudate in the food industry, mainly because of its ability to act as a natural emulsifier in beverage emulsions. It is derived from *Acacia senegal* and consists of a number of high-molecular-weight fractions. The surface-active fraction is believed to consist of branched arabinogalactan blocks attached to a polypeptide backbone. The hydrophobic polypeptide chain anchors the biopolymer to the droplet surface, while the hydrophilic polysaccharide blocks extend into solution and provide protection against aggregation through steric and electrostatic repulsion. Gum arabic has a high water solubility and a relatively low solution viscosity compared to other gums because of its relatively compact structure. Gum arabic tends to have a negative charge at relatively high pH values (pH > 3), which is important for its use as a structuring agent in the formation of delivery systems.

*Xanthan Gum*: Xanthan gum is a high-molecular-weight extracellular polysaccharide secreted by bacteria of the genus *Xanthomonas* (Cui 2005; Imeson 2010; Williams 2007). It primarily consists of a neutral β-(1–4)-D-glucose backbone with anionic trisaccharide side chains. These side chains typically consist of mannose–glucuronic acid–mannose, with a relatively high proportion of the terminal mannose units containing either pyruvate or acetate residues. In aqueous solutions at relatively low temperatures, xanthan gum exists as a stiff extended molecule with a helical structure, but at higher temperatures, it exists as a more flexible and disordered random coil molecule. The helix–coil transition temperature is highly sensitive to ionic strength and may range from around 40°C to >90°C. Under appropriate solution conditions, helical regions on different xanthan molecules associate with each other and form a weak gel. Xanthan gum is readily soluble in both hot and cold water and is stable over a wide range of solution and environmental conditions, for example, pH, ionic strength, heating, freeze–thaw cycling, and mixing. Xanthan gum forms highly viscous solutions at relatively low concentrations because it is a fairly stiff molecule that is highly extended in aqueous solutions.

*Gellan Gum*: Gellan gum is another extracellular polysaccharide produced as a fermentation product of a bacteria (*Pseudomonas elodea*) (Cui 2005; Imeson 2010; Williams 2007). It has an anionic linear chain consisting of a repeating unit of four saccharides: glucose, glucuronic acid, glucose, and rhamnose. In nature, there are approximately 1.5 substituents per repeating unit, composed mainly of glycerate or acetate. These side chains hinder the close association of the gellan gum molecules, which has a major impact on its gelling characteristics. Two forms of gellan gum that have different functional properties are commonly produced commercially: a low-acylated form that produces strong nonelastic brittle gels and a high-acylated form that produces soft elastic nonbrittle gels.

Gellan gums can be dissolved at ambient temperatures provided significant amounts of divalent ions are not present; otherwise, they have to be heated. They give solutions that are highly viscous and shear thinning. The solution viscosity decreases steeply with increasing temperature because of a reversible helix–coil transition that occurs upon heating (around 25°C–50°C). Gellan gums have good heat stability at neutral pH but are susceptible to thermal degradation under acidic conditions. They form gels when cooled from high temperatures owing to the formation of helical regions that associate with each other and form junction zones. Since they are electrically charged, their thickening and gelling properties are sensitive to salt type and concentration. Divalent ions usually promote gelation by forming salt bridges between negatively charged helical regions. Gels formed in the presence of monovalent ions are usually thermo-reversible, whereas those formed in the presence of multivalent ions may be thermo-irreversible. A variety of gel characteristics can be achieved by altering the degree of esterification of the gellan gum and the mineral composition.

*Cellulose and Its Derivatives*: Cellulose is the most abundant polysaccharide found in nature due to the fact that it is the major structural component of land plants (Cui 2005; Imeson 2010; Williams 2007). It is a linear polymer with a relatively high molecular weight consisting of D-glucose units joined together by D-β(1–4) linkages. In its natural state, cellulose is not usually suitable for utilization as a structure component because it forms strong intermolecular hydrogen bonds that make it insoluble in water. Nevertheless, it can be isolated and chemically modified in a number of ways to produce products that are useful as structural components for designing biopolymer-based delivery systems. The most common cellulose derivatives suitable for utilization in food products are methyl cellulose

(MC), carboxymethyl cellulose (CMC), hydroxypropyl cellulose (HPC), and methyl hydroxypropyl cellulose (MHPC). These ingredients consist of cellulose molecules that have been chemically modified by adding substituents (M, CM, HP, or MHP) to the cellulose backbone. These substituents provide a steric hindrance that helps prevent strong intermolecular associations between cellulose backbones.

MC, MHPC, and HPC are all soluble in cold water but tend to become insoluble when the solution is heated above a critical temperature (around 50°C–90°C). They all have some nonpolar groups attached to their polar backbone and therefore have surface activity. MC and MHPC both form reversible gels or highly viscous solutions upon heating, whereas HPC precipitates out of solution. The main driving force for the self-association of these types of amphiphilic cellulose derivatives at high temperatures has been attributed to hydrophobic attraction. MC, MHPC, and HPC are all nonionic polymers and therefore have good stability to pH and salt, as well as good compatibility with other ingredients. These ingredients also have good stability to pH (2 to 11), salt, and freeze–thaw cycling, which may be beneficial in a number of delivery applications.

CMC, also known as cellulose gum, is an anionic linear polymer, which is manufactured by chemically attaching carboxymethyl groups to the backbone of native cellulose. It is normally sold in the form of either sodium or calcium salts and is available in different molecular weights and degrees of substitution (DS). At a sufficiently high DS (>~0.4), it is readily soluble in water and forms viscous solutions. Because CMC is ionic, the viscosity of these solutions is sensitive to pH and ionic strength, as well as to the presence of other types of electrically charged molecules. CMC can form gels in the presence of multivalent ions because of electrostatic screening and bridging effects.

Another commonly used cellulose-based product in the food industry is microcrystalline cellulose (MCC). This product is manufactured by treating native cellulose with hydrochloric acid to dissolve the amorphous regions leaving crystalline regions as colloidal sized particles. MCC is water insoluble and so exists as small colloidal particles that are predominately dispersed in the aqueous phase. In aqueous solutions, MCC can form three-dimensional matrices of aggregated particles that form viscous solutions or gels depending on the concentration used. These solutions are shear thinning because the particle network breaks down upon application of shear forces, but the viscosity or gel strength is regained once the shearing stress is removed. MCC functions over a wide range of temperatures, providing freeze/thaw and heat stability to many products. This product is dispersible in water at relatively high pH (>3.8) but may need addition of protective hydrocolloids to disperse it at lower pH values. MCC may be a useful building block for the creation of certain types of biopolymer-based delivery systems.

### 7.2.2.7 Polysaccharide Selection

There are a number of factors that must be considered when selecting a suitable polysaccharide or combination of polysaccharides to fabricate a biopolymer-based delivery system (Jones and McClements 2010; Matalanis et al. 2011). It is important to establish the environmental and solution conditions where the polysaccharide molecules can associate with other polysaccharide or nonpolysaccharide structure-forming molecules. This usually requires knowledge of the physicochemical properties of the polysaccharides involved, such as helix–coil transition temperatures (for carrageenan, alginate, pectin), electrical properties ($pK_a$ values), amphiphilicity (for gum arabic, OSA-starch, and many modified celluloses), sensitivity to specific monovalent or multivalent ions, or susceptibility to enzyme or chemical reactions (Whistler and BeMiller 1997). For some applications, it is important to establish the electrical characteristics of the polysaccharide molecules used ($\zeta$-potential vs. pH), since electrostatic interactions may be used to assemble specific biopolymer structures, for example, coacervation, clustering, or multilayer formation. The electrical charge on polysaccharides depends on the nature of the ionic groups along the chain, as well as solution conditions. Some polysaccharides are neutral (starch, cellulose, seed gums), some are anionic (alginate, carrageenan, xanthan, gum arabic, gellan) and some are cationic (chitosan). The magnitude of the electrical charge on ionic polysaccharides depends on the pH relative to the $pK_a$ value of the charge groups. Anionic polysaccharides tend to be neutral at pH values sufficiently below their $pK_a$ value but negative above, whereas cationic polysaccharides tend to be neutral at pH values sufficiently above their $pK_a$ value but positive below. The most common charged groups on polysaccharides are sulfate groups (e.g., carrageenan), carboxyl groups (e.g., pectin, alginate, xanthan, carboxymethylcellulose), and amino groups (e.g., chitosan): $-SO_4H \leftrightarrow -SO_4^- (pK_a \approx 2)$; $-CO_2H \leftrightarrow -CO_2^- (pK_a \approx 3.5)$; $-NH_3^+ \leftrightarrow -NH_2 (pK_a \approx 6.5)$. The electrical charge on polysaccharides may also be altered by interactions with other ionic species in their environment. These interactions typically involve monovalent or multivalent ions such as sodium or calcium that bind to oppositely charged groups on the biopolymer chain, altering overall charge characteristics.

It is usually important to establish the nature of the biopolymer particles that can be formed by the polysaccharides used, such as their morphology, density, refractive index, size, charge, and stability to pH, salt, temperature, and enzymes (Jones and McClements 2010; Matalanis et al. 2011). In addition, knowledge of the type of environmental and solution conditions present within a particular food is often important for selecting the most appropriate polysaccharide building blocks for creating an appropriate delivery system. For example, pectin can start to depolymerize when exposed to neutral or alkaline conditions, leading to a loss of functionality (Sila et al. 2009).

## 7.3 Molecular Interactions

Biopolymer molecules interact with each other and with other molecules through a variety of physical and chemical interactions (Figure 7.3) (Israelachvili 2011;

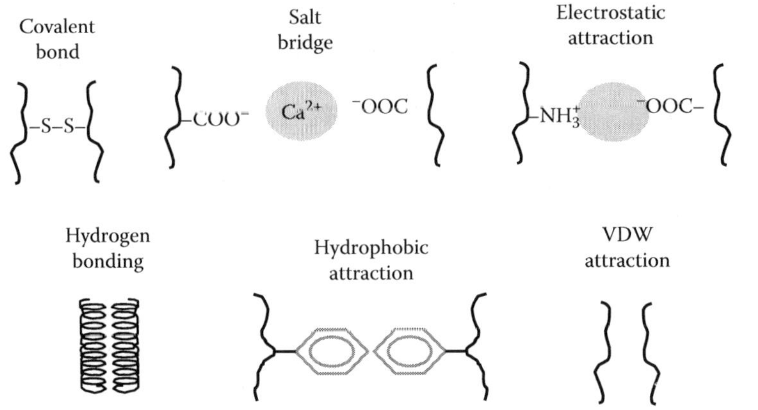

**Figure 7.3   Biopolymers may interact with each other through a variety of different kinds of molecular interactions.**

Jones and McClements 2010; Matalanis et al. 2011; Shewan and Stokes 2013). The sign, magnitude, strength, and direction of these interactions can often be modulated by changes in environmental conditions or solution composition, such as pH, ionic strength, temperature, and solvent type. These interactions are responsible for holding biopolymer-based delivery systems together and determining the way that they respond to changes in environmental and solution conditions, for example, whether the structures stay intact, swell, shrink, erode, or disintegrate. Understanding the nature of these interactions and the factors that affect them is therefore essential for designing colloidal delivery systems with specific functional attributes.

## 7.3.1  Electrostatic Interactions

Proteins and polysaccharides vary widely in their electrical characteristics and may be nonpolar, polar, anionic, cationic, or amphoteric. Many biopolymers have functional groups that are capable of becoming ionized under the appropriate solution conditions. The electrical charge on anionic (e.g., pectin, alginate, and carrageenan) or cationic (e.g., chitosan) polysaccharides depends on solution pH relative to the $pK_a$ values of their ionizable groups (Figure 7.2). The net electrical charge on proteins varies from positive, to neutral, to negative as the pH is increased from below to above their isoelectric points (pI). Nevertheless, it should be recognized that the electrical charge distribution on protein surfaces is heterogeneous, with some regions being negative and other regions being positive. Thus, proteins that have a net negative charge may still have substantial patches of positive charge on their surfaces (and vice versa), which has important implications for their ability to interact with other charged molecules (Cooper et al. 2005). The relative distribution of charges on a biopolymer molecule may also be altered by changes in its three-dimensional conformation; for example, the charge groups are often further apart in a random coil conformation than in a globular conformation. The charge status of functional groups may also depend on their specific local environments (e.g., dielectric constant) and may therefore be changed by biopolymer conformational changes (e.g., the movement of a group from within the hydrophobic core of a globular protein to its hydrophilic surface upon denaturation), location changes (e.g., the adsorption of a biopolymer to an interface so some groups can move from the water to oil phase), or solvent changes (e.g., addition of an organic solvent to an aqueous biopolymer solution). Changes in pH, ionic composition, solvent composition, and biopolymer conformation may therefore be used to control electrostatic interactions, and thereby control the assembly of biopolymer-based delivery systems.

Some of the major features of electrostatic interactions involving biopolymer molecules are discussed below (Jones and McClements 2010; Matalanis et al. 2011). Electrostatic interactions may be either attractive or repulsive depending on the signs of the electrical charges on the species involved: similar charges repel each other, whereas different charges attract each other. The magnitude of electrostatic interactions varies from strong to weak depending on the charge characteristics of the interacting species (e.g., charge density) and the properties of the solution separating them (e.g., ionic strength and dielectric constant). The charge characteristics of biopolymer molecules are determined by the total number of charged groups present, their sign and valance, and their distribution along the polymer chain. The electrical properties can be characterized by the *linear charge density* for linear polymers (i.e., the charge per unit length) or the *surface charge density* for globular polymers or biopolymer particles (i.e., the charge per unit surface area). The strength of any electrostatic interactions (attractive or repulsive) increases as the charge density of the molecular species involved increases.

The composition of the aqueous solution separating the charged species also plays an important role in determining the magnitude and range of electrostatic interactions. Counterions or dipolar molecules screen electrostatic interactions, which means the magnitude and range of electrostatic interactions decreases as the ionic strength or dielectric constant of a solution increases (Israelachvili 2011). The range of electrostatic interactions is usually characterized by the *Debye screening length* ($\kappa^{-1}$). For aqueous solutions at room temperatures, $\kappa^{-1} \approx 0.304/\sqrt{I}$ nm, where $I$ is the ionic strength expressed in moles per liter. As an example, for ionic strengths of 1, 10, 100, and 1000 mM, the Debye screening length is 9.6, 3, 0.96, and 0.3 nm, respectively. Increasing the ionic strength of a solution decreases the magnitude and range of electrostatic interactions through this screening effect. A multivalent ionic species may act as an *ionic bridge* between two similarly charged species, for example, A–C–A or C–A–C, where A and C represent anions and cations, respectively. Ion bridges are important for the formation of structures by many types of biopolymer molecules: $Ca^{2+}$ ions link anionic phosphate groups in casein micelles; $Ca^{2+}$ ions link anionic carboxylic acid or sulfate groups in alginate, pectin, and carrageenan; and anionic polyphosphate ions link cationic amino groups in chitosan. Decreasing the dielectric constant of a solvent (e.g., adding alcohol to water) increases the strength of any electrostatic interactions, which may be useful for forming certain types of structures. Temperature is believed to have little direct effect on the strength of electrostatic interactions over the temperature ranges typically found in foods, for example, –100°C to 200°C (Pink et al. 2006). However, it may have indirect effects on the electrostatic interactions in a system if it changes the conformation or charge distribution of biopolymer molecules. In summary, electrostatic interactions are one of the most powerful means of modulating the interactions between biopolymer molecules and are widely used to form specific structures in biopolymer-based delivery systems.

### 7.3.2 Hydrogen Bonding

Hydrogen bonding plays a major role in determining the structure, interactions, and functionality of many types of biopolymers (Jones and McClements 2010; Matalanis et al. 2011). A hydrogen bond is an attractive interaction between an electronegative atom and an electropositive hydrogen atom, for example, $O-H^{\delta+}...O^{\delta-}$. Hydrogen bonds may form within a particular biopolymer molecule or between different biopolymer molecules. They play a major role in stabilizing intramolecular structures in many proteins and polysaccharides (e.g., helices and sheets), as well as in the formation of intermolecular bonds between different biopolymer molecules (e.g., in gelatin, cellulose, and starch). Some of the most important characteristics of hydrogen bonds are discussed below.

Individual hydrogen bonds are relatively short-range and weak attractive interactions, but the combined impact of many hydrogen bonds is sufficient to hold together biopolymer structures, such as helices, sheets, and junction zones. Typically, pH and ionic strength have little direct impact on hydrogen bonds because they do not require fully ionized species. However, hydrogen bonding may be altered if solution conditions are adjusted so that the distance between polar groups is altered. Hydrogen bonds are weakened as the temperature is increased, because the thermal energy ($kT$) of the system increases, which favors entropy over enthalpy effects. Hydrogen bonds may also be weakened by changing the nature of the solvent surrounding them, for example, from water that can form strong hydrogen bonds to an organic solvent that cannot.

### 7.3.3 Hydrophobic Interactions

Hydrophobic interactions manifest themselves as attractive forces between nonpolar groups dispersed within water (Israelachvili 2011). However, their molecular origin is actually the fact that water–water attraction (mainly dipole–dipole) is much stronger than water–nonpolar group attraction (mainly van der Waals). For this reason, when a nonpolar group is introduced into water, the water molecules surrounding it change their organization so that they can maximize the number of hydrogen bonds formed with neighboring water molecules. Hydrophobic interactions play a major role in determining the conformation, interactions, and functionality of many biopolymer molecules (Jones and McClements 2010; Matalanis et al. 2011). First, they cause many types of protein to fold into compact globular structures to minimize the unfavorable contact area between nonpolar groups and water. Second, they cause amphiphilic biopolymers to absorb to oil–water or air–water interfaces. Third, they promote aggregation of proteins or polysaccharides that have nonpolar groups exposed to the water phase. Fourth, they promote binding of nonpolar molecules to hydrophobic patches on biopolymer surfaces, such as the nonpolar pockets in some globular proteins and in starch helices.

Hydrophobic interactions are medium-range, relatively strong, attractive interactions (Israelachvili 2011). Their strength depends on the interfacial tension between the water and nonpolar groups, as well as the surface area of nonpolar regions exposed to water: the higher the interfacial tension or exposed surface area, the stronger the hydrophobic attraction. Solution pH and ionic strength have little direct impact on hydrophobic interactions because they do not involve ionized species. Nevertheless, the strength of hydrophobic interactions may be altered if the normal tetrahedral structure of water is disrupted, for example, at pH extremes or high salt concentrations. Hydrophobic interactions are typically strengthened as the temperature is increased, because the entropic contribution to the interaction is increased. The strength of hydrophobic interactions may also be altered by changing the quality of the solvent surrounding the molecular species involved, for example, from water that promotes strong hydrophobic interactions to an organic solvent that does not. Any alterations in pH, ionic strength, temperature, or solvent quality that alter the amount of nonpolar groups exposed to water will also influence hydrophobic interactions by promoting changes in biopolymer conformation.

### 7.3.4 Excluded Volume Effects

The "excluded volume" or "steric exclusion" effect is the result of a competition for available volume within a system (Israelachvili 2011; Marenduzzo et al. 2010; Minton 1997). Different molecular species cannot occupy the same space in solution because of their physical dimensions (steric repulsion) or charge characteristics (electrostatic repulsion), which reduces the configuration entropy of the system by decreasing the number of arrangements possible. The free energy as a result of this steric exclusion effect increases as the concentration of molecular species in solution increases. Above a critical concentration, there is an osmotic driving force that favors separation of the system into phases with different molecular compositions (Schmitt et al. 1998; Turgeon et al. 2003). This type of *thermodynamic incompatibility* is the driving force for the segregative phase separation observed in many biopolymer mixtures (Tolstoguzov 1991), as well as the depletion flocculation that is sometimes observed when biopolymers are added to emulsions (Dickinson 2010).

Macroscopically, excluded volume interactions manifest themselves as an attraction between similar biopolymer molecules in solution; that is, they favor congregation of similar types of biopolymer into the same phase. The range of excluded volume interactions is of the order of the radius of gyration of the biopolymer molecules involved. The magnitude of these interactions increases as the molar concentration and radius of gyration of excluded biopolymers increases. If one or both of the biopolymers involved is electrically charged, then the excluded volume interactions depend on solution pH and ionic strength. In addition, the strength of the excluded volume effect depends on changes in solution conditions (such as temperature or solvent composition) that alter biopolymer conformation and therefore molecular volume.

### 7.3.5 Covalent Interactions

Biopolymers typically have a specific conformation in their native environment, which is determined by their biological function. However, this conformation may be considerably altered during the isolation, purification, processing, or utilization of food-grade biopolymer ingredients. Directed changes in the chemistry of biopolymer molecules can be utilized to change their molecular characteristics (e.g., charge, molecular weight, hydrophobicity, and chemical reactivity) and therefore functional properties (e.g., self-association, gelling, water holding, surface activity, and binding). Common covalent modifications of biopolymers involve oxidation/reduction, esterification/amidation, or Maillard reactions (Belitz et al. 2009; BeMiller and Huber 2007; Damodaran 1996). These reactions can be initiated using specific chemicals, enzymes, or physical treatments (e.g., drying, temperature, pressure, or mechanical forces). For example, Maillard reactions can be used to produce protein–polysaccharide conjugates by heating proteins and polysaccharides together under controlled conditions, for example, pH, relative humidity, holding temperature, and holding time (Oliver et al. 2006). Disulfide bonds may form between sulfhydrl groups that are exposed when globular proteins undergo thermal denaturation (Hoffman and van Mill 1997). Nonpolar side groups have been covalently attached to starches, celluloses, and other biopolymers to make them more surface active (Cizova et al. 2007; Erni et al. 2007). Proteins can be covalently cross-linked using certain chemicals (e.g., glutaraldehyde) or enzymes (e.g., transglutaminase) (Belitz et al. 2009; De Jong and Koppelman 2002; Gerrard 2002). Some polysaccharides can also be covalently cross-linked using suitable chemicals (e.g., starch by sodium trimetaphosphate) or enzymes (e.g., beet pectin by laccase) (Belitz et al. 2009; Minussi et al. 2002). The ability to form cross-links between biopolymer molecules is useful in stabilizing the structures formed in many types of biopolymer-based delivery systems.

## 7.3.6 Assembling Biopolymer Structures

The formation of biopolymer-based delivery systems usually involves assembling structures from numerous biopolymer molecules, which may be either the same kind or different (Jones and McClements 2010; Matalanis et al. 2011). Biopolymers can be made to associate using the various molecular interactions discussed in the previous section, such as electrostatic, hydrophobic, hydrogen bonding, and excluded volume interactions. When designing biopolymer-based delivery systems, it is advantageous to understand the influence of specific biopolymer, solution, and environmental characteristics on their tendency to assemble into particular structures. Most food-grade biopolymers and food matrices are so complex that it is difficult to develop detailed theoretical models to describe their behavior. Nevertheless, a number of theoretical concepts and mathematical models have been established for simpler systems that can provide important insights into structure formation by biopolymers and into the major factors that influence it. Reviews of the mathematical and computational models available to describe the phase behavior of biopolymers in solution are available in the literature (Mezzenga and Fischer 2013; Schmitt et al. 1998; Tolstoguzov 2003). In this section, a brief overview of one of the simplest theoretical approaches used to describe the interactions between biopolymers in solution is given to highlight some of the key factors involved. In addition, the phase separation of single and mixed biopolymer solutions is discussed, since phase separation methods are commonly used to fabricate biopolymer-based delivery systems.

Consider a system prepared by mixing two different molecular species together, for example, a biopolymer (species 2) and a solvent (species 1). The mixing free energy ($\Delta G_M$) depends on the net change in the molecular interactions and entropy in the system before and after mixing (Schmitt et al. 1998), which can be expressed by the following expression:

$$\frac{\Delta G_M}{VRT} = \frac{\phi_1 \ln \phi_1}{V_1} + \frac{\phi_2 \ln \phi_2}{V_2} + (\chi_{12}) \frac{\phi_1 \phi_2}{V_1} \tag{7.1}$$

Here $\phi_1$ and $\phi_2$ are the volume fractions of the two molecular species, $V_1$ and $V_2$ are their molar volumes, $V$ is the total volume of the solution, $R$ is the gas constant, $T$ is the absolute temperature, and $\chi_{12}$ is a *molecular interaction parameter*. This parameter depends on the net changes in the molecular interactions (1–1, 1–2, and 2–2) in the system associated with the mixing process and is usually expressed by the Flory–Huggins parameter:

$$\chi_{12} = \frac{z \Delta w_{12} r_1}{kT} \tag{7.2}$$

Here, $z$ is the lattice coordination number (which depends on the number of interactions an individual interacting unit participates in), $r_1$ is the number of interacting units on the biopolymer molecule that can interact with the solute, and $\Delta w_{12}$ is the free energy change associated with the various molecular interactions in the system: $\Delta w_{12} = 2 \times (w_{12} - 1/2 \, (w_{11} + w_{22}))$, which depends on changes in van der Waals interaction, electrostatic interaction, hydrogen bonding interaction, hydrophobic interaction, and so on (Schmitt et al. 1998). For an ideal solution, where the $w_{12}$ interactions are similar to the average of the $w_{11}$ and $w_{22}$ interactions, the Flory–Huggins parameters are equal to zero. For a system where the molecular interactions are unfavorable to mixing, the molecular interaction parameter is positive (i.e., $\chi_{12} > 0$), whereas for a system where they are favorable to mixing, it is negative (i.e., $\chi_{12} < 0$). This model provides some useful insights into the major parameters influencing the phase behavior of biopolymers in aqueous solutions. In the absence of any strong interactions (i.e., $\chi_{12} \approx 0$), the system will tend to be intimately mixed because the entropy of mixing term favors an even distribution of different molecules throughout the system. On the other hand, if there are strongly unfavorable interactions (i.e., $\chi_{12} \gg 0$) between the different types of molecules, then the system will phase separate so that similar molecules remain together (i.e., 1–1 and 2–2 interactions are favored). The above equation is only applicable for a single type of biopolymer in solution, but it can be modified to account for multiple biopolymer types (Schmitt et al. 1998). In the case of two biopolymers, one needs to take into account a variety of different molecular interaction parameters, such as $\chi_{12}$, $\chi_{13}$, and $\chi_{23}$, where the subscript 1 refers to the solvent, and the subscripts 2 and 3 refer to the two different types of biopolymers. In this case, $\chi_{23}$ is positive if there is a net repulsion between the two different biopolymers (which favors thermodynamic incompatibility) and $\chi_{23}$ is negative if there is an attraction (which favors coacervation). More detailed analysis of the theories available to understand the behavior of biopolymers in solutions is given elsewhere (Schmitt et al. 1998; Semenova and Dickinson 2010).

### 7.3.6.1 Single Biopolymer Systems

Phase separation in single biopolymer systems is governed by the relative balance of biopolymer–biopolymer, biopolymer–solvent, and solvent–solvent interactions. When the biopolymer–solvent interactions are strongly thermodynamically unfavorable (i.e., $\chi_{12} \gg 0$), then they overcome the favorable entropy of mixing contribution and phase separation occurs. The relative balance of the molecular interactions in single biopolymer systems can be altered in a number of ways, such as changing pH, ionic strength, or solvent type. This phenomenon is utilized in the formation of biopolymer particles using the antisolvent precipitation method. A biopolymer is initially dissolved in a good solvent ($\chi_{12} \ll 0$) or neutral solvent ($\chi_{12} \approx 0$). This solution is then titrated into a bad solvent ($\chi_{12} \gg 0$), so that biopolymer–biopolymer interactions are favored over biopolymer–solvent interactions, which leads to the formation of biopolymer particles.

### 7.3.6.2 Mixed Biopolymer Systems

Mixtures of different biopolymers are often used to assemble biopolymer-based delivery systems (Schmitt et al. 1998; Schmitt and Turgeon 2011; Turgeon et al. 2003). Consider what happens when two different types of biopolymers are mixed together in an aqueous solution (Figure 7.4). If there are relatively weak interactions between the two kinds of biopolymers (i.e., $\chi_{23} \approx 0$), then the solution exists as a single phase of uniformly dispersed molecules because entropy of mixing dominates molecular interactions. However, if there are relatively strong attractive (i.e., $\chi_{23} \ll 0$) or repulsive (i.e., $\chi_{23} \gg 0$) interactions between the two kinds of biopolymers, then the solution may separate into two phases with different compositions.

*Attractive Interactions*: When sufficiently strong attractive forces operate between different biopolymer types, they associate with each other. Under some circumstances, soluble molecular complexes with well-defined stoichiometry are formed that are evenly distributed throughout the solution. Under other circumstances, macroscopic phase separation occurs and a solution consisting of one phase

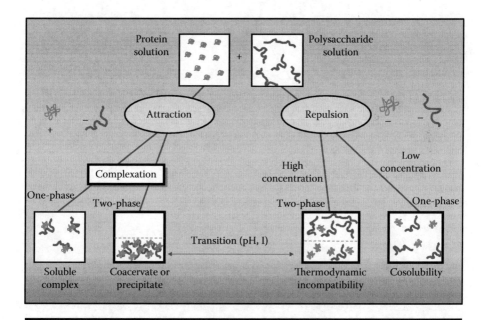

**Figure 7.4 A mixed biopolymer system may exist as a single phase or two phases depending on the sign and strength of the molecular interactions. Phase separation may occur because of thermodynamic incompatibility or complexation.**

rich in both biopolymers and another phase depleted in both biopolymers is formed. This process is usually known as *aggregative phase separation* or *complex coacervation* and is often driven by electrostatic attraction. A common example of electrostatic complex formation in the food industry is that based on the interactions of globular proteins and ionic polysaccharides. Consider a system consisting of a globular protein and an anionic polysaccharide that are mixed together at a pH value above the isoelectric point of the protein (Figure 7.5). At this pH, both the protein and polysaccharide are negatively charged and repel each other. When the pH is reduced, the number of positive charges on the protein increases ($-NH_2 \rightarrow -NH_3^+$), while the number of negative charges decreases ($-CO_2H \rightarrow -CO_2^-$) (Dickinson 1998). Below a critical pH, the anionic groups on the polysaccharide molecules associate with the cationic groups on the surface of the protein molecules leading to complexation. A number of key stages in the molecular interactions can be identified (Figure 7.5):

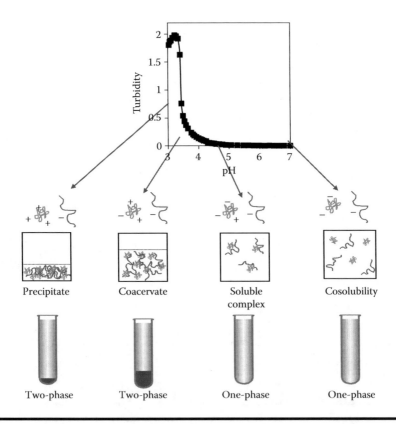

**Figure 7.5  Example of aggregative segregation (complex coacervation) in a solution containing a globular protein (such as β-lactoglobulin) and an anionic polysaccharide (such as pectin).**

■ *No complexation (pH >> pI)*. Initially, both protein and polysaccharide have strong negative charges so there is a strong electrostatic repulsion that prevents them from coming close enough together to associate.

■ *Soluble complexes (pH$_c$ < pH < pH$_s$)*. When the pH is reduced below a critical value, referred to as pH$_s$, the protein and polysaccharide associate with each other to form soluble complexes. Complexation often begins somewhat above the isoelectric point of the protein (even though the net protein charge is negative) owing to cationic patches on the protein surface (Turgeon et al. 2003; Xia et al. 1993).

■ *Coacervates (pH$_p$ < pH < pH$_c$)*. When the pH is reduced further, another critical pH is reached (pH$_c$) where the protein and polysaccharide molecules associate through electrostatic attraction to form neutral complexes or *coacervates*. Initially, the coacervates appear as droplets (*d* ~ 100–10,000 nm) that form turbid water-in-water (W/W) emulsions because of their ability to scatter light. These droplets are prone to coalescence because they have a low net charge and so they tend to separate into a two-phase system, with a dense lower phase rich in both biopolymers (the coacervate phase) and a lighter upper phase that is depleted in both biopolymers (serum phase). Typically, the coacervate phase is a highly viscous or gel-like material that contains large amounts of solvent (>70%). Coacervates are reversible phases that form and dissociate as the pH and ionic strength are adjusted to change the electrostatic interactions.

■ *Precipitates (pH < pH$_p$)*. When the electrostatic attraction between the protein and polysaccharide molecules is very strong, then precipitates are formed rather than coacervates. The biopolymer molecules within precipitates are packed more densely than in coacervates, thereby trapping less solvent. Sedimentation is fast and light scattering is high because of the large, dense structures formed (high refractive index).

■ *No complexation (pH << pK$_a$)*. If the pH decreases substantially below the pK$_a$ of the anionic polysaccharide, then the molecule will lose its charge and there will be a reduction in the electrostatic attraction. Consequently, any electrostatic complexes will dissociate. This effect is most relevant for anionic polysaccharides with relatively high pK$_a$ values (such as pectin and alginate with pK$_a$ = 3.5).

The composition and structure of the complexes formed at a particular pH value depend on the nature of the two biopolymers involved (e.g., molecular weight, conformation, flexibility, charge density, and charge distribution), solution conditions (such as ionic strength), and environmental conditions (such as temperature) (Cooper et al. 2005; Kayitmazer et al. 2013; Schmitt and Turgeon 2011).

*Repulsive Interactions*: In solutions where the interactions between two kinds of biopolymers are strongly unfavorable, the system tends to separate into two phases with one phase being rich in the first biopolymer type and depleted in the other, and another phase being rich in the second biopolymer type and depleted in the other (Schmitt et al. 1998; Schmitt and Turgeon 2011; Turgeon et al. 2003). This process is usually known as *segregative phase separation* or *thermodynamic incompatibility* and is often driven by excluded volume effects. The tendency for a given system to phase separate through this mechanism can be described by a *phase diagram* that shows the influence of initial system composition on the structural organization of the mixed system (Figure 7.6). A phase diagram is constructed by preparing a large number of mixed solutions containing different concentrations of two biopolymers ($B_1$ and $B_2$). The mixed system is allowed to come to equilibrium, and then the number of phases formed is determined (usually either one or two). The volume fraction ($\phi$) and composition ($[B_1]$ and $[B_2]$) of each phase can then be determined using suitable analytical methods for proteins (e.g., Biuret) and polysaccharides (e.g., phenol–sulfuric acid). At relatively low biopolymer concentrations, the solution exists as a single aqueous phase containing an intimate molecular dispersion of the two types of biopolymers, since entropy of mixing dominates the repulsive interactions. At relatively high biopolymer concentrations, the solution separates into two aqueous phases ($W_1$ and $W_2$) that have different biopolymer compositions: one of the phases is rich in $B_1$ and depleted in $B_2$, while the other is rich in $B_2$ and depleted in $B_1$. The composition and volume fraction of each phase can be established using tie lines, as described elsewhere (Norton and Frith 2001). If a phase-separated biopolymer mixture is blended, it will form a colloidal dispersion

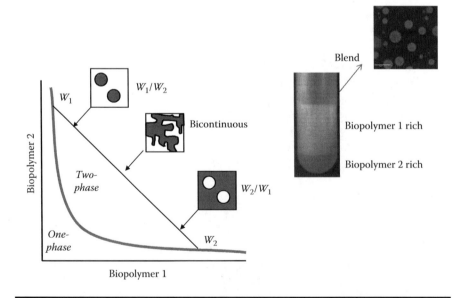

**Figure 7.6   The structural organization of a mixed biopolymer system that may undergo thermodynamic incompatibility can be described by a phase diagram.**

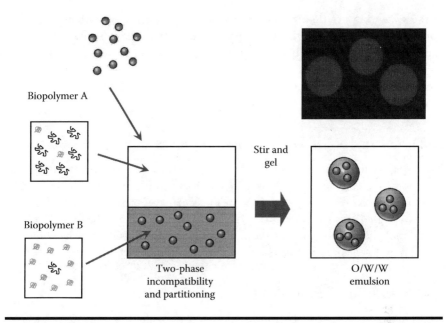

**Figure 7.7** **Schematic representation of production of an O/W/W emulsion from a two-phase system consisting of two aqueous phases.**

whose structure depends on the ratio of the two different phases (Figure 7.6). The aqueous phase that is present in excess tends to make up the continuous phase of the W/W emulsion. Thus, when the $W_1$ phase is in excess, a $W_2/W_1$ emulsion is formed, but when the $W_2$ phase is in excess, a $W_1/W_2$ emulsion is formed. When the two aqueous phases are present in approximately equal amounts, then a bicontinuous system is formed (Figure 7.6). These kinds of phase-separated systems tend to be unstable to coalescence and separate back into the two bulk aqueous phases if left, with the rate depending on the rheological properties of the two phases. The water droplets formed using this method can be stabilized by gelling them using a suitable physical or covalent cross-linking method (Norton and Frith 2001). This approach can be used to fabricate hydrogel particles suitable for encapsulating hydrophilic active ingredients or emulsified lipophilic active ingredients. In the case of lipophilic ingredients, an oil-in-water-in-water (O/W/W)-type emulsion is formed (Figure 7.7).

# 7.4 Physiochemical Methods for Biopolymer Particle Formation

Many of the approaches used to fabricate biopolymer structures suitable for creating delivery systems are based on manipulation of the physicochemical properties of the system, rather than on using mechanical processing operations. This section provides a brief overview of some of the most important physicochemical approaches for fabricating biopolymer-based delivery systems. A few examples of specific systems are given in each section, but this is a rapidly expanding area and there are many more examples of biopolymer-based delivery systems based on these principles described in the literature.

## 7.4.1 Formation of Molecular Complexes

A molecular complex is considered to be a relatively small entity with a well-defined stoichiometry, as opposed to a mixed system that forms a separate phase. Certain types of biopolymer molecules can form molecular complexes with specific types of active ingredients (Livney 2008). The active ingredients may be bound to individual biopolymer molecules or they may be incorporated into complexes formed from a single type or mixed types of biopolymers (Figure 7.8).

### 7.4.1.1 Individual Biopolymer Molecules

Molecular complexes can be formed between individual biopolymer molecules and certain types of active ingredients through various nonspecific or specific bonds, for example, van der Waals, electrostatic, hydrophobic, hydrogen bonding, and covalent interactions (Figure 7.3). The nature of the interactions involved depends on the molecular characteristics of the active ingredient and biopolymer (Livney 2008; Matalanis et al. 2011). Lipophilic active ingredients can bind to nonpolar patches on the surfaces of some biopolymer molecules through

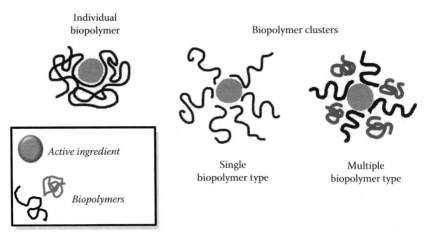

**Figure 7.8** **Highly schematic illustration of different types of molecular complexes that may be formed between active ingredients and biopolymers. (Pictures provided by Isobelle McClements.)**

hydrophobic interactions. For example, globular proteins, such as α-lactalbumin, β-lactoglobulin, and BSA, have hydrophobic pockets on their surfaces that can bind lipophilic molecules such as surfactants, resveratrol, α-tocopherol, docosahexaenoic acid, and conjugated linoleic acid (Kelley and McClements 2003; Liang and Subirade 2012; Liang et al. 2011; Zimet and Livney 2009). Flexible proteins, such as caseinate, have also been shown to bind certain kinds of lipophilic molecules through hydrophobic interactions and form molecular complexes that remain dispersed in aqueous solutions (Semo et al. 2007). Some polysaccharide molecules are also able to bind lipophilic active ingredients and form soluble complexes. For example, amylose and dextrin can form helices in aqueous solutions that have hydrophobic cores capable of incorporating lipophilic molecules, such as flavors, fatty acids, and nutraceuticals (Conde-Petit et al. 2006; Li and Zhang 2010; Nimz et al. 2004; Wulff et al. 2005; Xu et al. 2013). The inclusion of lipophilic molecules into these molecular complexes often improves their water dispersibility and chemical stability.

In some systems, electrostatic interactions play an important role in the binding of active ingredients to biopolymers. For example, electrostatic interactions play an important role in the binding of ionic surfactants and fatty acids to proteins and polysaccharides (Barbana et al. 2011; Kelley and McClements 2003; Livney 2010). They also play an important role in nonspecific and specific binding of mineral ions (such as calcium or iron) to biopolymers (Brock 2012; Carpenter and Mahoney 1992). The binding of bioactive peptides or proteins to other biopolymers is also often primarily driven by electrostatic interactions (Livney 2010).

In the pharmaceutical industry, the functional performance of many drugs is improved by covalently binding them to polymer molecules to form "pro-drugs" (Arpicco et al. 2011). The drugs are then released from the carrier molecule after they have been ingested. A similar approach can be used for the encapsulation and delivery of certain kinds of active ingredients in the food industry. For example, plant sterols are esterified with fatty acids to form plant sterol esters that have better physical and chemical stability in foods (Rocha et al. 2011). Similarly, vitamin E (α-tocopherol) is esterified to form vitamin E acetate (α-tocopherol acetate), which has better chemical stability in commercial products than the nonesterified form (Lauridsen et al. 2001). Bioactive peptides or proteins may be covalently bound to polysaccharides through the Maillard reaction, which can improve their water solubility, stability to environmental stresses (such as pH changes, high ionic strength, or thermal treatments), and biological activity (Dong et al. 2011, 2012; Livney 2010).

### 7.4.1.2 Biopolymer Molecular Clusters: Single Biopolymer Type

Active ingredients may also be encapsulated within clusters of biopolymer molecules (Figure 7.8). These clusters may be formed from a single biopolymer type or from different biopolymer types. Nanoscale clusters of casein molecules ("casein micelles") are capable of encapsulating and protecting various nonpolar active ingredients, including bioactive oils, vitamins, and drugs (Haham et al. 2012; Shapira et al. 2012; Zimet et al. 2011). Casein micelle-based delivery systems are formed by dissolving the nonpolar ingredients in a suitable organic solvent (such as ethanol) and then titrating the resulting solution into an aqueous caseinate solution. The casein micelles are then reassembled by adding calcium, phosphate, and citrate to promote clustering. The incorporation of vitamin D into casein micelles was shown to improve its stability to photochemical degradation, when compared to nonencapsulated vitamin D (Semo et al. 2007).

### 7.4.1.3 Biopolymer Molecular Clusters: Mixed Biopolymer Type

Molecular clusters may also be formed owing to the assembly of different types of biopolymers that are attracted to each other (Figure 7.8). Proteins and ionic polysaccharides form molecular clusters at pH values where there is a weak electrostatic attraction between them (Jones and McClements 2010; Livney 2010). Electrostatic complexes of proteins and ionic polysaccharides have been used to encapsulate and protect various forms of active lipophilic agents, including ω-3 fatty acids, oil-soluble vitamins, polyphenols, and flavors (Fang and Bhandari 2010; Madene et al. 2006; Ron et al. 2010; Zhang et al. 2009; Zimet and Livney 2009). These complexes may also be used to encapsulate hydrophilic or amphiphilic active ingredients, provided that there is a sufficiently strong attractive interaction between the active ingredients and one or more of the biopolymers in the molecular clusters. Thus, enzymes and other proteins can be encapsulated within these molecular clusters through electrostatic or hydrophobic interactions (Maculotti et al. 2009).

## 7.4.2 Formation of Hydrogel Particles

Biopolymers can be used as building blocks to form hydrogel particles that can be used to encapsulate active ingredients. These hydrogel particles can be formed from individual biopolymer types or from mixed biopolymer types depending on the mechanism used to produce them.

### 7.4.2.1 Hydrogel Particles: Single Biopolymer Type

Hydrogel particles can be formed from single biopolymer types by promoting their self-association under appropriate solution conditions, for example, where biopolymer–biopolymer interactions are stronger than biopolymer–solvent interactions. Hydrogel nanoparticles or microparticles can be formed when globular protein solutions are heated above their thermal denaturation temperature ($T_m$) under conditions where there is a weak attraction between the protein molecules (Bengoechea et al. 2011; Jones and McClements 2011). The size and charge of the biopolymer particles formed can be controlled by altering the initial biopolymer concentration, holding temperature, holding time, pH, and ionic strength. Recently, this approach was used to encapsulate and protect active ingredients in aqueous dispersions, for example, bioactive catechins (Shpigelman et al. 2010).

Hydrogel particles can also be formed from thermally denatured globular proteins using a cold-set particle formation approach (Ako et al. 2010; Zhang et al. 2012). In this method, a globular protein solution is first heated above its thermal denaturation temperature, causing the proteins to unfold and form thin filaments. Typical solution conditions that promote filament formation are pH values away from the pI and low ionic strength. Solution conditions are then altered to promote self-association of the protein filaments, for example, by adjusting the pH toward the pI or by adding counterions such as $Na^+$ and $Ca^{2+}$ at neutral pH (Bryant and McClements 1998; Mezzenga and Fischer 2013; Nicolai et al. 2011). The active ingredient to be encapsulated can be mixed with the heat-denatured protein solution prior to promotion of hydrogel particle formation.

Hydrogel particles can also be formed by changing the *quality* of the solvent surrounding the biopolymer molecules in solution. Many water-soluble biopolymers associate with each other when a sufficient quantity of alcohol (e.g., ethanol) is added to an aqueous solution, thereby leading to the formation of hydrogel particles. Hydrogel particles have been formed from various water-soluble biopolymers using this approach, including BSA, whey protein isolate, gelatin, and caseinate (Arroyo-Maya et al. 2012; Gulseren et al. 2012; Li et al. 2012). In some cases, it is necessary to stabilize the hydrogel particles formed so that they maintain their integrity when solution conditions are altered, for example, by cross-linking with glutaraldehyde, minerals, or enzymes.

Hydrogel particles can also be formed from water-insoluble biopolymers using an antisolvent precipitation approach. In this case, the biopolymer molecules are first dissolved within an organic solvent, and the resulting solution is then titrated into an aqueous solution to promote protein self-association. Hydrogel particles have been formed from wheat gliadin and corn zein (which are both water-insoluble proteins) using this approach. Gliadin particles have been used to encapsulate and protect retinoic acid and vitamin E (Duclairoir et al. 1999, 2002), whereas zein particles have been used to encapsulate and protect curcumin and quercetin (Patel et al. 2010, 2012).

Hydrogel particles can also be formed by cross-linking gelling biopolymers dissolved within an aqueous solution, provided the biopolymer concentration is kept below that required to form a macroscopic gel. Cross-linking can be induced in a number of ways: (i) adding chemical agents (such as glutaraldehyde), (ii) adding enzymes (such as transglutaminase and laccase), (iii) adding mineral counterions (such as potassium, calcium, or tripolyphosphate), (iv) adjusting pH (by adding acid or base), and (v) altering environmental conditions (such as temperature). The precise mechanism that is suitable for cross-linking the biopolymers depends on protein and polysaccharide type (Renard et al. 2006).

*Protein Gelation*: Globular proteins are commonly used to form hydrogel particles through controlled cross-linking. These particles are usually formed by heating globular proteins above their thermal denaturation temperature, which promotes protein self-association through hydrophobic attraction and disulfide formation (Jones and McClements 2010, 2011). The nature of the particles formed can be controlled by manipulating the intermolecular interactions, through controlling pH, ionic strength, and heating conditions. Hydrogel particles can also be formed by controlled aggregation of more flexible random coil proteins, such as casein and gelatin. Casein can be gelled by adjusting the pH to close to the proteins' isoelectric point, by adding multivalent counterions, or by adding rennet (Cooper et al. 2010). Gelatin can be gelled by cooling a biopolymer solution below the helix–coil transition temperature to promote the formation of helices and hydrogen bond cross-links (Djagny et al. 2001). Enzymes such as transglutaminase can be used to form covalent cross-links between amino acids on various protein substrates (De Jong and Koppelman 2002). Enzyme cross-linked systems have been used to stabilize hydrogel particles in various biopolymer-based delivery systems (Huppertz and de Kruif 2008; Matalanis and McClements 2013; Song and Zhang 2008).

*Polysaccharide Gelation*: Different gelation methods are required to cross-link different kinds of polysaccharides depending on the nature of the bonds between the molecules, such as cold-setting, heat-setting, and ion-setting gelation (Burey et al. 2008). Certain types of modified cellulose gel upon heating, which has been attributed to an increase in the strength of the hydrophobic attraction between nonpolar groups with increasing temperature. Other types of polysaccharides, such as agar, alginate, and carrageenan, gel upon cooling owing to formation of helices that associate with each other through hydrogen bonding. The enzyme laccase has been used to form gels from sugar beet pectin by forming covalent cross-links between phenolic groups (Minussi et al. 2002), which has been used to fabricate various types of biopolymer-based delivery systems (Chen et al. 2012; Chen et al. 2010; Zeeb et al. 2012).

### 7.4.2.2 Hydrogel Particles: Mixed Biopolymer Type

As discussed earlier, when two biopolymers are mixed together in aqueous solutions, they may form a one-phase or two-phase system depending on the nature of the biopolymers involved, solution composition, and environmental conditions (Figure 7.4). In a one-phase system, the two biopolymers can exist either as individual molecules or as soluble complexes that are distributed throughout the system. In a two-phase system, the solution separates into two phases with different biopolymer compositions through either aggregative or segregative mechanisms.

In *aggregative separation*, there is a relatively strong *attraction* between the two different kinds of biopolymers that causes them to associate with each other.

The most common example of this type of interaction for food biopolymers is the electrostatic attraction between molecules with opposite electrical charges. The resulting two-phase system consists of a phase that is rich in both biopolymers and a phase that is depleted in both biopolymers. The biopolymer-rich phase may form either a *coacervate* or a *precipitate*, depending on the strength of the attraction and the nature of the biopolymers involved (Section 7.3.6.2). In *segregative separation*, there is a relatively strong *repulsion* between the two different kinds of biopolymers; that is, there is a relatively high positive (unfavorable) free energy of mixing. This type of phase separation often occurs when one or both of the biopolymers are uncharged, or when both biopolymers have similar charges. At sufficiently low biopolymer concentrations, the two biopolymers are intimately mixed and form a one-phase solution (Figure 7.6). Once the biopolymer concentration exceeds a certain level, phase separation occurs and a two-phase solution is formed with each phase being rich in one type of biopolymer and depleted in the other type of biopolymer.

A variety of microstructures can be created in phase-separated biopolymer systems by varying the preparation conditions or by shearing the system (Norton and Frith 2001; Wolf et al. 2001, 2002). Examples of these systems include "W/W" emulsions or "O/W/W" emulsions (Figure 7.7). Once a particular microstructure has been formed, it is often possible to trap the system in a kinetically stable state and thus create novel food microstructures and rheological properties (Norton and Frith 2001). Kinetic trapping can be achieved by changing the solution or environmental conditions so that one or both of the phases thicken or gel. If this process is carried out in the presence of shear, it is possible to produce a wide variety of different microstructures such as spheres, teardrops, or fibers. Alternatively, it may be possible to adsorb another biopolymer around the internal water droplets that form the dispersed phase in a W/W emulsion, thereby stabilizing them.

# 7.5 Mechanical Methods for Biopolymer Particle Formation

Biopolymer-based delivery systems can also be formed using various mechanical methods, rather than relying primarily on the physicochemical (bottom–up) approaches described in the previous sections (Jones and McClements 2011; Matalanis et al. 2011). In this section, a brief overview of some of the mechanical methods available for creating biopolymer particles is given. A more general description of particle formation methods is given in Chapter 4. It should be noted that many mechanical methods actually use a combination of physicochemical and processing methods to form biopolymer particles; for example, a particle is first formed using a specific mechanical device and is then gelled using a physicochemical approach.

## 7.5.1 Extrusion Methods

Extrusion methods involve injection of a biopolymer solution into another solution that promotes gelation. One of the simplest means of fabricating biopolymer particles based on this principle is the simple injection device shown in Figure 7.9. Gelation may be induced by including an appropriate cross-linking agent into the gelling solution, such as mineral ions, glutaraldehyde, enzymes, or alcohol. Alternatively, the gelling solution could be heated or chilled to promote heat-set or cold-set gelation for thermally setting biopolymers, such as whey proteins or gelatin. A widely used example of the extrusion method is the formation of hydrogel particles using alginate as the biopolymer and calcium ions as the gelling agent (Li et al. 2011). Small drops of alginate solution are injected into a calcium solution, which promotes the formation of alginate hydrogel particles. The size and porosity of the hydrogel particles can be controlled by changing the alginate and calcium concentrations, as well as the extrusion conditions. This method has been used to control the digestibility and release of lipids under simulated gastrointestinal conditions and may be a useful method for developing colon-based delivery systems or for controlling the satiety response of foods (Li et al. 2011). The extrusion method can also be used to form hydrogel particles with different characteristics by using

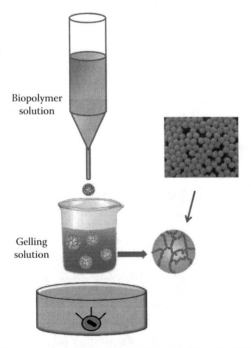

Biopolymer
solution

Gelling
solution

**Figure 7.9   Schematic representation of the injection method used to produce filled hydrogel beads. A biopolymer solution is injected into another aqueous solution containing a gelling agent or different environmental conditions.**

different kinds of biopolymer particles and gelling mechanisms. For example, pectin could be injected into a calcium solution (ionic gelation), chitosan could be injected into a tripolyphosphate solution (ionic gelation), whey protein could be injected into a hot liquid (heat-set gelation), or gelatin could be injected into a cold liquid (cold-set gelation).

The spray chilling methods discussed in Chapter 4 are examples of extrusion methods where multiple particles are formed simultaneously and then gelled (Oxley 2012b). In this approach, a biopolymer solution containing the active ingredient is sprayed through a nozzle into a cold chamber that promotes gelation. Membrane homogenization may also be seen as an example of an extrusion method that is capable of forming multiple particles simultaneously (Chapter 4), but this method is mainly used for forming emulsions (although it can be used to form biopolymer particles using the emulsion-templating method described later). Extrusion methods can be adopted to form core–shell structures by having a coaxial injection system with one fluid on the inside and another on the outside (Oxley 2012a).

## 7.5.2 Microfluidic Methods

Microfluidics is a particularly versatile method for creating hydrogel particles with well-defined dimensions and internal structures (Desmarais et al. 2012; Helgeson et al. 2011; Selimovic et al. 2012). In some respects, microfluidics methods are similar to extrusion methods since they involve injecting one solution into another solution to form hydrogel particles (Figure 7.10). Microfluidic devices can be designed to create hydrogel particles with different sizes, shapes, and internal structures (e.g., core–shell or dispersion) by carefully designing the nature of the various microchannels through which the fluids are made to flow. A simple microfluidic design for preparing hydrogel particles is shown in Figure 7.10. A biopolymer solution is made to flow though a narrow internal microchannel, while a solution of gelling

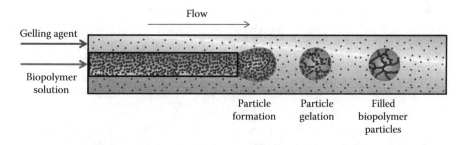

**Figure 7.10** **Schematic diagram of microfluidic method for forming biopolymer particles using a coaxial device. A biopolymer solution is passed through an inner cylinder, while a gelling agent solution is passed through an outer cylinder. Hydrogel particles are formed at the end of the inner cylinder, which are gelled by the gelling agent.**

agent is made to flow through an external microchannel. When the biopolymer exits the internal microchannel, it forms a particle, which comes into contact with a gelling agent that cross-links the biopolymer molecules, leading to the formation of hydrogel particles. Microfluidics are particularly suitable for basic research and development purposes because of the tight control one has over the characteristics of the hydrogel particles produced. Nevertheless, it is less suitable for widespread commercial applications within the food industry because of scale-up and economic problems.

## 7.5.3 Spray Drying and Other Drying Methods

The encapsulation of active food ingredients by spray drying leads to the formation of dried biopolymer particles when proteins or polysaccharides are used as wall materials (Burey et al. 2008). A number of steps are involved in the formation of biopolymer particles by spray drying (Fang and Bhandari 2012). First, the active ingredient is dissolved or dispersed into a liquid that is passed through a small nozzle, which leads to the formation of a mist of fine drops (Figure 4.10). The outlet of the nozzle is located within a heated chamber so that the volatile carrier liquid (usually water or organic solvent) quickly evaporates. The dried particles are then collected using a cyclone or filter bag. The temperature that the active ingredients experience is usually considerably less than that of the heated chamber because of the endothermic enthalpy change associated with solvent evaporation, which helps protect thermally labile substances from degradation. The diameters of the particles within the spray-dried powder are usually in the 10–100 µm range (Fang and Bhandari 2012). Spray drying can be used to convert a suspension of biopolymer particles into a powder that can be reconstituted prior to use. Hydrogel particles with different sizes can be prepared by controlling the relative rates of particle swelling and biopolymer cross-linking during reconstitution (Burey et al. 2008).

Spray drying has been used to encapsulate a wide variety of different active ingredients such as proteins, peptides, flavor oils, bioactive lipids, and probiotics (Fang and Bhandari 2012; Jafari et al. 2008; Vega and Roos 2006). The characteristics of the powder particles and the wall materials within them must be carefully controlled to prevent chemical degradation or release of the active ingredients during storage, to ensure good powder handling properties, and to ensure good powder dispersion characteristics after reconstitution.

Biopolymer particles may also be formed by other processing methods that involve drying a solution containing biopolymers and active ingredients. For example, an aqueous biopolymer solution may be converted into a solid form, which could then be ground using a mechanical device to form a powder. However, when the powder is reconstituted with water, the biopolymer particles may dissolve, unless there is a suitable cross-linking method that can be applied to the particles in the dry state or during rehydration.

## 7.5.4 Antisolvent Precipitation

The formation of biopolymer particles using antisolvent precipitation relies on changes in solvent quality (Horn and Rieger 2001; Joye and McClements 2013). In this process, the biopolymer and active component are first dissolved in a good solvent and then the conditions are altered so that it becomes a bad solvent, which leads to the spontaneous formation of biopolymer nanoparticles or microparticles. This method can be used to form biopolymer particles from either water-soluble or water-insoluble biopolymers. For water-insoluble biopolymers, an organic solvent is used to dissolve the biopolymer, and water is used as an antisolvent. For example, a water-insoluble biopolymer (such as zein) and active ingredient (such as curcumin or quercetin) are dissolved in an ethanol solution, which is then injected into an aqueous solution, leading to the formation of biopolymer particles (Patel et al. 2010, 2012). For water-soluble biopolymers, an aqueous solution is used to dissolve the biopolymer, and an organic liquid (such as ethanol) is used as the antisolvent. For example, a water-soluble biopolymer (such as protein) and active ingredient may be dissolved in water and then injected into an alcohol solution to promote biopolymer precipitation and biopolymer formation (Gulseren et al. 2012). Other components may also be added to a solution to cause it to become a bad solvent for a biopolymer, such as salts, cosolvents, acids, or bases. In some cases, it may be necessary to add a stabilizing agent to the antisolvent to improve the stability of the biopolymer particles once they have formed, for example, surfactants or polymers that adsorb to the particle surfaces (Joye and McClements 2013).

## 7.5.5 Emulsion Templating

The emulsion-templating method relies on the utilization of water-in-oil (W/O) emulsions as templates to create biopolymer particles with specific dimensions (Figure 7.11). An aqueous biopolymer solution is homogenized with an oil phase containing an oil-soluble emulsifier to form a W/O emulsion (Saglam et al. 2011, 2013; Zhang et al. 2006). Water droplet size can be controlled by varying homogenization conditions (pressure and number of passes), surfactant type, or system composition (oil-to-water ratio, surfactant-to-water ratio). The water droplets are then gelled using a mechanism appropriate for the particular biopolymer present, for example, temperature change, addition of cross-linking agent, or pH/ionic strength change. Finally, the biopolymer particles can be obtained by centrifuging/filtering the W/O emulsion, collecting the particles, and then washing them with an organic solvent to remove any residual oil. The resulting biopolymer particles can then be dispersed in an aqueous solution or dried.

This approach has been used to form biopolymer particles based on alginate (Reis et al. 2006). An aqueous solution containing alginate and insoluble calcium salt were dispersed into a continuous oil phase containing oil and surfactant under

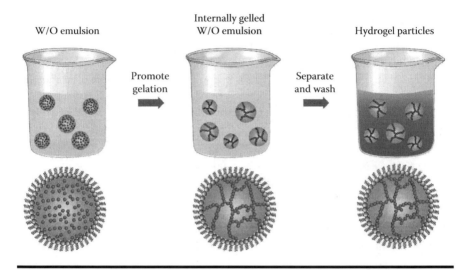

**Figure 7.11  Emulsion-templating method for forming biopolymer particles. A W/O emulsion is formed by homogenizing an oil phase (oil + lipophilic surfactant) and an aqueous phase (water + biopolymer). The water phase is then gelled by changing environmental conditions (such as temperature) or adding a gelling agent.**

constant agitation to form a W/O emulsion. An organic acid was then mixed with the oil phase, causing the insoluble calcium salt to slowly dissolve and release calcium ions into the aqueous phase. The cationic calcium ions then cross-linked the anionic alginate molecules, which led to the formation of biopolymer particles. Following gelation, these particles were separated from the oil phase by filtration/centrifugation and then washed using an organic solvent to remove any residual oil. This method has also been used to encapsulate riboflavin in alginate/whey protein microspheres (Chen and Subirade 2005) and to encapsulate probiotic bacteria in milk protein particles (Heidebach et al. 2009). This approach can be adopted to form filled biopolymer particles by mixing the biopolymer solution with an O/W emulsion prior to homogenizing it with an oil phase containing an oil-soluble surfactant. In this case, an oil-in-water-in-oil (O/W/O) emulsion is formed and then the internal water phase is gelled, and then the filled biopolymer particles are removed by centrifugation/filtration and washing as described previously (Egan et al. 2013).

## 7.5.6  Shearing Methods

There are a number of examples where biopolymer particles have been formed by applying shear to a solution to break down larger biopolymer particles into smaller ones (Burey et al. 2008; Matalanis et al. 2011; Norton and Frith 2001; Wolf et al.

2002). For example, it is possible to form a macroscopic gel using a particular type of biopolymer and then break it down by applying mechanical forces (Figure 7.12). The size and shape of the biopolymer particles formed depend on the nature of the biopolymers involved, the gelling mechanism, and the disruptive forces applied (e.g., intensity and duration).

It is also possible to form biopolymer particles in phase-separated biopolymer mixtures by shearing them to form a W/W emulsion and then adding a cross-linking agent or changing environmental conditions (such as temperature or solvent quality) to gel the particles. The size and morphology of the biopolymer particles formed depend on the magnitude of the applied shearing forces, that is, the stirring speed. The particle size typically decreases with increasing shear rate because of droplet disruption by the mechanical forces, but then it grows at higher shear rates because of increased droplet coalescence (Matalanis and McClements 2012a,b). Spherical biopolymer particles tend to be formed at relatively low shear rates, whereas ellipsoids or fibers are formed at higher shearing rates owing to elongation in the shear field (Wolf et al. 2001, 2002). The size and shape of the biopolymer particles formed depend on solution viscosity, applied shear rate, and interfacial tension. Suspensions of nonspherical hydrogel particles have higher viscosities than those containing equivalent amounts of spherical particles, which may be useful for the development of food products with novel textures and flow characteristics. By controlling the shear forces and particle cross-linking kinetics, it is possible to generate hydrogel particles with different sizes and morphologies (Norton and Frith 2001).

Biopolymer particles can also be fabricated from segregative phase-separated biopolymer mixtures. These systems can be made to form W/W emulsions by shearing and then the inner water phase can be gelled to create biopolymer particles (Stokes et al. 2001).

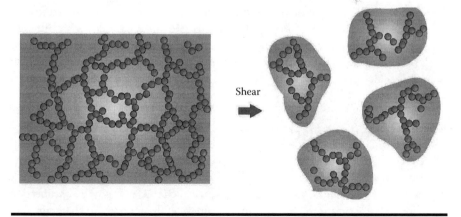

Shear

**Figure 7.12** **Biopolymer particles may be formed by applying shear forces to break down large particles into smaller ones.**

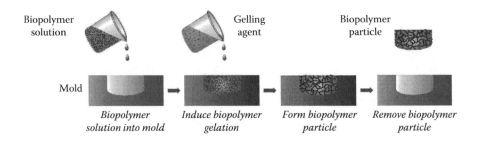

**Figure 7.13** **Schematic diagram of molding method for forming biopolymer particles. A biopolymer solution is poured into a mold. The gelation of the biopolymer molecules is then promoted by changing environmental conditions or adding a gelling agent. The biopolymer particles formed are then removed from the mold.**

### 7.5.7 Molding Techniques

Biopolymer particles with well-defined dimensions and morphologies can be formed using molding techniques (Jung and Yi 2012; Matalanis et al. 2011; McGuigan et al. 2008). A biopolymer solution is poured into a mold with a specific size and shape (Figure 7.13) and then the solution or environmental conditions are adjusted to promote biopolymer gelation (e.g., by changing temperature or adding a gelling agent). After gelation has occurred, the biopolymer particles are removed from the mold and used. The recent development of nanofabrication methods of producing molds with very small dimensions has meant that this method is now able to produce small biopolymer particles. Molds can be prepared containing multiple molding units so that many biopolymer particles can be formed simultaneously. The possibility of forming food-grade biopolymer particles with different morphologies and compositions using molding methods has been investigated (Malone and Appelqvist 2003). This method is particularly suitable for fundamental studies of the functional performance of biopolymer particles since particles with different sizes and morphologies can easily be prepared. Nevertheless, it may be less suitable for large-scale production of food biopolymer particles because of difficulties in economic scale-up.

## 7.6 Biopolymer Particle Properties

The effect of a biopolymer particle on the macroscopic properties of a food or beverage product, as well as on its functional performance (such as protection and release characteristics), depends on its structure, electrical charge, and physical properties. It is therefore important to control and measure biopolymer particle characteristics and understand how they influence the macroscopic properties of the commercial products that they are incorporated into.

## 7.6.1 Particle Structure

### 7.6.1.1 External Structure: Particle Dimension and Morphology

As mentioned earlier, the dimensions and morphology of biopolymer particles can be manipulated by controlling the building blocks and assembly procedures used to fabricate them (Jones and McClements 2010; Matalanis et al. 2011; Norton and Frith 2001). Depending on the fabrication method, biopolymer particles can range from a few nanometers to hundreds of micrometers in diameter. The size characteristics of a population of biopolymer particles are usually expressed as the full particle size distribution (PSD) or the mean particle diameter ($d$) and polydispersity index ($\sigma$) (Chapter 3). Analytical methods suitable for measuring the dimensions and morphology of biopolymer particles are discussed in Chapter 8.

### 7.6.1.2 Internal Structure

Biopolymer particles can be fabricated with various types of internal structures by controlling the building blocks and assembly conditions used to prepare them (Figure 7.14):

- *Homogeneous*: The interior of a biopolymer particle is composed of one or more biopolymer types that are intimately mixed with each other so that they appear to be homogeneous on the length scale considered, for example, particle diameter.
- *Heterogeneous—Bicontinuous*: The interior of a biopolymer particle consists of two or more biopolymer types that phase separate into regions with different compositions on the length scale considered. For example, one phase may be a protein-rich phase while another phase may be a polysaccharide-rich or solvent-rich phase.

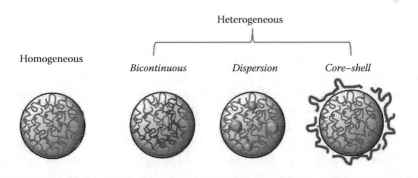

**Figure 7.14** **Examples of different kinds of internal structures that may be formed within hydrogel particles.**

■ *Heterogeneous—Dispersion*: The interior of a biopolymer particle consists of two or more discrete phases, with one or more of the phases being dispersed within the other as small particles. The composition, dimensions, shape, connectivity, interactions, and spatial organization of the various phases within the particle interior may vary. For example, the dispersed phase may be lipid droplets, solid particles, or air bubbles dispersed within a biopolymer matrix. In this case, the size, concentration, and location of the smaller dispersed particles may be important.

■ *Heterogeneous—Core–Shell*: A biopolymer particle may consist of two or more discrete phases, with at least one of the phases forming a shell around the other phase. This shell may vary in composition, thickness, and structure; for example, it may contain one or more layers. This kind of particle can be formed by coating a charged biopolymer particle with another oppositely charged biopolymer.

The internal structure of biopolymer particles may have a large impact on their physicochemical properties and functional performance. It is therefore possible to control the dimensions, shape, and internal structure of biopolymer particles to design delivery systems with different functional attributes, such as rheology, stability, encapsulation efficiency, loading capacity, permeability, integrity, environmental responsiveness, and digestibility. In particular, the porosity of a biopolymer particle may have a major influence on the release of an encapsulated bioactive, as well as on its exposure to other molecules within the surrounding aqueous phase. A highly porous structure will tend to facilitate rapid molecular transport, while a dense structure would limit it. This phenomenon has been demonstrated for alginate beads containing encapsulated lipids. Digestive enzymes (i.e., lipase) can diffuse into the alginate beads and hydrolyze the lipids more rapidly when the biopolymer matrix within the beads has a more open structure (Li et al. 2011).

## 7.6.2 Particle Electrical Characteristics

The electrical characteristics of biopolymer particles are determined by the electrical properties of the various building blocks used to fabricate them, the location of these building blocks within the particle (e.g., interior or exterior), and the surrounding solution conditions (such as pH, ionic composition, and dielectric constant) (Jones and McClements 2010; Matalanis et al. 2011). Usually, the charged species that are present at the exterior of the biopolymer particle are the most important in determining its overall electrical characteristics. Biopolymer particles may have electrical charges that range from highly positive to neutral to highly negative depending on their composition and environmental conditions.

Biopolymer particle charge is important for a number of reasons. The sign and magnitude of the charge on biopolymer particles influence the stability of biopolymer-based delivery systems to aggregation (McClements 2005). When the electrical charge on the particles is sufficiently high, then there will be a strong electrostatic repulsion that opposes aggregation. Usually, a charge magnitude greater than approximately 25 mV is taken to be a good indication of aggregation stability, but the actual value for a particular system depends on the other types of colloidal interactions operating, for example, van der Waals, steric, hydrophobic, depletion, and bridging (Israelachvili 2011). Particle charge also influences how biopolymer particles interact with other electrically charged species within their environment, such as mineral ions, proteins, polysaccharides, surfactants, and microorganisms. If a biopolymer particle has an opposite charge to another ionic ingredient within a food, then it may form an electrostatic complex that may precipitate and sediment (Guzey and McClements 2006). On the other hand, if it has a similar charge to another ionic ingredient, they should be prevented from coming into close contact, unless another oppositely charged species can act as an electrostatic bridge such as a multivalent mineral ion or polymer.

The electrical charge of biopolymer particles also determines how they interact with charged surfaces inside and outside of the human body. A food or beverage comes into contact with various solid surfaces during processing, storage, and utilization, such as processing equipment, packaging materials, cooking utensils, and storage containers. If any of these surfaces contain an electrical charge, then the biopolymer particles may interact with them through electrostatic interactions. Biopolymer particles may also interact with biological surfaces within the human GIT after ingestion of a food. A cationic biopolymer particle may bind to the anionic surface of the tongue, thereby causing perceived astringency (Needleman et al. 1997; Seo et al. 2011). Conversely, a cationic biopolymer particle could be designed to bind to a specific location within the GIT ("mucoadhesion") to delay its transit through the body and release its bioactive at a particular site (Sugihara et al. 2012; Varshosaz 2007). Finally, the electrical characteristics of the molecules within a biopolymer particle influence its changes in integrity and permeability in response to environmental changes such as pH or ionic strength (Thakur et al. 2012). For example, biopolymer particles held together by electrostatic attraction may dissociate when the ionic strength is increased owing to electrostatic screening effects or when the pH is changed so that the biopolymers no longer have opposite charges (Kizilay et al. 2011). When charged biopolymers within a hydrogel particle are cross-linked, they may maintain their overall physical integrity when solution conditions are changed, but they may still swell/shrink when pH or ionic strength is altered (Hebrard et al. 2013; Keppeler et al. 2009; Li et al. 2009). Swelling occurs when electrostatic repulsion between similarly charged biopolymer molecules is strong, for example, low ionic strength or pH values where biopolymer charge is high. Conversely, shrinking occurs when electrostatic repulsion between biopolymer molecules is weak, for example, high ionic strength or pH values where biopolymer charge is low. This phenomenon can be used to design biopolymer-based delivery systems that release active ingredients when they reach a particular environment, such as the mouth, stomach, or small intestine.

The electrical characteristics of biopolymer particles can be controlled by selecting one or more biopolymers with different charge versus pH profiles. Methods for measuring the electrical characteristics of particles in general are discussed in Chapter 8.

## 7.6.3 Particle Physicochemical Properties

The physicochemical properties of the biopolymer matrix that make up the interior of the particles also play an important role in determining their functional performance. These physicochemical properties include the density, refractive index, rheology, polarity, and permeability. Particle matrix properties influence the macroscopic properties of biopolymer-based delivery systems and foods they are incorporated into, such as their appearance, rheology, and stability. They also determine the molecular distribution of any ingredients originally present inside or outside of the particles, which is governed by physical properties such as equilibrium partition coefficients ($K_{OW}$), translational diffusion coefficients ($D$), and permeability characteristics ($P$). It is therefore important to carefully design, control, and measure the physicochemical properties of biopolymer particles. Some of the effects of biopolymer particles on the macroscopic properties of biopolymer-based delivery systems and food materials are discussed below.

### 7.6.3.1 Particle Integrity and Environmental Responsiveness

The "integrity" of a biopolymer particle is its ability to maintain its composition and structure under a given set of solution or environmental conditions, such as pH, ionic strength, temperature, and enzyme activity. Typically, biopolymer particles should maintain their integrity under one set of conditions so they can encapsulate and retain the active ingredients but then break down under another set of conditions so that they can release the active ingredients at the required site of action. Loss of biopolymer integrity may occur through several physicochemical mechanisms including simple diffusion, swelling, dissolution, erosion, or fragmentation (Figure 7.15). The precise mechanism involved depends on the nature of the active ingredient, type of biopolymers present, the molecular interactions holding them together within the particles, and the environmental conditions. Mathematical models for describing and predicting the release of active ingredients from colloidal particles are discussed in Chapter 10.

### 7.6.3.2 Optical Properties

Appearance is one of the most important attributes of any food or beverage product since it is the first sensory impression that a consumer receives to make a quality

Fragmentation—release by
particle breakup

Swelling—release by increase
in matrix pore size

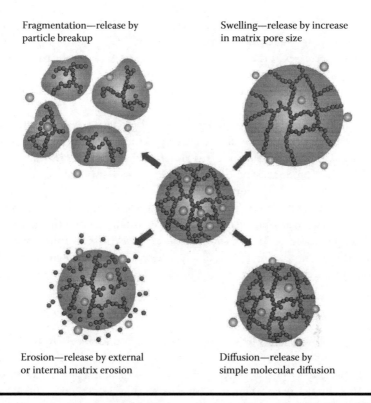

Erosion—release by external
or internal matrix erosion

Diffusion—release by
simple molecular diffusion

**Figure 7.15  Some common mechanisms for loss of particle integrity: diffusion, swelling, erosion, and fragmentation.**

judgment (McClements 2002). Biopolymer particles usually have a refractive index different from that of the surrounding medium and therefore they scatter light waves, which influence the optical properties of any material they are incorporated into. The overall appearance of a material can be separated into two major attributes: *opacity*—primarily determined by light scattering; *color*—primarily determined by light absorption. The overall impact of biopolymer particles on the optical properties of a biopolymer-based delivery system or food product depends on their size, concentration, and refractive index contrast (McClements 2002a,b). The impact of particle size and concentration on the optical properties of colloidal dispersions in general was discussed in Chapter 3. The turbidity or lightness of a particle suspension usually increases as the particle concentration increases, steeply at first and then more gradually at higher particle concentrations (Figure 3.7). The turbidity or lightness of a particle suspension goes from very low when the particle size is much smaller than the wavelength of light to a maximum when the particle size is fairly similar to the wavelength of light (Figure 3.8). Thus, a particle suspension appears clear when it contains very small particles ($r < 25$ nm), looks turbid when it contains somewhat larger particles ($25 < r < 100$ nm), appears cloudy or opaque when it contains intermediate-sized particles (100 nm < $r$ < 100 μm), and has particles that are discernible by the naked eye when it contains bigger particles. The turbidity decreases as the refractive index contrast between the particle and the surrounding medium decreases. Hence, biopolymer particles that contain relatively high concentrations of water tend to scatter light more weakly than similar sized particles containing less water.

The impact of biopolymer particles on optical properties has important consequences for their incorporation into commercial food and beverage products. For instance, some products should be transparent (such as some soft drinks and dressings), while other products should be opaque (such as many yogurts, soups, sauces, desserts, and dressings). The optical properties of food or beverage products are therefore altered by an amount that depends on the concentration, size, and refractive index contrast of any biopolymer particles incorporated into them. Thus, small biopolymer particles ($r$ < 25 nm) should be used for clear food and beverage applications as they do not scatter light strongly, whereas larger biopolymer particles ($r$ ≈ 100–1000 nm) should be used for opaque products as they scatter light strongly.

### 7.6.3.3 Rheological Properties

The textural characteristics of a food or beverage product may also be affected by incorporation of biopolymer particles (Jones and McClements 2010, Matalanis et al. 2011). The impact of particle properties on the rheology of delivery systems in general was discussed in Chapter 3. In this section, we therefore focus primarily on the impact of biopolymer particles on the rheology of aqueous dispersions. The impact of biopolymer particles on viscosity can be described by the following semiempirical equation:

$$\frac{\eta}{\eta_0} = \left(1 - \frac{\phi_{eff}}{\phi_c}\right)^{-2}. \tag{7.3}$$

Here, $\eta_0$ is the shear viscosity of the liquid surrounding the particles, $\phi_{eff}$ (= $R\phi$) is the *effective* volume fraction of the biopolymer particles, $\phi$ is the actual volume fraction occupied by the biopolymer chains, $R$ is the effective volume ratio, and $\phi_c$ is the critical packing parameter (≈0.6) where spherical particles become closely packed. $R$ is the effective volume occupied by the solvated biopolymer particles (biopolymer chains + trapped solvent) divided by the volume occupied by the biopolymer chains.

This equation can be used to predict how biopolymer concentration and effective volume ratio influence the viscosity of biopolymer particle suspensions (Figure 7.16). In general, the relative viscosity increases with increasing biopolymer concentration until the biopolymer particles become closely packed together; that is, $\phi_{eff}$ ≈ $\phi_c$ (Matalanis et al. 2011). Above $\phi_c$, the system becomes solid-like and can be

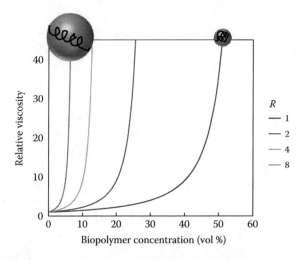

**Figure 7.16** **Theoretical prediction of the influence of biopolymer concentration and packing on the apparent viscosity of biopolymer colloidal suspensions. Here, $R$ is the effective volume ratio, that is, the particle volume (biopolymer + water) divided by the biopolymer volume. It is assumed that the biopolymer particles are spherical.**

characterized by parameters such as a yield stress, breaking stress, and elastic modulus (Quemada and Berli 2002). The effectiveness of biopolymer particles at increasing the viscosity increases as they entrap more water (higher $R$) within their structure (Figure 7.16). Thus, biopolymer matrices that have open structures tend to form more viscous solutions than those with more close packed structures. The rheology of a biopolymer particle suspension is also highly dependent on colloidal interactions between particles. If particles are attracted to each other, then a suspension tends to be much more viscous or even gel-like than if they exist as individual particles (Quemada and Berli 2002).

Biopolymer particles may be designed to provide desirable rheological attributes to a product (such as thickness or creaminess) or they may be designed to have negligible impact on product texture. The main characteristics of biopolymer particles that can be designed to control their impact on food rheology are their concentration, composition, shape, and interactions.

### 7.6.3.4 Stability

The physical and chemical stability of a delivery system before and after being incorporated into a food or beverage is important for many commercial applications (McClements 2010; McClements et al. 2009). The incorporation of biopolymer particles into a food product should not decrease its shelf life. Biopolymer particles may become physically unstable within a delivery system or food matrix through a variety of physicochemical mechanisms, including gravitational separation (creaming or sedimentation), aggregation (flocculation or coalescence), volumetric changes (swelling or shrinking), and dissociation (erosion or fragmentation). To ensure good stability, it is important that the major mechanisms responsible for particle instability in a particular system are identified. This knowledge can then be used to identify the most effective method of improving product stability. The dominant instability mechanism in a particular delivery system or food product depends on biopolymer particle characteristics (such as composition, size, charge, and structure), as well as environmental conditions (such as temperature, pH, ionic strength, and ingredient interactions).

The stability of colloidal delivery systems in general was discussed in Chapter 3, and so only a brief discussion of the major mechanisms influencing the stability of biopolymer particles is given here. The creaming rate of noninteracting rigid spherical particles dispersed within an ideal liquid is given by

$$U = -\frac{2gr^2(\rho_2 - \rho_1)}{9\eta_1}. \tag{7.4}$$

Here, $U$ is the creaming velocity (positive for creaming; negative for sedimentation), $g$ is the acceleration attributed to gravity, $r$ is the particle radius, $\rho$ is the density, $\eta$ is the shear viscosity, and the subscripts 1 and 2 refer to the continuous phase and particles, respectively. More sophisticated mathematical models that take into account polydispersity, nonspherical particles, particle fluidity, particle–particle interactions, and nonideal fluids are available (McClements 2005). The overall density of a biopolymer particle depends on the concentrations and densities of the different components within it. The particles in most biopolymer-based delivery systems contain biopolymer molecules and water, while in certain applications, they may also contain oil droplets (to encapsulate lipophilic active ingredients). The overall density of the particle is then given by (Matalanis et al. 2011)

$$\rho_{Particle} = \phi_B\rho_B + \phi_O\rho_O + (1 - \phi_B - \phi_O)\rho_W. \tag{7.5}$$

Here, the subscripts B, W, and O refer to the biopolymer, water, and oil phases, respectively. To a first approximation, the densities of the different components are as follows: $\rho_B \approx 1500$ kg m$^{-3}$, $\rho_W \approx 1000$ kg m$^{-3}$, and $\rho_O \approx 900$ kg m$^{-3}$. In some applications, biopolymer particles may also contain other components (such as minerals or solids) and so the above equation must be extended.

The above equations show that the rate of gravitational separation increases as the particle size increases, the density contrast increases, and the continuous phase viscosity decreases. In the absence of oil droplets, biopolymer particles are denser than water and tend to sediment. However, as more oil droplets are incorporated into the biopolymer particles, they become less dense, until eventually the filled biopolymer particles can have a density similar to that of water, which inhibits gravitational separation. The impact of particle composition (biopolymer, lipid, and water content) on the stability of an aqueous biopolymer particle suspension is shown in Figure 7.17. The rate and direction of gravitational separation depend on the composition of the biopolymer particles. At low oil droplet concentrations and high biopolymer concentrations, the particles tend to sediment ($-U$). Conversely, at high oil droplet concentrations and low biopolymer concentrations, they tend to cream ($+U$). There are particular oil and biopolymer combinations that provide density matching between the biopolymer particles and the surrounding aqueous phase, for example, 50% lipid and 10% biopolymer (Figure 7.17). The ability of oil droplets to delay sedimentation of biopolymer particles was recently demonstrated for systems consisting of casein-rich hydrogel particles containing casein-coated fat droplets (Matalanis and McClements 2013). Density-matched particles may be useful for preventing gravitational separation during the shelf life of low-viscosity products, such as soft drinks and other beverages. Gravitational separation is less of a concern for highly viscous or gelled products (such as many yogurts, desserts, dressings, and sauces) since the continuous phase inhibits particle movement.

The tendency for particle aggregation to occur in a suspension of particles can be predicted by calculating the various attractive and repulsive colloidal interactions acting between them, such as van der Waals, steric, electrostatic, hydrophobic, and depletion interactions (McClements 2005). When the attractive interactions dominate, the biopolymer particles have a tendency to aggregate, but when the repulsive forces dominate, they tend to be stable to aggregation. The tendency for swelling, shrinking, erosion, or fragmentation to occur is highly system specific and

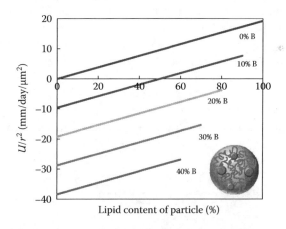

**Figure 7.17** **Theoretical prediction of the influence of biopolymer particle composition (lipid and biopolymer concentration) on the stability to gravitational separation. Assumed densities: biopolymer = 1500 kg m⁻³; water = 1000 kg m⁻³; oil = 900 kg m⁻³. B, biopolymer content of particle.**

depends on the type of bonds holding the biopolymer molecules together within the biopolymer particles, for example, covalent, electrostatic interactions, hydrogen bonds, and hydrophobic interactions. Hydrogel particles should be specifically designed to have stability characteristics that are suitable for the particular food or beverage product that they are going to be incorporated into. This requires knowledge of the specific solution conditions (e.g., pH, ionic strength, solvent type, and ingredients) and environmental conditions (e.g., heating, cooling, pressure, dehydration, or shearing) that the particles might be exposed to in the product.

### 7.6.3.5 Release Characteristics

An important attribute of many biopolymer-based delivery systems is their ability to control the release of encapsulated active ingredients, such as flavors, antimicrobials, antioxidants, vitamins, or nutraceuticals (Jones and McClements 2010; Matalanis et al. 2011). In this case, the biopolymer particles may have to be designed to release an active ingredient at a particular location outside or inside the human body (Chapter 3). The ingredient may have to be released at a controlled rate or in response to a specific environmental trigger (such as pH, ionic strength, temperature, or enzyme activity). The major mechanisms responsible for the release of encapsulated ingredients from colloidal particles in general are discussed in Chapter 10 and are only briefly summarized here in the context of biopolymer particles.

*Diffusion*: In this case, the active ingredient diffuses through the particle matrix and into the surrounding medium. The particle matrix may remain intact throughout this process or its properties may change (e.g., because of erosion, fragmentation, or dissolution). The release rate of an active ingredient from the particles depends on its relative solubility in the particle matrix and surrounding liquid and on its diffusion coefficient through the matrix. For hydrogel particles, the diffusion rate depends on the pore size of the biopolymer network compared to the dimensions of the diffusing active ingredient, as well as on any molecular interactions between the biopolymer network and the active ingredient (such as electrostatic or hydrophobic interactions). The change in the amount of an active ingredient within a biopolymer particle with time (assuming it does not interact with the biopolymer network) is given by the following expression (Crank 1975):

$$\Phi = \frac{M(t)}{M(\infty)} = 1 - \frac{6}{\pi^2} \sum_{n=1}^{\infty} \frac{1}{n^2} \exp\left(-\frac{D_{gel} n^2 \pi^2 t}{R^2}\right). \tag{7.6}$$

Here, $\Phi$ is the fraction of active ingredient that has diffused out of the hydrogel particle at time $t$, $M(t)$ and $M(\infty)$ are the concentrations of active ingredient present within the hydrogel particle at time $t$ and at equilibrium, $n$ is an integer, $R$ is the radius of the hydrogel beads, and $D_{gel}$ is the diffusion coefficient of the active ingredient through the hydrogel matrix. The following equation can be used to predict the diffusion coefficient of an active ingredient through a biopolymer network (Chan and Neufeld 2009):

$$D_{gel} = D_w \exp\left(-\pi\left(\frac{r_H - r_f}{\zeta + 2r_f}\right)\right). \tag{7.7}$$

Here, $D_w$ is the diffusion of the ingredient through pure water, $r_H$ is the hydrodynamic radius of the active ingredient, $r_f$ is the cross-sectional radius of the polymer chains in the biopolymer network, and $\zeta$ is the pore diameter of the biopolymer network. This equation indicates that diffusion of the active ingredient out of the particles will be retarded when its dimensions approach that of the pores (Li et al. 2011).

*Swelling*: An active ingredient may be released from a biopolymer particle when the particle experiences an environment that causes it to swell, for example, a change in pH, salt concentration, or temperature (Fang and Cathala 2011; Oh et al. 2008). For example, the active ingredient could be trapped within a biopolymer particle that initially has a pore size small enough to prevent it from moving out. When the biopolymer encounters an environment that causes swelling, the active ingredient diffuses out owing to the increased pore size (Equation 7.7). The active ingredient could be loaded into the biopolymer particles by initially swelling them in its presence and then changing the solution conditions to induce shrinkage.

*Dissolution*: An active ingredient may be released from a biopolymer particle owing to dissolution of the particle matrix (Siepmann and Siepmann 2013). In this case, the active ingredient is initially dissolved or dispersed within the particle matrix but is released when the particle encounters environmental conditions that weaken the physical interactions between the biopolymer molecules, causing them to dissociate from each other (e.g., pH, ionic strength, or temperature). As an example, biopolymer particles formed by electrostatic complexation of proteins and ionic polysaccharides will dissociate when the pH is altered so that there is no electrostatic attraction between them (Cooper et al. 2005; Kayitmazer et al. 2013).

*Erosion*: An active ingredient may be released from a biopolymer particle because of depolymerization of the biopolymer molecules that comprise the particle matrix (Siepmann and Siepmann 2013). The active ingredient may then be released into the surrounding medium as a result of biopolymer erosion processes that occur at the particle surface or throughout the entire particle volume. A number of physicochemical mechanisms may lead to depolymerization, with the precise mechanism depending on biopolymer type, for example, hydrolysis induced by strong acids, strong bases, specific enzymes, elevated temperatures, or mechanical treatments. Protein-based particles can be eroded by enzymatic hydrolysis within the stomach and small intestine, that is, by gastric and pancreatic proteases (Chen and Subirade 2005; Hurtado-Lopez and Murdan 2006). Starch-based particles can be eroded by enzymatic hydrolysis in the mouth and upper intestine, that is, by lingual or pancreatic amylases (Janssen et al. 2009; Rahmouni et al. 2001). Dietary fiber-based particles can be eroded by enzymes secreted by colonic bacteria in the large intestine (Kosaraju 2005; Shukla and Tiwari 2012). Pectin molecules are degraded by prolonged storage under highly acidic or basic conditions, particularly at elevated temperatures (Fraeye et al. 2007). Various types of biopolymers can also be depolymerized by high-intensity ultrasound treatment (Zhang et al. 2013). Knowledge of the breakdown mechanisms of different biopolymers is therefore important for designing biopolymer-based delivery systems with specific release characteristics.

*Fragmentation*: An active ingredient may be released from a biopolymer particle that is physically broken down into smaller parts when it experiences certain forces, for example, shearing, grinding, or mastication. The active ingredient still diffuses out of the particles, but the rate of release is faster than that for an intact particle because of the increased surface area and decreased diffusion path.

Mathematical models have been developed to describe various types of release mechanisms of active ingredients from particulate systems (Chapter 10). To select an appropriate release model, one must first know the origin of the release mechanism (e.g., diffusion, swelling, dissolution, erosion, or fragmentation). Some of the physical and chemical parameters required for these models include the initial PSD, the concentrations of the active component within the particle and surrounding medium, equilibrium partition coefficients, and the mass transport rate of the active ingredient through the biopolymer matrix (translational diffusion coefficients). While such empirical data are often available for drug delivery systems, the data for food delivery systems are only now starting to accumulate. The adoption of these models will help food scientists to rationalize the design, fabrication, and utilization of biopolymer particle systems with specific release characteristics (Lesmes and McClements 2009).

## 7.7 Potential Applications

In this section, a brief summary of some of the most important potential applications of biopolymer-based delivery systems within the food industry is given. A more general discussion of potential applications of different kinds of food-based colloidal delivery systems is given in Chapter 9.

### 7.7.1 Encapsulation and Protection

Biopolymer particles are commonly used to encapsulate and protect active ingredients, such as minerals, peptides, proteins, enzymes, carbohydrates, lipids, and probiotics (Chan et al. 2011; Chen et al. 2006; Desmarais et al. 2012; Martin and de Jong 2012; Matalanis et al. 2012; Sajeesh and Sharma 2006). Encapsulation may be carried out to improve the handling of an ingredient, to mask any undesirable flavors, or to protect it from chemical or physical degradation during storage or after ingestion (Matalanis et al. 2011). The active ingredients are typically dissolved or dispersed within one of the biopolymer phases used to prepare the delivery system. The biopolymer matrix can be specifically designed to retain the ingredient throughout storage and to inhibit any undesirable chemical reactions. For example, the porosity of the biopolymer network could be controlled to retard molecular diffusion, or substances could be incorporated into the biopolymer network to retard chemical degradation (such as antioxidants).

## 7.7.2 Controlled Release

It is possible to design biopolymer particles that release encapsulated active ingredients at a controlled rate or in response to some environmental trigger, such as pH, ionic strength, temperature, or enzyme activity (Baracat et al. 2012; Goldberg et al. 2007; Lima et al. 2012; Oh et al. 2008; Pridgen et al. 2007). Thus, the active ingredient can be delivered at the required site of action, such as the nose or mouth for a flavor molecule; the stomach, small intestine, or colon for a bioactive ingredient; or bacterial surfaces for an antimicrobial agent. Controlled release may be achieved using different kinds of proteins or polysaccharides to fabricate the particles, since different biopolymers have different susceptibilities to chemical or enzymatic degradation in the human GIT. Many starches are degraded in the mouth because of the action of amylases, many proteins are degraded in the stomach and small intestine because of the action of proteases, and many dietary fibers are only degraded once they reach the large intestine because of the action of enzymes secreted by colonic bacteria. The ability to control the release of an encapsulated ingredient may also be engineered into a particle by changing its dimensions, internal structure, or permeability. The rate of ingredient release from colloidal particles typically decreases as the particle size increases. The pore size of the biopolymer network within the particle matrix can also be controlled to modulate the release rate, or it can be made to change in response to specific environmental triggers, such as pH, ionic strength, or temperature. Typically, the release rate increases as the pore size increases relative to the size of the encapsulated active ingredients. For nonpolar active ingredients, it is possible to modulate the release rate by including oil droplets within the biopolymer particles since they alter the partitioning of the ingredients within the system.

## 7.7.3 Lightening Agents

Biopolymer particles are typically fabricated from proteins or polysaccharides that have higher refractive indices ($n \approx 1.5$) than water ($n \approx 1.33$) and are therefore capable of scattering light. In addition, the particle sizes of many biopolymer particles are on the same order as the wavelength of light ($d \approx \lambda$), which means that they scatter light strongly and make systems containing them appear turbid or opaque. Biopolymer particles may therefore be a useful means of increasing the opacity or lightness of commercial products, such as foods, beverages, sunscreens, or cosmetics. Indeed, studies have shown that biopolymer nanoparticles and microparticles can be used to modify the turbidity of model food products (Bengoechea et al. 2011; Chung et al. 2013; Jones and McClements 2011; Philippe et al. 2003). The light scattering theory predicts that the lightness of a material increases as the particle concentration and refractive index contrast increase and that it has an optimum value at a particular particle size (McClements 2002). Thus, biopolymer particles that have high refractive indices and intermediate diameters (200–2000 nm) can be used as effective lightening agents. The refractive index contrast increases as the biopolymer concentration within the particles increases or the water content decreases. On the other hand, the stability of biopolymer particles to gravitational separation may decrease as their biopolymer concentration increases. Consequently, it may be necessary to find a compromise between light scattering efficiency and particle stability within a product.

### 7.7.4 Texture Modification

Biopolymer particles increase the overall viscosity of solutions they are incorporated into by an amount that depends on their effective volume fraction (Figure 7.16). The increase in viscosity is relatively low at low particle concentrations but increases steeply above a certain particle concentration where they become closely packed. At sufficiently high particle concentrations, the overall system has solid-like characteristics. Biopolymer particles may therefore be incorporated into products to modify their rheological characteristics. The effective concentration of a biopolymer particle depends on how much water it traps within its structure and may be much higher than the concentration of the biopolymer chains themselves. Thus, biopolymer particles that trap high amounts of water tend to be the most effective texture modifiers. Numerous studies have shown that biopolymer nanoparticles or microparticles can be used to increase the viscosity or gel aqueous solutions. One of the most common examples of biopolymer particles that are used for this purpose are starch granules, which swell upon heating in the presence of water, resulting in a highly viscous or gel-like solution depending on starch concentration (Chung et al. 2013). However, other types of biopolymer particles can also be used for this purpose, such as whey, egg, or soy protein particles (Cheftel and Dumay 1993; Jong 2013) or protein/polysaccharide complexes (Schmitt et al. 1998; Schmitt and Turgeon 2011).

### 7.7.5 Fat Replacement

Biopolymer particles can be designed to have dimensions and surface characteristics similar to those of the lipid droplets found in many emulsion-based food products, such as milk, cream, beverages, sauces, dressings, and soups (Ma and Boye 2013; Matalanis et al. 2011; Shewan and Stokes 2013). Consequently, they may be used to mimic some of the desirable characteristics of lipid droplets, such as appearance, texture, or mouthfeel. Biopolymer particles could be used to replace some or all of the lipid droplets normally present in a food, beverage, or cosmetic product. The overconsumption of fat is a major problem in many developed countries, and so the availability of biopolymer particles that can act as fat mimetics may prove to be a valuable means of overcoming this problem. Ingredients based on protein particles are already commercially available for use as fat replacers in the food industry (Cheftel and Dumay 1993). These particles are used to replace some of the rheological, optical, and mouthfeel properties of food products that are normally lost when fat droplets are removed.

## 7.8 Summary

Biopolymer-based delivery systems can be used to encapsulate, protect, and release a wide range of different active ingredients. The physicochemical properties of bioactive particles (such as their size, shape, structure, charge, and permeability) can be controlled to give delivery systems with different physicochemical and functional properties (such as encapsulation, protection, and release). Biopolymer particles can be fabricated from a variety of different proteins and polysaccharides, using a wide range of different processing methods, giving broad scope for particle design for different functional attributes. Nevertheless, our understanding of creating biopolymer particles with specific functional attributes is still rather limited, and a better understanding of structure–function relationships is required to design functional biopolymer particles more systematically. Identification of the most appropriate ingredients and conditions required to create these particles requires knowledge of the molecular and functional characteristics of the biopolymers used, as well as of the physicochemical mechanisms underlying particle formation. From a practical point of view, the application of biopolymer-based delivery systems on an industrial scale is limited to those systems that can be manufactured economically and that are robust enough for commercial applications.

## References

Abd Elgadir, M., M. E. S. Mirghani and A. Adam (2013). "Fish gelatin and its applications in selected pharmaceutical aspects as alternative source to pork gelatin." *Journal of Food Agriculture and Environment* **11**(1): 73–79.

Aguilera, J. M. (2000). "Microstructure and food product engineering." *Food Technology* **54**(11): 56–60.

Akhtar, M. and E. Dickinson (2003). "Emulsifying properties of whey protein-dextran conjugates at low pH and different salt concentrations." *Colloids and Surfaces B: Biointerfaces* **31**(1–4): 125–132.

Akkermans, C., A. J. Van der Goot, P. Venema, H. Gruppen, J. M. Verijken, E. Van der Linden and R. M. Boom (2007). "Micrometer-sized fibrillar protein aggregates from soy glycinin and soy protein isolate." *Journal of Agricultural and Food Chemistry* **55**: 9877–9882.

Ako, K., T. Nicolai and D. Durand (2010). "Salt-induced gelation of globular protein aggregates: Structure and kinetics." *Biomacromolecules* **11**(4): 864–871.

Anton, M. (2013). "Egg yolk: Structures, functionalities and processes." *Journal of the Science of Food and Agriculture* **93**(12): 2871–2880.

Anton, M., F. Nau and V. Lechevalier (2009). Egg proteins. In *Handbook of Hydrocolloids*, 2nd Edition, G. O. Phillips and P. A. Williams (eds.). Cambridge, UK, Woodhead Publishing: 359–382.

Arpicco, S., F. Dosio, B. Stella and L. Cattel (2011). "Anticancer prodrugs: An overview of major strategies and recent developments." *Current Topics in Medicinal Chemistry* **11**(18): 2346–2381.

Arroyo-Maya, I. J., J. O. Rodiles-Lopez, M. Cornejo-Mazon, G. F. Gutierrez-Lopez, A. Hernandez-Arana, C. Toledo-Nunez, G. V. Barbosa-Canovas, J. O. Flores-Flores and H. Hernandez-Sanchez (2012). "Effect of different treatments on the ability of alpha-lactalbumin to form nanoparticles." *Journal of Dairy Science* **95**(11): 6204–6214.

Autio, K. (2006). Functional aspects of cereal cell-wall polysaccharides. In *Carbohydrates in Food*, A.-C. Eliasson (ed.). Boca Raton, FL, Taylor & Francis: 167–207.

Baracat, M. M., A. M. Nakagawa, R. Casagrande, S. R. Georgetti, W. A. Verri and O. de Freitas (2012). "Preparation and characterization of microcapsules based on biodegradable polymers: Pectin/Casein complex for controlled drug release systems." *AAPS Pharmscitech* **13**(2): 364–372.

Barbana, C., L. Sanchez and M. D. Perez (2011). "Bioactivity of alpha-lactalbumin related to its interaction with fatty acids: A review." *Critical Reviews in Food Science and Nutrition* **51**(8): 783–794.

Belitz, H. D., W. Grosch and P. Schieberle (2009). *Food Chemistry*. Berlin, Germany, Springer.

BeMiller, J. N. and K. C. Huber (2007). Carbohydrates. In *Food Chemistry*, S. Damodaran, K. L. Parkin and O. R. Fennema (eds.). Boca Raton, FL, CRC Press: 83–154.

Bengoechea, C., O. G. Jones, A. Guerrero and D. J. McClements (2011). "Formation and characterization of lactoferrin/pectin electrostatic complexes: Impact of composition, pH and thermal treatment." *Food Hydrocolloids* **25**(5): 1227–1232.

Bengoechea, C., I. Peinado and D. J. McClements (2011). "Formation of protein nanoparticles by controlled heat treatment of lactoferrin: Factors affecting particle characteristics." *Food Hydrocolloids* **25**(5): 1354–1360.

Benzaria, A., M. Maresca, N. Taieb and E. Dumay (2013). "Interaction of curcumin with phosphocasein micelles processed or not by dynamic high-pressure." *Food Chemistry* **138**(4): 2327–2337.

Brock, J. H. (2012). "Lactoferrin-50 years on." *Biochemistry and Cell Biology-Biochimie Et Biologie Cellulaire* **90**(3): 245–251.

Bryant, C. M. and D. J. McClements (1998). "Molecular basis of protein functionality with special consideration of cold-set gels derived from heat-denatured whey." *Trends in Food Science and Technology* **9**(4): 143–151.

Burey, P., B. R. Bhandari, T. Howes and M. J. Gidley (2008). "Hydrocolloid gel particles: Formation, characterization, and application." *Critical Reviews in Food Science and Nutrition* **48**(5): 361–377.

Campo, V. L., D. F. Kawano, D. B. da Silva, Jr. and I. Carvalho (2009). "Carrageenans: Biological properties, chemical modifications and structural analysis—A review." *Carbohydrate Polymers* **77**(2): 167–180.

Carpenter, C. E. and A. W. Mahoney (1992). "Contributions of heme and nonheme iron to human-nutrition." *Critical Reviews in Food Science and Nutrition* **31**(4): 333–367.

Chan, A. W. and R. J. Neufeld (2009). "Modeling the controllable pH-responsive swelling and pore size of networked alginate based biomaterials." *Biomaterials* **30**(30): 6119–6129.

Chan, E. S., S. L. Wong, P. P. Lee, J. S. Lee, T. B. Ti, Z. B. Zhang, D. Poncelet, P. Ravindra, S. H. Phan and Z. H. Yim (2011). "Effects of starch filler on the physical properties of lyophilized calcium-alginate beads and the viability of encapsulated cells." *Carbohydrate Polymers* **83**(1): 225–232.

Cheftel, J. C. and E. Dumay (1993). "Microcoagulation of proteins for development of creaminess." *Food Reviews International* **9**(4): 473–502.

Chen, B. C., H. J. Li, Y. P. Ding and H. Y. Suo (2012). "Formation and microstructural characterization of whey protein isolate/beet pectin coacervations by laccase catalyzed cross-linking." *Lwt-Food Science and Technology* **47**(1): 31–38.

Chen, B. C., D. J. McClements, D. A. Gray and E. A. Decker (2010). "Stabilization of soybean oil bodies by enzyme (laccase) cross-linking of adsorbed beet pectin coatings." *Journal of Agricultural and Food Chemistry* **58**(16): 9259–9265.

Chen, L. Y., G. E. Remondetto and M. Subirade (2006). "Food protein-based materials as nutraceutical delivery systems." *Trends in Food Science and Technology* **17**(5): 272–283.

Chen, L. Y. and M. Subirade (2005). "Chitosan/beta-lactoglobulin core-shell nanoparticles as nutraceutical carriers." *Biomaterials* **26**(30): 6041–6053.

Chevalier-Lucia, D., C. Blayo, A. Gracia-Julia, L. Picart-Palmade and E. Dumay (2011). "Processing of phosphocasein dispersions by dynamic high pressure: Effects on the dispersion physico-chemical characteristics and the binding of alpha-tocopherol acetate to casein micelles." *Innovative Food Science and Emerging Technologies* **12**(4): 416–425.

Chung, C., B. Degner and D. J. McClements (2013). "Physicochemical characteristics of mixed colloidal dispersions: Models for foods containing fat and starch." *Food Hydrocolloids* **30**(1): 281–291.

Cizova, A., A. Koschella, T. Heinze, A. Ebringerova and I. Srokova (2007). "Octenylsuccinate derivatives of carboxymethyl starch—Synthesis and properties." *Starch-Starke* **59**(10): 482–492.

Coffey, D. G., D. A. Bell and A. Henderson (2006). Cellulose and cellulose derivatives. In *Food Polysaccharides and Their Applications*, A. M. Stephen, G. O. Phillips and P. A. Williams (eds.). Boca Raton, FL, CRC Press: 147–180.

Conde-Petit, B., F. Escher and J. Nuessli (2006). "Structural features of starch-flavor complexation in food model systems." *Trends in Food Science and Technology* **17**(5): 227–235.

Cooper, C., M. Corredig and M. Alexander (2010). "Investigation of the colloidal interactions at play in combined acidification and rennet of different heat-treated milks." *Journal of Agricultural and Food Chemistry* **58**(8): 4915–4922.

Cooper, C. L., P. L. Dubin, A. B. Kayitmazer and S. Turksen (2005). "Polyelectrolyte-protein complexes." *Current Opinion in Colloid and Interface Science* **10**(1–2): 52–78.

Crank, J. (1975). *The Mathematics of Diffusion.* Oxford, UK, Oxford Science Publications.

Cui, S. W. (2005). *Food Carbohydrates: Chemistry, Physical Properties and Applications.* Boca Raton, FL, Taylor and Francis.

Damodaran, S. (1996). Amino acids, peptides, and proteins. In *Food Chemistry*, O. R. Fennema (ed.). New York, Marcel Dekker: 321–429.

Damodaran, S. (2007). Amino acids, peptides, and proteins. In *Food Chemistry*, S. Damodaran, K. L. Parkin and O. R. Fennema (eds.). New York, Marcel Dekker: 217–329.

Day, L. (2013). "Proteins from land plants—Potential resources for human nutrition and food security." *Trends in Food Science and Technology* **32**(1): 25–42.

De Jong, G. A. H. and S. J. Koppelman (2002). "Transglutaminase catalyzed reactions: Impact on food applications." *Journal of Food Science* **67**(8): 2798–2806.

Desmarais, S. M., H. P. Haagsman and A. E. Barron (2012). "Microfabricated devices for biomolecule encapsulation." *Electrophoresis* **33**(17): 2639–2649.

Dickinson, E. (1998). "Stability and rheological implications of electrostatic milk protein-polysaccharide interactions." *Trends in Food Science and Technology* **9**: 347–354.

Dickinson, E. (2003). "Hydrocolloids at interfaces and the influence on the properties of dispersed systems." *Food Hydrocolloids* **17**(1): 25–39.

Dickinson, E. (2010). "Flocculation of protein-stabilized oil-in-water emulsions." *Colloids and Surfaces B-Biointerfaces* **81**(1): 130–140.

Djagny, K. B., Z. Wang and S. Y. Xu (2001). "Gelatin: A valuable protein for food and pharmaceutical industries: Review." *Critical Reviews in Food Science and Nutrition* **41**(6): 481–492.

Dong, S. Y., A. Panya, M. Y. Zeng, B. C. Chen, D. J. McClements and E. A. Decker (2012). "Characteristics and antioxidant activity of hydrolyzed beta-lactoglobulin-glucose Maillard reaction products." *Food Research International* **46**(1): 55–61.

Dong, S. Y., B. B. Wei, B. C. Chen, D. J. McClements and E. A. Decker (2011). "Chemical and antioxidant properties of casein peptide and its glucose Maillard reaction products in fish oil-in-water emulsions." *Journal of Agricultural and Food Chemistry* **59**(24): 13311–13317.

Duclairoir, C., J. M. Irache, E. Nakache, A. M. Orecchioni, C. Chabenat and Y. Popineau (1999). "Gliadin nanoparticles: Formation, all-trans-retinoic acid entrapment and release, size optimization." *Polymer International* **48**(4): 327–333.

Duclairoir, C., A. M. Orecchioni, P. Depraetere and E. Nakache (2002). "Alpha-tocopherol encapsulation and in vitro release from wheat gliadin nanoparticles." *Journal of Microencapsulation* **19**(1): 53–60.

Egan, T., J. C. Jacquier, Y. Rosenberg and M. Rosenberg (2013). "Cold-set whey protein microgels for the stable immobilization of lipids." *Food Hydrocolloids* **31**(2): 317–324.

Einhorn-Stoll, U. and H. Kunzek (2009). "Thermoanalytical characterisation of processing-dependent structural changes and state transitions of citrus pectin." *Food Hydrocolloids* **23**(1): 40–52.

Eliasson, A.-C. (ed.) (2006). *Carbohydrates in Food.* Food Science and Technology. Boca Raton, FL, Taylor & Francis.

Erni, P., E. J. Windhab, R. Gunde, M. Graber, B. Pfister, A. Parker and P. Fischer (2007). "Interfacial rheology of surface-active biopolymers: Acacia senegal gum versus hydrophobically modifed starch." *Biomacromolecules* **8**(11): 3458–3466.

Evans, M., I. Ratcliffe and P. A. Williams (2013). "Emulsion stabilisation using polysaccharide-protein complexes." *Current Opinion in Colloid and Interface Science* **18**(4): 272–282.

Fang, A. P. and B. Cathala (2011). "Smart swelling biopolymer microparticles by a micro-fluidic approach. Synthesis, in situ encapsulation and controlled release." *Colloids and Surfaces B-Biointerfaces* **82**(1): 81–86.

Fang, Z. and B. Bhandari (2012). Spray drying, freeze drying, and related processes for food ingredient and nutraceutical encapsulation. In *Encapsulation Technologies and Delivery Systems for Food Ingredients and Nutraceuticals*, N. Garti and D. J. McClements (eds.). Oxford, UK, Woodhead Publishing: 73–108.

Fang, Z. X. and B. Bhandari (2010). "Encapsulation of polyphenols—A review." *Trends in Food Science and Technology* **21**(10): 510–523.

Fox, P. F. (2003). Milk proteins: General and historical aspects. In *Advanced Dairy Chemistry*, P. F. Fox and P. L. H. McSweeney (eds.). New York, Kluwer Academic/Plenum. **1**: Proteins: 1–48.

Fraeye, I., A. De Roeck, T. Duvetter, I. Verlent, M. Hendrickx and A. Van Loey (2007). "Influence of pectin properties and processing conditions on thermal pectin degradation." *Food Chemistry* **105**(2): 555–563.

Gerrard, J. A. (2002). "Protein-protein crosslinking in food: Methods, consequences, applications." *Trends in Food Science and Technology* **13**(12): 391–399.

Gidley, M. J. and J. S. G. Reid (2006). Galactomannans and other cell wall storage polysaccharides in seeds. In *Food Polysaccharides and Their Applications*, A. M. Stephen, G. O. Phillips and P. A. Williams (eds.). Boca Raton, FL, CRC Press: 181–216.

Goldberg, M., R. Langer and X. Q. Jia (2007). "Nanostructured materials for applications in drug delivery and tissue engineering." *Journal of Biomaterials Science-Polymer Edition* **18**(3): 241–268.

Gomez-Guillen, M. C., B. Gimenez, M. E. Lopez-Caballero and M. P. Montero (2011). "Functional and bioactive properties of collagen and gelatin from alternative sources: A review." *Food Hydrocolloids* **25**(8): 1813–1827.

Graveland-Bikker, J. F., R. I. Koning, H. K. Koerten, R. B. J. Geels, R. M. A. Heeren and C. G. de Kruif (2009). "Structural characterization of alpha-lactalbumin nanotubes." *Soft Matter* **5**(10): 2020–2026.

Gulseren, I., Y. Fang and M. Corredig (2012). "Complexation of high methoxyl pectin with ethanol desolvated whey protein nanoparticles: Physico-chemical properties and encapsulation behaviour." *Food and Function* **3**(8): 859–866.

Guzey, D. and D. J. McClements (2006). "Formation, stability and properties of multilayer emulsions for application in the food industry." *Advances in Colloid and Interface Science* **128**: 227–248.

Haham, M., S. Ish-Shalom, M. Nodelman, I. Duek, E. Segal, M. Kustanovich and Y. D. Livney (2012). "Stability and bioavailability of vitamin D nanoencapsulated in casein micelles." *Food and Function* **3**(7): 737–744.

Hebrard, G., V. Hoffart, J. M. Cardot, M. Subirade and E. Beyssac (2013). "Development and characterization of coated-microparticles based on whey protein/alginate using the Encapsulator device." *Drug Development and Industrial Pharmacy* **39**(1): 128–137.

Heidebach, T., P. Först and U. Kulozik (2009). Microencapsulation of probiotic cells by means of rennet-gelation of milk proteins. *Food Hydrocolloids*, **23**(7), 1670–1677.

Helgeson, M. E., S. C. Chapin and P. S. Doyle (2011). "Hydrogel microparticles from lithographic processes: Novel materials for fundamental and applied colloid science." *Current Opinion in Colloid and Interface Science* **16**(2): 106–117.

Hoffman, M. A. M. and P. J. J. M. van Mill (1997). "Heat-induced aggregtion of beta-lactoglobulin: Role of the free thiol group and disulfide bonds." *Journal of Agricultural and Food Chemistry* **45**: 2942–2948.

Hoffmann, M. A. M. and P. J. J. M. van Mill (1999). "Heat induced aggregation of β-lactoglobulin as a function of pH." *Journal of Agricultural and Food Chemistry* **47**(5): 1898–1905.

Horn, D. and J. Rieger (2001). "Organic nanoparticles in the aqueous phase—Theory, experiment, and use." *Angewandte Chemie-International Edition* **40**(23): 4331–4361.

Huppertz, T. and C. G. de Kruif (2008). "Structure and stability of nanogel particles prepared by internal cross-linking of casein micelles." *International Dairy Journal* **18**(5): 556–565.

Hurtado-Lopez, P. and S. Murdan (2006). "Zein microspheres as drug/antigen carriers: A study of their degradation and erosion, in the presence and absence of enzymes." *Journal of Microencapsulation* **23**(3): 303–314.

Imeson, A. (2010). *Food Stabilizers, Thickeners and Gelling Agents*. Chichester, UK, John Wiley & Sons.

Israelachvili, J. (2011). *Intermolecular and Surface Forces*, 3rd Edition. London, Academic Press.

Jafari, S. M., E. Assadpoor, Y. H. He and B. Bhandari (2008). "Encapsulation efficiency of food flavours and oils during spray drying." *Drying Technology* **26**(7): 816–835.

Janssen, A. M., A. M. van de Pijpekamp and D. Labiausse (2009). "Differential saliva-induced breakdown of starch filled protein gels in relation to sensory perception." *Food Hydrocolloids* **23**(3): 795–805.

Jones, O. G. and D. J. McClements (2010). "Functional biopolymer particles: Design, fabrication, and applications." *Comprehensive Reviews in Food Science and Food Safety* **9**(4): 374–397.

Jones, O. G. and D. J. McClements (2011). "Recent progress in biopolymer nanoparticle and microparticle formation by heat-treating electrostatic protein–polysaccharide complexes." *Advances in Colloid and Interface Science* **167**(1–2): 49–62.

Jong, L. (2013). "Characterization of soy protein nanoparticles prepared by high shear microfluidization." *Journal of Dispersion Science and Technology* **34**(4): 469–475.

Joye, I. J. and D. J. McClements (2013). "Production of nanoparticles by anti-solvent precipitation for use in food systems." *Trends in Food Science and Technology*. In press.

Jung, J.-M., G. Savin, M. Pouzot, C. Schmitt and R. Mezzenga (2008). "Structure of heat-induced b-lactoglobulin aggregates and their complexes with sodium-dodecyl sulfate." *Biomacromolecules* **9**: 2477–2486.

Jung, S. and H. Yi (2012). "Fabrication of chitosan-poly(ethylene glycol) hybrid hydrogel microparticles via replica molding and its application toward facile conjugation of biomolecules." *Langmuir* **28**(49): 17061–17070.

Kayitmazer, A. B., D. Seeman, B. B. Minsky, P. L. Dubin and Y. S. Xu (2013). "Protein-polyelectrolyte interactions." *Soft Matter* **9**(9): 2553–2583.

Kelley, D. and D. J. McClements (2003). "Interactions of bovine serum albumin with ionic surfactants in aqueous solutions." *Food Hydrocolloids* **17**(1): 73–85.

Keppeler, S., A. Ellis and J. C. Jacquier (2009). "Cross-linked carrageenan beads for controlled release delivery systems." *Carbohydrate Polymers* **78**(4): 973–977.

Kizilay, E., A. B. Kayitmazer and P. L. Dubin (2011). "Complexation and coacervation of polyelectrolytes with oppositely charged colloids." *Advances in Colloid and Interface Science* **167**(1–2): 24–37.

Kosaraju, S. L. (2005). "Colon targeted delivery systems: Review of polysaccharides for encapsulation and delivery." *Critical Reviews in Food Science and Nutrition* **45**(4): 251–258.

Kurita, K. (2006). "Chitin and chitosan: Functional biopolymers from marine crustaceans." *Marine Biotechnology* **8**(3): 203–226.

LaClair, C. E. and M. R. Etzel (2010). Ingredients and pH are key to clear beverages that contain whey protein. Journal of Food Science, **75**(1), C21–C27.

Lauridsen, C., M. S. Hedemann and S. K. Jensen (2001). "Hydrolysis of tocopheryl and retinyl esters by porcine carboxyl ester hydrolase is affected by their carboxylate moiety and bile acids." *The Journal of Nutritional Biochemistry* **12**(4): 219–224.

Lesmes, U. and D. J. McClements (2009). "Structure–function relationships to guide rational design and fabrication of particulate food delivery systems." *Trends in Food Science and Technology* **20**(10): 448–457.

Li, B. G. and L. M. Zhang (2010). "Inclusion complexation of amylose." *Progress in Chemistry* **22**(6): 1161–1168.

Li, J. L., Y. G. Zu, X. H. Zhao, D. M. Zhao, X. Q. Chen, Z. G. An and K. L. Wang (2012). Preparation, characterization, and in vitro evaluation of resveratrol-loaded bovine serum album nanoparticles. In *Smart Materials and Nanotechnology in Engineering*, J. L. Zhong (ed.). Stafa-Zurich, Trans Tech Publications Ltd. **345**: 349–354.

Li, X. Y., X. G. Chen, D. S. Cha, H. J. Park and C. S. Liu (2009). "Microencapsulation of a probiotic bacteria with alginate-gelatin and its properties." *Journal of Microencapsulation* **26**(4): 315–324.

Li, Y., M. Hu, Y. M. Du, H. Xiao and D. J. McClements (2011). "Control of lipase digestibility of emulsified lipids by encapsulation within calcium alginate beads." *Food Hydrocolloids* **25**(1): 122–130.

Liang, L. and M. Subirade (2012). "Study of the acid and thermal stability of beta-lactoglobulin-ligand complexes using fluorescence quenching." *Food Chemistry* **132**(4): 2023–2029.

Liang, L., V. Tremblay-Hebert and M. Subirade (2011). "Characterisation of the beta-lactoglobulin/alpha-tocopherol complex and its impact on alpha-tocopherol stability." *Food Chemistry* **126**(3): 821–826.

Lima, A. C., P. Sher and J. F. Mano (2012). "Production methodologies of polymeric and hydrogel particles for drug delivery applications." *Expert Opinion on Drug Delivery* **9**(2): 231–248.

Livney, Y. D. (2008). Complexes and conjugates of biopolymers for delivery of bioactive ingredients via food. In *Delivery and Controlled Release of Bioactives in Foods and Nutraceuticals*, N. Garti (ed.). Boca Raton, FL, CRC Press: 234–250.

Livney, Y. D. (2010). "Milk proteins as vehicles for bioactives." *Current Opinion in Colloid and Interface Science* **15**(1–2): 73–83.

Ma, Z. and J. I. Boye (2013). "Advances in the design and production of reduced-fat and reduced-cholesterol salad dressing and mayonnaise: A review." *Food and Bioprocess Technology* **6**(3): 648–670.

Maculotti, K., E. M. Tira, M. Sonaggere, P. Perugini, B. Conti, T. Modena and F. Pavanetto (2009). "In vitro evaluation of chondroitin sulphate-chitosan microspheres as carrier for the delivery of proteins." *Journal of Microencapsulation* **26**(6): 535–543.

Madene, A., M. Jacquot, J. Scher and S. Desobry (2006). "Flavour encapsulation and controlled release—A review." *International Journal of Food Science and Technology* **41**(1): 1–21.

Majhi, P. R., R. R. Ganta, R. P. Vanam, E. Seyrek, K. Giger and P. L. Dubin (2006). "Electrostatically driven protein aggregation: beta-lactoglobulin at low ionic strength." *Langmuir* **22**(22): 9150–9159.

Malone, M. E. and I. A. M. Appelqvist (2003). "Gelled emulsion particles for the controlled release of lipophilic volatiles during eating." *Journal of Controlled Release* **90**(2): 227–241.

Marenduzzo, D., C. Micheletti and E. Orlandini (2010). "Biopolymer organization upon confinement." *Journal of Physics-Condensed Matter* **22**(28): 1–16.

Martin, A. H. and G. A. H. de Jong (2012). "Impact of protein pre-treatment conditions on the iron encapsulation efficiency of whey protein cold-set gel particles." *European Food Research and Technology* **234**(6): 995–1003.

Matalanis, A., E. A. Decker and D. J. McClements (2012). "Inhibition of lipid oxidation by encapsulation of emulsion droplets within hydrogel microspheres." *Food Chemistry* **132**(2): 766–772.

Matalanis, A., O. G. Jones and D. J. McClements (2011). "Structured biopolymer-based delivery systems for encapsulation, protection, and release of lipophilic compounds." *Food Hydrocolloids* **25**(8): 1865–1880.

Matalanis, A. and D. J. McClements (2012a). "Factors influencing the formation and stability of filled hydrogel particles fabricated by protein/polysaccharide phase separation and enzymatic cross-linking." *Food Biophysics* **7**(1): 72–83.

Matalanis, A. and D. J. McClements (2012b). "Impact of encapsulation within hydrogel microspheres on lipid digestion: An in vitro study." *Food Biophysics* **7**(2): 145–154.

Matalanis, A. and D. J. McClements (2013). "Hydrogel microspheres for encapsulation of lipophilic components: Optimization of fabrication and performance." *Food Hydrocolloids* **31**(1): 15–25.

McClements, D. J. (2002a). "Colloidal basis of emulsion color." *Current Opinion in Colloid and Interface Science* **7**(5–6): 451–455.

McClements, D. J. (2002b). "Theoretical prediction of emulsion color." *Advances in Colloid and Interface Science* **97**(1–3): 63–89.

McClements, D. J. (2004). "Protein-stabilized emulsions." *Current Opinion in Colloid and Interface Science* **9**(5): 305–313.

McClements, D. J. (2005). *Food Emulsions: Principles, Practice, and Techniques.* Boca Raton, FL, CRC Press.

McClements, D. J. (2010). "Emulsion design to improve the delivery of functional lipophilic components." *Annual Review of Food Science and Technology* **1**: 241–269.

McClements, D. J., E. A. Decker, Y. Park and J. Weiss (2009). "Structural design principles for delivery of bioactive components in nutraceuticals and functional foods." *Critical Reviews in Food Science and Nutrition* **49**(6): 577–606.

McGuigan, A. P., D. A. Bruzewicz, A. Glavan, M. Butte and G. M. Whitesides (2008). "Cell encapsulation in sub-mm sized gel modules using replica molding." *Plos One* **3**(5): 1–11.

Mezzenga, R. and P. Fischer (2013). "The self-assembly, aggregation and phase transitions of food protein systems in one, two and three dimensions." *Reports on Progress in Physics* **76**(4): 1–43.

Mine, Y. (2002). "Recent advances in egg protein functionality in the food system." *Worlds Poultry Science Journal* **58**(1): 31–39.

Minton, A. P. (1997). "Influence of excluded volume upon macromolecular structure and associations in 'crowded' media." *Current Opinion in Biotechnology* **8**(1): 65–69.

Minussi, R. C., G. M. Pastore and N. Duran (2002). "Potential applications of laccase in the food industry." *Trends in Food Science and Technology* **13**(6–7): 205–216.

Monahan, F. J., J. B. German and J. E. Kinsella (1995). "Effect of pH and temperature on protein unfolding and thiol-disulfide interchange reactions during heat-induced gelation of whey proteins." *Journal of Agricultural and Food Chemistry* **43**: 46–52.

Moure, A., J. Sineiro, H. Dominguez and J. C. Parajo (2006). "Functionality of oilseed protein products: A review." *Food Research International* **39**(9): 945–963.

Muralikrishna, G. and M. V. S. S. T. S. Rao (2007). "Cereal non-cellulosic polysaccharides: Structure and function relationships: An overview." *Critical Reviews in Food Science and Nutrition* **47**(6): 599–610.

Necas, J. and L. Bartosikova (2013). "Carrageenan: A review." *Veterinarni Medicina* **58**(4): 187–205.

Needleman, I. G., F. C. Smales and G. P. Martin (1997). "An investigation of bioadhesion for periodontal and oral mucosal drug delivery." *Journal of Clinical Periodontology* **24**(6): 394–400.

Nicolai, T., M. Britten and C. Schmitt (2011). "Beta-Lactoglobulin and WPI aggregates: Formation, structure and applications." *Food Hydrocolloids* **25**(8): 1945–1962.

Nimz, O., K. Gessler, I. Uson, G. M. Sheldrick and W. Saenger (2004). "Inclusion complexes of V-amylose with undecanoic acid and dodecanol at atomic resolution: X-ray structures with cycloamylose containing 26 D-glucoses (cyclohexalcosaose) as host." *Carbohydrate Research* **339**(8): 1427–1437.

Norton, I. T. and W. J. Frith (2001). "Microstructure design in mixed biopolymer composites." *Food Hydrocolloids* **15**(4–6): 543–553.

Oh, J. K., R. Drumright, D. J. Siegwart and K. Matyjaszewski (2008). "The development of microgels/nanogels for drug delivery applications." *Progress in Polymer Science* **33**(4): 448–477.

Oliver, C. M., L. D. Melton and R. A. Stanley (2006). "Creating proteins with novel functionality via the Maillard reaction: A review." *Critical Reviews in Food Science and Nutrition* **46**: 337–350.

Oxley, J. D. (2012a). Coextrusion for food ingredients and nutraceutical encapsulation: Principles and technologies. In *Encapsulation Technologies and Delivery Systems for Food Ingredients and Nutraceuticals*, N. Garti and D. J. McClements (eds.). Oxford, UK, Woodhead Publishing: 131–149.

Oxley, J. D. (2012b). Spray cooling and spray chilling for food ingredient and nutraceutical encapsulation. In *Encapsulation Technologies and Delivery Systems for Food Ingredients and Nutraceuticals*, N. Garti and D. J. McClements (eds.). Oxford, UK, Woodhead Publishing: 110–130.

Patel, A., Y. C. Hu, J. K. Tiwari and K. P. Velikov (2010). "Synthesis and characterisation of zein-curcumin colloidal particles." *Soft Matter* **6**(24): 6192–6199.

Patel, A. R., P. C. M. Heussen, J. Hazekamp, E. Drost and K. P. Velikov (2012). "Quercetin loaded biopolymeric colloidal particles prepared by simultaneous precipitation of quercetin with hydrophobic protein in aqueous medium." *Food Chemistry* **133**(2): 423–429.

Perez, S. and E. Bertoft (2010). "The molecular structures of starch components and their contribution to the architecture of starch granules: A comprehensive review." *Starch-Starke* **62**(8): 389–420.

Philippe, M., F. Gaucheron, Y. Le Graet, F. Michel and A. Garem (2003). "Physicochemical characterization of calcium-supplemented skim milk." *Lait* **83**(1): 45–59.

Piculell, L. (1995). Gelling carrageenans. In *Food Polysaccharides and Their Applications*, A. M. Stephen (ed.). New York, Marcel Dekker: 205–244.

Pink, D. A., C. B. Hanna, B. E. Quinn, V. Levadny, G. L. Ryan, L. Filion and A. T. Paulson (2006). "Modelling electrostatic interactions in complex soft systems." *Food Research International* **39**(10): 1031–1045.

Pridgen, E. M., R. Langer and O. C. Farokhzad (2007). "Biodegradable, polymeric nanoparticle delivery systems for cancer therapy." *Nanomedicine* **2**(5): 669–680.

Quemada, D. and C. Berli (2002). "Energy of interaction in colloids and its implications in rheological modeling." *Advances in Colloid and Interface Science* **98**(1): 51–85.

Rahmouni, M., F. Chouinard, F. Nekka, V. Lenaerts and J. C. Leroux (2001). "Enzymatic degradation of cross-linked high amylose starch tablets and its effect on in vitro release of sodium diclofenac." *European Journal of Pharmaceutics and Biopharmaceutics* **51**(3): 191–198.

Reis, C. P., R. J. Neufeld, S. Vilela, A. J. Ribeiro and F. Veiga (2006). Review and current status of emulsion/dispersion technology using an internal gelation process for the design of alginate particles. *Journal of Microencapsulation*, **23**(3): 245–257.

Renard, D., F. van de Velde and R. W. Visschers (2006). The gap between food gel structure, texture and perception. *Food Hydrocolloids*, **20**(4): 423–431.

Ridley, B. L., M. A. O'Neill and D. Mohnen (2001). "Pectins: Structure, biosynthesis, and oligogalacturonide-related signaling." *Phytochemistry* **57**: 929–967.

Rinaudo, M. (2006). "Chitin and chitosan: Properties and applications." *Progress in Polymer Science* **31**(7): 603–632.

Rinaudo, M. (2008). "Main properties and current applications of some polysaccharides as biomaterials." *Polymer International* **57**: 397–430.

Rocha, M., C. Banuls, L. Bellod, A. Jover, V. M. Victor and A. Hernandez-Mijares (2011). "A review on the role of phytosterols: New insights into cardiovascular risk." *Current Pharmaceutical Design* **17**(36): 4061–4075.

Ron, N., P. Zimet, J. Bargarum and Y. D. Livney (2010). "Beta-lactoglobulin-polysaccharide complexes as nanovehicles for hydrophobic nutraceuticals in non-fat foods and clear beverages." *International Dairy Journal* **20**(10): 686–693.

Saglam, D., P. Venema, R. de Vries, L. M. C. Sagis and E. van der Linden (2011). "Preparation of high protein micro-particles using two-step emulsification." *Food Hydrocolloids* **25**(5): 1139–1148.

Saglam, D., P. Venema, R. de Vries and E. van der Linden (2013). "The influence of pH and ionic strength on the swelling of dense protein particles." *Soft Matter* **9**(18): 4598–4606.

Saiz-Abajo, M. J., C. Gonzalez-Ferrero, A. Moreno-Ruiz, A. Romo-Hualde and C. J. Gonzalez-Navarro (2013). "Thermal protection of beta-carotene in re-assembled casein micelles during different processing technologies applied in food industry." *Food Chemistry* **138**(2–3): 1581–1587.

Sajeesh, S. and C. P. Sharma (2006). "Novel pH responsive polymethacrylic acid-chitosan-polyethylene glycol nanoparticles for oral peptide delivery." *Journal of Biomedical Materials Research Part B-Applied Biomaterials* **76B**(2): 298–305.

Saulnier, L. and J.-F. Thibault (1999). "Ferulic acid and diferulic acids as components of sugar-beet pectins and maize bran heteroxylans." *Journal of the Science of Food and Agriculture* **79**: 396–402.

Sava, N., I. Van der Plancken, W. Claeys and M. Hendrickx (2005). "The kinetics of heat-induced structural changes of b-lactoglobulin." *Journal of Dairy Science* **88**: 1646–1653.

Schmitt, C., C. Sanchez, S. Desobry-Banon and J. Hardy (1998). "Structure and techno-functional properties of protein-polysaccharide complexes: A review." *Critical Reviews in Food Science and Nutrition* **38**(8): 689–753.

Schmitt, C. and S. L. Turgeon (2011). "Protein/polysaccharide complexes and coacervates in food systems." *Advances in Colloid and Interface Science* **167**(1–2): 63–70.

Selimovic, S., J. Oh, H. Bae, M. Dokmeci and A. Khademhosseini (2012). "Microscale strategies for generating cell-encapsulating hydrogels." *Polymers* **4**(3): 1554–1579.

Semenova, M. G. and E. Dickinson (2010). *Biopolymers in Food Colloids: Thermodynamics and Molecular Interactions*. Boca Raton, FL, CRC Press.

Semo, E., E. Kesselman, D. Danino and Y. D. Livney (2007). "Casein micelle as a natural nano-capsular vehicle for nutraceuticals." *Food Hydrocolloids* **21**(5–6): 936–942.

Seno, F. and A. Trovato (2007). "Minireview: The compact phase in polymers and proteins." *Physica A: Statistical Mechanics and its Applications* **384**(1): 122–127.

Seo, M. H., Y. H. Chang, S. Lee and H. S. Kwak (2011). "The physicochemical and sensory properties of milk supplemented with ascorbic acid-soluble nano-chitosan during storage." *International Journal of Dairy Technology* **64**(1): 57–63.

Shapira, A., I. Davidson, N. Avni, Y. G. Assaraf and Y. D. Livney (2012). "Beta-casein nanoparticle-based oral drug delivery system for potential treatment of gastric carcinoma: Stability, target-activated release and cytotoxicity." *European Journal of Pharmaceutics and Biopharmaceutics* **80**(2): 298–305.

Shewan, H. M. and J. R. Stokes (2013). "Review of techniques to manufacture micro-hydrogel particles for the food industry and their applications." *Journal of Food Engineering* **119**(4): 781–792.

Shpigelman, A., G. Israeli and Y. D. Livney (2010). Thermally-induced protein-polyphenol co-assemblies: Beta lactoglobulin-based nanocomplexes as protective nanovehicles for EGCG. *Food Hydrocolloids*, **24**(8): 735–743.

Shukla, R. K. and A. Tiwari (2012). "Carbohydrate polymers: Applications and recent advances in delivering drugs to the colon." *Carbohydrate Polymers* **88**(2): 399–416.

Siepmann, J. and F. Siepmann (2013). "Mathematical modeling of drug dissolution." *International Journal of Pharmaceutics* **453**(1): 12–24.

Sila, D. N., S. Van Buggenhout, T. Duvetter, I. Fraeye, A. D. Roeck, A. Van Loey and M. Hendrickx (2009). Pectins in Processed Fruits and Vegetables: Part II—Structure–Function Relationships. *Comprehensive Reviews in Food Science and Food Safety*, **8**(2): 86–104.

Singh, J., A. Dartois and L. Kaur (2010). "Starch digestibility in food matrix: A review." *Trends in Food Science and Technology* **21**(4): 168–180.

Siswoyo, T. A. and N. Morita (2003). "Physicochemical studies of defatted wheat starch complexes with mono and diacyl-sn-glycerophosphatidylcholine of varying fatty acid chain lengths." *Food Research International* **36**(7): 729–737.

Song, F. and L. M. Zhang (2008). "Enzyme-catalyzed formation and structure characteristics of a protein-based hydrogel." *Journal of Physical Chemistry B* **112**(44): 13749–13755.

Stephen, A. M., G. O. Phillips and P. A. Williams (2006). *Food Polysaccharides and Their Applications*. Boca Raton, FL, CRC Press.

Stokes, J. R., B. Wolf and W. J. Frith (2001). "Phase-separated biopolymer mixture rheology: Prediction using a viscoelastic emulsion model." *Journal of Rheology* **45**(5): 1173–1191.

Sugihara, H., H. Yamamoto, Y. Kawashima and H. Takeuchi (2012). "Effects of food intake on the mucoadhesive and gastroretentive properties of submicron-sized chitosan-coated liposomes." *Chemical and Pharmaceutical Bulletin* **60**(10): 1320–1323.

Sweedman, M. C., M. J. Tizzotti, C. Schafer and R. G. Gilbert (2013). "Structure and physicochemical properties of octenyl succinic anhydride modified starches: A review." *Carbohydrate Polymers* **92**(1): 905–920.

Taggart, P. and J. R. Mitchell (2009). Starch. In *Handbook of Hydrocolloids*, 2nd Edition. G. O. Phillips and P. A. Williams (eds.). Cambridge, UK, Woodhead Publishing: 108–141.

Tarte, R. (2011). Meat protein ingredients. In *Handbook of Food Proteins*, G. O. Phillips and P. A. Williams (eds.). Cambridge, UK, Woodhead Publishing: 56–91.

Thakur, G., M. A. Naqvi, D. Rousseau, K. Pal, A. Mitra and A. Basak (2012). "Gelatin-based emulsion gels for diffusion-controlled release applications." *Journal of Biomaterials Science-Polymer Edition* **23**(5): 645–661.

Tolstoguvoz, V. B. (2007). Ingredient interactions in complex foods: Aggregation and phase separation. In *Understanding and Controlling the Microstructure of Complex Foods*, D. J. McClements (ed.). Boca Raton, FL, CRC Press: 185–206.

Tolstoguzov, V. (2003). "Some thermodynamic considerations in food formulation." *Food Hydrocolloids* **17**(1): 1–23.

Tolstoguzov, V. B. (1991). "Functional properties of food proteins and role of protein-polysaccharide interaction." *Food Hydrocolloids* **4**(6): 429–468.

Turgeon, S. L., M. Beaulieu, C. Schmitt and C. Sanchez (2003). "Protein–polysaccharide interactions: Phase-ordering kinetics, thermodynamics and structural aspects." *Current Opinion in Colloid and Interface Science* **8**: 401–414.

Varshosaz, J. (2007). "The promise of chitosan microspheres in drug delivery systems." *Expert Opinion on Drug Delivery* **4**(3): 263–273.

Vega, C. and Y. H. Roos (2006). "Invited review: Spray-dried dairy and dairy-like—Emulsions compositional considerations." *Journal of Dairy Science* **89**(2): 383–401.

Whistler, R. L. and J. N. BeMiller (1997). *Carbohydrate Chemistry for Food Scientists*. St. Paul, MN, Eagan Press.

Williams, P. A. (2007). Gelling agents. In *Handbook of Industrial Water Soluble Polymers*, P. A. Williams (ed.). Oxford, UK, Blackwell Publ.: 73–97.

Wolf, B., W. J. Frith and I. T. Norton (2001). "Influence of gelation on particle shape in sheared biopolymer blends." *Journal of Rheology* **45**(5): 1141–1157.

Wolf, B., W. J. Frith and I. T. Norton (2002). Morphology control in disperse biopolymer mixtures with at least one gelling component. In *Gums and Stabilisers for the Food Industry 11*, P. A. Williams and G. O. Phillips (eds.). London, Royal Society of Chemistry: 112–119.

Wolf, W. J. (1993). "Sulfhydral content of glycinin: Effect of reducing agents." *Journal of Agricultural and Food Chemistry* **41**: 168–176.

Wulff, G., G. Avgenaki and M. S. P. Guzmann (2005). "Molecular encapsulation of flavours as helical inclusion complexes of amylose." *Journal of Cereal Science* **41**(3): 239–249.

Wurzburg, O. B. (2006). Modified starches. In *Food Polysaccharides and Their Applications*, A. M. Stephen, G. O. Phillips and P. A. Williams (eds.). Boca Raton, FL, CRC Press: 87–118.

Xia, J., P. L. Dubin, Y. Kim, B. B. Muhoberac and V. J. Klimkowski (1993). "Electrophoretic and quasi-elastic light scattering of soluble protein-polyelectrolyte complexes." *Journal of Physical Chemistry* **97**: 4528–4534.

Xu, J., W. X. Zhao, Y. W. Ning, M. Bashari, F. F. Wu, H. Y. Chen, N. Yang et al. (2013). "Improved stability and controlled release of omega 3/omega 6 polyunsaturated fatty acids by spring dextrin encapsulation." *Carbohydrate Polymers* **92**(2): 1633–1640.

Yan, Y. F., D. Seeman, B. Q. Zheng, E. Kizilay, Y. S. Xu and P. L. Dubin (2013). "pH-dependent aggregation and disaggregation of native beta-lactoglobulin in low salt." *Langmuir* **29**(14): 4584–4593.

Zeeb, B., M. Gibis, L. Fischer and J. Weiss (2012). "Crosslinking of interfacial layers in multilayered oil-in-water emulsions using laccase: Characterization and pH-stability." *Food Hydrocolloids* **27**(1): 126–136.

Zhang, H., E. Tumarkin, R. Peerani, Z. Nie, R. M. A. Sullan, G. C. Walker and E. Kumacheva (2006). "Microfluidic production of biopolymer microcapsules with controlled morphology." *Journal of the American Chemical Society* **128**(37): 12205–12210.

Zhang, J., L. Liang, Z. Tian, L. Chen and M. Subirade (2012). "Preparation and in vitro evaluation of calcium-induced soy protein isolate nanoparticles and their formation mechanism study." *Food Chemistry* **133**(2): 390–399.

Zhang, L. F., X. Q. Ye, T. Ding, X. Y. Sun, Y. T. Xu and D. H. Liu (2013). "Ultrasound effects on the degradation kinetics, structure and rheological properties of apple pectin." *Ultrasonics Sonochemistry* **20**(1): 222–231.

Zhang, W., C. Yan, J. May and C. J. Barrow (2009). "Whey protein and gum arabic encapsulated Omega-3 lipids The effect of material properties on coacervation." *Agro Food Industry Hi-Tech* **20**(4): 18–21.

Zhu, D., S. Damodaran and J. A. Lucey (2008). "Formation of whey protein isolate (WPI) and dextran conjugates in aqueous solutions." *Journal of Agricultural and Food Chemistry* **56**(16): 7113–7118.

Zimet, P. and Y. D. Livney (2009). "Beta-lactoglobulin and its nanocomplexes with pectin as vehicles for omega-3 polyunsaturated fatty acids." *Food Hydrocolloids* **23**(4): 1120–1126.

Zimet, P., D. Rosenberg and Y. D. Livney (2011). "Re-assembled casein micelles and casein nanoparticles as nano-vehicles for omega-3 polyunsaturated fatty acids." *Food Hydrocolloids* 25(5): 1270–1276.

# Chapter 8

# Delivery System Characterization Methods

## 8.1 Introduction

This chapter provides a brief overview of some of the most important analytical instruments and testing methods available for the characterization of food-grade colloidal delivery systems. These include methods to measure particle characteristics (such as size, structure, charge, and physical state) and methods to measure the physicochemical and functional properties of delivery systems (such as optical, rheological, stability, and release properties). The identification of appropriate analytical instruments and testing protocols is essential for the rational design and development of food-grade colloidal delivery systems for specific applications. These techniques are needed for research and development purposes to elucidate the relationship between particle characteristics and the bulk physicochemical and functional properties of delivery systems. They are also needed in quality control laboratories and manufacturing facilities to monitor the properties of the delivery systems before, during, and after production so as to ensure that their properties conform to predefined quality criteria or to predict how the final product will behave during storage and utilization (Aguilera et al. 1999).

## 8.2 Particle Characteristics

Initially, we consider the different analytical tools that can be used to provide information about the particles present within colloidal delivery systems, such as their concentration, morphology, organization, size, charge, and physical state. Many of the analytical instruments and testing protocols for colloidal systems are similar to those for emulsions, which have been reviewed elsewhere (McClements 2007).

### 8.2.1 Particle Concentration

The concentration of particles in a colloidal delivery system influences its physicochemical properties and functional performance, for example, appearance, texture, stability, mouthfeel, and release characteristics. Manufacturers of colloidal delivery systems usually know the concentration of particles present since they have close control over overall product composition. Similarly, the manufacturers of food or beverage products often know the amount of a delivery system that has been incorporated into a commercial product since this is controlled during the manufacturing process. In cases where this information is unknown or unavailable, it may be necessary to determine the particle concentration. If the particles can be isolated from the food matrix (e.g., by gravitational separation, centrifugation, or filtration), then they can be collected and weighed either before (wet basis) or after (dry basis) dehydration. If they cannot be isolated, then it is necessary to measure their concentration more indirectly, for example, using particle counting, scattering, or microscopy methods (see later sections). If the colloidal particles contain a specific component that is not present in the food matrix and that can be quantified using suitable analytical methods, then it may be possible to infer the particle concentration from this information. For example, the protein concentration can be determined using the Kjeldahl method, the lipid concentration can be determined using the Soxhlet method, and the carbohydrate content can be determined by the phenol–sulfuric acid method (Nielsen 2010).

### 8.2.2 Particle Morphology and Organization

The morphology and organization of the particles within a colloidal delivery system play an important role in determining its physicochemical and functional properties, and therefore it is important to have suitable analytical methods to characterize these attributes. The unaided human eye can only resolve objects greater than approximately 100 μm, whereas the particles in colloidal delivery systems are usually much smaller than this and so cannot be seen directly (Aguilera et al. 1999; Russ 2004). Numerous microscopy methods are suitable for characterizing structures in the colloidal size range, with the most commonly used being optical, electron, and atomic force microscopy (AFM) (Morris et al. 1999; Murphy 2012). Microscopy provides information about the structure of systems in the form of *images* that can be interpreted by human beings (Russ 2011). Each microscopy method works on different physical principles and can be used to examine different levels and types of structural organizations within delivery systems. Nevertheless, any microscopy method must have three attributes if it is going to be successfully used to examine the structure of small objects: resolution, magnification, and contrast (Aguilera et al. 1999). *Contrast* is the ability to distinguish specific objects within an image from the background and from other objects. *Resolution* is the ability to distinguish two objects that are in proximity. *Magnification* is the number of times the image appears larger than the actual object being examined. Most modern microscopes are attached to personal computers that can capture and store a digital version of the image, which can then be processed using various digital processing programs to obtain information about microstructure, for example, particle size distribution (PSD), particle organization, and sometimes chemical composition (Russ 2004, 2011).

## 8.2.2.1 Optical Microscopy

Optical microscopy uses light waves to provide images of samples (Mertz 2009). Typically, an optical microscope consists of a light source, a series of lenses, and an eyepiece or digital camera. The lenses direct the light waves through the sample and magnify the resulting image so that it can be observed or recorded. The *resolution* of an optical microscope is determined by the wavelength of light used and the mechanical design of the instrument. Theoretically, the resolution of an optical microscope is around 200 nm, but in practice, it is often difficult to obtain reliable measurements below approximately 1000 nm because of technical difficulties associated with the design and manufacture of the optical components and Brownian motion of small particles. Optical microscopes therefore have limited application to colloidal delivery systems containing relatively small particles ($d$ < 1000 nm). Nevertheless, they can provide valuable information about particle shape and size distribution for delivery systems containing larger particles and can be used to establish whether aggregation of smaller particles has occurred. Optical microscopy images of filled biopolymer particles are shown in Figure 8.1—these systems consist of small fat droplets dispersed within hydrogel particles that are themselves dispersed in an aqueous solution.

The *contrast* between the major components in colloidal delivery systems is often relatively poor because they have fairly similar refractive indices, which often makes it difficult to reliably distinguish them from each other using conventional bright-field optical microscopy. This problem can often be overcome using more specialized optical microscopy methods that are designed to enhance contrast, improve image quality, and distinguish specific components (Murphy 2012). The contrast between different components can often be enhanced by modifying the design of the optics within a light microscope, for example, using *phase contrast* or *differential interference contrast* microscopy (Murphy 2012). These techniques

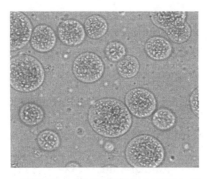

**Figure 8.1** **Optical microscopy image of an O/W$_1$/W$_2$ emulsion. This system consists of fish oil (O) droplets embedded within a whey protein-rich aqueous phase (W$_1$) that is contained within a pectin-rich aqueous phase W$_2$ (pH 7).**

improve the contrast by using specially designed lenses that convert small differences in refractive index in the sample into appreciable differences in light intensity in the final image. Structures of optically anisotropic components, such as fat crystals, liquid crystals, or native starch granules, can be studied using *polarization* light microscopy (Aguilera et al. 1999). In this case, the anisotropic components appear as white objects against a black background and can therefore be clearly distinguished.

The location of particular components within a sample (e.g., proteins, polysaccharides, or lipids) can be established using stains or dyes that bind specifically to the component of interest or that preferentially partition into a particular phase (e.g., oil or water). Specific structural features within a colloidal delivery system can thus be examined by selection of appropriate stains or dyes. Nevertheless, the stains/dyes must be carefully incorporated into a system so that they do not disturb the structures being examined.

One of the most powerful methods of providing information about the location of different components within a delivery system is laser scanning confocal microscopy (LSCM). LSCM allows one to obtain high-quality two-dimensional images of a sample in a particular plane or to get three-dimensional images of samples by taking a series of two-dimensional slices in the vertical direction without the need to physically section the specimen (Pawley 2006; Plucknett et al. 2001). These three-dimensional images can be used to determine the spatial location of different components within a delivery system. LSCM is also particularly powerful at examining the relative location and mass transport of specific components within delivery systems because of the possibility of using fluorescent probes to tag specific molecules or phases (Loren et al. 2007). LSCM images of colloidal delivery systems formed from fat, protein, and polysaccharide are shown in Figure 8.2.

Numerous other advanced forms of optical microscopy have been developed that can provide images of samples based on differences in their chemical composition,

(a)                                    (b)

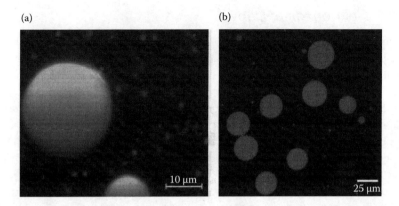

**Figure 8.2    Confocal fluorescent micrograph images of structures formed from oil, caseinate, and pectin (pH 5). Oil is dyed green, while caseinate is stained red. (a) Large oil droplets (green) surrounded by small caseinate/pectin particles (red). (b) Filled hydrogel particles consisting of fat droplets (green) trapped within casein-rich particles (red) that are dispersed within a pectin-rich continuous aqueous phase. (Image supplied by Alison Matalanis.) The original colors can be seen in the article by Matalanis and McClements (2013).**

such as Fourier transform infrared (FTIR) imaging, surface-enhanced resonance spectroscopy (SERS) imaging, and x-ray microscopy (Byelov et al. 2013; Cialla et al. 2012; Hertz et al. 2012; Levin and Bhargava 2005; Stewart et al. 2012). These methods can produce images of the location of different components (such as protein, lipids, or carbohydrates) within samples. SERS microscopy images of fat droplets surrounded by water are shown in Figure 8.3.

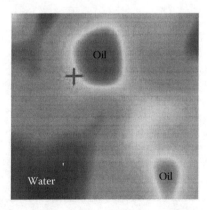

**Figure 8.3    SERS image of fat droplets surrounded by water. Each element of the image can be analyzed to obtain spectral and chemical information.**

## 8.2.2.2 Electron Microscopy

Electron microscopy is widely used to characterize the structure and organization of the particles in colloidal delivery systems, especially those containing structural elements too small to see using optical microscopes, that is, $d < 1000$ nm (Klang et al. 2012, 2013). The most powerful electron microscopes are large pieces of equipment that are relatively expensive to purchase and maintain and are therefore only available at large research and development institutions. Nevertheless, there are now bench-top electron microscopes available that are less powerful but that are cheaper and easier to use, which are finding more widespread utilization for routine analysis.

Electron microscopes use beams of electrons, rather than beams of light, to provide information about the structure of materials. The electron beams are directed through a series of magnetic fields, rather than optical lenses (Figure 8.4). A major advantage of using electron beams is that they have much smaller wavelengths than light beams and so they can be used to examine much smaller structures. In principle, the *resolution* of an electron beam is around 0.2 nm, but in practice, it is usually around 1 nm even with the most powerful instruments owing to mechanical limitations in the performance of magnetic lenses, and may be considerably larger for bench-top instruments. The *contrast* in the images obtained by electron microscopy is mainly due to differences in electron density between different components. The

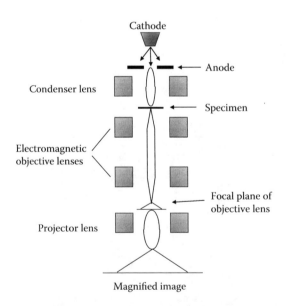

**Figure 8.4  Schematic diagram of a transmission electron microscope that can be used to provide images of the structure of colloidal delivery systems. An image is formed by passing a beam of electrons through the specimen.**

magnification of electron microscopy images depends on the type of instrument used and can be as high as several million times for the most powerful instruments.

Two types of electron microscopy are commonly used to examine the structure of delivery systems: transmission electron microscopy (TEM) and scanning electron microscopy (SEM) (Egerton 2008; Murphy 2012). Traditionally, electron microscopes could only work under high vacuum because electrons are easily scattered by the molecular species in gases, which would lead to poor image quality. Consequently, specimens had to undergo extensive preparation (e.g., fixation, dehydration) prior to imaging so as to ensure that they were free of volatile components that would evaporate and interfere with measurements, for example, water and some organic molecules. Nevertheless, environmental SEM instruments are now widely available that are capable of obtaining images of materials in a gaseous environment, and so there are far fewer problems associated with artifact generation during sample preparation (Donald 1998; Donald et al. 2000).

*Transmission Electron Microscopy*: TEM is particularly useful for characterizing very fine structures in colloidal delivery systems, for example, small particles or interfaces (Danino 2012; Dudkiewicz et al. 2011; Klang et al. 2013). In this case, a cloud of electrons, produced by a cathode, is accelerated through a small aperture in a positively charged plate to form an electron beam (Figure 8.4). This beam is focused and directed through the specimen by a series of magnetic lenses. Part of the electron beam is absorbed or scattered by the sample, but the rest is transmitted. The transmitted electron beam is magnified by a magnetic lens and then projected onto a screen to create an image of the sample. The fraction of electrons transmitted by a substance depends on its electron density: the lower the electron density, the greater the fraction of electrons transmitted and the darker the image. Components with different electron densities therefore appear as regions of different intensity on the image. The magnification factor for TEM is typically in the range of approximately $10^2$ to $10^6$ times, which enables very fine structures to be observed.

Electrons are highly attenuated by most materials and therefore samples must be extremely thin to allow enough of the electron beam to be detected. The samples used in TEM are usually much thinner (~50–100 nm) than those used in light microscopy (~1 μm to a few millimeters). The electron density contrast is often quite small for many of the major constituents in colloidal delivery systems, and so it is difficult to distinguish them. The density contrast can be enhanced by selectively staining the sample with heavy metal salts that have high electron densities, such as lead, tungsten, or uranium. These salts selectively bind to specific components, thereby enabling them to be distinguished from the rest of the sample. The need to use very thin dehydrated specimens that often require staining means that sample preparation is more time consuming and labor intensive than other forms of microscopy and may lead to appreciable artifacts in the images obtained. On the other hand, the level of detail that can be observed cannot be obtained by other forms of microscopy. A representative TEM image of lipid nanoparticles (chylomicrons) produced by Caco-2 cells when incubated with emulsion-based delivery system digestion products is shown in Figure 8.5.

*Scanning Electron Microscopy*: SEM is commonly used to provide images of the surface topography of specimens (Egerton 2008). It relies on the measurement of secondary electrons generated by a specimen when it is bombarded by an electron beam, rather than on measurement of the electrons that have traveled through the specimen (as in TEM). A focused electron beam is directed at a particular point on the surface of a specimen. Some of the energy associated with the electron beam is absorbed by the material and causes it to generate secondary electrons, which leave the surface of the sample and are recorded by a detector. An image of the specimen is obtained by scanning the electron beam in an $x$–$y$ direction over its surface and recording the intensity of the electrons generated at each location, which leads to an image with a three-dimensional appearance (Figure 8.6).

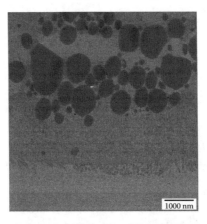

1000 nm

**Figure 8.5 TEM of lipoproteins (dark gray spheroids) produced in Caco-2 cell monolayers after incubation with lipid digestion products. (Image kindly supplied by Olivia Yao and Hang Xiao.)**

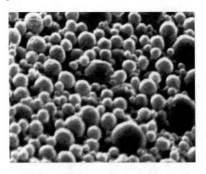

**Figure 8.6 SEM images of oil droplets in a hydrocarbon oil-in-water emulsion. (Image kindly supplied by John Coupland.)**

Sample preparation for SEM is considerably easier and tends to produce fewer artifacts than for TEM (Egerton 2008). Because an image is produced by secondary electrons generated at the surface of a specimen, rather than by an electron beam that travels through a specimen, it is not necessary to use ultrathin samples. Even so, the specimens used in conventional SEM often have to be cut, fractured, fixed, and dehydrated, which may alter their structures. Nevertheless, many of these problems have been overcome with the introduction of environmental SEM (Donald 1998; Donald et al. 2000). The resolving power of a good analytical SEM instrument is around 3 to 4 nm (but it is higher for bench-top instruments), which is an order of magnitude worse than TEM, but about two or three orders of magnitude better than optical microscopy. Another major advantage of SEM over optical microscopy is the large *depth of field*, which means that images of relatively large structures are all in-focus. An image of a colloidal delivery system (oil-in-water emulsion) acquired using SEM is shown in Figure 8.6.

## 8.2.2.3 Atomic Force Microscopy

AFM has the ability to provide information about the structure of materials at the atomic and molecular level (Morris et al. 1999; Sitterberg et al. 2010) and is therefore often complementary to the other forms of microscopy mentioned above. Commercial instruments based on this technology are widely available and are finding increasing utilization for providing information about the microstructure and organization of the structural entities present within colloidal delivery systems. An atomic force microscope creates an image by scanning a tiny probe, across the surface of the specimen being analyzed (Figure 8.7). When the probe is held extremely close to the surface of a material, it experiences a repulsive force, which causes the cantilever to which it is attached to bend away from the surface. The extent of bending is measured using an optical system. By measuring the deflection of the probe as it is moved over the surface of the material, it is possible to obtain an image of its structure. In practice, it is more common to measure the force required to keep the deflection of the probe constant, as this reduces the possible damage caused by the probe as it moves across the surface of the material. The resolution of AFM depends on the size and shape of the probe and the accuracy to which it can be positioned relative to the sample. AFM instruments typically enable lateral resolution of structures on the order of nanometers and vertical resolutions on the order of tenths of nanometers. AFM can be used to analyze both wet and dry samples. Despite its ability to examine very fine structures, AFM is fairly difficult to use for routine analysis of colloidal delivery systems, and it is therefore only used if the information cannot be provided by other methods.

## 8.2.2.4 Practical Considerations

Like any analytical tool, it is important that microscopy instruments are used correctly to obtain useful and meaningful results. Samples should be carefully

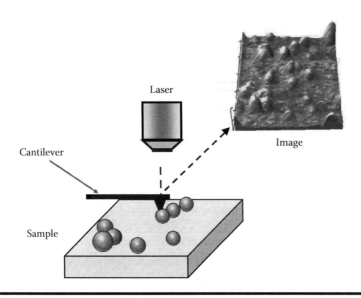

**Figure 8.7 Schematic diagram of an atomic force microscope that can be used to provide images of the structure of colloidal delivery systems. The AFM image shown is of biopolymer nanoparticles produced by heating protein and polysaccharide complexes together. (Image kindly supplied by Owen Jones.)**

prepared to avoid introducing artifacts into the images that are unrepresentative of the structures being analyzed. In addition, image acquisition and analysis should also be carried out carefully to avoid introducing artifacts and to ensure high-quality representative images. Different microscopy methods have advantages and disadvantages for particular applications. Optical microscopy is the simplest and cheapest method to use and is probably the most suitable technique for samples containing relatively large structures (>1000 nm). It usually requires the least sample preparation and can be used for a wide range of different materials. It also has the ability to stain or dye different components within a system (e.g., fats, proteins, and carbohydrates), which is useful for determining the structural organization and location of these components within a sample. In addition, specialist measurement cells can be purchased for application with optical microscopes that can be useful for characterizing colloidal delivery systems. Temperature-controlled cells that can be used to analyze a sample at a particular temperature or over a range of temperatures are available. These cells are particularly suitable for examining temperature-induced changes in the structure or properties of colloidal delivery systems, such as denaturation, helix–coil transitions, crystallization, or melting. Measurement cells can also be purchased to examine changes in the microstructure of a sample when it is subjected to controlled mechanical stresses (van der Linden et al. 2003). It is also possible to purchase micromanipulation techniques that enable one to measure the forces required to pull structures apart.

Electron microscopy is most suitable for examining samples with small structures that are too small to observe using optical microscopy, such as micelles, microemulsions, nanoemulsions, solid lipid nanoparticles, and interfacial layers (<1000 nm). TEM is particularly suitable for examining very fine structures in a material (<10 nm). The major disadvantage of electron microscopy techniques is the extensive sample preparation that is required, which is often time consuming and labor intensive and may introduce artifacts into the images (Aguilera et al. 1999). AFM is a more specialist method for analyzing very fine structures in materials, which has proved useful for examining fine structures in some colloidal systems (Sitterberg et al. 2010).

## 8.2.3 Particle Size

The size of the particles within a colloidal delivery system is one of the most important factors influencing its overall physicochemical properties and functional performance, for example, optical properties, rheology, stability, release characteristics, and biological fate (Lesmes and McClements 2009; McClements et al. 2009a,b; McClements and Li 2010b). It is therefore important to be able to reliably measure and report the size of the particles present within a colloidal delivery system. As discussed in Chapter 3, particle dimensions are usually represented as a PSD or by a measure of the central tendency and width of the distribution (such as the mean and polydispersity index).

A variety of analytical instruments are available commercially to determine the PSD of colloidal delivery systems (McClements 2007). The microscopy methods discussed in the previous section can be used to measure PSDs, but this approach is often laborious and time consuming. More specialist instruments ("particle sizers") have therefore been specifically designed to measure PSDs. Many of these instruments are automated, easy to use, rapid, and precise. A number of different kinds of particle sizers have been developed based on different physical principles, for example, measurements of light scattering patterns, particle diffusion rates, individual particle counting, and so on. Different types of particle sizers are suitable for analyzing colloidal dispersions with different particle sizes and concentrations. For example, static light scattering can be used to analyze systems with particle diameters between approximately 100 nm and 1000 μm and droplet concentrations between approximately 0.001 and 0.1 wt%, whereas dynamic light scattering can be used to analyze systems with particle diameters between approximately 1 nm and 3 μm and particle concentrations between approximately 0.001 and 30 wt%. This means that concentrated systems have to be diluted for analysis by static light scattering but can be directly analyzed by dynamic light scattering (in reflection mode). Each type of particle sizing technology also requires that specific information is provided to the instrument so that it can accurately calculate a PSD from the measured data. For example, one needs to input the refractive index and absorption coefficient of the continuous and dispersed phases to analyze the scattering pattern measured by a static light scattering instrument. Finally, it should be stressed that all particle sizing instruments are designed to operate under a specific set of conditions and that reliable measurements can only be obtained if these conditions are met in practice.

### 8.2.3.1 Static Light Scattering

Particle sizers based on static light scattering (also called laser diffraction) are one of the most widely used instruments to characterize the PSD of colloidal dispersions (McClements 2007). They are based on the principle that the light scattering pattern (intensity vs. scattering angle) produced when a laser beam propagates through a dilute particle suspension depends on the PSD. Particle sizers based on this principle come with specialized software that contains a mathematical model (usually the "Mie theory") that relates the measured scattering pattern of a colloidal dispersion to the characteristics of the particles it contains (concentration, diameter, and refractive index). The software finds the PSD that gives the best fit between the measured scattering pattern and a theoretically predicted one and then reports the data as a table or plot of particle concentration versus particle size (e.g., Figure 8.8). These mathematical models assume that the particles are homogeneous spheres with well-defined refractive indices. To calculate the PSD, the model requires information about the complex refractive index of both the particles and the dispersing medium (usually water).

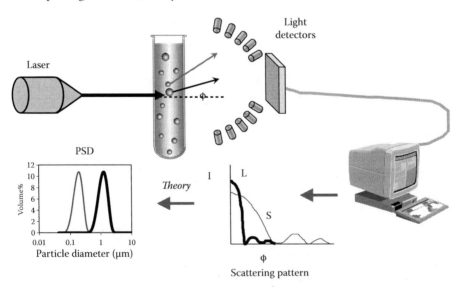

**Figure 8.8 PSD of colloidal dispersions can be measured using static light scattering, which relies on determining the scattering pattern (intensity vs. scattering angle), and using a mathematical model to convert the data into a PSD.**

Commercial static light scattering instruments are capable of determining particle diameters within the range of approximately 100 nm to 1000 μm. These instruments normally require that the droplet concentration is relatively low (<0.1 wt%) so that the laser beam can pass through and to avoid multiple scattering effects (since the mathematical model used assumes single scattering). Consequently, most delivery systems need to be diluted considerably prior to analysis, which must be done carefully to avoid altering the microstructure (see Section 8.2.3.7).

## 8.2.3.2 Dynamic Light Scattering

Dynamic light scattering is another particle sizing technology widely used to measure the PSD of colloidal dispersions (McClements 2007). It is particularly useful for analysis of colloidal delivery systems containing relatively small particles ($d < 1000$ nm). Particle sizers based on this principle measure fluctuations in the intensity of scattered light over time that occurs when the particles in a colloidal dispersion periodically change their relative spatial location owing to Brownian motion (Figure 8.9). The frequency of these intensity fluctuations depends on the velocity at which the particles move and hence on their particle size: smaller particles move more rapidly than larger ones and therefore give more rapid intensity fluctuations.

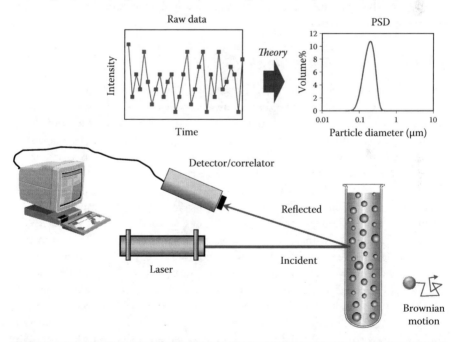

**Figure 8.9** **PSDs of colloidal dispersions can be measured using dynamic light scattering, which relies on measuring fluctuations in scattered light (intensity vs. time), and using a mathematical model to convert the data into a PSD.**

A sample is analyzed by measuring the change in intensity of the scattered wave over time at a particular scattering angle and then using a suitable mathematical model to convert the intensity fluctuations into a PSD.

Commercial dynamic light scattering instruments are capable of determining particle diameters within the range of approximately 1 nm to 3 μm. The particle concentration range that can be analyzed depends on the approach used to measure the scattered light intensity fluctuations. Some instruments measure the light that has been transmitted through a suspension and are therefore only suitable for analysis of dilute dispersions (<0.1 wt%), whereas other instruments measure the back-scattered light and are therefore suitable for analysis of both dilute and concentrated dispersions (0.001 to 10 wt%). Dynamic light scattering can also be used to measure the microrheology experienced by small particles of known size, based on analysis of their diffusion coefficients.

Diffusive wave spectroscopy (DWS) methods that are also capable of providing information about particle size in relatively concentrated systems have become available (Alexander and Dalgleish 2006). The operating principle of DWS is related to that of dynamic light scattering, except that it relies on the light waves being multiply scattered by the particles so that each photon takes a diffusive path through the system (rather than undergoing single scattering). DWS is particularly suited for analysis of mean particle size, aggregation phenomenon, and microrheology of concentrated systems.

### 8.2.3.3 Optical Pulse Counting

Particle sizers are available based on the principle of measuring the size of particles in a colloidal dispersion one at a time using a suitable optical arrangement (Zhang et al. 2009). A dilute colloidal dispersion is placed within a container and then a small volume is pulled through a narrow tube (Figure 8.10). The colloidal dispersion is made to flow past a laser beam, which is scattered when any particles are present, and a photodetector measures the intensity of scattered or nonscattered light. Nonscattered light is that fraction of the laser beam that passes directly through the measurement chamber (which decreases as particles pass through), whereas scattered light is that fraction of the laser beam that is scattered at a particular angle (which increases as the particles pass through). Changes in light intensity signals are recorded as a known volume of colloidal dispersion passes through the measurement chamber. The magnitude of each signal depends on the size of the scattering particles: the larger the size, the higher the signal. The concentration of the colloidal dispersion must be kept sufficiently low so that only one particle flows through the laser beam at a given time. The concentration of the particles within a colloidal dispersion can be determined from the number of light pulses per unit volume, whereas the size distribution of the particles can be determined from the intensities of the individual pulses. Commercial instruments can analyze particles from approximately 500 nm to 2500 μm and require highly

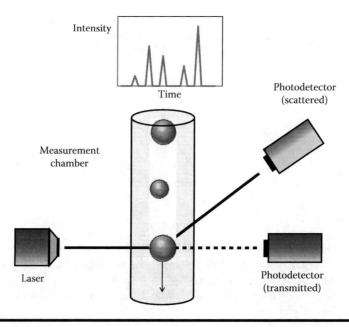

**Figure 8.10 PSDs of colloidal dispersions can be measured using optical pulse counting, which relies on measuring the number and intensity of the scattered or transmitted waves produced when single particles pass through a laser beam.**

diluted colloidal dispersions (<0.1% particles). A potential advantage of these systems is that they are highly sensitive to outliers in a distribution (i.e., a few very small or large particles), which may be important for certain applications of colloidal delivery systems.

## 8.2.3.4 Electrical Pulse Counting

Electrical pulse counting techniques are based on measurements of changes in the electrical conductivity across a small orifice when a dilute colloidal dispersion is pulled through it (McClements 2007). The colloidal dispersion is placed within a container that has two electrodes dipping into it (an anode and a cathode). One of the electrodes is located in the container, while the other is located within a glass tube that has a small orifice in it (Figure 8.11). The colloidal dispersion is pulled through the small orifice by applying a vacuum to the glass tube. When a particle passes through the orifice, it causes a decrease in the electrical current between the two electrodes because the particles usually have a lower electrical conductivity than water. Each time a particle passes through the orifice, the instrument records a decrease in current, which it converts into an electrical pulse. The particle concentration is determined from the number of pulses per unit volume of colloidal dispersion that passes through the orifice. The PSD is determined by measuring the

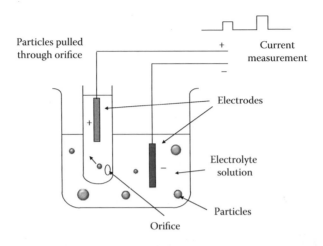

**Figure 8.11  PSDs of colloidal dispersions can be determined using electrical pulse counting, which relies on measuring the number and amplitude of the electrical pulses produced as individual colloidal particles pass through an orifice separating two electrodes.**

height of each individual electrical pulse, since the height of a pulse is proportional to the particle volume ($\propto d^3$).

Commercial electrical pulse counting techniques are capable of determining the size distribution of particles with diameters between approximately 0.4 and 1200 μm. Nevertheless, to cover this entire range, it is necessary to use a series of glass tubes with different sized orifices. Electrical pulse counting instruments normally require that the particle concentration be relatively low (<0.1 wt%) so that only a single particle passes through the orifice at a time. Consequently, many types of colloidal delivery systems need to be diluted considerably prior to analysis using this method.

### 8.2.3.5  Gravitational Settling and Centrifugation

These techniques obtain information about the particle size of colloidal dispersions by measuring the velocity at which particles move in a gravitational or centrifugal field (McClements 2007). Larger particles move more rapidly than smaller ones, and therefore measurements of particle velocity can be used to determine the particle size using a suitable mathematical model. These techniques typically measure the time dependence of the particle concentration versus sample height. The particle concentration profile can be measured using numerous methods, including light scattering, electrical conductivity, ultrasonic velocity, NMR, or x-rays. Commercially available instruments typically use light scattering (transmission or reflection) to measure particle concentration profiles, although more specialized laboratory instruments may use the other methods mentioned.

Commercial instruments are capable of determining the size distribution of particles with diameters between approximately 1 nm and 1 mm, although a number of different instruments are usually needed to cover the entire range. For example, the size of large particles can be determined by measuring their sedimentation or creaming velocity owing to gravity, whereas that of smaller particles can be determined by measuring their velocity in a strong centrifugal field. These measurements can be carried out on dilute or concentrated particle suspensions, but the theories used to interpret the data become more complicated when extensive particle–particle interactions occur.

## 8.2.3.6 Miscellaneous Techniques

There are a variety of other particle sizing technologies that have been developed that are suitable for application to delivery systems, such as NMR, ultrasonic spectrometry, electroacoustics, x-ray scattering, neutron scattering, and dielectric spectroscopy (McClements 2007). These instruments are not as widely used as the ones mentioned above, but they may be useful for applications with special requirements. For example, ultrasonic spectrometry and NMR methods can be used to analyze concentrated colloidal dispersions without dilution, but they are often only suitable for relatively simple systems that do not have complex compositions or structures. The principles behind many of these techniques have been reviewed elsewhere (McClements 2007) or are available from the companies that manufacture and market them.

## 8.2.3.7 Practical Considerations

A major advantage of most instrumental particle sizers is that they are simple to use and rapidly provide full PSDs. However, to obtain reliable data from any commercial particle sizing instrument, it is important to understand the underlying physical principles, to operate it properly, and to be aware of and eliminate any potential sources of error (McClements 2005). Errors may come from a variety of sources: an inappropriate mathematical model is used to interpret the experimental data; the particles in the colloidal dispersion do not conform to the assumptions made in the mathematical model (e.g., they are nonspherical or nonhomogeneous); the appropriate physical parameters required in the mathematical model are unknown (e.g., refractive indices in light scattering); the preparation procedure alters the structure of the sample (e.g., dilution and stirring); the size or concentration of the particles in the colloidal dispersion is not appropriate for the instrument used.

Many commercial particle sizers require some form of sample preparation prior to carrying out the measurements, for example, stirring to ensure homogeneity, dilution to prevent multiple scattering, or temperature control. These sample preparation procedures may alter the structure of the colloidal dispersion being analyzed so that the measured PSD is different from that in the original sample.

For example, dilution and stirring may cause appreciable alterations in the PSD, particularly in suspensions containing weakly flocculated particles or those that are sensitive to aggregation (McClements 2007). When dilution is necessary prior to analysis, it is important to carry it out using a buffer solution that has the same properties as the continuous phase of the original sample, for example, pH, ionic strength, and temperature.

Most commercial particle sizing instruments use some form of mathematical model to convert the measured data into a PSD. These models are usually based on some simplifying assumptions about the system, such as the particles are spherical, homogeneous, and noninteracting. If reliable quantitative information about a PSD is required, then it is important to ensure that the properties of the sample being analyzed conform to the assumptions underlying the mathematical model. In many colloidal delivery systems, the particles are not simple homogeneous spheres but have a more complex structure, such as core–shell or dispersion, for example, nanoemulsions, multilayer emulsions, multiple emulsions, and filled hydrogel particles. If the particles in a colloidal dispersion are flocculated, then the mathematical models used are unlikely to work correctly, because the scatterers will be nonspherical, inhomogeneous, and interacting. Consequently, the results obtained may only provide a qualitative indication of the actual PSD, or they may be completely unreliable.

It is always important to ensure that any particle sizer is used correctly by carefully following the instructions provided by the instrument manufacturer. In particular, it is usually important to thoroughly clean particle sizing instruments regularly to ensure that they work correctly and to calibrate them periodically with standards (i.e., particles with known sizes). In addition, it is important to ensure that the colloidal dispersion analyzed, falls within the particle size and concentration range that the instrument is sensitive to.

## 8.2.4 Particle Charge

The particles in many delivery systems have an electrical charge because of the presence of ionized substances at their surfaces, such as ionic surfactants, phospholipids, free fatty acids, proteins, ionic polysaccharides, and mineral ions (McClements 2007). The electrical characteristics of colloidal particles depend on the type, concentration, and location of any ionized species present at their surfaces, as well as the ionic composition and physical properties of the surrounding liquid. The charge on a colloidal particle is important because it determines the way it interacts with other charged species and surfaces, which influences the physicochemical properties, stability, and functional performance of colloidal delivery systems. The particles in many colloidal delivery systems are prevented from aggregating as a result of electrostatic repulsion. Colloidal particles may interact with oppositely charged polymers, particles, or mineral ions in the surrounding aqueous phase, thereby altering their physical or chemical stability, for example, through aggregation, gelation, or oxidation. The ability of colloidal particles to adhere to solid surfaces on processing equipment or product containers (such as cans, bottles, cups, or caps) influences their effective concentration in the product, as well as the overall appearance of the product. The ability of particles to adhere to biological surfaces after ingestion (e.g., mouth, stomach, small intestine, and colon) influences the biological fate of encapsulated substances (e.g., flavor perception, release profile, and bioavailability). For these reasons, it is often important to have analytical instruments to measure the electrical charge on the particles within colloidal delivery systems.

In general, the electrical characteristics of a particle are characterized by its surface charge density, surface electrical potential, and $\zeta$-potential (Hunter 1986; McClements 2005). The surface charge density is related to the number of electrical charges per unit surface area. The surface electrical potential is the free energy required to increase the surface charge density from zero to a finite value, by bringing charges from an infinite distance to the surface through the surrounding medium. The surface electrical potential therefore depends on the ionic composition of the surrounding medium owing to electrostatic screening effects, usually decreasing as the ionic strength of the aqueous phase increases. The $\zeta$-potential is the electrical potential at the "shear plane," which is defined as the distance away from the particle surface below which any counterions remain strongly attached to the particle when it moves in an electrical field. Practically, the $\zeta$-potential is a better representation of the electrical characteristics of a particle because it inherently accounts for the adsorption of any charged counterions. In addition, the $\zeta$-potential is much easier to measure than the electrical potential or surface charge density. Typically, the electrical properties of a particle are characterized by measuring the $\zeta$-potential versus pH profile under solution conditions that mimic the final application (e.g., ionic composition and temperature). A variety of analytical instruments that can be used to measure the electrical charge ($\zeta$-potential) on particles in colloidal suspensions are commercially available.

## 8.2.4.1 Microelectrophoresis

Microelectrophoresis instruments measure the velocity and direction that charged particles move in a well-defined applied electrical field (McClements 2007). A colloidal dispersion is placed into a measurement cell that contains a pair of electrodes, and then a well-defined electrical field is applied (Figure 8.12). The electric field causes any charged particles to move toward the oppositely charged electrode at a speed that depends on the magnitude of their charge and the viscosity of the surrounding liquid. The instrument calculates the electrophoretic mobility from measurements of particle velocity and then uses a mathematical model in the instrument software to convert this information into a $\zeta$-potential. The movement of the particle in a measurement cell can be followed using various techniques, with

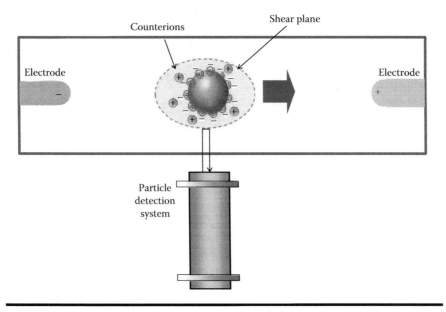

**Figure 8.12  The particle charge of colloidal dispersions can be determined using microelectrophoresis. The direction and velocity of particle movement are measured using a suitable technique in a well-defined electrical field.**

the most common being laser light scattering. Most commercial ζ-potential instruments based on light scattering require that the droplet concentration be relatively low (<0.1 wt%) so that light can be transmitted through the system and multiple scattering effects can be avoided. Consequently, many delivery systems have to be diluted prior to analysis, which must be done carefully to avoid changes in the electrical characteristics of the particles (see Section 8.2.4.3).

### 8.2.4.2 Electroacoustics

Electroacoustic instruments are commercially available to measure the size, concentration, and ζ-potential of the particles in suspensions (McClements 2007). These instruments work by applying an oscillating electric signal to a sample and measuring the resulting acoustic signal generated by the oscillating particles or by applying an oscillating acoustic signal to the sample and measuring the resulting electric signal generated by the oscillating particles. The resulting signals can be converted to the ζ-potential of the particles using a suitable mathematical model.

The major advantage of electroacoustic techniques is that they can be applied to concentrated colloidal dispersions ($\phi \leq 50\%$) without the need for sample dilution. Nevertheless, the strength of the signal received depends on the density contrast between the particles and surrounding liquid, and therefore electroacoustic measurements may have limited application for systems with low density contrast. In addition, there may be particle–particle interactions that occur at higher concentrations that make the measurements unreliable.

### 8.2.4.3 Practical Considerations

To obtain reliable particle charge measurements, it is important to operate $\zeta$-potential instruments properly and to be aware of any potential sources of error (McClements 2005, 2007). Errors may come from a variety of sources, including the sample itself, the sample preparation procedure, the mechanical operation of the instrument, and the mathematical model used to convert the measured signal into a particle charge. One of the major potential sources of error in $\zeta$-potential measurements carried out by the widely used microelectrophoresis method is the need for concentrated emulsions to be diluted prior to analysis. Dilution can cause a significant alteration in the $\zeta$-potential of particles owing to the change in pH, ionic strength, or composition of the solution surrounding them. In addition, dilution may cause some of the adsorbed material to be displaced from the droplet surfaces. Ideally, one should dilute a colloidal dispersion with its own continuous phase to minimize changes in particle charge caused by dilution. This can be achieved by making up a continuous phase containing no particles with the same composition as that in the colloidal dispersion or by centrifuging a colloidal dispersion and collecting the serum phase. Otherwise, one should ensure that the pH and ionic strength of the diluting fluid is similar to that of the continuous phase of the original suspension. Even if the pH and ionic strength of the continuous phase is kept the same, there is still the possibility that dilution will change partitioning of a charged substance between the interfacial region and the continuous phase, thereby altering the $\zeta$-potential of the particles; for example, an adsorbed polyelectrolyte or surfactant may be partially displaced after dilution.

### 8.2.5 Particle Physical State

The physical state of the particles in colloidal delivery systems may have a major influence on their stability and functional performance. Lipids may undergo solid–liquid (melting/crystallization) or polymorphic transitions. Proteins and polysaccharides may undergo conformational changes (such as unfolding or helix–coil transitions) or sol–gel transitions. Surfactants and phospholipids may undergo a variety of phase transitions between different states at surfaces or in the bulk. Solid particles may undergo glass–rubbery transitions in response to changes in temperature or water activity. It is therefore important to have analytical tools to determine the physical state of the particles in colloidal delivery systems. Numerous analytical techniques are available to provide this kind of information, with the most appropriate method for a particular application depending on the nature of the system being studied (McClements 2007). A number of analytical approaches that are useful for characterizing the physical state of colloidal particles are highlighted below.

### 8.2.5.1 Dilatometry

This method is based on measuring changes in the volume of a fixed mass of sample in response to some environmental change (usually temperature). It can be used to monitor phase transitions that cause an appreciable change in the density of a material, for example, solid–liquid phase transitions. This method is particularly suitable for studying the crystallization and melting of particles in colloidal delivery systems (McClements 2007).

### 8.2.5.2 Differential Scanning Calorimetry

Differential scanning calorimetry (DSC) is based on measurements of the enthalpy changes (heat released or absorbed) when a sample is subjected to a controlled temperature scan (i.e., cooling or heating). This method can be used to monitor a variety of phase transitions in wet and dry colloidal dispersions, including melting, crystallization, polymorphic changes, and glass–rubbery transitions (Qian et al. 2012c; Shrestha et al. 2007; Shukat et al. 2012; Silalai and Roos 2010).

### 8.2.5.3 Nuclear Magnetic Resonance

This method is based on measurements of the response of a sample to an applied pulse of radiofrequency electromagnetic radiation (McClements 2005). The hydrogen nuclei within a sample move to an excited state after application of the pulse, but then they decay back to their unexcited states. The decay rate can be measured using magnetic coils and depends on the physical state of the substance: the signal from solids decays more rapidly than that from liquids. Thus, the physical state of a sample can be determined by measuring the decay rate of the NMR signal. This method is commonly used to measure the solid content of particles in colloidal delivery systems.

### 8.2.5.4 Ultrasonics

This method is based on precise measurements of the ultrasonic velocity or attenuation of a material, which may change appreciably when one or more components within the material undergoes a phase transition (Awad et al. 2012; McClements 1997). This method may be used to nondestructively monitor phase transitions such as melting, crystallization, polymorphism, and glass–rubbery transitions.

### 8.2.5.5 X-Ray Analysis

X-ray diffraction can be used to provide information about changes in the spatial arrangement of molecules within a material and can therefore be used to provide information about their physical state, for example, solid state (crystalline vs. amorphous) and polymorphic form (Hartel 2013).

### 8.2.5.6 Microscopy

Various forms of microscopy can be used to provide information about the physical state of the particles in colloidal dispersions (Section 8.2.2). Optical microscopy combined with crossed polarizers can be used to determine the presence and location of crystalline or liquid crystal regions in a sample. TEM can be used to provide information about molecular packing within colloidal particles (Bunjes et al. 2007; Kuntsche et al. 2011).

### 8.2.5.7 Practical Considerations

There are a number of practical considerations that must be taken into account when identifying a suitable analytical tool to provide information about the physical state of the particles in a colloidal delivery system. First, it is important that the analytical instrument is sensitive to the particular kind of phase transition occurring within that system, for example, solid–liquid, polymorphic form, sol–gel, or glass–rubbery. Second, the instrument should be capable of covering the temperature range over which this transition occurs. Third, it is important that any sample preparation procedure does not alter the physical state of the particles being analyzed. Fourth, in multicomponent systems, it may be important to distinguish different kinds of phase transitions associated with different constituents.

## 8.3 Bulk Physicochemical Properties

The bulk physicochemical properties of delivery systems, such as their optical properties, rheology, and stability, play an important role in determining the types of products that they can be successfully incorporated into. In this section, a brief overview of the different methods for providing information about the bulk physicochemical properties of delivery systems is given.

### 8.3.1 Optical Properties

The most important optical properties of a colloidal delivery system are the opacity, color, and surface sheen (Hutchings 1999), which depend on the scattering properties of the particles and the absorption properties of any chromophores present (McClements 2002a,b, 2005). The opacity of a delivery system is usually characterized by measuring the transmittance or reflectance of visible light. Transmission measurements are typically used to measure the turbidity of relatively clear systems, whereas reflectance measurements are used to measure the lightness of relatively opaque systems. The transmittance is determined by measuring the reduction in intensity of a beam of light propagated through a sample: the lower the transmittance, the higher the turbidity. Transmittance measurements are most suitable for samples that are only slightly cloudy (e.g., suspensions with low particle sizes, concentrations, or refractive index contrasts). The reflectance is determined by measuring the fraction of a light beam that is reflected from the surface of a sample: the greater the reflectance, the lighter the material. It is therefore more suitable for optically opaque materials (e.g., suspensions with large particle sizes, concentrations, or refractive index contrasts). The color of a delivery system is determined by measuring the selective absorption of visible light in either transmission or reflectance modes. For samples that are only slightly cloudy, the fraction of light transmitted at different wavelengths is measured by passing light through the sample. For samples that are opaque, the fraction of light reflected from the surface of the sample is measured as a function of wavelength.

The opacity and color of delivery systems can be quantitatively determined using tristimulus color coordinates, such as the $L^*a^*b^*$ system measured using a colorimetry instrument (McClements 2002a,b). $L^*$ represents lightness, where low $L^*$ (0) is dark (black) and high $L^*$ (100) is light (white). The $a^*$ and $b^*$ values are color coordinates, where $+a^*$ is the red direction, $-a^*$ is the green direction, $+b^*$ is the yellow direction, and $-b^*$ is the blue direction. The opacity of a colloidal dispersion can therefore be characterized by the lightness ($L^*$), while the overall color intensity can be characterized by the chroma: $C = (a^{*2} + b^{*2})^{1/2}$. Theoretical models are available to mathematically relate particle characteristics (size, concentration, refractive index contrast) and dye characteristics (concentration and absorption spectrum) to tristimulus color coordinates (McClements 2002a,b).

The optical properties of a colloidal delivery system can be measured using various analytical instruments, which vary in their physical principles, range of applications, cost, and ease of use (Hutchings 1999). The color of slightly cloudy samples can be determined by measuring the transmittance spectra using most bench-top UV-visible spectrophotometers. The transmittance is measured across the range of wavelengths of visible light (approximately 380 to 780 nm), and then a suitable mathematical model is used to convert the spectra into $L^*a^*b^*$ values (McClements 2002a,b). Some of these spectrophotometers also have specially designed measurement cells that enable the user to measure the reflectance spectra and can therefore also be applied to opaque samples. Again, mathematical models are used to convert the measured spectra into $L^*a^*b$ values, but it is important to calibrate the cells properly to take into account changes caused by the interaction of the light waves with cuvette walls. There are also a number of handheld, bench-top, and online instruments that are specifically designed to measure the color coordinates of samples. Typically, these colorimeters measure the full reflectance spectra as a function of wavelength, or they measure the relative intensity of red, green, and blue light reflected from a sample. Measurements are made under well-controlled optical conditions using either a black or white plate behind the sample. The surface sheen of a colloidal dispersion can be measured by analyzing the angular dependence of the light scattered from the surface, which may vary from shiny (specular reflection) to matt (diffuse reflection).

## 8.3.2 Rheology

A variety of instrumental methods are available to test the rheological properties of suspensions (Barnes 1994; Genovese et al. 2007; Pal et al. 1992; Tadros 1994, 2004). These instruments vary according to the type of samples they can analyze (liquids, solids, or viscoelastic materials), the type of deformation they apply to the sample (shear, compression, or some combination), the property measured (viscosity, yield stress, elastic modulus, or fracture behavior), their cost, range of applications, ease of operation, and data analysis software (McClements 2005). For industrial applications, it is important to have instruments that are inexpensive and easy to use and that make rapid and reliable measurements of some textural parameter related to product quality. For research and development applications, it is often more important to use instruments that provide fundamental information about the rheological constants of the material being tested (such as viscosity, elastic modulus, or yield stress), so that data can be compared with theoretical predictions or measurements made on other samples or in other laboratories.

The instruments used to characterize the rheological properties of colloidal suspensions typically apply either well-defined shear forces or compression forces to the samples being tested, although some instruments are capable of using a combination of both types of forces (Bourne 2002; Norton et al. 2011; Rao 1999). The most common instruments utilizing shear forces are shear viscometers and dynamic shear rheometers, which are primarily used for testing fluid and viscoelastic materials. The most common instruments that utilize compression forces are universal testing machines, which are primarily used for testing viscoelastic, plastic, or solid-like materials. The rheology of fluid and semisolid materials is typically characterized by measuring the shear stress versus shear stress profile, from which parameters such as apparent viscosity, consistency, flow index, and yield stress can be obtained. The rheology of semisolid materials is often characterized by measuring their complex shear modulus ($G'$ and $G''$), which simultaneously provides information about their fluid-like and solid-like characteristics. The complex shear modulus is typically measured as a function of applied frequency, strain, or temperature to provide information about material properties. The rheology of solid materials is typically characterized by measuring their elastic modulus, fracture stress, and fracture strain using either shear or compression tests.

### 8.3.2.1 Shear Testing

Rheology instruments that apply shear forces typically consist of a temperature-controlled measurement cell where the material to be analyzed is placed (Bourne 2002; Rao 1999). The sample is then subjected to a controlled *shear stress* and the resulting *shear strain* (solids) or *rate of shear strain* (fluids) is measured. The rheological properties of the sample are then determined by analyzing the resulting stress–strain relationship. The measurement cells may have different designs depending on the nature of the sample and the sensitivity required, with the most common being concentric cylinder, parallel plate, cone and plate, and vane. The surfaces of the measurement cells may need to be serrated or roughened to avoid slip effects. The type of rheological test carried out depends on whether the sample is liquid, solid, or viscoelastic. For a solid, the *elastic modulus, yield stress,* and *fracture stress* can be determined by measuring the stress–strain relationship. For a fluid, the *apparent viscosity* can be determined by applying a constant stress and measuring the resulting rate of strain. For a viscoelastic material, the *complex shear modulus* can be determined by applying a sinusoidal stress and measuring the resulting sinusoidal strain. These tests are often carried out as a function of applied stress, temperature, or time to provide more information about material properties. Experiments must be carried out carefully to avoid errors, such as gap effects, wall slip effects, gravitational separation, and sample history (Larson 1999; Macosko 1994; McClements 2005).

### 8.3.2.2 Compression Testing

This type of test is usually carried out on solid or semisolid materials that are capable of supporting their own weight without flowing (Bourne 1997; Rao 1999; Walstra 2003). A sample is placed between a fixed plate and a probe, which is capable of being moved vertically (upward or downward) at a controlled speed. The instrument has a pressure sensor in either the probe or the plate that measures the force exerted on the sample when it is deformed. The instrument also records the distance that the probe moves upward or downward. The force–distance relationship is used to generate a stress–strain curve, where stress is the force per unit surface area and strain is the fractional deformation. This type of curve can be used to calculate the elastic modulus, fracture stress, fracture strain, and yield stress of the material (Bourne 2002). A number of different probes are available to simulate different mechanical forces a material may experience, such as flat plates, spikes, blades, or sets of teeth. As with all rheological experiments, compression tests must be carried out carefully to avoid erroneous results. Some of the factors to consider are the parallelism of the plate and probe, the friction between plate/probe and sample, changes in sample cross-sectional area during compression, changes in material properties during sample preparation, and time effects for viscoelastic or plastic materials.

### 8.3.2.3 Compression–Shear Testing

In some situations, it is convenient to apply both compression and shear forces to the material to obtain fundamental rheological parameters or to simulate certain types of mechanical forces that the material may experience (Bourne 2002; van Aken 2007). For example, after ingestion, delivery systems and the foods containing them undergo a complex flow profile during mastication that involves both compression and shear movements. Rheological methods may therefore be specifically designed to simulate the complex processes occurring during oral processing (Le Reverend et al. 2010; Lillford 2000). One such method involves mixing samples with saliva and then subjecting them to a sequence of compression–shear–elongation cycles to mimic the movement of the jaw and tongue during mastication (Chung et al. 2012, 2013). The change in the sample rheology and microstructure can be measured during each cycle.

## 8.3.3 Flavor

The incorporation of a delivery system into a food or beverage product may alter its desirable flavor profile by changing the distribution of volatile or nonvolatile flavor molecules or by providing rheological and thin film properties that are detected by the tongue (mouthfeel) (Le Reverend et al. 2010; van Vliet et al. 2009). In addition, one of the most common applications of delivery systems in the food industry is controlling flavor release profiles during consumption (Madene et al. 2006). It is therefore important to have analytical techniques to characterize the flavor release profiles of delivery systems.

### 8.3.3.1 Analysis of Volatile Flavors

The interaction of volatile flavor compounds from a food or beverage product with the receptors within the nose is responsible for its aroma (Madene et al. 2006). Analytical methods are therefore required to measure the change in concentration of aroma compounds in the headspace above a product. These measurements can be made under static, dynamic, or simulated in vivo methods (McClements 2005).

*Static Measurements*: Static headspace analysis is primarily used to determine the equilibrium partition coefficient of a flavor compound between a sample and the gaseous headspace above it (Balasubramanian and Panigrahi 2011; Jelen et al. 2012). The sample to be analyzed is placed in a sealed container stored under fixed temperature and pressure (Figure 8.13). Samples of the gas phase are then removed from the headspace using a syringe (or other device) inserted through the lid of the sealed container, and the flavor concentration is measured, usually by gas chromatography (GC) or high-performance liquid chromatography (HPLC) (Wang et al. 2008). Information about the type of flavor molecules present within the headspace can be determined after they have been separated by GC or HPLC, either by comparing the retention times of the chromatogram peaks with known standards, or by analyzing each of the peaks using an analytical technique that provides information about chemical structure, for example, mass spectrometry or nuclear magnetic resonance.

General information about the overall flavor profile above a static sample can be obtained using analytical instruments called "electronic noses" (Baldwin et al.

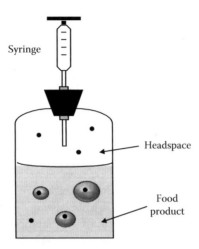

**Figure 8.13 The concentration of volatile molecules above a food product containing a colloidal delivery system can be measured using static headspace analysis.**

2011; Deisingh et al. 2004). Electronic noses consist of an array of sensors that generate electronic signals when they interact with flavors that the human nose is sensitive to. Each individual sensor responds to many different kinds of flavor molecules, but the intensity of the signal produced depends on molecular type and concentration. The pattern of signal intensities produced by the different sensors within an electronic nose provides a means of discriminating samples based on differences in their flavor profiles.

*Dynamic Measurements*: Dynamic headspace analysis involves measuring the concentration of flavor molecules in the headspace over time to monitor the kinetics of flavor release from a sample contained in a measurement cell (McClements 2005; Snow and Slack 2002; Wang et al. 2008). Similar analytical instruments can be used to measure the type and concentration of flavor compounds in the headspace in dynamic measurements as in static measurements, for example, GC, HPLC, mass spectrometry, NMR, or electronic nose. A number of powerful analytical instruments based on mass spectrometry have been developed to provide real-time measurements of changes in flavor profile over time (Deleris et al. 2013; Linforth et al. 2010). The headspace is continuously collected and fed to a mass spectrometer for continuous analysis of flavor type and concentration. Dynamic methods are particularly useful for measuring flavor release from colloidal delivery systems or to determine how delivery systems influence the flavor profiles of food or beverage products. Delivery systems with different particle characteristics can be formulated (such as particle composition, structure, or size) and then the flavor release profile can be measured (Arancibia et al. 2011; Linforth et al. 2010).

*Simulated In Vivo Analysis*: Analytical instruments that can actually measure the headspace concentration of volatiles during mastication by human beings are available (Deleris et al. 2013; Linforth et al. 2007, 2010). In these instruments, the gas phase is collected from the nose of a human subject before, during, and after consumption of a food or beverage by attaching a collection tube to one of the nostrils. As with dynamic methods, aliquots of gas phase can be collected periodically for later analysis or the gas phase can be analyzed continuously using specially designed mass spectrometers. These simulated in vivo methods provide a more accurate representation of the complex dynamic processes occurring during food ingestion than conventional dynamic methods. Application of these techniques has shown that different flavor molecules are released at different times, depending on their physicochemical characteristics, the structure and composition of the food matrix, and the mastication conditions (Arancibia et al. 2011; Linforth et al. 2010). There is considerable emphasis in the field of flavor research on correlating the results of in vivo headspace analysis with sensory time–intensity measurements.

### 8.3.3.2 Analysis of Nonvolatile Flavors

The taste of a food or beverage is determined by specific types of nonvolatile molecules present within the saliva after ingestion (Bigiani and Prandi 2011; McClements 2005). The analytical characterization of taste therefore involves measurements of the type and concentration of nonvolatile flavors present within aqueous solutions. A variety of analytical methods have been developed to measure flavor concentrations depending on the type of flavor molecule, such as chemical methods, ion-selective electrodes, spectrophotometry, chromatography, mass spectrometry, NMR, radiolabeling, and electrophoresis. Measurements can be made after a sample has come to equilibrium or it can be measured over time to provide dynamic information. Simulated in vivo methods have been developed to measure the release of nonvolatile flavor molecules from foods and beverages during mastication. Aliquots of saliva can be periodically collected from the mouth of a human subject during mastication and then analyzed using conventional methods. Samples can be processed in the mouth for a given period and then spat out and analyzed. Alternatively, sensors can be placed in the mouth to measure changes in the type and concentration of nonvolatile flavor molecules over time. Recently, electronic tongues have been developed to monitor changes in flavor profiles in the mouth during mastication (Escuder-Gilabert and Peris 2010; Smyth and Cozzoino 2013; Vlasov et al. 2010). These methods may be used to monitor the release of a flavor component from a delivery system or to determine how incorporation of a delivery system influences the flavor profile of a product.

Recently, there have been major advances in the development of biochemical and animal models for measuring certain kinds of nonvolatile flavor compounds, such as acid, sweet, salt, bitter, and umami (Morini et al. 2011; Ozeck et al. 2004;

Zhang et al. 2010). Taste-receptor assays that provide a readout when a specific flavor molecule interacts with a receptor have been developed. These methods allow high-throughput screening to identify flavors, flavor enhancers, or flavor suppressors from various plant, animal, or microbial sources.

### 8.3.3.3 Sensory Analysis

Ultimately, the influence of a delivery system on the flavor profile of a food or beverage product is assessed by food consumers. The analytical instruments carried out in a laboratory are useful for identifying the most important factors that influence flavor, such as particle composition, structure, and size. However, they cannot be used to model the complexity of the mastication experience, which involves dynamic physicochemical, physiological, and psychological responses (Le Reverend et al. 2010; van Aken 2007; van Vliet et al. 2009). It is therefore important to have analytical protocols that involve human beings to test foods or model food products. Consequently, sensory analysis by human subjects is one of the most important methods of assessing the overall flavor profile of foods and beverages (Meilgaard et al. 2006; Piggott et al. 1998; Stone and Sidel 2004). In general, sensory methods can be divided into two general categories: discriminant and descriptive methods. In discriminant methods, panelists are requested to identify whether there is a sensory difference in specified properties between two or more food samples. In descriptive methods, the panelists are requested to assess and rank specified properties of food samples based on previously established sensory descriptors. Sensory analysis can also be categorized according to whether a trained or nontrained panel is used to carry out the evaluation. In some situations, sensory evaluation is carried out by specialists that have previously been trained to recognize particular flavors or to detect slight differences in flavor profiles. In other situations, sensory evaluation is carried out using relatively large panels of untrained individuals that are more representative of the general population or some specific segment of the general population, and the resulting data are statistically analyzed to ascertain significant differences between samples. Sensory analysis may involve a panelist giving an overall impression of some specified characteristic of a food or beverage sample after smelling or tasting it. Alternatively, impressions of the sample may be provided over time before, during, and after mastication. For example, a panelist may be asked to rank the intensity of some flavor characteristic over time to provide an intensity–time profile. Usually, the flavor intensity starts from a low value, reaches some maximum value during mastication, and then fades away (Figure 8.14). Rather than report the full intensity versus time profile, it is often convenient to report specific parameters derived from the time–intensity profile: onset time ($t_{onset}$), time to reach maximum intensity ($t_{max}$), total duration ($t_{dur}$), maximum intensity ($I_{max}$), area under the curve (AUC), and initial release rate ($dI/dt$).

Despite their practical importance in the food industry, sensory methods do have a number of limitations for routine flavor analysis. Sensory analysis is often

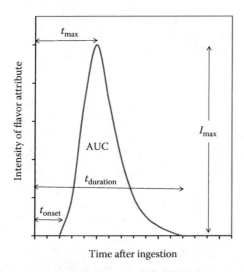

**Figure 8.14** **A typical flavor release profile can be characterized by the area under the curve (AUC), onset time ($t_{onset}$), time to reach maximum intensity ($t_{max}$), overall duration of effect ($t_{duration}$), and maximum intensity ($I_{max}$).**

time consuming and expensive to carry out, and individuals vary widely in their evaluation of food flavors. For this reason, there is considerable emphasis on the development of quantitative analytical procedures that correlate with the results of sensory analysis. When designing a delivery system, the major factors influencing the sensory properties of foods are usually established first using analytical methods, and then the final versions of the product are tested using sensory analysis.

## 8.3.4 Stability

A colloidal delivery system may break down through a variety of physical and chemical mechanisms (McClements 2005; McClements and Decker 2000). Physical instability mechanisms usually involve a change in the location, dimensions, or structure of the particles, whereas chemical instability mechanisms involve a change in the chemistry of the molecules present (e.g., as a result of hydrolysis, oxidation, or reduction). In this section, some of the commonly used analytical methods that are used to assess the physical stability of particle suspensions are described. Methods for measuring the chemical stability of active ingredients in colloidal dispersions are discussed in Section 8.4.

### 8.3.4.1 Particle Location Changes: Gravitational Separation

Gravitational separation involves the upward (creaming) or downward (sedimentation) movement of the particles in a colloidal dispersion caused by the forces of gravity (McClements 2005). Colloidal particles that are denser than the surrounding fluid tend to move downward, whereas those that are lighter tend to move upward. The rate of gravitational separation increases with increasing density contrast, increasing particle size, and decreasing continuous phase viscosity (Section 3.4.3.1). In this section, a brief overview of some of the methods available to determine the stability of a colloidal delivery system is given.

*Mathematical Prediction*: To a first approximation, the stability of a colloidal dispersion to gravitational separation can be predicted from Stokes' law (McClements 2005). To do this, it is necessary to have information about the densities of the particles and surrounding fluid, the particle size, and the rheological properties of the continuous phase. The density of the different phases can often be measured using suitable analytical techniques, such as density bottles, hydrometers, and oscillating U-tube density meters (Pomeranz and Meloan 1994). The particle size can be measured using the various methods described earlier (Section 8.2). The rheological properties of the continuous phase can be characterized using various types of rheometers (Rao 1999). In principle, the long-term stability of a colloidal delivery system can therefore be estimated from knowledge of these physicochemical properties and Stokes' law. In practice, this approach has limited use because Stokes' law does not take into account the inherent complexity of most food dispersions, such as nonspherical particles, particle–particle interactions, and nonideal fluids. In addition, the particle size may change during storage (e.g., owing to aggregation, dissociation, or dissolution), which would change the rate of gravitational separation. If these effects are ignored, then the rate of gravitational separation predicted by Stokes' law for the original sample will be different from the observed rate. For these reasons, it is often more appropriate to directly measure gravitational separation in a colloidal dispersion.

*Visual Observation*: The simplest and cheapest method of following gravitational separation of a colloidal dispersion is through visual observation. The colloidal dispersion is placed in a transparent test tube, gently mixed to ensure that it is initially homogeneous, and stored for a certain period (or exposed to specific environmental stresses), and then the height of any boundaries formed between different layers is measured (Figure 8.15). For example, in a colloidal dispersion containing particles with a lower density than the surrounding liquid (such as an oil-in-water emulsion), it is often possible to observe two or three layers after creaming has occurred (Figure 8.15). In the initial stages of creaming, three layers are formed, a lower particle-depleted serum layer ($\phi < \phi_{initial}$), an intermediate layer ($\phi = \phi_{initial}$), and an upper particle-rich cream layer ($\phi > \phi_{initial}$) are observed. At the later stages of creaming, a two-layer system is formed consisting of a lower serum layer and an upper cream layer. Ideally, the serum layer is optically transparent, the intermediate layer has an appearance similar to that of the original colloidal dispersion, and the cream layer is more optically opaque (lighter) than the original

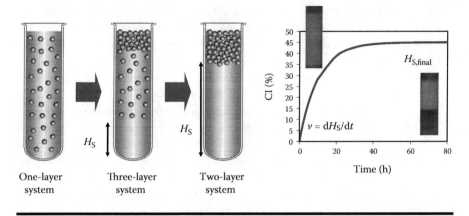

**Figure 8.15** **The stability of a colloidal delivery system to gravitational separation can often be characterized by simple visual observation. The height of the serum layer ($H_S$) can be measured as a function of time to determine the creaming velocity and final height.**

colloidal dispersion. The extent of creaming can then be simply characterized by a creaming index (*CI*):

$$CI = 100 \times \frac{H_S}{H_T} \tag{8.1}$$

where $H_T$ is the height of the total sample and $H_S$ is the height of the serum layer (Figure 8.15). Typically, the creaming index starts at zero and increases over time as the particles move upward until a fairly constant value is reached when all the particles are tightly packed in the cream layer. The initial slope of a plot of the creaming index versus time is related to the creaming velocity: $v = \mathrm{d}(H_S)/\mathrm{d}(t)$. The value of $H_S$ at the end of the process depends on the initial particle concentration in the overall colloidal dispersion, as well as the packing efficiency of the particles in the cream layer (McClements 2005). In the case of particles that are heavier than the continuous phase, a similar approach can be used, but a *sedimentation index* is defined rather than a creaming index. In this case, the height of the serum layer is measured from the top of the sample.

Creaming or sedimentation can be monitored visually by recording the change in the height of the serum layer over time, or it can be monitored using video imaging instruments. These instruments can be coupled with image analysis software to quantify changes in sample appearance during storage. A simpler version of this approach is to periodically take photographs of a sample using a digital camera. The photographs can then be used as a record of the stability, or they can be further analyzed using image analysis software to calculate a creaming index or creaming velocity.

Two major problems associated with determining the extent of gravitational separation by visual inspection are as follows: (i) it is only possible to obtain information about the location of the boundaries between the different layers, rather than about the full vertical particle concentration profile, and (ii) in some systems, it is difficult to clearly locate the boundaries between the different layers because they are diffuse or both adjacent layers are optically opaque.

*Physical Sectioning*: The change in particle concentration with time and height in a colloidal dispersion can be determined by physically sectioning a sample in the vertical direction (Pal 1994). A colloidal dispersion is placed in a container and then stored for a certain period. The system is then rapidly frozen and the sample is removed from the container and cut into a series of vertical sections. The particle concentration in each section can then be measured using an appropriate analytical technique, such as solvent evaporation, solvent extraction, density, spectroscopy, or turbidity measurements (Pal 1994). The vertical resolution of a particle concentration profile depends on the thickness of the sections. It is also important to ensure that the method used to physically trap the particles prior to sectioning (e.g., freezing) does not alter their vertical position. An alternative approach is to store a colloidal dispersion within a cylindrical burette for a certain period and then to carefully collect successive aliquots of the sample and analyze their particle concentration. Again, it is important that the relative vertical position of the particles is not altered as the aliquots are collected.

The major disadvantage of these techniques is that they cause destruction of the sample being analyzed and therefore cannot be used to monitor gravitational separation in the same sample over time. Instead, a large number of similar samples have to be prepared and each one has to be analyzed separately at a different time. Nevertheless, it is possible to determine the composition of the different components in a sample at each vertical location using this technique, provided appropriate analytical methods are available to measure the concentrations of the components of interest. In addition, it is possible to determine differences in other particle characteristics as a function of sample height, such as PSD, charge, aggregation, or physical state. This technique is also suitable for analyzing optically opaque colloidal dispersions that may be difficult to analyze using techniques that require light transmission.

*Particle Profiling*: A number of instrumental methods that can provide information about the change in particle concentration within a colloidal dispersion as a function of vertical height and storage time are available. One of the most commonly used is an optical profiling method based on light scattering (Chanamai and McClements 2000; Mengual et al. 1999a,b). A colloidal dispersion is placed in a vertical flat-bottomed transparent tube and a monochromatic beam of near-infrared light is directed through it (Figure 8.16). The percentage of transmitted and back-scattered light is measured as a function of height using one or two detectors by scanning the light beam up and down the sample. The fraction of transmitted light decreases as the particle concentration increases, whereas the fraction of

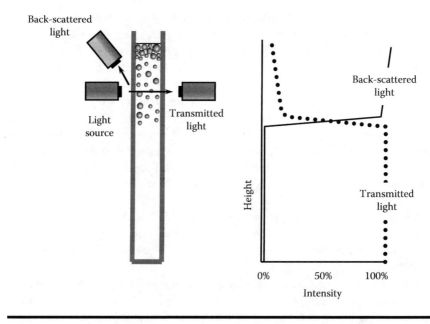

**Figure 8.16   The stability of a colloidal delivery system to gravitational separation can be characterized by optical scanning methods. The reflectance or transmittance of light through a sample is measured as a function of height.**

back-scattered light increases. The variation of particle concentration with sample height can sometimes be deduced from the percentage of transmitted and back-scattered light using a suitable theory or calibration curve. In practice, it is difficult to accurately measure the particle concentration as a function of sample height because the intensities of scattered and transmitted light do not change appreciably with the amount of particles present at higher particle concentrations, that is, >5% to 10% (Chantrapornchai et al. 1999, 2001; McClements 2002a,b). In addition, the measured intensities depend on the particle size, as well as the particle concentration. In principle, optical profiling techniques can be used to measure both the size and concentration of the particles at any vertical location by measuring the angular dependence of the intensity of the scattered light. Nevertheless, this is typically only possible for dilute colloidal dispersions where multiple scattering effects are not appreciable. If the size of the particles within a colloidal dispersion does not change throughout the course of an experiment, then the PSD can be estimated from the creaming profile, for example, using Stokes' law. Optical profiling has found widespread utilization within the food industry for the characterization of gravitational separation in colloidal dispersions because analytical instruments based on this principle are commercially available (Mengual et al. 1999a,b). These instruments are capable of rapid, quantitative, and reproducible measurements and can often detect instability in a colloidal dispersion before it is visible to the human eye.

There are also commercial instruments available that measure changes in the particle concentration with sample height based on applying a centrifugal force. The speed of particle movement is accelerated by applying a well-defined centrifugal force. These accelerated optical profiling methods may therefore be useful for predicting the long-term stability of colloidal dispersions from measurements made over relatively short times, provided that the accelerated tests correlate with the long-term stability tests (see below).

A number of other nondestructive methods have also been developed to provide information about particle concentration profiles in colloidal dispersions without disturbing the sample, for example, electrical conductivity, ultrasound, and NMR (McClements 2005; Pal 1994; Robins et al. 2002). Particle concentration profiles can be determined by inserting electrodes into a colloidal dispersion and measuring changes in electrical conductivity at different heights. A theoretical model or empirical calibration curve can then be used to convert the electrical conductivity at a particular height into a particle concentration (Bury et al. 1995; Gundersen et al. 2001).

Ultrasonic profiling devices are very similar to optical profile devices, except that ultrasonic waves are propagated through a colloidal dispersion rather than light waves (Basaran et al. 1998; Dickinson and Golding 1998; Pinfield et al. 1994, 1996). The ultrasonic velocity and attenuation are measured as a function of sample height by scanning an ultrasonic transducer up and down the colloidal dispersion. These ultrasonic measurements can then be converted into a particle concentration (and sometimes a particle size also) as a function of sample height.

Nondestructive NMR imaging techniques have also been used to monitor gravitational separation in colloidal dispersions (McDonald et al. 1999; Newling et al. 1997; Simoneau et al. 1993). NMR techniques are based on differences in the relaxation times of different components within a colloidal dispersion after a radiofrequency pulse is applied. NMR imaging enables one to obtain a three-dimensional image of the particle concentration within a concentrated system without the need for dilution, but they are expensive to purchase and require highly skilled operators, which has limited their application. Nevertheless, NMR imaging techniques do provide some exciting possibilities for analyzing the biological fate of colloidal delivery systems after ingestion, for example, monitoring the location of fat droplets in the stomach after consumption of a fatty food (Marciani et al. 2001, 2003, 2004).

### Accelerated Stability Tests

It is often important to be able to predict the long-term stability of a colloidal dispersion to gravitational separation from measurements made over relatively short times, so as to avoid waiting for a long time before a product defect is detected. If a defect is detected early, then it is often possible to change the ingredients or processing operations to rectify the problem before too much product is lost. The stability of a colloidal dispersion to gravitational separation can often be rapidly assessed by centrifuging it at a fixed speed for a certain length of time (Sherman 1995). The extent of gravitational separation can then be measured after centrifugation using one of the analytical methods outlined above, for example, visual inspection, physical sectioning, or concentration profiling. Accelerated tests are useful for predicting the long-term storage stability of colloidal dispersions provided the particle size does not change during storage and the continuous phase has a Newtonian viscosity. Nevertheless, the use of accelerated tests for predicting the long-term stability of colloidal dispersions to gravitational separation should be treated with caution because the factors that determine particle movement in a gravitational field may differ from those that are important in a centrifugal field. For example, the continuous phase may have a *yield stress* that is exceeded in a centrifuge, but which would never be exceeded under normal storage conditions. Alternatively, the mean particle size in a colloidal dispersion may change during storage (e.g., because of droplet coalescence, flocculation, or Ostwald ripening), which then promotes more rapid creaming. This kind of instability may not be observed when a sample is tested over a relatively short time using an accelerated test, but it may be observed during long-term storage. For this reason, it is always advisable to carefully compare the results of accelerated tests with those made during long-term storage before routinely using an accelerated test.

### 8.3.4.2 Particle Size Changes: Aggregation, Ostwald Ripening, and Dissociation

The size of the particles in a colloidal delivery system may change after it has been fabricated through a variety of physicochemical and physiological processes, such as flocculation, coalescence, partial coalescence, Ostwald ripening, swelling, fragmentation, erosion, and dissolution (Chapter 3). It is therefore important to be able to measure changes in particle size during the manufacture, storage, or utilization of colloidal delivery systems. The most common analytical methods used to measure the PSD and microstructure of colloidal dispersions were covered earlier in this chapter (Section 8.2).

### 8.3.4.3 Environmental Stress Tests

The colloidal delivery systems used in the food industry must be capable of functioning under a range of different environmental conditions, which are highly dependent on food product type. For example, there may be appreciable variations in pH, ionic strength, solution composition, temperature, and mechanical forces from product to product. It is therefore important to establish the range of environmental conditions under which a particular colloidal delivery system can successfully operate (McClements 2007). A number of environmental stress tests that can be used to establish the performance of colloidal delivery systems under different conditions are outlined below. Typically, a sample is prepared, subjected to one or more of these stresses, and then its stability is tested using one or more of the methods described previously (e.g., microstructure, particle size, and gravitational separation).

- *Extended Storage.* Many colloidal delivery systems must remain stable for relatively long periods before they are utilized. The resistance of a delivery system to long-term storage can be assessed by keeping it in a temperature-, humidity-, or light-controlled environment for a certain time and then testing its properties. The precise conditions used will depend on the expected shelf life and storage conditions of the final product (e.g., 6 months at 30°C in a dark room).

- *Mechanical Forces.* Many colloidal delivery systems must remain stable when they are subjected to mechanical forces, for example, flow through a pipe, mixing, stirring, or vibration during transport. The resistance of delivery systems to mechanical forces can be determined by placing them in containers and then subjecting them to a well-defined mechanical force for a specified period. For example, delivery systems may be placed in sealed containers and then rotated at a fixed speed in a temperature-controlled shaker, for example, 120 rpm at 30°C for 1 week. Alternatively, samples may be placed in a high-speed blender and sheared at a fixed speed for a certain time, for example, 1000 rpm at 30°C for 1 min. The device used will depend on the type of mechanical forces that one is trying to simulate.

- *Freeze–Thaw Cycling.* The ability of delivery systems to remain stable during freezing and thawing is important for certain types of applications, for example, frozen foods. The resistance of delivery systems to freeze–thaw cycling can be established by placing a sample in a freezer at a fixed temperature and time (e.g., –20°C for 22 h) and then thawing it at ambient temperature (e.g., +20°C for 2 h) prior to testing. This procedure can be repeated a number of times to establish the freeze–thaw stability, with the samples being tested after each cycle. In some cases, it is more appropriate to assess the influence of freezing rate on stability, for example, fast, intermediate, or slow cooling.

- *Thermal Processing.* Many colloidal delivery systems are subjected to some kind of thermal processing during their manufacture or utilization, for example, sterilization, pasteurization, or cooking. The stability of delivery systems to thermal processing can be tested by placing them in a temperature-controlled water bath at different temperatures (e.g., 30°C to 100°C for 20 min) or different holding times (e.g., 90°C for 0 to 30 min). Alternatively, one can simulate a specific time–temperature profile that a food or beverage product may experience, such as sterilization (e.g., 30 min at 110°C) or pasteurization (e.g., 72°C for 16 s). The samples can then be cooled to ambient temperature and tested.

- *Dehydration.* Some delivery systems are dehydrated during the manufacturing process to increase their shelf life, facilitate transport, and improve convenience. The resistance of materials to dehydration can be assessed by drying them, storing them for a fixed period under standardized conditions (e.g., light exposure, oxygen content, relative humidity, temperature), and then testing them after they have been rehydrated.
- *pH and Minerals.* Food products vary considerably in the pH and mineral composition of the aqueous phase, which may have a pronounced impact on their stability. The resistance of colloidal delivery systems to these effects can be characterized by preparing samples with a range of pH values (e.g., 3–8) and mineral concentrations (e.g., 0–500 mM NaCl or 0–100 mM $CaCl_2$) and then testing their properties after they have been stored for a fixed period (e.g., 24 h or 1 week at 30°C).

If a delivery system is going to be used in a particular food or beverage product, then it is important to be sure that it is capable of functioning over the range of environmental stresses that it will encounter in this product. In addition, there may be other environmental stress tests that are important for certain products, for example, UV-visible light exposure, chilling, vibration, aeration, microwave, or high-pressure treatment.

## 8.4 Protection, Retention, and Release Characteristics

An overview of the analytical methods that can be used to measure the protection, retention, and release of active ingredients within colloidal delivery systems is given in this section. Changes in the stability or location of active ingredients within colloidal delivery systems may be measured as a function of time or as a function of exposure to specific environmental conditions (such as pH, ionic strength, temperature, or enzyme activity). The release of active ingredients from colloidal delivery systems after exposure to gastrointestinal tract (GIT) conditions is discussed in Section 8.5. In this section, the focus is on general methods for measuring the chemical stability and retention/release of active ingredients.

### 8.4.1 Protection

A number of active ingredients utilized within the food and other industries are chemically unstable and therefore have a tendency to degrade within a product (e.g., during preparation, transport, or storage) or after a product has been ingested (e.g., during passage through the GIT). Carotenoids, such as β-carotene and lycopene, are natural colorants and nutraceuticals that are highly susceptible to chemical degradation caused by oxidation, leading to color fading and loss of biological activity (Boon et al. 2010). Polyunsaturated fatty acids, such as ω-3 fatty acids, are also highly unstable to lipid oxidation, leading to the formation of undesirable off-flavors, loss of biological activity, and production of potentially toxic reaction products (McClements and Decker 2000; Waraho et al. 2011). Some natural flavors, such as citral, are unstable to acid-catalyzed degradation, leading to loss of desirable flavor notes and formation of undesirable ones (Choi et al. 2009). Many proteins and peptides may be chemically degraded within the stomach because of the high acidity and the presence of digestive enzymes (Moreno 2007; Wickham et al. 2009). The development of effective strategies to inhibit the chemical degradation of active ingredients relies on the availability of appropriate analytical tools to monitor the rate and extent of their degradation. Numerous analytical tools are available to monitor changes in the chemical properties of active ingredients, with the most suitable for a particular application depending on the nature of the active ingredient, delivery system, and food matrix involved. These analytical tools include methods such as chemical, titrimetric, potentiometric, colorimetric, spectrophotometric, chromatography, mass spectrometry, scattering, and microscopy methods (Nielsen 2010). Some of the most widely used methods are spectrophotometric methods (such as UV-visible and fluorescence spectroscopy, infrared spectroscopy, atomic absorption/emission spectroscopy, nuclear magnetic resonance, and mass spectrometry) and chromatographic methods (such as thin-layer chromatography, GC, and HPLC). Often different types of methods are used in combination in order to improve the selectivity and sensitivity of the analysis. For example, a chromatography method may be used to separate a complex mixture of molecules, and then a spectrophotometric method is used to determine the chemical identity of the molecules.

In some cases, the analysis of an active ingredient can be carried out nondestructively and in situ, that is, without altering the properties of the delivery system or having to extract the active ingredient. For example, simple colorimetric methods can be used to nondestructively monitor chemical degradation of active ingredients that undergo color changes, for example, color fading of β-carotene (Qian et al. 2012a,b). In this case, changes in the tristimulus color coordinates (e.g., $L^*a^*b^*$ values) are measured as a function of time by reflecting light waves off the surface of the sample. Some fluorescence, infrared, and Raman spectroscopy methods can also be used to nondestructively monitor chemical degradation by reflecting electromagnetic waves from their surfaces. However, in many cases, analysis can only be carried out by altering the properties of the sample (e.g., drying) or after the active ingredient has been isolated from the delivery system (e.g., by centrifugation, filtration, or solvent extraction). The level of active ingredient remaining or of any degradation products formed can then be measured using a suitable analytical method. For example, the degradation of flavors (such as citral) can be measured using chromatographic methods once the flavors have been isolated from the delivery system by solvent extraction (Choi et al. 2009). Similarly, the degradation of polyunsaturated lipids and the formation of primary or secondary reaction products can be measured using spectrophotometric (e.g., UV-visible) and chromatographic (e.g., GC) methods once the lipids have been extracted from the delivery system by solvent extraction (Matalanis et al. 2012; Waraho et al. 2011). Typically, an appropriate analytical methodology has to be developed for each situation, which depends on the nature of the substance being analyzed and the nature of the surrounding food matrix.

## 8.4.2 Retention and Release

The amount of active ingredient retained within a colloidal particle during storage and then released at the appropriate site of action is important for many applications of colloidal delivery systems. It is therefore useful to have analytical methods that can measure the fraction of active ingredient present within the colloidal particles and the fraction released into the surrounding medium. There are a wide number of different analytical tools that can be used to achieve this goal, with the most appropriate one for a particular application depending on the nature of the active ingredient, delivery system, and food matrix involved. For the sake of convenience, the different methods available are classified according to the amount of sample preparation needed.

### 8.4.2.1 Nondestructive Methods

Ideally, an analytical method should be able to measure the relative concentration of an active ingredient inside and outside of the colloidal particles in situ and nondestructively. The retention or release profile of an active ingredient could then be monitored over time using the same sample without altering its properties. There are only a limited number of analytical methods capable of this type of measurement. Nuclear magnetic resonance methods are one of the most powerful methods of determining the fraction of compounds in different physical environments within multiphase materials and can be used to monitor release kinetics of certain types of active ingredients and food matrices (Heins et al. 2007; Hey and Alsagheer 1994; Mantle 2013). If a suitable spin probe is available, then electron paramagnetic resonance can be used to determine the relative concentrations of model active ingredients in different physical environments within colloidal systems (Berton-Carabin et al. 2012, 2013).

Front-face fluorescence methods can be used to provide in situ information about changes in the local environment of certain active ingredients within colloidal systems (Castelain and Genot 1996; Panya et al. 2012). Fluorescence spectroscopy can also be used to determine the release of nutraceuticals (such as vitamin D) from biopolymer complexes (Diarrassouba et al. 2013). In this case, the active ingredient must exhibit fluorescence and the fluorescent spectra must change when the local environment of the active ingredient is altered.

Fiber optic systems can be used for in situ measurement of the concentration of active ingredients that absorb UV, visible, or infrared radiation (Guillot et al. 2013; McFearin et al. 2011). In these instruments, a fiber optic probe is dipped into the sample to be analyzed and measures either transmission or absorption spectra, which can be used to determine the concentration of active ingredient in the immediate vicinity of the probe. Certain forms of microscopy enable one to determine the change in concentration of a substance as a function of distance and time, which enables release kinetics to be monitored, for example, fluorescence, Raman, or FTIR microscopy (Everall 2010; Mauricio-Iglesias et al. 2009; Perry et al. 2006). In cases where the particles in a delivery system sediment to the bottom of the container, it is possible to simply measure the release of an active ingredient in the supernatant using colorimetry or spectrophotometry (Janaswamy and Youngren 2012).

Ion-selective electrodes can be used to nondestructively measure the concentration of specific ions (such as $Na^+$ and $Ca^{2+}$) within aqueous solutions (Kotanen et al. 2012; Pflaum et al. 2013) and can therefore be used to monitor the release of these ions from colloidal particles. Electrical conductivity measurements can also be used to measure the release of ions from colloidal particles into the surrounding aqueous phase, for example, from the internal aqueous phase of W/O/W emulsions (Lutz et al. 2009).

## 8.4.2.2 Destructive Methods

In many cases, the amount of active ingredient retained or released by a colloidal delivery system can only be ascertained by isolating the particles from the surrounding matrix. The concentration of active ingredient in the colloidal particles or matrix can then be measured using appropriate analytical tools. A number of approaches can be used to isolate the colloidal particles from the matrix, including dehydration, gravitational separation, centrifugation, filtration, and dialysis (Nielsen 2010; Stockmann and Schwarz 1999). In cases where the matrix surrounding the colloidal particles is a simple fluid (such as an aqueous solution), it is often possible to directly measure the concentration of active ingredient present within it after the particles have been removed. For example, colorimetry or spectrophotometry could be used to detect a substance in solution that absorbs UV-visible radiation, whereas ion-selective electrodes could be used to detect the presence of specific mineral ions (such as sodium or calcium). In other cases, it is necessary to further process the matrix containing the active ingredient before it can be analyzed, for example, using concentration, evaporation, or solvent extraction methods (Nielsen 2010). Organic solvents are commonly used to isolate lipophilic active ingredients from delivery systems or food matrices prior to analysis by chromatographic or spectrophotometric methods. Rotary evaporation is often used to concentrate the amount of active ingredient present in a sample before analysis by removing water or organic solvent. If the active ingredient is a mineral (such as sodium or calcium), then the colloidal delivery system or food matrix may need to be converted into an ash before it can be measured by analytical methods such as atomic absorption or emission spectroscopy (Nielsen 2010).

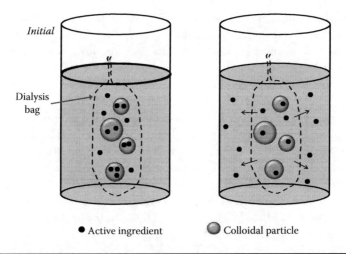

*Initial*

Dialysis bag

● Active ingredient  ◉ Colloidal particle

**Figure 8.17   The release of an active ingredient from a colloidal delivery system can be measured using a dialysis bag or other cell containing a semipermeable membrane.**

One simple approach to measuring the release of an active ingredient that has some water solubility from a colloidal delivery systems is to use dialysis (Baker 1987; Moreno-Bautista and Tam 2011). The delivery system is placed within a dialysis bag that has pores smaller than the colloidal particles but larger than the active ingredient (Figure 8.17). The concentration of active ingredient released from the colloidal particles and into the liquid surrounding the dialysis bag is then measured over time. Rather than using a simple dialysis bag, more specialized semipermeable membrane cells can also be used to carry out this type of experiment. In these experiments, it is usually important to carefully control the conditions in the system so that quantitative results about release kinetics can be obtained, for example, by ensuring that perfect sink conditions are maintained throughout the experiment. This often requires the use of flow through cells that maintain a high concentration gradient of active ingredient between the surfaces of the colloidal particles and the surrounding liquid (Baker 1987).

## 8.5  Biological Fate

The biological fate of colloidal delivery systems is one of their most important functional characteristics (McClements and Li 2010a,b; Porter and Charman 2001). A delivery system may be specifically designed to release an active ingredient at a specific site or at a controlled rate within the GIT, or it may be designed to increase its oral bioavailability. Even delivery systems that are primarily designed to encapsulate and protect bioactives within a food product during storage, or to mask undesirable flavor profiles of bioactives, should release the bioactives after ingestion so that they can be absorbed by the human body. In addition, it is possible that encapsulation of an ingredient may alter its potential toxicity (McClements and Rao 2011). It is therefore important to establish the biological fate of active ingredients after they have been consumed. There have been important developments in the establishment of analytical approaches for characterizing the biological fate of delivery systems in recent years. These approaches include in vitro screening tools, cell culture models, animal feeding studies, and human feeding studies that can broadly be divided into in vitro and in vivo methods. Numerous detailed review articles have been written on the analytical approaches available to characterize the potential biological fate of colloidal delivery systems (Butts et al. 2012; Fatouros and Mullertz 2008; Macfarlane and Macfarlane 2007; McClements and Li 2010a,b; Porter and Charman 2001; Wickham et al. 2009). In this section, a brief overview of the general approaches that can be used is given.

## 8.5.1 In Vitro Approaches

In vitro approaches are most suitable for rapidly screening different formulations during the early development of delivery systems, as well as to identify the key physicochemical attributes of the colloidal particles in delivery systems that determine their biological fate. A number of in vitro approaches have been developed to study the potential biological fate of delivery systems, which vary in their sophistication and accuracy. Some of these approaches focus on one particular region of the GIT (such as the small intestine or colon), whereas others are based on a number of sequential steps to more accurately mimic the entire GIT (Yoo and Chen 2006). Typically, a sample is prepared and then subjected to one or more treatments designed to simulate the human GIT, for example, mouth, stomach, small intestine, and colon. These treatments typically involve mixing the sample with simulated gastrointestinal fluids with specific compositions (e.g., pH, mineral composition, enzyme activity, etc.) under fluid flow conditions and incubation times that mimic specific regions of the GIT. The properties of the particles present in a colloidal delivery system are typically changed appreciably when they are exposed to different GIT environments. In addition, any encapsulated active ingredients may be retained, released, or chemically modified in different GIT regions. It is therefore important to have analytical tools and protocols to measure both the fate of the delivery system and the active ingredients they contain.

### 8.5.1.1 Passage through GIT

The development of a suitable testing protocol requires an understanding of the influence of different GIT regions on the properties of a delivery system (McClements and Li 2010a,b; van Aken 2010). For this reason, the passage of a delivery system through the GIT is often simulated by sequentially exposing it to a series of artificial GIT fluids with different compositions at body temperature, that is, ≈37°C (Figure 8.18).

*Mouth*: The sample containing the delivery system to be tested is mixed with simulated saliva fluid (SSF) for a specified time (e.g., a few seconds to a few minutes depending on food type) at a specified speed (e.g., 60 s$^{-1}$). In the case of semisolid or solid foods, it may also be necessary to have some kind of mechanical device to simulate the mastication process. Typically, the SSF is a neutral aqueous solution (pH ≈ 7) that contains various components that simulate human saliva, such as acids, buffers, minerals, biopolymers (mucin), and enzymes (amylase). A volume ratio of 1:1 of the sample to SSF is often used to simulate dilution of the sample with saliva in the mouth. The material collected at the end of the oral processing stage is often referred to as the "bolus."

*Stomach*: The bolus sample is mixed with simulated gastric fluid (SGF) for a specified time (e.g., 2 h) at a specified speed (e.g., 60 s$^{-1}$). For semisolid or solid foods, it may again be necessary to apply some kind of mechanical device to simulate complex mechanical forces and fluid flows associated with gastric motility (Ferrua et al. 2011; Kong and Singh 2010). Typically, the SGF is a highly acidic aqueous solution (pH ≈ 2) with a composition that simulates gastric conditions, for example, acids, buffers, salts, and digestive enzymes (proteases and lipases). A volume ratio of 1:1 of the bolus sample to SGF is often used to simulate dilution in gastric juices. The material collected at the end of the gastric processing stage is often referred to as "chyme."

*Small Intestine*: The chyme is mixed with simulated small intestinal fluid (SSIF) for a specified time (e.g., 2 h) with controlled agitation (e.g., shearing at 60 s$^{-1}$). Typically,

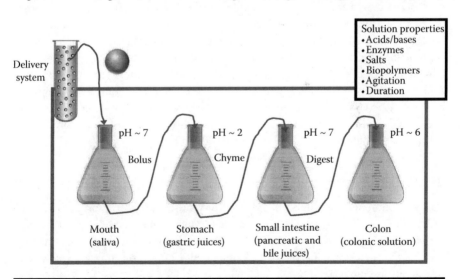

**Figure 8.18 Schematic diagram of a multiple-step in vitro digestion model to simulate the GIT. A sample is passed through a series of fluids that simulate the mouth, stomach, small intestine, and colon.**

the SSIF is a neutral aqueous solution (pH ≈ 7) containing various components that simulate small intestine composition, such as bile salts, phospholipids, bases, buffers, mineral salts, and digestive enzymes (e.g., amylases, proteases, and lipases). A volume ratio of 1:1 of the chyme to SSIF is often used to simulate dilution in small intestinal fluids. The material collected at the end of the small intestinal processing stage is often referred to as the "digest." The digest can be passed on to a simulated colon stage, or it can be used to analyze the release and absorption of bioactive components (Figure 8.18).

*Colon*: The digest is typically incubated in simulated colonic fluids (SCFs) for a specified time and under agitation conditions. However, the colon is one of the most difficult parts of the GIT to simulate in the laboratory because it requires the cultivation of colonic bacteria under anaerobic conditions. In vitro testing methods designed to simulate the human large intestine range from simple static batch microbial cultures to multiple-stage continuous cultures (Macfarlane and Macfarlane 2007; Ouwehand et al. 2009; Rummey and Rowland 1992). A sample is typically incubated in one or more SCFs that contain populations of bacteria representative of those normally found in the human colon. These bacteria are typically cultivated from animal or human feces. One difficulty in accurately simulating the human colon is the considerable variations in bacterial populations that exist between individuals. Rather than using bacteria, it is possible to formulate SCFs that contain a mixture of enzymes typically produced by colonic bacteria, for example, glycosidases to degrade dietary fibers and proteases to degrade proteins (Souto-Maior et al. 2009; Yang 2008). Because of the difficulties in setting up and maintaining in vitro colonic models, many researchers prefer to go directly to animal models (Macfarlane and Macfarlane 2007). Alternatively, if the active ingredient encapsulated within a delivery system is fully absorbed within the small intestine, then this step may be omitted.

Researchers can establish their own in vitro GIT models in the laboratory using appropriate reagents (such as acids, bases, buffers, minerals, bile salts, phospholipids, biopolymers, and enzymes) and equipment (such as glassware, stirrers, temperature control units, pH meters, and titration units). Alternatively, a number of companies have developed sophisticated analytical instruments specifically designed to simulate the GIT process, for example, the TIM system from TNO Quality of Life, the Netherlands (Minekus et al. 1999; Venema et al. 2009).

### 8.5.1.2 Absorption

Various components within a colloidal delivery system may be absorbed from GIT fluids: intact colloidal particles, digestion products from colloidal particles, and active ingredients. Absorption typically occurs at the epithelium cells located in the walls of the small intestine and can be simulated using various physical, biological, or cell culture models.

*Physical Models*: For some components, it is assumed that their concentration within the SSIF in a soluble form is representative of the amount absorbed by the epithelium cells; that is, the rate-limiting step is solubility, rather than permeability.

The amount of active ingredients solubilized within the SSIF can be determined by measuring their aqueous phase concentration, either directly or after they have been isolated from the undigested material by centrifugation, filtration, or dialysis (Christensen et al. 2004; Do Thi et al. 2009). Sometimes it is convenient to place the sample being analyzed within a dialysis bag or cell with a suitable pore size (to prevent undigested material from exiting) and then measuring the amount of released material that moves across the semipermeable membrane as a function of time using an appropriate method (Figure 8.17).

*Ex Vivo Permeation Methods*: In these methods, a section of the GIT is cut from an animal after it has been sacrificed and is then washed to remove residual components. Part of the intestine is then clamped between two chambers: one of which contains the sample to be analyzed (donor chamber), while the other contains only buffer solution (receiver chamber) (Dahan and Hoffman 2007). The transport of colloidal particles, digestion products, or active ingredients across the chamber is then measured over time using suitable analytical methods. Alternatively, the sample to be tested can be placed inside an intact section of GIT, which is then placed in an appropriate buffer solution. The amount of bioactive component or digestion product that moves across the intestinal walls and into the surrounding buffer solution is measured using an appropriate analytical method.

*Cell Culture Methods*: Cell culture models offer a convenient means of simulating the epithelium cells coating the inner surface of the small intestine, where the absorption of bioactive components and digestion products normally occurs. One of the most commonly used cell culture models are Caco-2 cells (Dhuique-Mayer et al. 2007; Reboul et al. 2006; Versantvoort et al. 2002, 2003). A layer of Caco-2 cells is normally grown directly on the surface of a plastic plate or on a semipermeable membrane that is placed in the plastic plate (Figure 8.19). An aqueous solution of the sample (which may have previously been exposed to simulated GIT

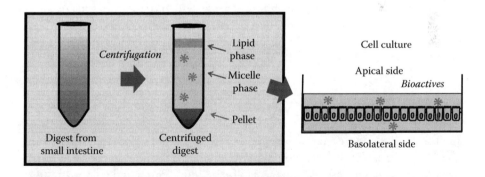

**Figure 8.19 Schematic diagram of the method used to measure the bioaccessibility and absorption of bioactive components. The sample is centrifuged after the small intestine phase and the micelle phase are placed on top of the cells in the cell culture model.**

conditions) is placed in contact with the surface of the cells, and the amount of the component of interest that is absorbed by the cells or that passes through the layer of cells is measured. A suitable analytical technique is used to measure the location or quantity of absorbed material, for example, microscopy, chromatography, spectrometry, electrophoresis, or chemical methods. An example of a microscopy approach to study the formation of chylomicrons in Caco-2 cells after exposure to lipid digestion products is shown in Figure 8.5.

### 8.5.2 In Vivo Approaches

In vitro methods are useful for rapidly screening many different samples and for establishing physicochemical mechanisms (McClements and Li 2010a,b). However, they are insufficiently sophisticated to mimic all of the complex processes that occur within the human GIT. A more realistic indication of the potential performance of a delivery system can be obtained using in vivo studies with animals or humans (McClements et al. 2009a,b). A number of different kinds of information can be collected from animal feeding studies. An animal may be fed the delivery system being tested, either alone or as part of a diet, over a specified period. Changes in the whole animal (such as behavior or body weight) or changes in specific internal organs (such as liver, pancreas, kidneys, spleen, lungs, heart, brain, subcutaneous fats, and GIT) may be measured, typically after the animal is sacrificed. Alternatively, changes in the composition of the feces, urine, blood, or breath of an animal may be measured to monitor the absorption, metabolism, or excretion of a particular component. In some cases, it is possible to use invasive tubes to collect samples of delivery systems as they pass through the digestive tract of live animals. Alternatively, whole-body imaging methods, such as those based on magnetic resonance, ultrasound, x-ray, or fluorescence, can be used to visualize the processes occurring within the GIT as a food passes through. The in vivo approach typically provides a more accurate picture of how delivery systems perform in practice, but they have many limitations since they are expensive, time consuming, and have ethical and legal implications. In addition, when using animals, it is important that their GIT behaves in a similar fashion to the human GIT to obtain meaningful results.

### 8.5.3 In Vitro versus In Vivo Correlations

In vitro studies offer several advantages over in vivo studies, because they are usually faster, less expensive, and more versatile, and provide more details about physicochemical mechanisms (Dahan and Hoffman 2006; Porter and Charman 2001; Yoo and Chen 2006). Nevertheless, it is extremely difficult to accurately mimic the complex physicochemical and physiological processes that occur within the human GIT. For this reason, it is often useful to use a combination of in vitro studies and in vivo studies when developing an appropriate delivery system. The in vitro method is used to screen a range of initial samples, and then the in vivo method is used to test those candidates that look the most promising. If this approach is used, it is important to establish robust in vitro–in vivo correlations to ensure that any in vitro method used to test a sample gives a reliable prediction of its performance in vivo (Dahan and Hoffman 2008; Porter and Charman 2001; Pouton and Porter 2006). Typically, the rate or extent of absorption of an active ingredient is measured for similar test samples using an in vitro method and an in vivo method, and then the results are correlated to one another. It is also useful to have computational models that can predict the real-life performance of delivery systems on the basis of knowledge of their composition and structure, as well as physiological conditions (Dressman et al. 2011; Mathias and Crison 2012).

## 8.5.4 Measurement of Changes in Delivery System Properties

The composition, structure, or physicochemical properties of delivery systems can be measured at specific points in actual or simulated GIT environments to establish the major factors influencing their performance. Some of the key factors that may be measured are highlighted below:

*Particle Composition*: The chemical composition of a particle may change as it passes through the GIT for a number of reasons: enzyme or chemical degradation of specific components, dissociation of specific structures, and exchange of molecules with the surrounding fluids. In some cases, it is possible to isolate particles from GIT fluids and measure their composition using traditional methods, such as chemical methods, chromatography, or spectroscopy. Alternatively, the same techniques can be used to measure changes in the chemical composition of the surrounding fluids, for example, to determine accumulation of digestion products or other components released from the particles. Fluorescence microscopy is a valuable tool for providing qualitative information about changes in particle composition in situ by tagging specific components with fluorescent dyes and measuring their location throughout the GIT. Alternatively, indirect information about particle composition can be obtained by measuring particle properties such as the electrical charge (which provides information about changes in interfacial composition).

*Particle Dimensions and Structural Organization*: The size and structure of the particles in a delivery system may change appreciably as they pass through the GIT owing to various physicochemical and physiological processes: flocculation, coalescence, erosion, fragmentation, dissolution, and digestion. Information about particle size can be obtained using the various particle sizing technologies discussed earlier in this chapter, such as static and dynamic light scattering. Nevertheless, the particles may have to be isolated from the GIT fluids first to avoid interference from other particulate matter within them. Information about the structure and spatial organization of the particles in a delivery system can be obtained using the microscopy methods discussed earlier, such as optical and electron microscopy. Confocal fluorescence microscopy is particularly suitable for this purpose because different molecules within the delivery system can be tagged to ascertain changes in their locations (Figure 8.20).

*Physical State*: The physical state of various components in a delivery system may change within the GIT, such as melting or crystallization of lipids, glass–rubbery phase transitions of biopolymer matrices, and dissolution of lipid, sugar, or salt crystals. A number of analytical techniques that can be used to provide information about the physical state of specific food components are available (Section 8.2). Polarized light microscopy is particularly suitable for providing structural information about the presence of crystalline or liquid crystalline material within a sample. X-ray diffraction and infrared methods can be used to provide information about the packing of molecules within crystals. DSC can be used to provide information about crystal type and amount and about the temperatures for melting, crystallization, and polymorphic transitions.

*Bioactive Release*: The release of an active ingredient from a colloidal delivery system within the GIT can also be established using appropriate analytical techniques. If the active ingredient can be fluorescently labeled, then its release can be directly monitored in situ using fluorescent microscopy. Other types of chemical microscopy can also be used for this purpose such as infrared or Raman microscopy. Alternatively, it may be possible to measure the type and concentration of specific molecules present within the GIT fluids surrounding the particles over time. In some cases, the particles may have to be isolated from the GIT fluids first, whereas in other cases, it may be possible to measure chemical composition directly. Analytical techniques such as chemical analysis, spectroscopy, chromatography, and electrophoresis can be used depending on the nature of the component being analyzed (Nielsen 2010).

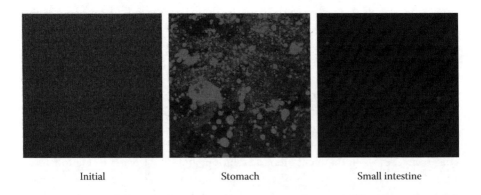

|  |  |  |
|:---:|:---:|:---:|
| Initial | Stomach | Small intestine |

**Figure 8.20  Confocal fluorescent microscopy images of changes in the structure of samples collected from different locations of a rat's GIT after feeding an emulsion-based delivery system. The lipid phase appears red in these diagrams as a result of staining with a fluorescent dye. (Images kindly provided by Yan Li.)**

## 8.6 Summary

This chapter has highlighted some of the most important analytical tools and experimental protocols that can be used to characterize the properties of colloidal delivery systems. Analysis is needed to determine the initial characteristics of a delivery system; to monitor any changes in its properties during storage, transport, and utilization; and to establish its potential biological fate. Numerous analytical tools have therefore been developed to characterize colloidal dispersions, and many of these can be purchased commercially. These instruments are often fully automated, user friendly, and capable of rapid measurements, which has greatly increased their utilization for the characterization of colloidal dispersions. Nevertheless, it is always important to be familiar with the operating principles of an instrument. This will enable one to be sure that the instrument is suitable for the sample of interest, that the sample is prepared appropriately for analysis, and that any theoretical models used to interpret the data are used correctly. It is always important to remember that any analytical instrument will only give reliable data if it is used properly.

## References

Aguilera, J. M., D. W. Stanley and G. V. Barbosa-Cánovas (1999). *Microstructural Principles of Food Processing Engineering*. Gaithersburg, MD, Aspen Publishers.

Alexander, M. and D. G. Dalgleish (2006). "Dynamic light scattering techniques and their applications in food science." *Food Biophysics* **1**(1): 2–13.

Arancibia, C., L. Jublot, E. Costell and S. Bayarri (2011). "Flavor release and sensory characteristics of o/w emulsions. Influence of composition, microstructure, and rheological behavior." *Food Research International* **44**(6): 1632–1641.

Awad, T. S., H. A. Moharram, O. E. Shaltout, D. Asker and M. M. Youssef (2012). "Applications of ultrasound in analysis, processing and quality control of food: A review." *Food Research International* **48**(2): 410–427.

Baker, R. W. (1987). *Controlled Release of Biologically Active Agents*. New York, John Wiley and Sons.

Balasubramanian, S. and S. Panigrahi (2011). "Solid-phase microextraction (SPME) techniques for quality characterization of food products: A review." *Food and Bioprocess Technology* **4**(1): 1–26.

Baldwin, E. A., J. H. Bai, A. Plotto and S. Dea (2011). "Electronic noses and tongues: Applications for the food and pharmaceutical industries." *Sensors* **11**(5): 4744–4766.

Barnes, H. A. (1994). "Rheology of emulsions—A review." *Colloids and Surfaces A-Physicochemical and Engineering Aspects* **91**: 89–95.

Basaran, T. K., K. Demetriades and D. J. McClements (1998). "Ultrasonic imaging of gravitational separation in emulsions." *Colloids and Surfaces A-Physicochemical and Engineering Aspects* **136**(1–2): 169–181.

Berton-Carabin, C. C., J. N. Coupland, C. Qian, D. J. McClements and R. J. Elias (2012). "Reactivity of a lipophilic ingredient solubilized in anionic or cationic surfactant micelles." *Colloids and Surfaces A-Physicochemical and Engineering Aspects* **412**: 135–142.

Berton-Carabin, C. C., R. J. Elias and J. N. Coupland (2013). "Reactivity of a model lipo-philic ingredient in surfactant-stabilized emulsions: Effect of droplet surface charge and ingredient location." *Colloids and Surfaces A-Physicochemical and Engineering Aspects* **418**: 68–75.

Bigiani, A. and S. Prandi (2011). "Functional diversity of taste cells. A review." *Flavour and Fragrance Journal* **26**(4): 214–217.

Boon, C. S., D. J. McClements, J. Weiss and E. A. Decker (2010). "Factors influencing the chemical stability of carotenoids in foods." *Critical Reviews in Food Science and Nutrition* **50**(6): 515–532.

Bourne, M. C. (1997). *Food Texture and Viscosity: Concept and Measurement.* New York, Academic Press.

Bourne, M. C. (2002). *Food Texture and Viscosity: Concept and Measurement.* London, Academic Press.

Bunjes, H., F. Steiniger and W. Richter (2007). "Visualizing the structure of triglyceride nanoparticles in different crystal modifications." *Langmuir* **23**(7): 4005–4011.

Bury, M., J. Gerhards, W. Erni and A. Stamm (1995). "Application of a new method based on conductivity measurements to determine the creaming stability of O/W emul-sions." *International Journal of Pharmaceutics* **124**(2): 183–194.

Butts, C. A., J. A. Monro and P. J. Moughan (2012). "In vitro determination of dietary protein and amino acid digestibility for humans." *British Journal of Nutrition* **108**: S282–S287.

Byelov, D. V., J. M. Meijer, I. Snigireva, A. Snigirev, L. Rossi, E. van den Pol and A. Kuijk (2013). "In situ hard x-ray microscopy of self-assembly in colloidal suspensions." *RSC Advances* **3**(36): 15670–15677.

Castelain, C. and C. Genot (1996). "Partition of adsorbed and nonadsorbed bovine serum albumin in dodecane-in-water emulsions calculated from front-face intrin-sic fluorescence measurements." *Journal of Agricultural and Food Chemistry* **44**(7): 1635–1640.

Chanamai, R. and D. J. McClements (2000). "Creaming stability of flocculated monodis-perse oil-in-water emulsions." *Journal of Colloid and Interface Science* **225**(1): 214–218.

Chantrapornchai, W., F. Clydesdale and D. J. McClements (1999). "Theoretical and experi-mental study of spectral reflectance and color of concentrated oil-in-water emulsions." *Journal of Colloid and Interface Science* **218**(1): 324–330.

Chantrapornchai, W., F. M. Clydesdale and D. J. McClements (2001). "Influence of floc-culation on optical properties of emulsions." *Journal of Food Science* **66**(3): 464–469.

Choi, S. J., E. A. Decker, L. Henson, L. M. Popplewell and D. J. McClements (2009). "Stability of citral in oil-in-water emulsions prepared with medium-chain triacylglyc-erols and triacetin." *Journal of Agricultural and Food Chemistry* **57**(23): 11349–11353.

Christensen, J. O., K. Schultz, B. Mollgaard, H. G. Kristensen and A. Mullertz (2004). "Solubilisation of poorly water-soluble drugs during in vitro lipolysis of medium- and long-chain triacylglycerols." *European Journal of Pharmaceutical Sciences* **23**(3): 287–296.

Chung, C., B. Degner and D. J. McClements (2012). "Instrumental mastication assay for texture assessment of semi-solid foods: Combined cyclic squeezing flow and shear vis-cometry." *Food Research International* **49**(1): 161–169.

Chung, C., K. Olson, B. Degner and D. J. McClements (2013). "Textural properties of model food sauces: Correlation between simulated mastication and sensory evaluation methods." *Food Research International* **51**(1): 310–320.

Cialla, D., A. Marz, R. Bohme, F. Theil, K. Weber, M. Schmitt and J. Popp (2012). "Surface-enhanced Raman spectroscopy (SERS): Progress and trends." *Analytical and Bioanalytical Chemistry* **403**(1): 27–54.

Dahan, A. and A. Hoffman (2006). "Use of a dynamic in vitro lipolysis model to rationalize oral formulation development for poor water soluble drugs: Correlation with in vivo data and the relationship to intra-enterocyte processes in rats." *Pharmaceutical Research* **23**(9): 2165–2174.

Dahan, A. and A. Hoffman (2007). "The effect of different lipid based formulations on the oral absorption of lipophilic drugs: The ability of in vitro lipolysis and consecutive ex vivo intestinal permeability data to predict in vivo bioavailability in rats." *European Journal of Pharmaceutics and Biopharmaceutics* **67**(1): 96–105.

Dahan, A. and A. Hoffman (2008). "Rationalizing the selection of oral lipid based drug delivery systems by an in vitro dynamic lipolysis model for improved oral bioavailability of poorly water soluble drugs." *Journal of Controlled Release* **129**(1): 1–10.

Danino, D. (2012). "Cryo-TEM of soft molecular assemblies." *Current Opinion in Colloid and Interface Science* **17**(6): 316–329.

Deisingh, A. K., D. C. Stone and M. Thompson (2004). "Applications of electronic noses and tongues in food analysis." *International Journal of Food Science and Technology* **39**(6): 587–604.

Deleris, I., A. Saint-Eve, E. Semon, H. Guillemin, E. Guichard, I. Souchon and J. L. Le Quere (2013). "Comparison of direct mass spectrometry methods for the on-line analysis of volatile compounds in foods." *Journal of Mass Spectrometry* **48**(5): 594–607.

Dhuique-Mayer, C., P. Borel, E. Reboul, B. Caporiccio, P. Besancon and M. J. Amiot (2007). "Beta-cryptoxanthin from citrus juices: Assessment of bioaccessibility using an in vitro digestion/Caco-2 cell culture model." *British Journal of Nutrition* **97**(5): 883–890.

Diarrassouba, F., G. Remondetto, L. Liang, G. Garrait, E. Beyssac and M. Subirade (2013). "Effects of gastrointestinal pH conditions on the stability of the beta-lactoglobulin/vitamin D-3 complex and on the solubility of vitamin D-3." *Food Research International* **52**(1): 515–521.

Dickinson, E. and M. Golding (1998). "Influence of alcohol on stability of oil-in-water emulsions containing sodium caseinate." *Journal of Colloid and Interface Science* **197**(1): 133–141.

Do Thi, T., M. Van Speybroeck, V. Barillaro, J. Martens, P. Annaert, P. Augustijns, J. Van Humbeeck, J. Vermant and G. Van den Mooter (2009). "Formulate-ability of ten compounds with different physicochemical profiles in SMEDDS." *European Journal of Pharmaceutical Sciences* **38**(5): 479–488.

Donald, A. M. (1998). "Environmental scanning electron microscopy for the study of 'wet' systems." *Current Opinion in Colloid and Interface Science* **3**(2): 143–147.

Donald, A. M., C. B. He, C. P. Royall, M. Sferrazza, N. A. Stelmashenko and B. L. Thiel (2000). "Applications of environmental scanning electron microscopy to colloidal aggregation and film formation." *Colloids and Surfaces A-Physicochemical and Engineering Aspects* **174**(1–2): 37–53.

Dressman, J. B., K. Thelen and S. Willmann (2011). "An update on computational oral absorption simulation." *Expert Opinion on Drug Metabolism and Toxicology* **7**(11): 1345–1364.

Dudkiewicz, A., K. Tiede, K. Loeschner, L. H. S. Jensen, E. Jensen, R. Wierzbicki, A. B. A. Boxall and K. Molhave (2011). "Characterization of nanomaterials in food by electron microscopy." *Trac-Trends in Analytical Chemistry* **30**(1): 28–43.

Egerton, R. (2008). *Physical Principles of Electron Microscopy: An Introduction to TEM, SEM, and AEM*. New York, Springer Science.

Escuder-Gilabert, L. and M. Peris (2010). "Review: Highlights in recent applications of electronic tongues in food analysis." *Analytica Chimica Acta* **665**(1): 15–25.

Everall, N. J. (2010). "Confocal Raman microscopy: Common errors and artefacts." *Analyst* **135**(10): 2512–2522.

Fatouros, D. G. and A. Mullertz (2008). "In vitro lipid digestion models in design of drug delivery systems for enhancing oral bioavailability." *Expert Opinion on Drug Metabolism and Toxicology* **4**(1): 65–76.

Ferrua, M. J., F. B. Kong and R. P. Singh (2011). "Computational modeling of gastric digestion and the role of food material properties." *Trends in Food Science and Technology* **22**(9): 480–491.

Genovese, D. B., J. E. Lozano and M. A. Rao (2007). "The rheology of colloidal and noncolloidal food dispersions." *Journal of Food Science* **72**(2): R11–R20.

Guillot, A., M. Limberger, J. Kramer and C. M. Lehr (2013). "In situ drug release monitoring with a fiber-optic system: Overcoming matrix interferences using derivative spectrophotometry." *Dissolution Technologies* **20**(2): 15–19.

Gundersen, S. A., O. Saether and J. Sjoblom (2001). "Salt effects on lignosulfonate and Kraft lignin stabilized O/W-emulsions studied by means of electrical conductivity and video-enhanced microscopy." *Colloids and Surfaces A-Physicochemical and Engineering Aspects* **186**(3): 141–153.

Hartel, R. W. (2013). Advances in food crystallization. In *Annual Review of Food Science and Technology*, M. P. Doyle and T. R. Klaenhammer (eds.). Palo Alto, Annual Reviews. **4**: 277–292.

Heins, A., T. Sokolowski, H. Stockmann and K. Schwarz (2007). "Investigating the location of propyl gallate at surfaces and its chemical microenvironment by H-1 NMR." *Lipids* **42**(6): 561–572.

Hertz, H. M., M. Bertilson, O. von Hofsten, S. C. Gleber, J. Sedlmair and J. Thieme (2012). "Laboratory X-ray microscopy for high-resolution imaging of environmental colloid structure." *Chemical Geology* **329**: 26–31.

Hey, M. J. and F. Alsagheer (1994). "Interphase transfer rates in emulsions studied by NMR-spectroscopy." *Langmuir* **10**(5): 1370–1376.

Hunter, R. J. (1986). *Foundations of Colloid Science*. Oxford, UK, Oxford University Press.

Hutchings, J. B. (1999). *Food Color and Appearance*. Gaithersburg, MD, Aspen Publishers.

Janaswamy, S. and S. R. Youngren (2012). "Hydrocolloid-based nutraceutical delivery systems." *Food and Function* **3**(5): 503–507.

Jelen, H. H., M. Majcher and M. Dziadas (2012). "Microextraction techniques in the analysis of food flavor compounds: A review." *Analytica Chimica Acta* **738**: 13–26.

Klang, V., N. B. Matsko, C. Valenta and F. Hofer (2012). "Electron microscopy of nanoemulsions: An essential tool for characterisation and stability assessment." *Micron* **43**(2–3): 85–103.

Klang, V., C. Valenta and N. B. Matsko (2013). "Electron microscopy of pharmaceutical systems." *Micron* **44**: 45–74.

Kong, F. and R. P. Singh (2010). "A human gastric simulator (HGS) to study food digestion in human stomach." *Journal of Food Science* **75**(9): E627–E635.

Kotanen, C. N., A. N. Wilson, A. M. Wilson, K. Ishihara and A. Guiseppi-Elie (2012). "Biomimetic hydrogels gate transport of calcium ions across cell culture inserts." *Biomedical Microdevices* **14**(3): 549–558.

Kuntsche, J., J. C. Horst and H. Bunjes (2011). "Cryogenic transmission electron microscopy (cryo-TEM) for studying the morphology of colloidal drug delivery systems." *International Journal of Pharmaceutics* **417**(1–2): 120–137.

Larson, R. G. (1999). *The Structure and Rheology of Complex Fluids.* Oxford, UK, Oxford University Press.

Le Reverend, B. J. D., I. T. Norton, P. W. Cox and F. Spyropoulos (2010). "Colloidal aspects of eating." *Current Opinion in Colloid and Interface Science* **15**: 84–89.

Lesmes, U. and D. J. McClements (2009). "Structure-function relationships to guide rational design and fabrication of particulate food delivery systems." *Trends in Food Science and Technology* **20**(10): 448–457.

Levin, I. W. and R. Bhargava (2005). Fourier transform infrared vibrational spectroscopic imaging: Integrating microscopy and molecular recognition. In *Annual Review of Physical Chemistry*, M. P. Doyle and T. R. Klaenhammer (eds.). Palo Alto, Annual Reviews. **56**: 429–474.

Lillford, P. (2000). "The materials science of eating and food breakdown." *MRS Bulletin* **25**: 38–43.

Linforth, R., M. Cabannes, L. Hewson, N. Yang and A. Taylor (2010). "Effect of fat content on flavor delivery during consumption: An in vivo model." *Journal of Agricultural and Food Chemistry* **58**(11): 6905–6911.

Linforth, R. S. T., K. S. K. Pearson and A. J. Taylor (2007). "In vivo flavor release from gelatin-sucrose gels containing droplets of flavor compounds." *Journal of Agricultural and Food Chemistry* **55**(19): 7859–7863.

Loren, N., M. Langton and A. M. Hermansson (2007). Confocal fluorescence microscopy for structure characterization. In *Understanding and Controlling the Microstructure of Complex Foods*, D. J. McClements (ed.). Cambridge, UK, Woodhead Publishing.

Lutz, R., A. Aserin, L. Wicker and N. Garti (2009). "Release of electrolytes from W/O/W double emulsions stabilized by a soluble complex of modified pectin and whey protein isolate." *Colloids and Surfaces B-Biointerfaces* **74**(1): 178–185.

Macfarlane, G. T. and S. Macfarlane (2007). "Models for intestinal fermentation: Association between food components, delivery systems, bioavailability and functional interactions in the gut." *Current Opinion in Biotechnology* **18**(2): 156–162.

Macosko, C. W. (1994). *Rheology: Principles, Measurements and Applications.* New York, VCH Publishers.

Madene, A., M. Jacquot, J. Scher and S. Desobry (2006). "Flavour encapsulation and controlled release—A review." *International Journal of Food Science and Technology* **41**(1): 1–21.

Mantle, M. D. (2013). "NMR and MRI studies of drug delivery systems." *Current Opinion in Colloid and Interface Science* **18**(3): 214–227.

Marciani, L., C. Ramanathan, D. J. Tyler, P. Young, P. Manoj, M. Wickham, A. Fillery-Travis, R. C. Spiller and P. A. Gowland (2001). "Fat emulsification measured using NMR transverse relaxation." *Journal of Magnetic Resonance* **153**(1): 1–6.

Marciani, L., M. Wickham, B. P. Hills, J. Wright, D. Bush, R. Faulks, A. Fillery-Travis, R. C. Spiller and P. A. Gowland (2004). "Intragastric oil-in-water emulsion fat fraction measured using inversion recovery echo-planar magnetic resonance imaging." *Journal of Food Science* **69**(6): E290–E296.

Marciani, L., M. Wickham, J. Wright, D. Bush, R. Faulks, A. Fillery-Travis, P. Gowland and R. C. Spiller (2003). "Magnetic resonance imaging (MRI) insights into how fat emulsion stability alters gastric emptying." *Gastroenterology* **124**(4): A581.

Matalanis, A., E. A. Decker and D. J. McClements (2012). "Inhibition of lipid oxidation by encapsulation of emulsion droplets within hydrogel microspheres." *Food Chemistry* **132**(2): 766–772.

Matalanis, A. and D. J. McClements (2013). "Hydrogel microspheres for encapsulation of lipophilic components: Optimization of fabrication and performance." *Food Hydrocolloids* **31**(1): 15–25.

Mathias, N. R. and J. Crison (2012). "The use of modeling tools to drive efficient oral product design." *AAPS Journal* **14**(3): 591–600.

Mauricio-Iglesias, M., V. Guillard, N. Gontard and S. Peyron (2009). "Application of FTIR and Raman microspectroscopy to the study of food/packaging interactions." *Food Additives and Contaminants Part A-Chemistry Analysis Control Exposure and Risk Assessment* **26**(11): 1515–1523.

McClements, D. J. (1997). "Ultrasonic characterization of foods and drinks: Principles, methods, and applications." *Critical Reviews in Food Science and Nutrition* **37**(1): 1–46.

McClements, D. J. (2002a). "Colloidal basis of emulsion color." *Current Opinion in Colloid and Interface Science* **7**(5–6): 451–455.

McClements, D. J. (2002b). "Theoretical prediction of emulsion color." *Advances in Colloid and Interface Science* **97**(1–3): 63–89.

McClements, D. J. (2005). *Food Emulsions: Principles, Practices, and Techniques.* Boca Raton, FL, CRC Press.

McClements, D. J. (2007). "Critical review of techniques and methodologies for characterization of emulsion stability." *Critical Reviews in Food Science and Nutrition* **47**(7): 611–649.

McClements, D. J. and E. A. Decker (2000). "Lipid oxidation in oil-in-water emulsions: Impact of molecular environment on chemical reactions in heterogeneous food systems." *Journal of Food Science* **65**(8): 1270–1282.

McClements, D. J., E. A. Decker and Y. Park (2009a). "Controlling lipid bioavailability through physicochemical and structural approaches." *Critical Reviews in Food Science and Nutrition* **49**(1): 48–67.

McClements, D. J., E. A. Decker, Y. Park and J. Weiss (2009b). "Structural design principles for delivery of bioactive components in nutraceuticals and functional foods." *Critical Reviews in Food Science and Nutrition* **49**(6): 577–606.

McClements, D. J. and Y. Li (2010a). "Review of in vitro digestion models for rapid screening of emulsion-based systems." *Food and Function* **1**(1): 32–59.

McClements, D. J. and Y. Li (2010b). "Structured emulsion-based delivery systems: Controlling the digestion and release of lipophilic food components." *Advances in Colloid and Interface Science* **159**(2): 213–228.

McClements, D. J. and J. Rao (2011). "Food-grade nanoemulsions: Formulation, fabrication, properties, performance, biological fate, and potential toxicity." *Critical Reviews in Food Science and Nutrition* **51**(4): 285–330.

McDonald, P. J., E. Ciampi, J. L. Keddie, M. Heidenreich and R. Kimmich (1999). "Magnetic-resonance determination of the spatial dependence of the droplet size distribution in the cream layer of oil-in-water emulsions: Evidence for the effects of depletion flocculation." *Physical Review E* **59**(1): 874–884.

McFearin, C. L., J. Sankaranarayanan and A. Almutairi (2011). "Application of fiber-optic attenuated total reflection-FT-IR methods for in situ characterization of protein delivery systems in real time." *Analytical Chemistry* **83**(10): 3943–3949.

Meilgaard, M., G. V. Civille and B. T. Carr (2006). *Sensory Evaluation Techniques.* Boca Raton, FL, CRC.

Mengual, O., G. Meunier, I. Cayre, K. Puech and P. Snabre (1999a). "Characterisation of instability of concentrated dispersions by a new optical analyser: The TURBISCAN MA 1000." *Colloids and Surfaces A-Physicochemical and Engineering Aspects* **152**(1–2): 111–123.

Mengual, O., G. Meunier, I. Cayre, K. Puech and P. Snabre (1999b). "TURBISCAN MA 2000: Multiple light scattering measurement for concentrated emulsion and suspension instability analysis." *Talanta* **50**(2): 445–456.

Mertz, J. (2009). *Introduction to Optical Microscopy.* Greenwood Village, CO, Roberts and Company Publishers.

Minekus, M., M. Smeets-Peeters, A. Bernalier, S. Marol-Bonnin, R. Havenaar, P. Marteau, M. Alric, G. Fonty and J. Veld (1999). "A computer-controlled system to simulate conditions of the large intestine with peristaltic mixing, water absorption and absorption of fermentation products." *Applied Microbiology and Biotechnology* **53**(1): 108–114.

Moreno, F. J. (2007). "Gastrointestinal digestion of food allergens: Effect on their allergenicity." *Biomedicine and Pharmacotherapy* **61**(1): 50–60.

Moreno-Bautista, G. and K. C. Tam (2011). "Evaluation of dialysis membrane process for quantifying the in vitro drug-release from colloidal drug carriers." *Colloids and Surfaces A-Physicochemical and Engineering Aspects* **389**(1–3): 299–303.

Morini, G., A. Bassoli and G. Borgonovo (2011). "Molecular modelling and models in the study of sweet and umami taste receptors. A review." *Flavour and Fragrance Journal* **26**(4): 254–259.

Morris, V. J., A. P. Gunning and A. R. Kirby (1999). *Atomic Force Microscopy for Biologists.* London, Imperial College Press.

Murphy, D. B. (2012). *Fundamentals of Light Microscopy and Electronic Imaging.* New York, John Wiley & Sons.

Newling, B., P. M. Glover, J. L. Keddie, D. M. Lane and P. J. McDonald (1997). "Concentration profiles in creaming oil-in-water emulsion layers determined with stray field magnetic resonance imaging." *Langmuir* **13**(14): 3621–3626.

Nielsen, S. S. (2010). *Food Analsysis.* New York, Springer.

Norton, I. T., F. Spyropoulos and P. Cox (2011). *Practical Food Rheology: An Interpretive Approach.* New York, Wiley-Blackwell.

Ouwehand, A. C., K. Tiihonen, H. Makelainen, N. Rautonen, O. Hasselwander and G. Sworn (2009). Non-starch polysaccharides in the gastrointestinal tract. In *Designing Functional Foods: Measuring and Controlling Food Structure Breakdown and Nutrient Absorption,* D. J. McClements and E. A. Decker (eds.). Boca Raton, FL, CRC Press: 126–147.

Ozeck, M., P. Brust, H. Xu and G. Servant (2004). "Receptors for bitter, sweet and umami taste couple to inhibitory G protein signaling pathways." *European Journal of Pharmacology* **489**(3): 139–149.

Pal, R. (1994). "Techniques for measuring the composition (oil and water content) of emulsions—A state of the art review." *Colloids and Surfaces A-Physicochemical and Engineering Aspects* **84**: 141–160.

Pal, R., Y. Yan and J. Masliyah (1992). "Rheology of emulsions." *Advances in Chemistry Series* (231): 131–170.

Panya, A., M. Laguerre, C. Bayrasy, J. Lecomte, P. Villeneuve, D. J. McClements and E. A. Decker (2012). "An investigation of the versatile antioxidant mechanisms of action of rosmarinate alkyl esters in oil-in-water emulsions." *Journal of Agricultural and Food Chemistry* **60**(10): 2692–2700.

Pawley, J. (2006). *Handbook of Biological Confocal Microscopy*. New York, Springer.

Perry, P. A., M. A. Fitzgerald and R. G. Gilbert (2006). "Fluorescence recovery after photobleaching as a probe of diffusion in starch systems." *Biomacromolecules* **7**(2): 521–530.

Pflaum, T., K. Konitzer, T. Hofmann and P. Kochler (2013). "Analytical and sensory studies on the release of sodium from wheat bread crumb." *Journal of Agricultural and Food Chemistry* **61**(26): 6485–6494.

Piggott, J. R., S. J. Simpson and S. A. R. Williams (1998). "Sensory analysis." *International Journal of Science and Technology* **33**: 7–12.

Pinfield, V. J., E. Dickinson and M. J. W. Povey (1994). "Modeling of concentration profiles and ultrasound velocity profiles in a creaming emulsion—Importance of scattering effects." *Journal of Colloid and Interface Science* **166**(2): 363–374.

Pinfield, V. J., M. J. W. Povey and E. Dickinson (1996). "Interpretation of ultrasound velocity creaming profiles." *Ultrasonics* **34**(6): 695–698.

Plucknett, K. P., S. J. Pomfret, V. Normand, D. Ferdinando, C. Veerman, W. J. Frith and I. T. Norton (2001). "Dynamic experimentation on the confocal laser scanning microscope: Application to soft-solid, composite food materials." *Journal of Microscopy-Oxford* **201**: 279–290.

Pomeranz, Y. and C. E. Meloan (1994). *Food Analysis*. New York, Chapman & Hall.

Porter, C. J. H. and W. N. Charman (2001). "In vitro assessment of oral lipid based formulations." *Advanced Drug Delivery Reviews* **50**: S127–S147.

Pouton, C. W. and C. J. H. Porter (2006). Formulation of lipid-based delivery systems for oral administration: Materials, methods and strategies. *Annual Meeting of the American-Association-of-Pharmaceutical-Scientists, San Antonio, TX*.

Qian, C., E. A. Decker, H. Xiao and D. J. McClements (2012a). "Inhibition of beta-carotene degradation in oil-in-water nanoemulsions: Influence of oil-soluble and water-soluble antioxidants." *Food Chemistry* **135**(3): 1036–1043.

Qian, C., E. A. Decker, H. Xiao and D. J. McClements (2012b). "Physical and chemical stability of beta-carotene-enriched nanoemulsions: Influence of pH, ionic strength, temperature, and emulsifier type." *Food Chemistry* **132**(3): 1221–1229.

Qian, C., E. A. Decker, H. Xiao and D. J. McClements (2012c). "Solid lipid nanoparticles: Effect of carrier oil and emulsifier type on phase behavior and physical stability." *Journal of the American Oil Chemists Society* **89**(1): 17–28.

Rao, M. A. (1999). *Rheology of Fluids and Semisolid Foods: Principles and Applications*. New York, Springer.

Reboul, E., M. Richelle, E. Perrot, C. Desmoulins-Malezet, V. Pirisi and P. Borel (2006). "Bioaccessibility of carotenoids and vitamin E from their main dietary sources." *Journal of Agricultural and Food Chemistry* **54**(23): 8749–8755.

Robins, M. M., A. D. Watson and P. J. Wilde (2002). "Emulsions—Creaming and rheology." *Current Opinion in Colloid and Interface Science* **7**(5–6): 419–425.

Rummey, C. J. and I. R. Rowland (1992). "In vivo and in vitro models of the human colonic flora." *Critical Reviews in Food Science and Nutrition* **31**: 299–331.

Russ, J. C. (2004). *Image Analysis of Food Microstructure*. Boca Raton, FL, CRC Press.

Russ, J. C. (2011). *The Image Processing Handbook*. Boca Raton, FL, CRC Press.

Sherman, P. (1995). "A critique of some methods proposed for evaluating the emulsifying capacity and emulsion stabilizing performance of vegetable proteins." *Italian Journal of Food Science* **7**(1): 3–10.

Shrestha, A. K., T. Ua-Arak, B. P. Adhikari, T. Howes and B. R. Bhandari (2007). "Glass transition behavior of spray dried orange juice powder measured by differential scanning calorimetry (DSC) and thermal mechanical compression test (TMCT)." *International Journal of Food Properties* **10**(3): 661–673.

Shukat, R., C. Bourgaux and P. Relkin (2012). "Crystallisation behaviour of palm oil nano-emulsions carrying vitamin E." *Journal of Thermal Analysis and Calorimetry* **108**(1): 153–161.

Silalai, N. and Y. H. Roos (2010). "Roles of water and solids composition in the control of glass transition and stickiness of milk powders." *Journal of Food Science* **75**(5): E285–E296.

Simoneau, C., M. J. McCarthy and J. B. German (1993). "Magnetic-resonance-imaging and spectroscopy for food systems." *Food Research International* **26**(5): 387–398.

Sitterberg, J., A. Ozcetin, C. Ehrhardt and U. Bakowsky (2010). "Utilising atomic force microscopy for the characterisation of nanoscale drug delivery systems." *European Journal of Pharmaceutics and Biopharmaceutics* **74**(1): 2–13.

Smyth, H. and D. Cozzoino (2013). "Instrumental methods (spectroscopy, electronic nose, and tongue) as tools to predict taste and aroma in beverages: Advantages and limitations." *Chemical Reviews* **113**(3): 1429–1440.

Snow, N. H. and G. C. Slack (2002). "Head-space analysis in modern gas chromatography." *Trac-Trends in Analytical Chemistry* **21**(9–10): 608–617.

Souto-Maior, J. F. A., A. V. Reis, L. N. Pedreiro and O. A. Cavalcanti (2009). "Phosphated crosslinked pectin as a potential excipient for specific drug delivery: Preparation and physicochemical characterization." *Polymer International* **59**(1): 127–135.

Stewart, S., R. J. Priore, M. P. Nelson and P. J. Treado (2012). Raman imaging. In *Annual Review of Analytical Chemistry*, R. G. Cooks and E. S. Yeung (eds.). Palo Alto, Annual Reviews. **5**: 337–360.

Stockmann, K. and K. Schwarz (1999). "Partitioning of low molecular weight compounds in oil-in-water emulsions." *Langmuir* **15**(19): 6142–6149.

Stone, H. and J. L. Sidel (2004). *Sensory Evaluation Practices.* Amsterdam, Elsevier Academic Press.

Tadros, T. F. (1994). "Fundamental principles of emulsion rheology and their applications." *Colloids and Surfaces A-Physicochemical and Engineering Aspects* **91**: 39–55.

Tadros, T. F. (2004). "Application of rheology for assessment and prediction of the long-term physical stability of emulsions." *Advances in Colloid and Interface Science* **108–109**: 227–258.

van Aken, G. A. (2007). Relating food microstructure to sensory quality. In *Understanding and Controlling the Microstructure of Complex Foods*, D. J. McClements (ed.). Cambridge, UK, Woodhead Publishing Limited: 449–482.

van Aken, G. A. (2010). "Relating food emulsion structure and composition to the way it is processed in the gastrointestinal tract and physiological responses: What are the opportunities?" *Food Biophysics* **5**(4): 258–283.

van der Linden, E., L. Sagis and P. Venema (2003). "Rheo-optics and food systems." *Current Opinion in Colloid and Interface Science* **8**(4–5): 349–358.

van Vliet, T., G. A. van Aken, H. H. J. de Jongh and R. J. Hamer (2009). "Colloidal aspects of texture perception." *Advances in Colloid and Interface Science* **150**(1): 27–40.

Venema, K., R. Havenaar and M. Minekus (2009). Improving in vitro simulation of the stomach and intestines. In *Designing Functional Foods: Measuring and Controlling Food Structure Breakdown and Nutrient Absorption*, D. J. McClements and E. A. Decker (eds.). Boca Raton, FL, CRC Press: 314–339.

Versantvoort, C. H. M., R. C. A. Ondrewater, E. Duizer, J. J. M. Van de Sandt, A. J. Gilde and J. P. Groten (2002). "Monolayers of IEC-18 cells as an in vitro model for screening the passive transcellular and paracellular transport across the intestinal barrier: Comparison of active and passive transport with the human colon carcinoma Caco-2 cell line." *Environmental Toxicology and Pharmacology* **11**(3): 335–344.

Versantvoort, C. H. M., R. C. A. Ondrewater, E. Duizer, J. J. M. Van de Sandt, A. J. Gilde and J. P. Groten (2003). "Monolayers of IEC-18 cells as an in vitro model for screening the passive transcellular and paracellular transport across the intestinal barrier: Comparison of active and passive transport with the human colon carcinoma Caco-2 cell line (vol 11, pg 335, 2002)." *Environmental Toxicology and Pharmacology* **13**(1): 55.

Vlasov, Y. G., Y. E. Ermolenko, A. V. Legin, A. M. Rudnitskaya and V. V. Kolodnikov (2010). "Chemical sensors and their systems." *Journal of Analytical Chemistry* **65**(9): 880–898.

Walstra, P. (2003). *Physical Chemistry of Foods*. New York, Marcel Decker.

Wang, Y. W., J. McCaffrey and D. L. Norwood (2008). "Recent advances in headspace gas chromatography." *Journal of Liquid Chromatography and Related Technologies* **31**(11–12): 1823–1851.

Waraho, T., D. J. McClements and E. A. Decker (2011). "Mechanisms of lipid oxidation in food dispersions." *Trends in Food Science and Technology* **22**(1): 3–13.

Wickham, M., R. Faulks and C. Mills (2009). "In vitro digestion methods for assessing the effect of food structure on allergen breakdown." *Molecular Nutrition and Food Research* **53**(8): 952–958.

Yang, L. (2008). "Biorelevant dissolution testing of colon-specific delivery systems activated by colonic microflora." *Journal of Controlled Release* **125**(2): 77–86.

Yoo, J. Y. and X. D. Chen (2006). "GIT physicochemical modeling—A critical review." *International Journal of Food Engineering* **2**(4): 12.

Zhang, F., B. Klebansky, R. M. Fine, H. T. Liu, H. Xu, G. Servant, M. Zoller, C. Tachdjian and X. D. Li (2010). "Molecular mechanism of the sweet taste enhancers." *Proceedings of the National Academy of Sciences of the United States of America* **107**(10): 4752–4757.

Zhang, H. P., C. H. Chon, X. X. Pan and D. Q. Li (2009). "Methods for counting particles in microfluidic applications." *Microfluidics and Nanofluidics* **7**(6): 739–749.

# Chapter 9

## Selection of Delivery Systems: Case Studies

### 9.1 Introduction

The purpose of this chapter is to highlight some of the issues that should be considered when selecting an appropriate delivery system for a particular application. This aim will be achieved using a number of case studies involving the potential utilization of colloidal delivery systems for particular applications within the food and other industries. As highlighted throughout this book, there are a large number of different types of colloidal delivery systems available for encapsulating, protecting, and releasing active ingredients. The manufacturer of a commercial product must therefore decide the type of delivery system that is the most appropriate for their specific application, which depends on a variety of economic, marketing, legal, and physicochemical factors.

#### 9.1.1 Design Criteria for Colloidal Delivery Systems

In this section, a number of general criteria that should be considered when selecting any delivery system for application in foods, beverages, and other products are given. In the following sections, more specific criteria are given for particular applications.

#### 9.1.1.1 Economics

The ingredients and processing operations used to fabricate a delivery system should be economical so that the benefits gained from utilizing it outweigh any associated

**401**

additional costs. Many of the delivery systems developed in academic laboratories and reported in the scientific literature are unsuitable for commercial application because they are too expensive for large-scale commercial applications.

### 9.1.1.2 Ease and Reliability of Fabrication

Ideally, the manufacturing process used to fabricate a delivery system should be relatively simple to implement on a large scale and should reliably produce a final product with the desired properties and functionality. The development of any new processing procedures or new manufacturing equipment must be justified by the potential benefits of the delivery system. Ideally, a manufacturer wants to utilize existing equipment and production lines to produce any new product; otherwise, there may be high capital costs associated with purchasing and maintaining new equipment and buildings.

### 9.1.1.3 Transportation and Storage

Delivery systems can come in a variety of different forms, such as liquids, pastes, capsules, solids, or powders. The physical form of a delivery system determines its storage stability, the storage space required, the ease of transport, and the ease of utilization. Often a manufacturer wants an active ingredient to be present at a relatively high loading capacity, so as to reduce transport and storage costs. In this case, an ingredient may need to be produced in a highly concentrated form or in a powdered form. On the other hand, an active ingredient should also be prepared in a form that can easily be dispersed into the final product. In this case, it may be beneficial for the ingredient to be in a liquid form or another easily dispersible form.

### 9.1.1.4 Labeling and Marketing

It is often important that the ingredients and processing conditions used to produce a food product are perceived as being desirable to consumers so that a company can make claims on a product's label or in advertising campaigns. There is increasing interest in using ingredients that are perceived as healthy, natural, organic, vegan, vegetarian, Kosher, Halal, non-GMO (genetically modified organism), fair-trade, or allergen-free. In addition, there is a tendency toward using processing operations and supply chains that can be considered to be "green," ethical, or sustainable. Thus, it may be undesirable to use high levels of energy, water, or organic solvents to produce a delivery system.

### 9.1.1.5 Legislation

The type and amount of ingredients used to fabricate a delivery system should be legally acceptable in all of the countries where the final product is going to be marketed and sold. The legislation governing the use of food ingredients varies from place to place, and this issue should be addressed early in the development of a delivery system when identifying appropriate ingredients to assemble it.

### 9.1.1.6 Patent Status

A number of companies would like to develop innovative delivery systems that can be patented so that their competitors cannot use the same approach. On the other hand, a manufacturer may be unable to use a technology that has already been patented by another company. Knowledge of the multitude of patents that have been published on delivery systems is therefore important for developing and implementing any new technology in this field.

### 9.1.1.7 Shelf Life

A delivery system should be designed so that it will remain physically and chemically stable throughout the required shelf life of the product it is incorporated into and so that it can resist any environmental stresses that a product typically experiences, such as thermal processing, freezing, chilling, dehydration, dilution, high shear mixing, vibration, or flow through a pipe. Delivery systems may therefore have to be designed to inhibit physical instability mechanisms, such as flocculation, coalescence, Ostwald ripening, gravitational separation, and phase inversion. For certain active ingredients, it may also be necessary to inhibit chemical instability, such as oxidation or hydrolysis. The precise mechanism responsible for chemical degradation depends on the active ingredient and may be promoted by solution composition and environmental conditions, such as temperature, oxygen, catalysts, light, and pH. It is therefore important to establish the chemical degradation pathways involved, as well as the major factors that influence them, when developing effective strategies to improve the shelf life of products.

### 9.1.1.8 Matrix Compatibility

It is important that the delivery system is compatible with the food matrix into which it will be incorporated. The delivery system should not adversely affect the appearance, texture, or flavor profile of the product. For example, some food and beverage products should be optically transparent (e.g., fortified waters, soft drinks, and juices) and therefore any delivery system used should not cause them to become turbid or cloudy. In this case, the particles within the delivery system should be so small that they do not scatter light strongly. Some food and beverage products should have a relatively low viscosity and smooth mouthfeel and therefore the delivery system should not cause an appreciable increase in viscosity or alteration in mouthfeel.

### 9.1.1.9 Loading, Retention, and Release Characteristics

A delivery system should be capable of incorporating an appropriate amount of the active ingredient into the final product, which will depend on its potency and required function, for example, antioxidant, antimicrobial, color, flavor, or nutraceutical. For nutraceuticals, the amount of active ingredient in a serving of a food product is typically around 10% of the recommended daily allowance (RDA)—if this value has been established. For colors or flavors, the amount of active ingredient required depends on the desired appearance or flavor profile of the product. A manufacturer should therefore calculate the amount of active ingredient needed to obtain the desired effect and determine the loading capacity of the delivery system to ensure that it can deliver the required effect in the final product. The delivery system should also be designed to retain the active ingredient within the colloidal particles during storage and transport, which will depend on the nature of the environmental stresses the product experiences throughout its lifetime. Finally, a delivery system should be designed to release the active component at the appropriate site of action.

## 9.1.2 Identification of Appropriate Colloidal Delivery Systems

Initially, a manufacturer should clearly identify and describe the problem that the delivery system is intended to address. The problem to be resolved may include one or more different aspects, such as stabilization of an active ingredient against chemical degradation; conversion of an active ingredient into a form that facilitates transport, handling, or utilization; incorporation of an active ingredient into a food matrix that it is not normally compatible with; controlling the release profile of an active ingredient outside or inside the human body; masking an off-flavor or undesirable mouthfeel associated with an active ingredient; and isolating chemically reactive ingredients from one another. A manufacturer should then identify the various types of delivery systems that might be able to address the specified problem and select the one that is most appropriate on the basis of the various criteria outlined above, such as cost, ease of production, labeling and legal requirements, stability, matrix compatibility, loading characteristics, and so on. The process of selecting an appropriate delivery system for a particular application is highlighted in the following sections by using specific case studies.

## 9.2 Nutraceutical-Fortified Soft Drinks

There has been growing interest in the beverage industry in fortifying waters, soft drinks, and juices with lipophilic bioactive ingredients that are perceived to provide health benefits, such as ω-3 fatty acids, conjugated linoleic acid, carotenoids, phytosterols, and oil-soluble vitamins (e.g., A, D, E, and K) (Given 2009; Sagalowicz and Leser 2010; Velikov and Pelan 2008). These lipophilic nutraceuticals typically have very low water solubility and cannot simply be dispersed into an aqueous-based product. Instead, they must first be encapsulated within a suitable colloidal delivery system.

## 9.2.1 Design Criteria

The general criteria for selecting colloidal delivery systems were discussed in Section 9.1.1. In this section, the most important criteria for designing colloidal delivery systems for nutraceutical-fortified beverages that are optically clear are discussed.

### 9.2.1.1 Loading and Retention Characteristics

The delivery system should be capable of incorporating an appropriate amount of the lipophilic nutraceutical into the final product. For nutraceuticals, the amount of active ingredient in a serving of a food product is typically around 10% of the RDA. A manufacturer can therefore use the reported RDA for those lipophilic nutraceuticals where these values have been established. For other nutraceuticals, a manufacturer may have to use published data to estimate the amount required to produce a beneficial biological effect and then aim to fortify a product with 10% or more of this value. One of the potential limitations of nutraceutical-fortified beverages (such as waters, soft drinks, and juices) is that the overall lipid content is relatively low (typically <0.1%), which limits their use to nutraceuticals with low RDAs.

### 9.2.1.2 Influence on Physicochemical and Sensory Properties of Product

*Optical properties*: One of the most important characteristics of this type of beverage product is high optical clarity, and so any colloidal delivery system used should not noticeably increase the turbidity or cloudiness (McClements 2002a,b). It is therefore important to use a delivery system that contains colloidal particles that do not scatter light strongly. The three most important factors affecting the light scattering properties of a colloidal delivery system are the refractive index contrast, particle concentration, and particle size (McClements 2002a,b). The turbidity tends to increase with increasing refractive index contrast and particle concentration and has a maximum value when the particle size is similar to the wavelength of light (Figure 9.1). A colloidal delivery system suitable for use in an optically transparent beverage product should be designed to contain small particles ($d < 50$ nm), low particle concentration ($\phi < 0.1\%$), or small refractive index contrast ($\Delta n < 0.01$). An example of the influence of particle size on the appearance of orange oil delivery systems is shown in Figure 9.2 for nanoemulsions and conventional emulsions. These systems have the same overall orange oil content (5%) but have different particle sizes. The nanoemulsion ($d = 25$ nm) appears transparent, whereas the emulsion ($d = 150$ nm) appears opaque.

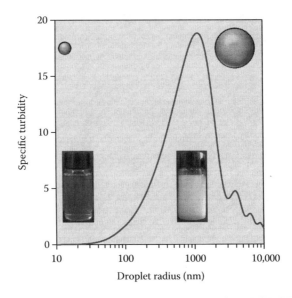

**Figure 9.1    The small size of the droplets in nanoemulsions means that they have different optical properties to conventional emulsions: they can be designed to have high optical clarity, which is important for some commercial applications. Plot shows change in emulsion turbidity per unit mass of droplets with changing droplet radius calculated using Mie theory.**

*Rheological properties*: This type of beverage product is usually expected to have a relatively low viscosity in its final form, and therefore any colloidal delivery system used should not cause a large increase in product viscosity. This can be achieved by ensuring that the effective particle concentration of the delivery system in the final product is relatively low ($\phi_{eff} < 5\%$). The effective particle concentration depends on the nature of the delivery system: for oil-in-water (O/W) emulsions, it is similar to the total oil content; for O/W nanoemulsions, it depends on the thickness of the interfacial layer relative to the oil core; for water-in-oil-in-water (W/O/W) emulsions, it depends on the amount of internal aqueous phase trapped within the oil droplets; for filled hydrogel particles, it depends on the amount of water and oil trapped within the biopolymer matrix; for microclusters, it depends on the amount of water trapped within the floc structures (Chapters 6 and 7).

## 9.2.1.3 Product Stability

The delivery system should not promote instability of the beverage product it is incorporated into, and it should itself remain stable throughout the expected shelf life of the product.

*Physical stability*: A colloidal delivery system may undergo changes in the spatial arrangement of the particles within a beverage product over time that have

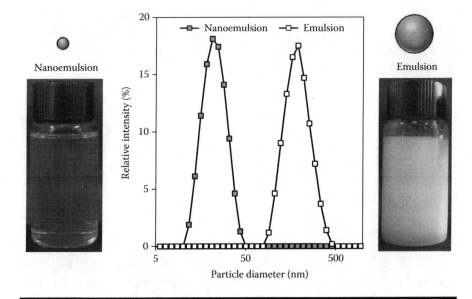

**Figure 9.2  Appearance and particle size distributions of two orange oil delivery systems suitable for use in acidic beverages (5% orange oil, 5% medium-chain triglycerides, 1% or 20% Tween 80, pH 3.5, 5 mM citrate buffer): the nanoemulsion ($d$ = 25 nm) contained 20% Tween 80, while the emulsion ($d$ = 150 nm) contained 1% Tween 80. The nanoemulsion appears transparent because the droplet size is much smaller than the wavelength of light. (Data kindly supplied by Yuhua Chang, University of Massachusetts.)**

adverse effects on product quality, such as droplet growth, aggregation, creaming, or sedimentation. It is important that the delivery system is carefully designed so that it is compatible with the food matrix being used. For example, positively charged particles within a delivery system may interact with negatively charged biopolymers or minerals within a food matrix, thereby promoting instability through electrostatic complexation. Ideally, the colloidal particles in a commercial product should maintain their size distribution and remain evenly distributed throughout the system during storage. However, the size and spatial organization of the particles may change over time through a variety of physicochemical mechanisms, including flocculation, coalescence, Ostwald ripening, and gravitational separation (Chapter 6). The particles in a delivery system may therefore have to be specifically designed to inhibit these different instability mechanisms (McClements 2005, 2010b). Flocculation and coalescence can often be prevented by ensuring that there is a sufficiently strong repulsive interaction operating between the particles, such as an electrostatic or steric repulsion. This can be achieved by coating the particles with an interfacial layer that has a sufficiently high electrical charge or that contains polymer chains that protrude into the aqueous phase. Ostwald ripening can be inhibited by using an oil phase that has low water solubility or by adding ripening inhibitors (McClements et al. 2012). Gravitational separation can be retarded by reducing the density contrast between the particles and surrounding liquid, by reducing the particle size, or by increasing the viscosity of the aqueous phase (McClements 2005). For most beverage systems, it is impractical to increase the aqueous phase viscosity since this would cause an undesirable change in the product texture and mouthfeel. Gravitational separation is usually inhibited in commercial beverage products by using very small particles ($d < 300$ nm) or by ensuring a low density contrast. For emulsions or nanoemulsions, a low density contrast is usually achieved by adding weighting agents to the oil phase so that the density of the particle and that of the surrounding aqueous phase are similar. Alternatively, the density of a lipid particle may be increased by coating it with a thick layer of dense biopolymer (McClements and Rao 2011). For filled hydrogel particles, the ratio of oil and biopolymer in the particles can be optimized to ensure that the overall density is close to that of the aqueous phase (Matalanis and McClements 2013).

*Chemical stability*: A number of lipophilic active ingredients that need to be incorporated into beverages are susceptible to chemical degradation during storage, such as ω-3 fatty acids, phytosterols, and carotenoids (Waraho et al. 2011). In this case, the colloidal delivery system should be designed to inhibit the chemical degradation of the encapsulated active ingredient within the product. This may be achieved in numerous ways depending on the chemical degradation mechanism involved, for example, controlling temperature, light exposure, oxygen levels, antioxidant levels, prooxidant levels, and the relative location of the various molecular species involved in the reactions (McClements and Decker 2000). Colloidal delivery systems can often be designed to enhance the chemical stability of labile active ingredients. For example, it may be possible to embed a chemically unstable active ingredient within a particle matrix or coating that has strong antioxidant capacity or that physically prevents it from interacting with reactants or catalysts within the product (Lesmes et al. 2010; Marze 2013; Matalanis et al. 2011).

### 9.2.1.4 Storage Form

A delivery system suitable for use in the beverage industry may be formulated in a variety of different forms, such as a fluid, a paste, a solid, or a powder. It is often beneficial for beverage manufacturers to store and transport an active ingredient in a highly concentrated form since this reduces storage and transport costs. For example, a highly concentrated flavor O/W emulsion may initially be prepared for storage purposes. This concentrated flavor delivery system is then diluted into an aqueous solution to form the final beverage product. In this case, the delivery system should be designed to remain physically and chemically stable in both the concentrated and diluted forms. A lipophilic nutraceutical could be incorporated into the concentrated flavor emulsion during its preparation, or it could be incorporated into the final beverage product during the dilution step.

## 9.2.2 Potential Delivery Systems

In this section, we highlight a number of delivery systems that might be suitable candidates to address the design criteria for fortifying clear beverages with lipophilic nutraceuticals. Two of the most important characteristics of these products that limit the type of delivery system that can be utilized are their high optical clarity and low viscosity. The need for high optical clarity means that the particles have to be very small ($d < 50$ nm), have low refractive index contrast, or have to be used at a low concentration. The maximum particle concentration that can be used before the system appears turbid depends on the size and refractive index contrast of the colloidal particles (McClements 2002a,b). The larger the particle size or refractive index contrast, the lower the amount that can be added before the system looks turbid. For lipophilic nutraceuticals that need to be incorporated at relatively high levels, it is necessary to use very fine particles since these can be added at fairly high concentrations before they make a product look cloudy. However, this limits the type of colloidal delivery system that can be used to those that contain very fine particles, such as microemulsions, nanoemulsions, and nanoparticles (McClements 2010, 2011). Systems such as emulsions, multiple emulsions, microparticles, and filled hydrogel particles are usually unsuitable because the particles they contain are so large they scatter light and make the system look turbid. In addition, the fact that the particles are relatively large makes them prone to gravitational separation during storage because beverages typically have a low viscosity and therefore particle movement is rapid. It may still be possible to use these types of delivery systems if they can be designed so that the refractive index and density contrasts are relatively low. However, these systems are usually considerably more complicated to fabricate than microemulsions, nanoemulsions, and nanoparticles, and therefore there must be some additional advantages to make them a worthwhile candidate for development.

### 9.2.2.1 Surfactant-Based Delivery Systems

The main advantages of using microemulsions for this particular application are that they contain very small particles that are thermodynamically stable (under a given set of conditions), and so they are optically transparent and have good stability characteristics (Flanagan and Singh 2006; Leser et al. 2006; Spernath and Aserin 2006). In addition, they are relatively easy to prepare and can often be fabricated in a concentrated form that can simply be diluted with an aqueous solution containing the other beverage ingredients (such as sweeteners, buffers, preservatives, and water-soluble flavors and colors) to form the final beverage product. Microemulsions therefore have many of the characteristics desired for application as a delivery system in transparent beverages, such as fortified waters, soft drinks, or juices. On the other hand, they also have a number of disadvantages that limit their commercial utilization in certain beverage products. For example, microemulsions usually require fairly high surfactant-to-oil ratios when compared to other delivery systems. In addition, the surfactants used tend to be synthetic ingredients that may cause problems because of their cost, undesirable flavor profile, potential toxicity, poor label friendliness, or regulatory restrictions.

## 9.2.2.2 Emulsion-Based Delivery Systems

O/W nanoemulsions consist of small ($r$ < 100 nm) oil droplets dispersed within an aqueous solution, with each droplet being coated by a layer of emulsifier to help stabilize them against aggregation (McClements 2011; McClements and Rao 2011). Nanoemulsions have some similar advantages as microemulsions: they contain very fine particles that can have high kinetic stability, and so they can be optically clear and have good long-term stability. However, stable nanoemulsions containing small particles are often more difficult to fabricate than microemulsions. Nanoemulsions may be fabricated from high-energy or low-energy homogenization methods depending on the nature of the emulsifier, oil, and aqueous phases used (Chapter 6). High-energy methods require the utilization of mechanical devices ("homogenizers") that produce intense disruptive forces that intermingle and break up the oil and water phases. These devices are often expensive to purchase, maintain, and operate and typically produce particles that are larger than those created using low-energy methods. On the other hand, a wider range of ingredients can be used to produce nanoemulsions, and the emulsifier-to-oil ratio is typically much lower than that needed for low-energy methods (McClements 2011; McClements and Rao 2011). Low-energy methods rely on the spontaneous formation of fine oil droplets when system composition or environmental conditions (such as temperature) are changed in a particular manner. They require simple inexpensive equipment to produce nanoemulsions, such as stirrers and metering units. The main disadvantages of low-energy methods are that they normally require the use of synthetic surfactants and high surfactant-to-oil ratios. Like microemulsions, nanoemulsions can be fabricated in a concentrated form that is diluted with an aqueous solution containing the other functional ingredients to form the final beverage product. Nanoemulsions therefore have characteristics that make them suitable for application as delivery systems in clear beverages. However, there are a number of potential disadvantages of using them in commercial products. Nanoemulsions often have better stability to flocculation and coalescence than conventional emulsions, but they may be prone to droplet growth through Ostwald ripening, particularly if the oil phase has some solubility in the aqueous phase. Ostwald ripening can be retarded by adding sufficient quantities of a ripening inhibitor to the oil phase prior to nanoemulsion formation (Lim et al. 2011; McClements et al. 2012). The droplets in some nanoemulsions may also undergo instability owing to flocculation and coalescence if the repulsive interactions (e.g., electrostatic and steric) are not sufficiently strong (Lee and McClements 2010). Problems with droplet aggregation can often be overcome by using an emulsifier or coating that increases the repulsive interactions between the droplets. Nanoemulsions stabilized by some small-molecule surfactants are prone to coalescence at relatively high temperatures because the curvature of the surfactant monolayer tends toward unity near the phase inversion temperature (Rao and McClements 2010). Nanoemulsions often have a higher loading capacity for lipophilic active ingredients than microemulsions, which can be an advantage in some applications.

### 9.2.2.3 Biopolymer-Based Delivery Systems

Filled hydrogel particles are suitable for encapsulating lipophilic nutraceuticals, but they are usually so large that they are susceptible to gravitational separation and may make a product look cloudy or opaque and therefore have limited application in optically transparent low-viscosity beverages. However, biopolymer nanoparticles can be produced from proteins or polysaccharides using a variety of fabrication methods (Hu and Huang 2013; Matalanis et al. 2011; Thies 2012). These nanoparticles can be designed to encapsulate lipophilic active ingredients so that they can be incorporated into clear beverages. Biopolymer nanoparticles can be created that have sizes so small that they do not scatter light strongly and are therefore suitable for incorporation into clear products. This type of nanoparticle is prone to aggregation and so it is important that there is a relatively strong repulsive interaction (steric or electrostatic) between the particles. It may also be important to ensure that gravitational separation is inhibited by using small particles that have a low density contrast (which may be achieved by controlling the oil-to-biopolymer ratio) (Matalanis and McClements 2013). A potential advantage of using biopolymer nanoparticles is that they can be created from natural ingredients, such as proteins and polysaccharides. However, there are some potential disadvantages of using nanoparticles in beverage products, such as their relatively high susceptibility to aggregation and gravitational separation. In addition, some proteins may be allergens, non-Kosher, nonvegan, or nonvegetarian and so manufacturers do not want to include them in commercial products.

## 9.3 Dairy-Based Functional Beverages Designed to Enhance Nutraceutical Bioavailability

Many lipophilic nutraceuticals have a low oral bioavailability when consumed in their conventional or isolated forms, for example, as part of natural foods (such as carotenoids in fruits or vegetables) or as part of powders isolated from natural foods (such as crystalline carotenoid powders). There is therefore considerable interest in the design of colloidal delivery systems to improve the oral bioavailability of lipophilic nutraceuticals (Hu and Huang 2013; Marze 2013; McClements et al. 2009). In this section, the focus will be on the design of dairy-based beverages for delivery of lipophilic nutraceuticals. It will be assumed that the beverage product should have similar physicochemical and sensory characteristics as milk, that is, a relatively low viscosity fluid with a white creamy appearance that should be stable to phase separation during storage and utilization. It will also be assumed that the manufacturer would like to create the product from "label-friendly" ingredients.

## 9.3.1 Design Criteria

In this section, the most important criteria that should be considered when selecting an appropriate colloidal delivery system for incorporating a lipophilic nutraceutical into a dairy-based beverage designed to increase its oral bioavailability are considered.

### 9.3.1.1 Loading and Retention Characteristics

Some of the major challenges that need to be overcome when designing this kind of delivery system are the fact that many lipophilic nutraceuticals have a low water solubility and high melting point, for example, carotenoids, curcumoids, phytosterols, and flavonoids (McClements 2012b; Thies 2012). These lipophilic substances must therefore be incorporated into some kind of colloidal delivery system before they can be successfully introduced into an aqueous-based beverage product. In the case of crystalline nutraceuticals, it may be necessary to dissolve them in an oil phase prior to creation of the delivery system (McClements 2012b). The amount that can be incorporated in a soluble form is then limited by the equilibrium saturation concentration of the nutraceutical in the delivery system. Alternatively, it may be possible to encapsulate crystalline nutraceuticals into a delivery system without dissolving them (Muller et al. 2011). In this case, it may be important to ensure that the crystals are not too large; otherwise, they may cause problems with appearance, sedimentation, and mouthfeel (Fredrick et al. 2010). The amount of nutraceutical that should be incorporated into a dairy-based beverage to give a specific beneficial effect depends on its bioactivity and may be available in published studies or in government regulations. Only those delivery systems that have a loading capacity that is sufficiently high to achieve this amount can therefore be utilized in commercial applications. Dairy-based beverages can have relatively high lipid contents (typically up to 5%) and therefore it may be possible to incorporate relatively high levels of a lipophilic nutraceutical into them.

### 9.3.1.2 Influence on Physicochemical and Sensory Properties of Product

In general, a suitable colloidal delivery system for lipophilic nutraceuticals should not adversely affect the appearance, texture, flavor, or mouthfeel of a functional food or beverage, and it should maintain its desirable characteristics throughout the shelf life of the product (Lesmes and McClements 2009; McClements 2012b). In the case of a dairy-based beverage product, the colloidal delivery system needs to be incorporated into a product that has a relatively low viscosity and that is optically opaque. Since the product is opaque, it is not important that the delivery system contains very small particles from an optical property perspective. Indeed, it may actually be advantageous to have particles that have dimensions similar to the wavelength of light so that they contribute to the creamy appearance of the product. On the other hand, the size of the particles may have a pronounced influence on the storage stability of the product. Since dairy-based beverage products typically have a low viscosity, then the particles may have to be relatively small ($d < 500$ nm) to inhibit creaming or sedimentation. Alternatively, it may be possible to inhibit or retard gravitational separation by designing the particles so that they have a similar density as the surrounding aqueous phase (Matalanis and McClements 2013). Adding thickening or gelling agents to inhibit creaming may not be an option since they would have an undesirable influence on the perceived texture and mouthfeel of the product. The delivery system itself should also not cause any undesirable changes in the textural characteristics of the product.

## 9.3.1.3 Product Stability

*Physical stability*: In general, it is important that a colloidal delivery system remains stable to various physical degradation mechanisms that may occur within a commercial product, such as particle aggregation, gravitational separation, Ostwald ripening, and phase separation. For a dairy-based beverage, the colloidal delivery system may have to operate in a compositionally complex environment that undergoes specific processing operations, such as high-temperature, short-time processing or retorting. It is important that the delivery system remains stable under the pH and ionic strength of the aqueous phase in the product and that there are no adverse ingredient interactions with other components present. For example, if the particles in the delivery system had an opposite charge to some of the other ingredients used in the beverage (such as biopolymers or multivalent minerals), then undesirable particle aggregation and sedimentation may occur. In the case of a dairy-based product, the aqueous phase typically has a pH around neutral and contains a complex mixture of proteins, minerals, and sugars.

*Chemical stability*: Some lipophilic nutraceuticals are susceptible to chemical transformations during food processing and storage, as well as during passage through the gastrointestinal tract (GIT) (e.g., carotenoids and ω-3 fatty acids). These chemical transformations may alter their oral bioavailability and biological activity, thereby affecting their potential health benefits. The precise chemical transformation mechanism involved depends on the nature of the nutraceutical and of the food matrix and has to be established for each system. In general, some common factors that influence the chemical transformation of lipophilic nutraceuticals are pH, temperature, oxygen levels, exposure to UV and visible light, catalysts, and inhibitors (Boon et al. 2010; McClements and Decker 2000; Waraho et al. 2011). The delivery system should therefore be carefully designed to control any potential chemical degradation reactions of nutraceuticals that may occur within a dairy-based beverage matrix. This may involve adding antioxidants that are located in regions where the reaction occurs or creating physical barriers that prevent reactive species from coming into contact with each other.

### 9.3.1.4 Bioavailability Enhancement

A major problem limiting the effectiveness of many lipophilic nutraceuticals is their low oral bioavailability. An overview of the most important factors affecting the oral bioavailability of lipophilic nutraceuticals is therefore given in this section. Bioavailability can be defined as the amount of a bioactive component that eventually reaches the site of action in an active state (Rein et al. 2013). The overall oral bioavailability ($F$) depends on a number of factors, which can be represented by the following expression (Lentz et al. 2007):

$$F = F_L \times F_A \times F_D \times F_M \times F_E \quad (9.1)$$

$F_L$ is the fraction of the bioactive lipophilic component *liberated* into the lumen of the GIT to become bioaccessible. $F_A$ is the fraction of the released lipophilic component that is *absorbed* by the epithelial cells. $F_D$ is the fraction of the absorbed lipophilic component that reaches the site of action after *distribution* within the body. $F_M$ is the fraction of the absorbed lipophilic component that reaches the site of action in a *metabolically* active form and depends on any chemical or enzymatic modifications that take place before and after ingestion. $F_E$ is the fraction of the absorbed lipophilic component that has not been *excreted* by the body. In practice, each of these parameters varies over time after a bioactive component is ingested to give a profile of bioavailability *versus* time at a specified site of action. Some of the key processes that influence these parameters are outlined below (Bauer et al. 2005; Bermudez et al. 2004; Fave et al. 2004):

- *Liberation*: A lipophilic nutraceutical must be *liberated* from the food matrix and solubilized within mixed micelles in the small intestinal fluids before it is bioaccessible (Figure 9.3). Mixed micelles are assembled from bile salts and phospholipids secreted by the body, as well as the products of lipid digestion such as monoacylglycerides and free fatty acids (FFAs) (Mullertz et al. 2012). The fraction of a lipophilic bioactive solubilized within the mixed micelle phase of the small intestine can often be used as a measure of the fraction liberated ($F_L$) from the food matrix in a bioaccessible form.

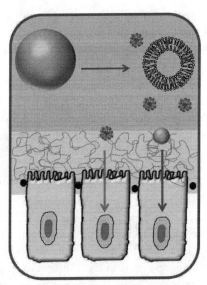

**Figure 9.3 Lipid nanoparticles may penetrate into the mucus layer and become trapped (mucoadhesion) or transported across it. Nanoparticles that are not digested and that travel through the mucus layer may be directly adsorbed by the epithelium cells.**

- *Absorption*: Mixed micelles transport solubilized lipophilic nutraceuticals across the intestinal lumen, through the mucous layer, and to the apical side of the intestinal enterocyte cells. The lipophilic nutraceuticals may then be *absorbed* by enterocyte cells through either passive or active transfer mechanisms (Singh et al. 2009). The cell permeation step determines the fraction of the liberated nutraceutical eventually absorbed ($F_A$) by the body.
- *Distribution*: After absorption, the lipophilic nutraceuticals are *distributed* among various tissues throughout the body, such as the systemic circulation, liver, kidney, adipose tissue, lungs, heart, brain, and so on. The molecular and physicochemical characteristics of the lipophilic nutraceutical, as well as those of any coingested food components in the food matrix, may alter the distribution of the bioactive nutraceuticals among the different tissues. The target tissue for a nutraceutical depends on the nature of the biological response required, such as prevention of chronic disease, maintenance of general well-being, enhanced performance, or treatment of specific acute diseases.
- *Metabolism*: Lipophilic nutraceuticals may be molecularly transformed before or after ingestion owing to chemical or biochemical transformation processes within the food or body, such as oxidation, hydrolysis, or enzyme activity. The transformations of a lipophilic nutraceutical as it passes through the body therefore determine the fraction that arrives in a metabolically active state at the site of action ($F_M$).

■ *Excretion*: Lipophilic nutraceuticals and their metabolites are eventually excreted from the human body, for example, in the feces, urine, sweat, or breath. The rate and extent of the various excretion processes of a nutraceutical determine the fraction remaining ($F_E$).

The oral bioavailability of ingested lipophilic components can therefore be improved by designing delivery systems and food matrices that increase the fraction liberated ($F_L$) and absorbed ($F_A$) and reaching the site of action ($F_D$) in a metabolically active form ($F_M$). This goal can be achieved by manipulating the composition and structure of delivery systems based on knowledge of the impact of specific food properties on the biological fate of lipophilic bioactives (McClements 2013).

One of the major factors limiting the oral bioavailability of many lipophilic nutraceuticals is their low solubility in gastrointestinal juices. It is well known that the absorption of lipophilic bioactive agents from the GIT can be enhanced in the presence of certain coingested lipids, which has mainly been attributed to their ability to form mixed micelles within the small intestine that are capable of solubilizing them and transporting them to the enterocyte cells where they are absorbed (Charman et al. 1997; Porter et al. 2007; Pouton and Porter 2008; Rein et al. 2013; Yeap et al. 2013). The extent of the increase in bioavailability depends on the nature of the coingested lipids, such as their chain length and degree of unsaturation. Typically, mixed micelle phases containing long-chain fatty acid digestion products have higher solubilization capacities than those containing short- or medium-chain ones. In addition, the absorption route (portal vein vs. lymphatic system) depends on the type and amount of lipids ingested since this influences chylomicron formation in the epithelium cells. Chylomicrons are capable of incorporating lipophilic bioactive agents and transporting them to the systemic circulation via the lymphatic system, thereby avoiding first passing through the liver. On the other hand, lipophilic agents that are transported via the portal vein route undergo first pass metabolism in the liver, which may alter their bioactivity appreciably. Other types of food components may also play an important role in either increasing or decreasing the absorption of lipophilic nutraceuticals, such as certain carbohydrates, proteins, minerals, surfactants, and phytochemicals (Gu et al. 2007). These components may operate through various mechanisms, such as altering enzyme activity (e.g., lipase or protease), altering mass transport processes (e.g., by increasing viscosity), binding to components involved in the digestion and absorption process (such as bile salts and calcium), altering cell permeability (e.g., by altering membrane flexibility or tight junctions), and inhibiting efflux mechanisms. The dependence of oral bioavailability on the type and amount of ingredients present means that there is great scope for designing colloidal delivery systems to improve the oral bioavailability of lipophilic nutraceuticals in functional foods and beverages.

## 9.3.2 Potential Delivery Systems

In this section, we highlight a number of delivery systems that might be suitable candidates for increasing the oral bioavailability of lipophilic nutraceuticals on the basis of the information given in the previous section. An appropriate delivery system should contain a sufficient quantity of a carrier lipid that favors the solubilization, transport, and absorption of the lipophilic nutraceuticals within the GIT. A delivery system may also contain various other food ingredients specifically designed to increase bioavailability, such as permeation enhancers or efflux inhibitors (McClements 2013).

### 9.3.2.1 Surfactant-Based Delivery Systems

Microemulsions are widely used in the pharmaceutical industry to increase the bioavailability of lipophilic substances, particularly in the form of self-microemulsifying drug delivery systems, that is, SMEDDS (Porter et al. 2006, 2007). In this case, the lipophilic drug is usually mixed with a surfactant and carrier lipid and placed within a capsule that breaks down within the stomach or small intestine, leading to the spontaneous formation of microemulsions containing the released drug in the GIT. The ingested surfactants usually mix with the bile salts and phospholipids secreted by the human body, which increases their solubilization capacity (Rozner et al. 2010). This approach is less useful for food applications because of the high levels of surfactant required to form microemulsions and because it is often difficult to form microemulsions from long-chain triglycerides (LCTs) (which give high solubilization capacities for lipophilic nutraceuticals). Some researchers have shown that stable microemulsions can be formed using food-grade ingredients that are suitable for encapsulating and delivering lipophilic nutraceuticals, and these may be useful for increasing the oral bioavailability of nutraceuticals that do not need to be present at high levels in foods (Amar-Yuli et al. 2009; Flanagan and Singh 2006; Spernath and Aserin 2006). Nevertheless, microemulsions are unlikely to be suitable in the dairy-based beverage product considered in this application because the high levels of synthetic surfactants required to fabricate them may cause taste problems and would not lead to a "clean" label.

Liposomes are primarily formed from phospholipids and can be used to encapsulate certain types of lipophilic nutraceuticals between the nonpolar regions formed by the phospholipid bilayers (Constantinides et al. 2006; Liu et al. 2013; Maherani et al. 2011; Taylor et al. 2005). Phospholipids are natural surfactants that can be obtained from various food sources, including soy, milk, eggs, and sunflowers, and therefore they are more suitable for production of dairy-based beverages that require clean labels. In addition, phospholipids normally form part of the mixed micelles present in the small intestinal fluids and may therefore be able to increase the solubilization capacity of the mixed micelle phase for the lipophilic nutraceutical. One of the major limitations to using liposome-based delivery systems is their relatively poor stability under conditions normally encountered in food and beverage products. The proteins, minerals, and fat droplets in a complex dairy-based beverage product may interact with the liposomes and promote their instability. Nevertheless, there are a number of strategies that can be used to improve their stability, such as controlling their initial composition or preparation method (Isailovic et al. 2013; Lu et al. 2011b) or coating them after they have been formed with biopolymer layers to increase the steric or electrostatic repulsion between them (Laye et al. 2008; Liu et al. 2013a). Coating liposomes with biopolymers may also be useful for controlling their fate within the GIT after ingestion (Liu et al. 2013a).

### 9.3.2.2 Emulsion-Based Delivery Systems

Emulsion-based delivery systems, such as emulsions, nanoemulsions, and solid lipid nanoparticles (SLNs), are particularly suitable for increasing the bioavailability of lipophilic substances, since their composition, size, interfacial properties, and physical state can easily be controlled (McClements 2010, 2012b). These delivery systems can be prepared using low-energy or high-energy homogenization methods from a variety of lipids and emulsifiers. Typically, the lipophilic nutraceutical to be encapsulated is dissolved in an oil phase containing an appropriate carrier lipid. The concentration of nutraceutical used should be below the saturation level; otherwise, it may crystallize and be expelled from the lipid particles during storage (McClements 2012b). As discussed earlier, a carrier lipid that will be highly digested within the stomach and small intestine and produce a mixed micelle phase with a high solubilization capacity should be selected. Typically, LCTs are most suitable for this purpose (Figure 9.4). However, LCTs tend to have relatively high viscosities, high interfacial tensions, and high phase inversion temperatures, which limit the type of homogenization method that can be used to prepare them. For example, it is often difficult to produce nanoemulsions with very fine droplets using low-energy or high-energy methods using LCTs. The total amount of carrier lipid present in an emulsion-based delivery system is also important as the bioaccessibility of a lipophilic nutraceutical usually increases with increasing lipid content (Salvia-Trujillo et al. 2013). In addition, the size of the droplets in an emulsion-based delivery system can be controlled by controlling the homogenization conditions, for example, the duration or intensity for high-energy methods (Qian and McClements 2011) or the surfactant-to-oil ratio for low-energy methods (Saberi et al. 2013). Some studies have shown that the smaller the droplets in an emulsion, the higher the bioavailability of an encapsulated lipophilic bioactive (Figure 9.5). This effect has been attributed to the increase in surface area of lipid exposed to the digestive enzymes as the droplet size decreases, which increases the rate at which mixed micelles capable of solubilizing the lipophilic nutraceuticals are produced.

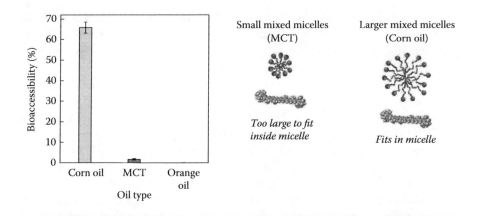

**Figure 9.4** **The bioaccessibility of a lipophilic compound is often influenced by the nature of the carrier oil. In this study, we examined the bioaccessibility of β-carotene encapsulated within nanoemulsions made from long-chain triglycerides (corn oil), medium-chain triglycerides (MCT), or an indigestible oil (orange oil). (Data from C. Qian et al.,** *Food Chemistry*, **135(3), 1440–1447, 2012.)**

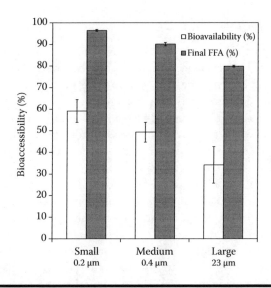

**Figure 9.5** **Influence of the diameter of initial engineered lipid nanoparticles in corn oil-in-water nanoemulsions on the extent of fatty acid release and bioaccessibility of β-carotene after 2 h digestion by lipase under simulated small intestine conditions. The initial mean droplet diameters were as follows: large, $d_{43} \approx$ 23 μm; medium, $d_{43} \approx$ 0.4 μm; and small, $d_{43} \approx$ 0.2 μm. (From L. Salvia-Trujillo et al.,** *Food Chemistry*, **139(1–4), 878–884, 2013.)**

Since a dairy-based product is optically opaque, it is possible to utilize either nanoemulsions or emulsions as delivery systems. Emulsions have the advantage that they are easier to prepare than nanoemulsions and that they contain droplets that scatter light strongly so they would contribute to the desirable creamy appearance of the product (McClements 2002a, b). On the other hand, nanoemulsions have the advantage that the small particle size may lead to a higher bioavailability and a better stability than emulsions (McClements 2011; McClements and Rao 2011). Emulsions and nanoemulsions are probably the best option for increasing the bioavailability of lipophilic nutraceuticals in dairy-based beverages. They can be prepared using natural food-grade ingredients (such as triglyceride oils and dairy protein emulsifiers) specifically selected to increase bioavailability: (i) by facilitating lipid digestion and release, (ii) by increasing the formation of mixed micelles with a high solubilization capacity, and (iii) by stimulating the formation of chylomicrons that transport the lipophilic active ingredients via the lymphatic route (McClements 2013). Other components could also be added to the aqueous phase of these systems to increase the bioavailability of the lipophilic components, such as efflux inhibitors or permeation enhancers (McClements 2013).

### 9.3.2.3 Biopolymer-Based Delivery Systems

Filled hydrogel particles may be suitable for incorporating lipophilic nutraceuticals into dairy-based beverages and increasing their bioavailability (Matalanis et al. 2011). These systems could be designed to protect the lipophilic nutraceutical from degradation within the product, but to release it in the stomach or small intestine after it has been ingested. For example, a biopolymer matrix could be designed to physically and chemically protect the lipid droplets within a product during storage, but to dissociate and release the droplets when the hydrogel particles encounter gastric or small intestinal fluids (McClements 2010; McClements and Li 2010). Filled hydrogel particles can be designed to have similar densities to the surrounding aqueous phase, which is useful for inhibiting gravitational separation in low-viscosity beverages (Matalanis and McClements 2013).

Biopolymer nanoparticles may also be suitable for delivering certain types of lipophilic nutraceuticals (Hu et al. 2012; Patel et al. 2012). These nanoparticles are typically formed using an antisolvent precipitation approach (Chapter 7). For example, a hydrophobic biopolymer (such as zein) and a lipophilic nutraceutical are dissolved in a concentrated alcohol solution. The resulting mixture is then injected into an aqueous solution to promote the formation of biopolymer nanoparticles containing the lipophilic nutraceutical. Additives, such as emulsifiers or thickening agents, may have to be included in the aqueous phase to ensure the formation and stability of the biopolymer nanoparticles.

Any biopolymer-based system has to be carefully designed so that it does not adversely influence product quality (appearance, texture, stability, and flavor profile) and so that it ensures high bioavailability of the nutraceutical after ingestion.

## 9.4 Delivery of Probiotics to the Colon

There is growing evidence that the nature of the microbial population within the gut plays a critical role in determining the health and wellness of human beings (Parvez et al. 2006; Roberfroid et al. 2010). Research has identified certain types of microbial populations that are associated with promoting human health (Playne 2002). These bacterial populations are claimed to have a variety of beneficial effects, including stimulating the immune system, improving gastrointestinal health, inhibiting harmful bacteria, improving lipid metabolism, synthesizing essential nutrients (such as vitamins), decreasing allergies, and reducing the incidences of certain kinds of cancer (Parvez et al. 2006; Roberfroid et al. 2010; Teitelbaum and Walker 2002). Food manufacturers are therefore attempting to incorporate these types of live microbes (probiotics) into functional foods and beverages so as to increase their potential health benefits. One of the major challenges is that the viability and bioactivity of probiotics may be lost before they reach the colon owing to microbial degradation within the food or within the GIT (Burgain et al. 2011; Corona-Hernandez et al. 2013). The viability of probiotics may be increased by encapsulating them within delivery systems, and therefore there has been considerable interest in the development of delivery systems suitable for use in the food industry (Burgain et al. 2011; Cook et al. 2012; Heidebach et al. 2012; Islam et al. 2010). In this section, we will consider the development of colloidal delivery systems to encapsulate and protect probiotic bacteria within a fruit yogurt. This type of product is typically an optically opaque product that has paste-like rheological properties; that is, it remains solid at low applied stresses but flows when a critical applied stress (the "yield stress") is exceeded.

Initially, it should be noted that there are a number of ways of improving the viability of probiotics in foods and beverages that do not involve using delivery systems (Ranadheera et al. 2010; Sanders and Marco 2010). First, the viability of a probiotic depends on the strain of bacteria, with some strains having good resistance to stresses encountered in foods or the GIT, such as high acidity and bile salts. Second, the nature of the food matrix surrounding the probiotics in a food or beverage product can be optimized to improve their viability. For example, foods (like yogurt) that contain high levels of protein may be able to reduce the high acidity of the stomach through their buffering capacity, thereby causing a less severe stress on the probiotics. Third, the environmental conditions experienced by a food or beverage product could be carefully controlled to prevent loss of probiotic viability, such as nutrient levels, pH, ionic strength, temperature, and oxygen. Nevertheless, encapsulation is often required to ensure that probiotics reach the large intestine in sufficient quantities.

## 9.4.1 Design Criteria

### 9.4.1.1 Loading and Retention Characteristics

It has been estimated that there should be approximately $10^6$ to $10^7$ viable probiotic bacteria per gram of a product to get a beneficial health effect (Krasaekoopt and Bhandari 2012). A delivery system should therefore be capable of encapsulating at least this amount of bacteria in a food or beverage product. In addition, the dimensions of the particles in the delivery system must be sufficiently greater than that of the probiotic bacteria. Typically, probiotic bacteria have dimensions in the order of a few micrometers, which mean that delivery systems containing smaller particles are unsuitable (such as nanoemulsions or microemulsions). To maintain their protective effect, it is important that the particles in a delivery system maintain their integrity within the food product and within regions of the GIT where probiotic viability may be lost.

### 9.4.1.2 Influence on Physicochemical and Sensory Properties of Product

A colloidal delivery system for probiotics should not adversely affect product quality, and it should maintain its protective characteristics throughout the shelf life of the product. In the case of a fruit yogurt, the delivery system needs to be incorporated into an optically opaque product with paste-like characteristics. The fact that the product is opaque means that it is not necessary to use a delivery system that contains very small particles (which would not be possible anyway, because small particles could not contain the probiotics). As discussed in previous sections, it may actually be advantageous to use particles with dimensions similar to the wavelength of light so that they contribute to the desirable creamy appearance of the product. Typically, yogurt should have specific rheological characteristics, such as elastic modulus, yield stress, and apparent viscosity, which are normally a result of the network of aggregated biopolymers in the system. The introduction of a colloidal delivery system into a yogurt should not therefore disrupt this gel network and adversely affect the textural characteristics of the product. Finally, the delivery system should not influence the desirable mouthfeel of a product. For example, the particles should not be so large that they give a grainy or gritty texture within the mouth during mastication.

### 9.4.1.3 Product Stability

The delivery system should be designed so that it remains physically stable within the product throughout storage. In general, the colloidal particles that may be used to encapsulate probiotics may be susceptible to instability through a variety of mechanisms, such as particle aggregation, gravitational separation, and phase separation. For a yogurt-based product, many of these instability mechanisms will be prevented because the yield stress and high viscosity of the product inhibits particle movement.

### 9.4.1.4 Maintain Viability within Foods

Probiotics are living organisms that may be destroyed by adverse changes in solution properties or environmental conditions, such as pH extremes, high ionic strengths, thermal processing, freezing, organic solvents, and dehydration (Krasaekoopt and Bhandari 2012; Riaz and Masud 2013). Numerous studies have shown a substantial decrease in the number of viable probiotic microorganisms in food products during storage or after exposure to specific processing treatments (Huq et al. 2013; Karimi et al. 2011; Kosin and Rakshit 2006). There are only a limited number of food and beverage products that are particularly suitable for use with probiotics, that is, those with appropriate solution conditions (e.g., pH and ionic composition) and that only experience relatively mild processing and storage conditions. Yogurts appear to be particularly suitable for this purpose since they meet many of the desirable criteria for probiotic foods (Sanders and Marco 2010). Delivery systems have been shown to increase the viability of probiotics when exposed to certain processing or storage conditions, thereby increasing the range of products they can be successfully incorporated into (Burgain et al. 2011; Cook et al. 2012; Heidebach et al. 2012; Krasaekoopt and Bhandari 2012).

### 9.4.1.5 Maintain Viability within the GIT

Studies on probiotics have shown that they may undergo appreciable loss of viability during transit through the human GIT. For example, it has been reported that a $10^8$- to $10^9$-fold decrease in the number of viable probiotic bacteria occurred from the initial food to the point when the food reached the small intestine (Priya et al. 2011). A delivery system has to be designed to withstand the variable challenges it encounters along the length of the GIT, such as high acidity, enzyme activity, bile salt levels, antimicrobial activity, and oxygen levels. The stomach is probably the harshest environment that probiotics encounter on their journey through the GIT because of the high acidity (typically pH 1 to 3), high enzyme activity (e.g., lipases and proteases), bile salts, and other stressors. To overcome these challenges, researchers have devised many ways to manipulate delivery system properties to ensure that a system remains viable along the length of the GIT, for example, by encapsulating bacteria in particles designed to protect them from harsh environmental conditions (Figure 9.6).

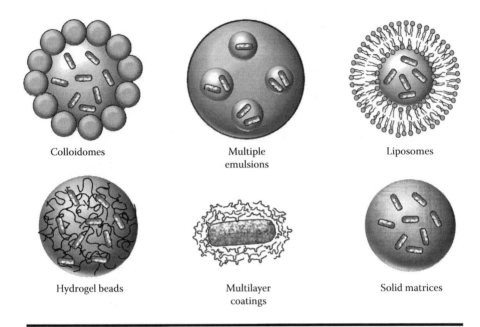

**Figure 9.6 Probiotic microorganisms can be encapsulated within a variety of different colloidal delivery systems to protect them from degradation in foods and within the GIT.**

### 9.4.1.6 Release Probiotics within the Large Intestine

Ingested probiotics should reach the colon in a viable state if they are going to exhibit their beneficial effects by altering the colonic microbial population (Huq et al. 2013; Teoh et al. 2011). Having said this, there is some evidence that even non-viable probiotics may have some beneficial health effects. A delivery system should therefore be designed so that it protects the bacteria in the mouth, stomach, and small intestine, but then releases them within the large intestine. A number of different trigger mechanisms may be used to stimulate the release of probiotics within the large intestine, including pH changes, transit time, and colonic microflora. A delivery system may be designed so that it breaks down under the approximately neutral pH conditions within the small intestine and colon. The delivery system would be incorporated into an acidic product that maintained its low pH in the mouth and stomach but started to break down in the small intestine thereby releasing the probiotics. Alternatively, the fact that a delivery system reaches the large intestine after a certain time and remains there for a prolonged period may be used to release the probiotics in the colon. A delivery system could be designed so that it slowly breaks down as it travels through the GIT, thereby releasing most of the probiotics within the colon. Finally, the most suitable method probably is to use the presence of the microbial population in the colon as a means to release the probiotics. Dietary fibers (such as pectin, alginate, and carrageenan) are largely indigestible within the mouth, stomach, and small intestine, but they are digested by enzymes released by certain colonic bacteria. These dietary fibers are therefore particularly suitable for use as building blocks for fabricating delivery systems to protect and release probiotics in the GIT.

## 9.4.2 Potential Delivery Systems

In this section, a number of delivery systems that might be suitable candidates for encapsulating and protecting probiotics in foods and beverages are discussed. An appropriate delivery system should be capable of encapsulating sufficient quantities of probiotics without adversely affecting food quality or shelf life. In addition, it should protect the probiotics within the food product and within the GIT and then release them within the colon.

### 9.4.2.1 Surfactant-Based Delivery Systems

Micelles and microemulsions are unsuitable delivery systems for probiotics because the particles are too small ($d < 100$ nm) to encapsulate the bacteria ($d > 1000$ nm). In principle, large unilamellar vesicles can be used to trap bacteria within the internal aqueous phase and therefore protect them from components in the external aqueous phase that may cause loss of viability (Figure 9.6). In practice, the liposomes are sensitive to solution conditions and environmental changes and therefore must be carefully designed to remain stable in food products. It is also difficult to ensure that a large fraction of the probiotics is actually trapped within the internal aqueous phase of the liposomes. Finally, liposomes typically have a relatively poor stability in commercial food products. For these reasons, there are very few published examples of the encapsulation and delivery of probiotics using liposomes in foods. The stability of liposomes outside and inside the human body may be modulated by altering their composition or preparation method (Isailovic et al. 2013; Lu et al. 2011b) or by coating them with dietary fiber layers after they have been formed (Chun et al. 2013; Laye et al. 2008; Liu et al. 2013a). However, there is still the problem of obtaining a sufficiently high loading capacity.

### 9.4.2.2 Emulsion-Based Delivery Systems

Emulsions, nanoemulsions, and SLNs are unsuitable for encapsulation of probiotics because the internal phase of the particles is hydrophobic, whereas the bacteria are hydrophilic. Probiotics can be encapsulated within the internal aqueous phase of W/O/W emulsions and therefore protected from any components in the external aqueous phase that may cause loss of viability (Jimenez-Colmenero 2013). Initially, the probiotics are dispersed within the internal aqueous phase and then homogenized with an oil phase containing an oil-soluble surfactant to form a stable water-in-oil (W/O) emulsion. The water droplets in this emulsion must be quite large ($d > 5000$ nm) in order to accommodate the probiotics. In addition, the homogenization conditions and ingredients selected should not cause loss of viability of the bacteria during the emulsion preparation process. The internal aqueous phase may contain components that facilitate the survival of the probiotics during manufacture, storage, transport, and utilization, such as nutrients and cryo-protectants. A W/O/W emulsion is then formed by homogenizing the W/O emulsion with an external aqueous phase that contains a water-soluble emulsifier. Again, the homogenization conditions and ingredients should be selected so that they do not cause loss of probiotic viability. Multiple emulsions are often highly unstable when incorporated into food products and would therefore need to be carefully designed to prevent flocculation, coalescence, gravitational separation, and osmotic swelling (Chapter 6). In the case of probiotics, the droplets in W/O/W emulsions would have to be rather large to ensure that the bacteria were encapsulated. Large droplets are often associated with a high degree of physical instability in emulsions; for example, droplet aggregation and gravitational separation usually increase with increasing particle size. The multiple emulsions would probably be most suitable for highly viscous or semisolid food products (such as yogurts, desserts, spreads, cheeses, and sauces) or for solid products after the delivery system has been dried (such as breads, crackers, cookies, and cakes). They may therefore be suitable for the yogurt-based product considered in this section, but their utilization may be limited by their relatively high production costs and low stability in commercial products.

### 9.4.2.3 Biopolymer-Based Delivery Systems

The most suitable delivery systems for encapsulating, protecting, and releasing probiotics in the colon are biopolymer-based ones (Huq et al. 2013; Islam et al. 2010; Teoh et al. 2011). Probiotics can be trapped within hydrogel particles that can be formed by many of the preparation methods described in Chapter 7, such as extrusion, coacervation, and thermodynamic incompatibility. The probiotics are mixed with an aqueous solution containing a gelling biopolymer, and then solution conditions or environmental conditions are changed to induce bead formation and gelation. A commonly used example of this approach is the encapsulation of probiotics in calcium alginate beads (Huq et al. 2013; Riaz and Masud 2013). The probiotics are mixed with an aqueous solution containing alginate, and then this mixture is dripped or injected into an aqueous solution containing calcium (Krasaekoopt and Bhandari 2012).

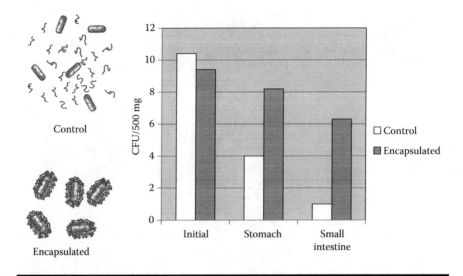

**Figure 9.7   Delivery systems can be designed to protect probiotics in the GIT.** This figure shows data from a study where *Lactobacillus acidophilus* was coated with layers of chitosan and carboxymethyl cellulose. (Data adapted from A.J. Priya et al., *Journal of Agricultural and Food Chemistry*, 59(21), 11838–11845, 2011.)

The cationic calcium ions promote gelation of the anionic alginate molecules through electrostatic bridging, which leads to the formation of bacteria trapped within hydrogel beads (Chandramouli et al. 2004). Coacervation involves mixing the probiotics with a mixture of positive and negative biopolymers to form a hydrogel bead with probiotics trapped inside (Oliveira et al. 2007a, b). An alternative approach is to coat the bacteria with multiple layers of biopolymers using the electrostatic deposition method (similar to multilayer emulsion formation) (McClements 2010a). Indeed, a dramatic improvement in the fraction of viable bacteria reaching the colon upon encapsulating the bacteria in multilayer polymer coatings formed from CMC and chitosan has been reported (Priya et al. 2011) (Figure 9.7). It is also possible to use the electrostatic deposition method to form a biopolymer coating around probiotic-containing hydrogel beads to obtain improved functional properties (Chen et al. 2013). There are numerous examples in the literature of the utilization of biopolymer-based delivery systems for encapsulating and protecting probiotics (Burgain et al. 2011; Cook et al. 2012; Heidebach et al. 2012; Krasaekoopt and Bhandari 2012; Sanders and Marco 2010; Teoh et al. 2011). A biopolymer-based delivery system should be designed to protect the encapsulated bacteria within the food, mouth, stomach, and small intestine, but then release them within the colon (Figure 9.8). This goal may be achieved by controlling changes in matrix integrity, swelling, or

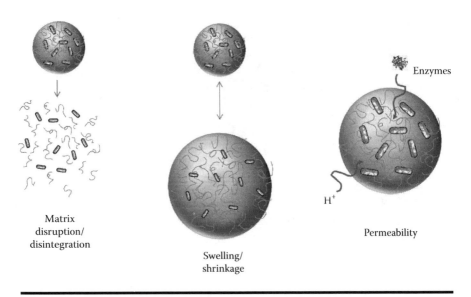

**Figure 9.8** **A probiotic delivery system should protect the encapsulated bacteria within the food, mouth, stomach, and small intestine but release them within the colon. This may be achieved by controlling the matrix integrity, swelling, or permeability in response to different solution conditions.**

permeability in response to changes in solution conditions, such as pH, ionic strength, or enzyme activity.

## 9.5 Controlled Flavor Release

Food and beverage manufacturers often want to create products with specific flavor release profiles, such as burst, sustained, or sequential release (Augustin et al. 2001; Madene et al. 2006). Colloidal delivery systems provide a potential means of creating products with well-defined flavor release profiles. In these applications, the delivery system must be designed to retain and protect the flavor molecules during storage, but then release them at a controlled rate after the product is prepared or consumed. There are a number of potential triggers for flavor release from a colloidal delivery system, such as hydration, dilution, temperature changes, enzyme activity, and pH alterations. In this section, we will consider controlling the release of flavor from a snack product such as a potato chip (crisp). In this example, the delivery system will initially be in the form of a powder that is sprayed onto the surface of the product. After ingestion, the flavors are released as the product is masticated within the oral cavity as a result of changes such as fragmentation, hydration, dilution, pH, temperature, and enzyme activity.

## 9.5.1 Design Criteria

### 9.5.1.1 Loading and Retention Characteristics

Typically, the overall flavor profile of a food or beverage product is determined by a complex mixture of various nonvolatile and volatile components. Flavor molecules differ according to their oil–water partition coefficients, solubilities, volatilities, and flavor intensities. Each flavor component should be released at a specific rate and level after food ingestion to achieve the desired flavor profile. A suitable colloidal delivery system must therefore be capable of incorporating appropriate amounts of the various flavor molecules, controlling their initial location within the system (e.g., oil phase, water phase, and headspace), preventing their loss during storage, and controlling their release rate after ingestion. If the desired flavor contains both polar and nonpolar components, then it may be necessary to use a delivery system that has both oil and aqueous phases. As mentioned earlier, a delivery system for a flavor on a snack product will usually be in a powdered form that coats the surface of the product. Consequently, the delivery system must be capable of being converted into a powdered form, for example, by spray drying. The properties of the solid matrix in a powder can often be controlled to inhibit loss of a flavor component during storage. For example, ingredients and processing operations that create a solid matrix in a glassy state can be chosen, since this will slow down molecular diffusion processes and inhibit flavor loss (Uhlemann and Reiss 2010).

### 9.5.1.2 Influence on Physicochemical and Sensory Properties of Product

In this example, the delivery system will form part of a powder that coats the surfaces of the snack product. This powder should remain attached to the snack surfaces during storage and give a desirable appearance to the product once it is opened and consumed. It may therefore be necessary to also encapsulate an appropriate mixture of dyes within the delivery system to give a pleasant visual appearance. It may also be necessary to control the size of the powder particles to control their appearance, mouthfeel, and dissolution characteristics in the mouth. This may be achieved by controlling dehydration conditions and using postdrying methods, such as agglomeration.

### 9.5.1.3 Product Stability

The flavor molecules should be retained within the food or beverage product throughout storage and transport and should remain stable to chemical degradation. Flavor molecules may leach out of the delivery system into the snack product or into the gaseous phase above the food during storage. The tendency for flavor molecules to move from one environment to another is determined by both thermodynamics (equilibrium partition coefficients) and kinetics (mass transport processes) (Uhlemann and Reiss 2010). The thermodynamic driving force for flavor release is governed by particle–water ($K_{PW}$) and particle–gas ($K_{PG}$) partitioning coefficients, as well as changes in system composition (such as dilution after ingestion). In addition, some flavor molecules are susceptible to chemical degradation during storage, such as citral and limonene (Uhlemann and Reiss 2010). The rate, pathway, and extent of degradation depend on the nature of the flavor molecule and food matrix, as well as on environmental conditions such as temperature, oxygen levels, light exposure, prooxidants, water activity, and pH. A delivery system should therefore be designed so that it will retain and protect the flavor molecules within the product during storage, for example, by ensuring they remain within the core of the colloidal particles. This objective may be achieved using a thermodynamic approach, such as selecting materials that control the equilibrium partition coefficients (i.e., particle–water and particle–gas) so that the flavor molecules tend to remain within the particles. Alternatively, it may be achieved using a kinetic approach, such as using core or shell materials that slow down the diffusion of the flavor molecules out of the particles, for example, a glassy wall matrix (Uhlemann and Reiss 2010). It may also be necessary to control food matrix composition and environmental conditions to inhibit chemical degradation, such as light exposure, oxygen concentration, pH, antioxidant addition, transition ion levels, and so on.

### 9.5.1.4 Flavor Release Profile

The flavor release profile desired by the food or beverage manufacturer (e.g., burst, sustained, or sequential) will determine the nature of the delivery system used (Madene et al. 2006). A burst release profile may be obtained by designing a system that rapidly breaks down in the mouth during mastication, whereas a sustained release profile may be obtained by designing a delivery system that slowly dissolves or disintegrates within the mouth or one that inhibits the diffusion of the flavor molecules out of intact particles. A sequential release profile may be obtained by encapsulating different types of flavor molecules in different types of colloidal particles (e.g., one type of flavor in rapid-release particles and another in slow-release particles) or in different regions of colloidal particles (e.g., one type of flavor in the core and another in the shell). The composition, structure, and physicochemical properties of the colloidal particles in a delivery system will ultimately determine the release profile.

## 9.5.2 Potential Delivery Systems

In this section, a number of delivery systems that might be suitable candidates for encapsulating and protecting flavors are reviewed. An appropriate delivery system should be capable of encapsulating sufficient quantities of flavor molecules, retaining them during storage, and then releasing them during mastication at an appropriate rate and extent.

## 9.5.2.1 Surfactant-Based Delivery Systems

In general, micelles and microemulsions are suitable delivery systems for lipophilic flavors for certain food and beverage applications because they can be used to solubilize them within their hydrophobic interiors (Garti and Aserin 2012; Rao and McClements 2011; Ziani et al. 2012). The fraction of the flavor molecules that remains within the nonpolar core of surfactant-based delivery systems depends on their solubilization capacity and the equilibrium partition coefficients of the system, as well as the total amount of micelle or microemulsion particles present (McClements 2005). One major advantage of using this type of delivery system is that they can be used in optically transparent products (such as clear drinks) because the particles are so small ($d < 100$ nm) that they do not scatter light strongly. In addition, they are thermodynamically stable systems and should therefore remain stable indefinitely, provided the solution and environmental conditions are not changed (McClements 2012c). On the other hand, high surfactant levels are undesirable because of high costs, off-flavors, and potential toxicity. Liposomes may also be used to encapsulate lipophilic flavor molecules; however, they are typically less stable than micelles or microemulsions and may give a cloudy appearance in aqueous-based products because of their larger dimensions (Liu et al. 2013b; Mozafari et al. 2008; Uhlemann and Reiss 2010). Nevertheless, they can be converted into a powdered form (Misra et al. 2009) so that they can be utilized in dried products such as snacks. Another potential problem with micelles, microemulsions, and liposomes is that they have little ability to slow down the release of flavor molecules because of their very small dimensions.

In a snack application, a surfactant-based delivery system would have to be dehydrated prior to application onto the surface of the snack product. It would therefore need to be mixed with a suitable wall material (e.g., carbohydrate or protein) that would form the solid matrix around the colloidal particles in the powdered flavor formulation. The composition and physicochemical properties of this solid matrix have to be carefully controlled to restrict molecular diffusion and thereby increase the retention and stability of the encapsulated flavor molecules (Carolina et al. 2007). As mentioned already, one of the major problems associated with using this kind of delivery system is that relatively high amounts of surfactant are needed to encapsulate nonpolar flavor molecules, which might provide an unpleasant taste after ingestion.

## 9.5.2.2 Emulsion-Based Delivery Systems

Emulsions, nanoemulsions, multiple emulsions, and SLNs are suitable for encapsulation of both hydrophilic and hydrophobic flavors because they have polar (water) and nonpolar (oil) phases in the same system (McClements 2010; Sagalowicz and Leser 2010; Velikov and Pelan 2008). Initially, the flavors would be incorporated into a suitable emulsion-based delivery system and then the system would need to be dehydrated, for example, by spray drying (Desai and Park 2005; Fang and Bhandari 2012; Jafari et al. 2007, 2008). Prior to dehydration, a suitable wall material (e.g., carbohydrate or protein) would need to be dispersed in the aqueous phase of the emulsion-based system. This wall material would form the solid matrix around the lipid particles in the powdered flavor formulation. The composition of the water material and the processing operations used during dehydration have to be carefully controlled to ensure that a solid matrix is formed that gives the powder appropriate properties, such as flow characteristics, stickiness, dissolution properties, inhibition of chemical degradation, and high flavor retention (Carolina et al. 2007; Murrieta-Pazos et al. 2012; Vega and Roos 2006). The rate of flavor release from a powdered form of an emulsion-based delivery system may be controlled in a number of ways:

- *Matrix dissolution*: Initially, the oil droplets containing the flavor molecules are trapped within a solid matrix (wall material) that typically consists of proteins or carbohydrates. The dissolution behavior of this solid matrix within the mouth will determine the flavor release profile. Thus, by selecting different types of proteins or carbohydrates to form the solid matrix, it may be possible to control the dissolution behavior and flavor release profile.
- *Particle size*: The release of flavor molecules from a colloidal particle occurs more slowly as the particle dimensions increase. Thus, the flavor release profile can be controlled by using differently sized oil droplets in an emulsion-based delivery system. Nevertheless, relatively large particles ($d > 10$ μm) are typically needed to cause an appreciable delay (a few seconds or more) of flavor release within emulsions (McClements 2005).
- *Oil phase properties*: The oil phase used to prepare an emulsion-based delivery system determines oil droplet polarity, which will in turn determine the solubility and partitioning of hydrophobic and hydrophilic flavor molecules within them (McClements 2005). In addition, the rheological properties of the oil phase will determine the rate at which flavor molecules can diffuse through the particles—the higher the viscosity, the slower the diffusion. Thus, it may be possible to control the flavor release profile by altering the nature of the oil phase used to prepare the initial emulsion.
- *Particle physical state*: The release of a nonpolar flavor molecule from a lipid particle may be controlled by controlling its physical state (Ghosh et al. 2006, 2007, 2008). For example, a flavor molecule may be trapped within a solid lipid particle and then released when the particle melts in the mouth. In this case, it is important to ensure that the flavor molecules are trapped throughout the solid matrix, rather than expelled to the surface of the lipid particles once the fat phase crystallizes (Mei et al. 2010).
- *Particle structure*: Hydrophilic flavor molecules may be trapped within the internal aqueous phase of W/O/W emulsions, for example, NaCl (Pawlik and Norton 2012; Sapei et al. 2012). The release of these molecules may

then depend on the breakdown of these double emulsions within the oral cavity. If the W/O droplets can be designed so that they break down at different rates, it may be possible to control the release of hydrophilic flavors, such as salt, bitter, sweet, or acid compounds. Alternatively, W/O/W emulsions may be used to encapsulate bitter components so that they are not released in the mouth, which may be important for some bioactive components, such as bitter peptides and astringent proteins (Jimenez-Colmenero 2013).

■ *Particle surface characteristics*: The ability of lipid droplets to adhere to the tongue and oral cavity during mastication depends on their surface characteristics, such as interfacial composition, electrical charge, and hydrophobicity (van Aken and de Hoog 2009; van Aken et al. 2005, 2007). For example, cationic fat droplets may stick to the tongue more strongly than anionic ones, which will alter their retention time within the oral cavity and therefore their flavor release profiles.

■ *Particle stability*: The stability of lipid droplets to flocculation, coalescence, and phase separation within the oral cavity may also be utilized to control the release of flavors from them. Certain types of fat droplets stay intact within the mouth and do not interact strongly with the oral cavity, whereas others may flocculate, coalesce, or spread on the tongue. The behavior of fat droplets within the mouth may therefore have an important impact on the flavor release profile (Dresselhuis et al. 2007, 2008). A fat droplet that breaks down and spreads across the tongue will release flavor molecules differently from one that remains intact within the mouth.

### 9.5.2.3 Biopolymer-Based Delivery Systems

Biopolymer-based delivery systems offer considerable scope for controlling the release of flavor molecules in food and beverage products (Burey et al. 2008; Desai and Park 2005). Flavors can be trapped within hydrogel particles using preparation methods such as extrusion, coacervation, thermodynamic incompatibility, and emulsion templating (Chapter 7). Filled hydrogel particles are particularly suitable for this purpose because they are capable of incorporating both hydrophobic and hydrophilic flavor molecules in the same system. The flavor release profile of a filled hydrogel particle system can be controlled in numerous ways: the composition, dimensions, location, interfacial properties, and physical state of the lipid droplets can be varied; the composition, dimensions, interactions, and pore size of the hydrogel particles can be varied. For example, studies have shown that the release of nonpolar flavor molecules can be delayed by increasing the size of the hydrogel particles where the fat droplets are embedded in (Malone and Appelqvist 2003). This phenomenon was used to create reduced-fat products that have similar flavor release profiles as high-fat products. Emulsified flavor oils have also been trapped within hydrogel beads formed using coacervation (Bouquerand et al. 2012; Buldur and Kok 2011; Weinbreck et al. 2004). Interestingly, studies of the release characteristics of flavors from complex coacervate beads showed that the release rate did not depend strongly on cross-linking or core-to-wall ratio but was mainly determined by the equilibrium partition coefficients of the flavor molecules (Leclercq et al. 2009). After formation, biopolymer-based delivery systems may be dehydrated (e.g., by spray or freeze drying) to form powders suitable for utilization as coatings on snack products (Desai and Park 2005; Fang and Bhandari 2012; Jafari et al. 2008). As mentioned above, a suitable wall material needs to be incorporated into the product prior to drying to ensure that the final powder has suitable physico-chemical characteristics.

# 9.6 Protection of Lipophilic Active Agents against Oxidation

There is growing interest in incorporating various kinds of bioactive lipids into commercial food and beverage products because of their potential health benefits, for example, ω-3 oils, conjugated linoleic acid, oil-soluble vitamins, and carotenoids (Given 2009; McClements et al. 2009; Sagalowicz and Leser 2010; Velikov and Pelan 2008). This can be a challenge because many of these bioactive lipids are polyunsaturated molecules that are highly susceptible to oxidative degradation during product preparation, transport, and storage (McClements and Decker 2000; Waraho et al. 2011). The chemical stability of these bioactive lipids can often be improved by encapsulating them within colloidal delivery systems that inhibit their oxidative degradation. In this section, we will focus on the development of delivery systems to incorporate and protect ω-3-rich oils (such as fish, flax, or algae oils) in salad dressings. These products are optically opaque, highly viscous liquids that normally have an acidic aqueous phase.

It should be noted that there are a number of strategies that can be used to protect ω-3-rich oils from oxidation that do not require encapsulation. These strategies may be more economically viable than encapsulation for some commercial applications, or they may be used in combination with encapsulation to further improve product stability. Typically, the rate of lipid oxidation in aqueous-based foods increases with increasing oxygen concentration, temperature, light exposure, and prooxidant levels (McClements and Decker 2000; Waraho et al. 2011). The susceptibility of a food product to oxidation can therefore be reduced by decreasing oxygen levels, storing at cold temperatures, avoiding exposure to light, reducing prooxidant levels (e.g., by using purified ingredients or adding chelating agents), and adding antioxidants. Nevertheless, the utilization of colloidal delivery systems can often provide additional protective strategies that can improve the shelf life and quality of a commercial product.

## 9.6.1 Design Criteria

### 9.6.1.1 Loading and Retention Characteristics

The amount of bioactive lipids that must be ingested to have a beneficial biological effect depends on the lipid. For ω-3 fatty acids, it has been suggested that a level of at least 250 mg/day is required to demonstrate any health benefits (Mozaffarian and Wu 2011). A food or beverage product must therefore be fortified with a significant fraction of this amount of bioactive lipid per serving to have a beneficial effect. For high-fat products, this level is relatively easy to achieve, but for low-fat products, it may be necessary to have a high loading capacity in the delivery system. Salad dressings typically have a relatively high fat content, and therefore it is relatively easy to incorporate sufficient quantities of ω-3 oils into these products (Let et al. 2007a, b). It is possible to deliver ω-3 fatty acids into foods in a number of different forms, such as FFAs, monoacyglycerols (MGs), diacylglycerols (DGs), triacylglycerols (TGs), or phospholipids. These different molecular forms typically have a very low water solubility and therefore the retention of the ω-3 oils within colloidal particles is usually relatively high.

### 9.6.1.2 Influence on Physicochemical and Sensory Properties of Product

In general, a suitable colloidal delivery system for ω-3 oils should not adversely affect the appearance, texture, flavor, or mouthfeel of the food product it is incorporated into, and it should maintain its desirable characteristics throughout the shelf life of the product (Lesmes and McClements 2009; McClements 2012a). In the case of a salad dressing, the colloidal delivery system needs to be incorporated into a cloudy or opaque product that has a relatively high viscosity (McClements 2005; Sikora et al. 2008). As discussed earlier, it is not important to use a delivery system that contains very small particles in a cloudy or opaque product, and there may be some advantages because the delivery system itself may contribute to the desirable creamy appearance of the product. Incorporating an ω-3 oil delivery system into a commercial salad dressing should also not cause any undesirable changes in its texture or mouthfeel.

### 9.6.1.3 Product Stability

The delivery system used must remain physically stable within the food product throughout its lifetime, and hence it should be designed to prevent instability because of gravitational separation, flocculation, coalescence, or oiling off. Salad dressings typically have an acidic aqueous phase that contains relatively high levels of salts and other components, and so the delivery system must be designed to operate in this environment (McClements 2005; Sikora et al. 2008). Ostwald ripening is usually not a problem for ω-3 oils because of their very low water solubilities. Probably the biggest challenge to incorporating ω-3 oils into food products, such as salad dressings, is their high susceptibility to chemical degradation (McClements and Decker 2000; Waraho et al. 2011). Polyunsaturated lipids are highly susceptible to lipid oxidation, which leads to the formation of undesirable off-flavors ("rancidity"), loss of nutritional value, and formation of potentially toxic reaction products. The rate and extent of oxidation depend on the nature of the lipid molecules (such as degree of unsaturation and molecular form), solution conditions (such as pH and ionic composition), light exposure, oxygen levels, temperature, and the presence of catalysts or inhibitors (Decker et al. 2011; Kargar et al. 2011; McClements and Decker 2000). The development of a suitable delivery system that can be used to inhibit lipid oxidation requires an understanding of the chemical degradation pathway and the factors that influence it.

Lipid oxidation involves a complex sequence of chemical changes that result from the interaction of lipids with oxygen active species (McClements and Decker 2000; Waraho et al. 2011). The pathway and rate of lipid oxidation depend on the nature of the reactive species involved, the surrounding food matrix, and environmental conditions. Lipid oxidation can be conveniently divided into three distinct stages: initiation, propagation, and termination stages (Figure 9.9) (McClements and Decker 2008).

- *Initiation* involves abstraction of a hydrogen atom from a fatty acid to form an *alkyl radical* (L˙) that is stabilized by delocalization over the double bond(s).
- *Propagation* involves addition of oxygen to an *alkyl radical* resulting in the formation of a *peroxyl radical* (LOO˙). Peroxyl radicals are highly reactive and promote abstraction of a hydrogen atom from another molecule. The C–H

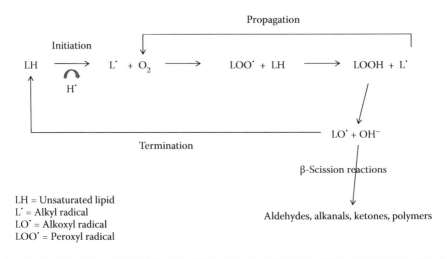

**Figure 9.9 Schematic representation of the major steps involved in lipid oxidation. (Diagram modified from the one kindly provided by Leqi Cui.)**

bond of unsaturated fatty acids is relatively weak and is therefore highly susceptible to attack from LOO˙ radicals. Hydrogen addition to a LOO˙ radical results in the formation of a lipid hydroperoxide (LOOH), as well as a new alkyl radical (L˙) on the other fatty acid molecule that was attacked, thereby propagating the reaction from one fatty acid to another.

■ *Termination* involves the combination of two free radicals to form nonradical species, thereby terminating the lipid oxidation reaction.

Lipid hydroperoxides formed in the above process decompose into *alkoxyl* radicals (LO˙), leading to numerous fragmentation and polymerization reactions that generate a wide range of reaction products. Some of these reaction products are small volatile molecules that contribute to the undesirable flavor profile of rancid fats, whereas others are large nonvolatile polymers that increase the viscosity.

Various factors influence the rate and extent of lipid oxidation in food products (Jacobsen 2008; McClements and Decker 2000; Waraho et al. 2011). The rate of lipid oxidation tends to increase with the following: increasing number of double bonds in the lipids, increasing oxygen concentration, increasing prooxidants, decreasing antioxidants, increasing light exposure, increasing temperature, and extreme water activities. These factors must therefore be carefully controlled to limit lipid oxidation in commercial products. In addition, a number of other antioxidant strategies can be designed into colloidal delivery systems, such as engineering the interface so that prooxidant transition metals are electrostatically or sterically repelled from the droplet surface, or that the antioxidants are located in the most effective location where oxidation occurs (e.g., within the oil droplets or at the oil–water interface).

## 9.6.2 Potential Delivery Systems

An overview of various delivery systems that can potentially be used to protect ω-3 oils from oxidation in food products such as salad dressing is given in this section. A suitable delivery system should be capable of carrying a biologically significant level of ω-3 oil, should not adversely affect food quality and shelf life, should protect the polyunsaturated lipids from oxidation, and should release the bioactive fatty acids within the GIT so that they can be absorbed.

### 9.6.2.1 Surfactant-Based Delivery Systems

Micelles or microemulsions are suitable for incorporating bioactive lipids into optically transparent food or beverage products (such as fortified waters or soft drinks) because of the weak light scattering by the very small colloidal particles they contain. However, salad dressings are typically optically opaque and therefore this attribute is not particularly important in this product. On the other hand, microemulsions containing FFAs have been reported to have unusual oxidation properties: the oxidative stability of encapsulated FFAs increased as their degree of unsaturation increased (Miyashita et al. 1993). This phenomenon was attributed to differences in the physical orientation of the fatty acid double bonds within the microemulsion particles: the double bonds in more unsaturated fatty acids appear to be more effectively shielded from interactions with aqueous phase prooxidants. Nevertheless, it is typically undesirable to use FFAs (rather than TGs) as a source of ω-3 oils in foods because of their poor taste profiles. Studies have shown that ω-3-rich TG oils (fish oil) can be incorporated into microemulsions (Zheng et al. 2011), and these systems could be used as delivery systems for these bioactive lipids in foods. Nevertheless, there are a number of limitations of micelles and microemulsions that may limit their widespread commercial application: (i) they are often difficult to fabricate from long-chain TG oils (such as most sources of ω-3 fatty acids); (ii) the loading capacity is usually relatively low (<10%); (iii) light can penetrate into them (which may increase oxidation); and (iv) they require high levels of surfactants that may be unacceptable for use in many foods because of taste, legal, and labeling concerns. Therefore, there appears to be few advantages to using micelles or microemulsions to deliver polyunsaturated oils in salad dressing-type products.

Liposomes may also be suitable delivery systems for polyunsaturated lipids in some food products (Lu et al. 2011a). Liposomes may be fabricated from phospholipids that have ω-3 fatty acid tails so that the delivery system itself serves as the source of bioactive lipids. Alternatively, MG, DG, or TG forms of ω-3 fatty acids can be incorporated into the liposome structure by solubilizing them in the lipid bilayers. In one study where polyunsaturated fatty acids were part of the phospholipid molecule, it was reported that the number of double bonds had little impact on the rate of lipid oxidation, which is different from that reported for bulk systems where the oxidation rate typically increases with increasing unsaturation (Araseki et al. 2002). One of the major limitations of using liposomes is that they tend to have a relatively poor physical stability in commercial products, unless they are stabilized, for example, by incorporating cholesterol into the bilayer to decrease membrane fluidity or by using biopolymer coatings that help prevent aggregation (Chun et al. 2013; Gibis et al. 2012). Liposomes may be suitable for application in salad dressing products because they can be formulated from label-friendly ingredients and the degree of unsaturation does not appear to have a major impact on the rate of oxidation; however, they must be carefully designed to remain physically stable within the product, which is often a challenge.

### 9.6.2.2 Emulsion-Based Delivery Systems

Conventional O/W emulsions offer one of the simplest means of incorporating polyunsaturated lipids into food products because they are simple to fabricate and have a high loading capacity (Jacobsen 2008; McClements and Decker 2000; Salminen et al. 2013; Waraho et al. 2011). However, they can only be used for products that are cloudy or opaque owing to the relatively strong light scattering efficiency of the droplets they contain. They also have to be carefully designed to ensure that they remain physically stable throughout product processing, storage, and utilization (McClements 2005). Numerous strategies have been developed to improve the oxidative stability of polyunsaturated lipids in conventional emulsions based on knowledge of the physicochemical mechanisms of oxidation in these heterogeneous systems (Figure 9.10). The presence of transition metals (which act as powerful catalysts of lipid oxidation) has been identified as one of the most important factors influencing the rate of lipid oxidation in emulsions (McClements and Decker 2000). Consequently, strategies that remove or inactivate transition metals are particularly effective at inhibiting lipid oxidation, such as using ingredients with low transition metal concentrations, removing them, chelating them, or preventing them from accumulating at droplet surfaces. Studies have shown that emulsions stabilized with anionic surfactants oxidize faster than those stabilized by cationic surfactants since negative interfaces attract positively charged metal ions to the lipid droplet surfaces, whereas positive interfaces repel them (Hu et al. 2003; Mei et al. 1999). In addition, increasing the thickness or decreasing the permeability of the interfacial layer also decreases the ability of transition metals from interacting with encapsulated lipids by forming a physical barrier that separates prooxidants from lipid substrates (McClements and Decker 2000).

The physical and chemical stability of emulsions can be altered by coating the droplet interfaces with one or more layers of biopolymers (Gudipati et al. 2010; Lomova et al. 2010). Multilayer emulsions may inhibit lipid oxidation by producing a thicker interfacial layer that limits interactions between the encapsulated lipid and prooxidants in the aqueous phase components or by making the lipid droplets

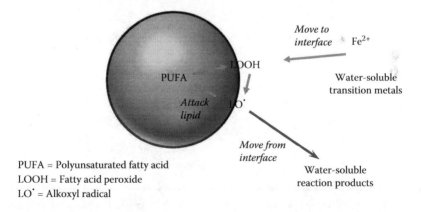

**Figure 9.10  Proposed mechanism of lipid oxidation in O/W emulsions. The rate and extent of oxidation depend on the location of the different reactants in the system.**

positively charged so that they repel transition metals. In addition, antioxidants (such as some proteins) can be incorporated into the interfacial layer surrounding the lipid droplets, thereby improving their oxidative stability (Lesmes et al. 2010). The stability of emulsified polyunsaturated fats to oxidation is compared for droplets coated by caseinate (primary) or caseinate and lactoferrin (secondary) in Figure 9.11. The secondary emulsion was much more stable to oxidation than the primary emulsion, which was attributed to differences in the thickness, charge, and antioxidant activity of the interfacial coatings (Lesmes et al. 2010). Multiple (W/O/W) emulsions may also be useful at inhibiting the oxidation of polyunsaturated lipids by incorporating water-soluble antioxidants or chelating agents in the interior or exterior aqueous phases and oil-soluble surfactants in the lipid phase (Poyato et al. 2013).

Conventional emulsions appear to be particularly suitable for application in salad dressing products because they are relatively easy to fabricate from food-grade ingredients and simple processing operations and because they contribute to the desirable creamy texture and appearance of this kind of product. Other types of emulsion-based delivery systems may have certain advantages over conventional emulsions for certain situations. Nanoemulsions may have greater physical stability and oral bioavailability than conventional emulsions (Deshpande et al. 2013; Dey et al. 2012), but they are more difficult to prepare, and the rate of oxidation may be faster because of their higher surface areas. SLNs or nanostructured lipid carriers may also be

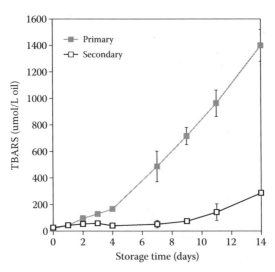

**Figure 9.11 Delivery systems can be designed to protect encapsulated bioactive lipids from chemical degradation. This example compares lipid oxidation (increase in TBARS) during storage for fat droplets coated by caseinate (primary) or lactoferrin + caseinate (secondary). (Data from U. Lesmes et al., *Food Chemistry*, 123(1), 99–106, 2010.)**

suitable for stabilizing ω-3 oils against oxidation, provided they can be designed correctly (Lacatusu et al. 2013). Some studies have shown that unsaturated lipids may be expelled from lipid nanoparticles when they are cooled to a temperature where the fat phase crystallizes (Okuda et al. 2005). Under these circumstances, the rate of lipid oxidation may be worse in an SLN than in a liquid lipid nanoparticle because the unsaturated molecules are located at the particle surface close to prooxidants in the aqueous phase. On the other hand, recent studies have shown that SLN suspensions that consist of lipid particles that have a liquid core of ω-3 oils and a solid shell of saturated lipids can be prepared (Helgason et al. 2009). In this case, the ω-3 oils are protected from oxidation because the solid shell acts as a barrier that inhibits the contact between reactants.

## 9.6.2.3 Biopolymer-Based Delivery Systems

There is considerable scope for controlling the rate and extent of lipid oxidation in food products using biopolymer-based delivery systems, such as hydrogel particles, filled hydrogel particles, and biopolymer nanoparticles. Filled hydrogel particles are probably the most suitable biopolymer-based delivery system for controlling lipid oxidation in salad dressing-type products. These systems can be formed using a number of different preparation methods (Chapter 7). Typically, an O/W emulsion that consists of an ω-3-rich oil dissolved within emulsifier-coated lipid droplets is formed. The lipid droplets would then be trapped within a suitable biopolymer matrix. Filled hydrogel particles may be designed to inhibit lipid oxidation in a number of ways. First, the properties of the lipid droplets could be controlled, such as their composition, size, interfacial characteristics, or physical state as described in the previous section. Second, the physicochemical properties of the hydrogel matrix could be controlled, for example, its composition, dimensions, or porosity. Antioxidants could be incorporated into the lipid droplets or into the hydrogel matrix to inhibit oxidation. The hydrogel matrix could be designed to inhibit the contact of the encapsulated ω-3 lipids with any prooxidants in the aqueous phase (such as transition metals) by creating a physical barrier or by binding them. The filled hydrogel particles would also have to be fabricated so that they could easily be incorporated into commercial products, such as salad dressings. As mentioned earlier, a salad dressing is typically opaque and has a high viscosity, and so there should be few problems in incorporating relatively large filled hydrogel particles into them. The main challenge will be to fabricate the hydrogel particles using commercially viable production methods that are capable of producing large volumes of material.

A number of studies have explored the possibility of using filled hydrogel particles to protect polyunsaturated lipids from oxidation. Emulsified fish oil has been encapsulated within caseinate-rich hydrogel particles fabricated using a multistep biopolymer phase separation method (Matalanis et al. 2012). Caseinate is known to have good antioxidant properties and was reported to stabilize the encapsulated oils from oxidation. Emulsified flaxseed and fish oils have been encapsulated within gelatin–gum arabic hydrogel particles fabricated using complex coacervation (Liu et al. 2010; Tamjidi et al. 2012). The encapsulated form of these polyunsaturated lipids was reported to be more stable than the nonencapsulated form. A review of different biopolymer-based delivery systems suitable for inhibiting oxidation has been published (Drusch et al. 2008).

## 9.7 Summary

This chapter has used a number of case studies to highlight some of the factors that need to be considered when selecting an appropriate delivery system for a particular application. The choice of a delivery system depends on the nature of the active ingredient to be encapsulated, the nature of the food matrix that it will be incorporated into, and the particular problem that is being addressed (such as improved dispersibility, increased bioavailability, controlled flavor release, or better chemical stability). Consequently, there is no single colloidal delivery system that can be used for every application. Instead, care must be taken to clearly define the problem to be addressed, the challenges that need to be overcome, and the specific criteria required for an appropriate delivery system. An appropriate delivery system that meets these requirements can then be identified based on knowledge of their ease of fabrication, stability to environmental conditions, physicochemical properties, and functional performance.

## References

Amar-Yuli, I., D. Libster, A. Aserin and N. Garti (2009). "Solubilization of food bioactives within lyotropic liquid crystalline mesophases." *Current Opinion in Colloid and Interface Science* **14**(1): 21–32.

Araseki, M., K. Yamamoto and K. Miyashita (2002). "Oxidative stability of polyunsaturated fatty acid in phosphatidylcholine liposomes." *Bioscience Biotechnology and Biochemistry* **66**(12): 2573–2577.

Augustin, M. A., L. Sanguansri, C. Margetts and B. Young (2001). "Microencapsulation of food ingredients." *Food Australia* **53**(6): 220–223.

Bauer, E., S. Jakob and R. Mosenthin (2005). "Principles of physiology of lipid digestion." *Asian-Australasian Journal of Animal Sciences* **18**(2): 282–295.

Bermudez, B., Y. M. Pacheco, S. Lopez, R. Abia and F. J. G. Muriana (2004). "Digestion and absorption of olive oil." *Grasas Y Aceites* **55**(1): 1–10.

Boon, C. S., D. J. McClements, J. Weiss and E. A. Decker (2010). "Factors influencing the chemical stability of carotenoids in foods." *Critical Reviews in Food Science and Nutrition* **50**(6): 515–532.

Bouquerand, P. E., G. Dardelle, P. Erni and V. Normand (2012). An industry perspective on the advantages and disadvantages of different flavor delivery systems. In *Encapsulation Technologies and Delivery Systems for Food Ingredients and Nutraceuticals*, N. Garti and D. J. McClements (eds.). Cambridge, UK, Woodhead Publ Ltd: 453–487.

Buldur, P. M. and F. N. Kok (2011). "Encapsulation of food flavors via coacervation method." *Current Opinion in Biotechnology* **22**: S96.

Burey, P., B. R. Bhandari, T. Howes and M. J. Gidley (2008). "Hydrocolloid gel particles: Formation, characterization, and application." *Critical Reviews in Food Science and Nutrition* **48**(5): 361–377.

Burgain, J., C. Gaiani, M. Linder and J. Scher (2011). "Encapsulation of probiotic living cells: From laboratory scale to industrial applications." *Journal of Food Engineering* **104**(4): 467–483.

Carolina, B. C., S. Carolina, M. C. Zamora and C. Jorge (2007). "Glass transition temperatures and some physical and sensory changes in stored spray-dried encapsulated flavors." *Lwt-Food Science and Technology* **40**(10): 1792–1797.

Chandramouli, V., K. Kailasapathy, P. Peiris and M. Jones (2004). "An improved method of microencapsulation and its evaluation to protect Lactobacillus spp. in simulated gastric conditions." *Journal of Microbiological Methods* **56**(1): 27–35.

Charman, W. N., C. J. H. Porter, S. Mithani and J. B. Dressman (1997). "Physicochemical and physiological mechanisms for the effects of food on drug absorption: The role of lipids and pH." *Journal of Pharmaceutical Sciences* **86**(3): 269–282.

Chen, S., Y. Cao, L. R. Ferguson, Q. Shu and S. Garg (2013). "Evaluation of mucoadhesive coatings of chitosan and thiolated chitosan for the colonic delivery of microencapsulated probiotic bacteria." *Journal of Microencapsulation* **30**(2): 103–115.

Chun, J.-Y., M.-J. Choi, S.-G. Min and J. Weiss (2013). "Formation and stability of multiple-layered liposomes by layer-by-layer electrostatic deposition of biopolymers." *Food Hydrocolloids* **30**(1): 249–257.

Constantinides, P. P., M. V. Chaubal and R. Shorr (2006). Advances in lipid nanodispersions for parenteral drug delivery and targeting. *Annual Meeting of the American-Association-of-Pharmaceutical-Scientists*, San Antonio, TX.

Cook, M. T., G. Tzortzis, D. Charalampopoulos and V. V. Khutoryanskiy (2012). "Microencapsulation of probiotics for gastrointestinal delivery." *Journal of Controlled Release* **162**(1): 56–67.

Corona-Hernandez, R. I., E. Alvarez-Parrilla, J. Lizardi-Mendoza, A. R. Islas-Rubio, L. A. de la Rosa and A. Wall-Medrano (2013). "Structural stability and viability of microencapsulated probiotic bacteria: A review." *Comprehensive Reviews in Food Science and Food Safety* **12**(6): 614–628.

Decker, E. A., B. Chen, A. Panya and R. J. Elias (2011). Understanding antioxidant mechanisms in preventing oxidation in foods. In *Oxidation in Foods and Beverages and Antioxidant Applications, Vol 1: Understanding Mechanisms of Oxidation and Antioxidant Activity*, E. A. Decker, R. J. Elias and D. J. McClements (eds.). Cambridge, UK, Woodhead Publishing, 225–248.

Desai, K. G. H. and H. J. Park (2005). "Recent developments in microencapsulation of food ingredients." *Drying Technology* **23**(7): 1361–1394.

Deshpande, D., D. R. Janero and M. Amiji (2013). "Engineering of an omega-3 polyunsaturated fatty acid-containing nanoemulsion system for combination C6-ceramide and 17 beta-estradiol delivery and bioactivity in human vascular endothelial and smooth muscle cells." *Nanomedicine-Nanotechnology Biology and Medicine* **9**(7): 885–894.

Dey, T. K., S. Ghosh, M. Ghosh, H. Koley and P. Dhar (2012). "Comparative study of gastrointestinal absorption of EPA & DHA rich fish oil from nano and conventional emulsion formulation in rats." *Food Research International* **49**(1): 72–79.

Dresselhuis, D. M., H. J. Klok, M. A. C. Stuart, R. J. de Vries, G. A. van Aken and E. H. A. de Hoog (2007). "Tribology of O/W emulsions under mouth-like conditions: Determinants of friction." *Food Biophysics* **2**(4): 158–171.

Dresselhuis, D. M., M. A. C. Stuart, G. A. van Aken, R. G. Schipper and E. H. A. de Hoog (2008). "Fat retention at the tongue and the role of saliva: Adhesion and spreading of 'protein-poor' versus 'protein-rich' emulsions." *Journal of Colloid and Interface Science* **321**(1): 21–29.

Drusch, S., S. Benedetti, M. Scampicchio and S. Mannino (2008). "Stabilisation of omega-3 fatty acids by microencapsulation." *Agro Food Industry Hi-Tech* **19**(4): 31–32.

Fang, Z. and B. Bhandari (2012). Spray drying, freeze drying, and related processes for food ingredient and nutraceutical encapsulation. In *Encapsulation Technologies and Delivery Systems for Food Ingredients and Nutraceuticals*, N. Garti and D. J. McClements (eds.). Oxford, UK, Woodhead Publishing: 73–108.

Fave, G., T. C. Coste and M. Armand (2004). "Physicochemical properties of lipids: New strategies to manage fatty acid bioavailability." *Cellular and Molecular Biology* **50**(7): 815–831.

Flanagan, J. and H. Singh (2006). "Microemulsions: A potential delivery system for bioactives in food." *Critical Reviews in Food Science and Nutrition* **46**(3): 221–237.

Fredrick, E., P. Walstra and K. Dewettinck (2010). "Factors governing partial coalescence in oil-in-water emulsions." *Advances in Colloid and Interface Science* **153**(1–2): 30–42.

Garti, N. and A. Aserin (2012). Micelles and microemulsions as food ingredient and nutraceutical delivery systems. In *Encapsulation Technologies and Delivery Systems for Food Ingredients and Nutraceuticals*, N. Garti and D. J. McClements (eds.). Cambridge, UK, Woodhead Publishing, 211–251.

Ghosh, S., D. G. Peterson and J. N. Coupland (2006). "Effects of droplet crystallization and melting on the aroma release properties of a model oil-in-water emulsion." *Journal of Agricultural and Food Chemistry* **54**(5): 1829–1837.

Ghosh, S., D. G. Peterson and J. N. Coupland (2007). "Aroma release from solid droplet emulsions: Effect of lipid type." *Journal of the American Oil Chemists Society* **84**(11): 1001–1014.

Ghosh, S., D. G. Peterson and J. N. Coupland (2008). "Temporal aroma release profile of solid and liquid droplet emulsions." *Food Biophysics* **3**(4): 335–343.

Gibis, M., E. Vogt and J. Weiss (2012). "Encapsulation of polyphenolic grape seed extract in polymer-coated liposomes." *Food and Function* **3**(3): 246–254.

Given, P. S. (2009). "Encapsulation of flavors in emulsions for beverages." *Current Opinion in Colloid and Interface Science* **14**(1): 43–47.

Gu, C. H., H. Li, J. Levons, K. Lentz, R. B. Gandhi, K. Raghavan and R. L. Smith (2007). "Predicting effect of food on extent of drug absorption based on physicochemical properties." *Pharmaceutical Research* **24**(6): 1118–1130.

Gudipati, V., S. Sandra, D. J. McClements and E. A. Decker (2010). "Oxidative stability and in vitro digestibility of fish oil-in-water emulsions containing multilayered membranes." *Journal of Agricultural and Food Chemistry* **58**(13): 8093–8099.

Heidebach, T., P. Forst and U. Kulozik (2012). "Microencapsulation of probiotic cells for food applications." *Critical Reviews in Food Science and Nutrition* **52**(4): 291–311.

Helgason, T., T. S. Awad, K. Kristbergsson, E. A. Decker, D. J. McClements and J. Weiss (2009). "Impact of surfactant properties on oxidative stability of beta-carotene encapsulated within solid lipid nanoparticles." *Journal of Agricultural and Food Chemistry* **57**(17): 8033–8040.

Hu, B. and Q. R. Huang (2013). "Biopolymer based nano-delivery systems for enhancing bioavailability of nutraceuticals." *Chinese Journal of Polymer Science* **31**(9): 1190–1203.

Hu, D. D., C. C. Lin, L. Liu, S. N. Li and Y. P. Zhao (2012). "Preparation, characterization, and in vitro release investigation of lutein/zein nanoparticles via solution enhanced dispersion by supercritical fluids." *Journal of Food Engineering* **109**(3): 545–552.

Hu, M., D. J. McClements and E. A. Decker (2003). "Lipid oxidation in corn oil-in-water emulsions stabilized by casein, whey protein isolate, and soy protein isolate." *Journal of Agricultural and Food Chemistry* **51**(6): 1696–1700.

Huq, T., A. Khan, R. A. Khan, B. Riedl and M. Lacroix (2013). "Encapsulation of probiotic bacteria in biopolymeric system." *Critical Reviews in Food Science and Nutrition* **53**(9): 909–916.

Isailovic, B. D., I. T. Kostic, A. Zvonar, V. B. Dordevic, M. Gasperlin, V. A. Nedovic and B. M. Bugarski (2013). "Resveratrol loaded liposomes produced by different techniques." *Innovative Food Science and Emerging Technologies* **19**: 181–189.

Islam, M. A., C. H. Yun, Y. J. Choi and C. S. Cho (2010). "Microencapsulation of live probiotic bacteria." *Journal of Microbiology and Biotechnology* **20**(10): 1367–1377.

Jacobsen, C. (2008). "Omega-3s in food emulsions: Overview and case studies." *Agro Food Industry Hi-Tech* **19**(5): 9–12.

Jafari, S. M., E. Assadpoor, Y. H. He and B. Bhandari (2008). "Encapsulation efficiency of food flavours and oils during spray drying." *Drying Technology* **26**(7): 816–835.

Jafari, S. M., Y. H. He and B. Bhandari (2007). "Encapsulation of nanoparticles of d-limonene by spray drying: Role of emulsifiers and emulsifying techniques." *Drying Technology* **25**(4–6): 1069–1079.

Jimenez-Colmenero, F. (2013). "Potential applications of multiple emulsions in the development of healthy and functional foods." *Food Research International* **52**(1): 64–74.

Kargar, M., F. Spyropoulos and I. T. Norton (2011). Microstructural design to reduce lipid oxidation in oil-in-water emulsions. In *11th International Congress on Engineering and Food*, G. Saravacos, P. Taoukis, M. Krokida, V. Karathanos, H. Lazarides, N. Stoforos, C. Tzia and S. Yanniotis (eds.). Elsevier, Amsterdam, Procedia Food Science, **1**: 104–108.

Karimi, R., A. M. Mortazavian and A. G. Da Cruz (2011). "Viability of probiotic microorganisms in cheese during production and storage: A review." *Dairy Science and Technology* **91**(3): 283–308.

Kosin, B. and S. K. Rakshit (2006). "Microbial and processing criteria for production of probiotics: A review." *Food Technology and Biotechnology* **44**(3): 371–379.

Krasaekoopt, W. and B. Bhandari (2012). Properties and applications of different probiotic delivery systems. In *Encapsulation Technologies and Delivery Systems for Food Ingredients and Nutraceuticals*, N. Garti and D. J. McClements (eds.). Cambridge, UK, Woodhead Publishing: 541–594.

Lacatusu, I., E. Mitrea, N. Badea, R. Stan, O. Oprea and A. Meghea (2013). "Lipid nanoparticles based on omega-3 fatty acids as effective carriers for lutein delivery. Preparation and in vitro characterization studies." *Journal of Functional Foods* **5**(3): 1260–1269.

Laye, C., D. J. McClements and J. Weiss (2008). "Formation of biopolymer-coated liposomes by electrostatic deposition of chitosan." *Journal of Food Science* **73**(5): N7–N15.

Leclercq, S., C. Milo and G. A. Reineccius (2009). "Effects of cross-linking, capsule wall thickness, and compound hydrophobicity on aroma release from complex coacervate microcapsules." *Journal of Agricultural and Food Chemistry* **57**(4): 1426–1432.

Lee, S. J. and D. J. McClements (2010). "Fabrication of protein-stabilized nanoemulsions using a combined homogenization and amphiphilic solvent dissolution/evaporation approach." *Food Hydrocolloids* **24**(6–7): 560–569.

Lentz, K. A., M. Quitko, D. G. Morgan and J. E. Grace (2007). "Development and validation of a preclinical food effect model." *Journal of Pharmaceutical Sciences* **96**(2): 459–472.

Leser, M. E., L. Sagalowicz, M. Michel and H. J. Watzke (2006). "Self-assembly of polar food lipids." *Advances in Colloid and Interface Science* **123**: 125–136.

Lesmes, U. and D. J. McClements (2009). "Structure-function relationships to guide rational design and fabrication of particulate food delivery systems." *Trends in Food Science and Technology* **20**(10): 448–457.

Lesmes, U., S. Sandra, E. A. Decker and D. J. McClements (2010). "Impact of surface deposition of lactoferrin on physical and chemical stability of omega-3 rich lipid droplets stabilised by caseinate." *Food Chemistry* **123**(1): 99–106.

Let, M. B., C. Jacobsen and A. S. Meyer (2007a). "Ascorbyl palmitate, gamma-tocopherol, and EDTA affect lipid oxidation in fish oil enriched salad dressing differently." *Journal of Agricultural and Food Chemistry* **55**(6): 2369–2375.

Let, M. B., C. Jacobsen and A. S. Meyer (2007b). "Lipid oxidation in milk, yoghurt, and salad dressing enriched with neat fish oil or pre-ernulsified fish oil." *Journal of Agricultural and Food Chemistry* **55**(19): 7802–7809.

Lim, S. S., M. Y. Baik, E. A. Decker, L. Henson, L. M. Popplewell, D. J. McClements and S. J. Choi (2011). "Stabilization of orange oil-in-water emulsions: A new role for ester gum as an Ostwald ripening inhibitor." *Food Chemistry* **128**(4): 1023–1028.

Liu, S., N. H. Low and M. T. Nickerson (2010). "Entrapment of flaxseed oil within gelatin-gum arabic capsules." *Journal of the American Oil Chemists Society* **87**(7): 809–815.

Liu, W. L., J. H. Liu, W. Liu, T. Li and C. M. Liu (2013a). "Improved physical and in vitro digestion stability of a polyelectrolyte delivery system based on layer-by-layer self-assembly alginate-chitosan-coated nanoliposomes." *Journal of Agricultural and Food Chemistry* **61**(17): 4133–4144.

Liu, W. L., A. Q. Ye, W. Liu, C. M. Liu and H. Singh (2013b). "Liposomes as food ingredients and nutraceutical delivery systems." *Agro Food Industry Hi-Tech* **24**(2): 68–71.

Lomova, M. V., G. B. Sukhorukov and M. N. Antipina (2010). "Antioxidant coating of microsize droplets for prevention of lipid peroxidation in oil-in-water emulsion." *ACS Applied Materials and Interfaces* **2**(12): 3669–3676.

Lu, F. S. H., N. S. Nielsen, M. Timm-Heinrich and C. Jacobsen (2011a). "Oxidative stability of marine phospholipids in the liposomal form and their applications." *Lipids* **46**(1): 3–23.

Lu, Q., D. C. Li and J. G. Jiang (2011b). "Preparation of a tea polyphenol nanoliposome system and its physicochemical properties." *Journal of Agricultural and Food Chemistry* **59**(24): 13004–13011.

Madene, A., M. Jacquot, J. Scher and S. Desobry (2006). "Flavour encapsulation and controlled release—A review." *International Journal of Food Science and Technology* **41**(1): 1–21.

Maherani, B., E. Arab-Tehrany, M. R. Mozafari, C. Gaiani and M. Linder (2011). "Liposomes: A review of manufacturing techniques and targeting strategies." *Current Nanoscience* **7**(3): 436–452.

Malone, M. E. and I. A. M. Appelqvist (2003). "Gelled emulsion particles for the controlled release of lipophilic volatiles during eating." *Journal of Controlled Release* **90**(2): 227–241.

Marze, S. (2013). "Bioaccessibility of nutrients and micronutrients from dispersed food systems: Impact of the multiscale bulk and interfacial structures." *Critical Reviews in Food Science and Nutrition* **53**(1): 76–108.

Matalanis, A., E. A. Decker and D. J. McClements (2012). "Inhibition of lipid oxidation by encapsulation of emulsion droplets within hydrogel microspheres." *Food Chemistry* **132**(2): 766–772.

Matalanis, A., O. G. Jones and D. J. McClements (2011). "Structured biopolymer-based delivery systems for encapsulation, protection, and release of lipophilic compounds." *Food Hydrocolloids* **25**(8): 1865–1880.

Matalanis, A. and D. J. McClements (2013). "Hydrogel microspheres for encapsulation of lipophilic components: Optimization of fabrication and performance." *Food Hydrocolloids* **31**(1): 15–25.

McClements, D. J. (2002a). "Colloidal basis of emulsion color." *Current Opinion in Colloid and Interface Science* **7**(5–6): 451–455.

McClements, D. J. (2002b). "Theoretical prediction of emulsion color." *Advances in Colloid and Interface Science* **97**(1–3): 63–89.

McClements, D. J. (2005). *Food Emulsions: Principles, Practice, and Techniques.* Boca Raton, FL, CRC Press.

McClements, D. J. (2010a). "Design of nano-laminated coatings to control bioavailability of lipophilic food components." *Journal of Food Science* **75**(1): R30–R42.

McClements, D. J. (2010b). "Emulsion design to improve the delivery of functional lipophilic components." *Annual Review of Food Science and Technology* **1**(1): 241–269.

McClements, D. J. (2011). "Edible nanoemulsions: Fabrication, properties, and functional performance." *Soft Matter* **7**(6): 2297–2316.

McClements, D. J. (2012a). "Advances in fabrication of emulsions with enhanced functionality using structural design principles." *Current Opinion in Colloid and Interface Science* **17**(5): 235–245.

McClements, D. J. (2012b). "Crystals and crystallization in oil-in-water emulsions: Implications for emulsion-based delivery systems." *Advances in Colloid and Interface Science* **174**: 1–30.

McClements, D. J. (2012c). "Nanoemulsions versus microemulsions: Terminology, differences, and similarities." *Soft Matter* **8**(6): 1719–1729.

McClements, D. J. (2013). "Utilizing food effects to overcome challenges in delivery of lipophilic bioactives: Structural design of medical and functional foods." *Expert Opinion on Drug Delivery* 1–12.

McClements, D. J. and E. A. Decker (2000). "Lipid oxidation in oil-in-water emulsions: Impact of molecular environment on chemical reactions in heterogeneous food systems." *Journal of Food Science* **65**(8): 1270–1282.

McClements, D. J. and E. A. Decker (2008). Lipids. In *Food Chemistry*, 4th Edition, S. Damodaran, K. L. Parkin and F. O.R. (eds.). Boca Raton, FL, CRC Press: 155–216.

McClements, D. J., E. A. Decker, Y. Park and J. Weiss (2009). "Structural design principles for delivery of bioactive components in nutraceuticals and functional foods." *Critical Reviews in Food Science and Nutrition* **49**(6): 577–606.

McClements, D. J., L. Henson, L. M. Popplewell, E. A. Decker and S. J. Choi (2012). "Inhibition of Ostwald ripening in model beverage emulsions by addition of poorly water soluble triglyceride oils." *Journal of Food Science* **77**(1): C33–C38.

McClements, D. J. and Y. Li (2010). "Structured emulsion-based delivery systems: Controlling the digestion and release of lipophilic food components." *Advances in Colloid and Interface Science* **159**(2): 213–228.

McClements, D. J. and J. Rao (2011). "Food-grade nanoemulsions: Formulation, fabrication, properties, performance, biological fate, and potential toxicity." *Critical Reviews in Food Science and Nutrition* **51**(4): 285–330.

Mei, L. Y., S. J. Choi, J. Alamed, L. Henson, M. Popplewell, D. J. McClements and E. A. Decker (2010). "Citral stability in oil-in-water emulsions with solid or liquid octadecane." *Journal of Agricultural and Food Chemistry* **58**(1): 533–536.

Mei, L. Y., D. J. McClements and E. A. Decker (1999). "Lipid oxidation in emulsions as affected by charge status of antioxidants and emulsion droplets." *Journal of Agricultural and Food Chemistry* **47**(6): 2267–2273.

Misra, A., K. Jinturkar, D. Patel, J. Lalani and M. Chougule (2009). "Recent advances in liposomal dry powder formulations: Preparation and evaluation." *Expert Opinion on Drug Delivery* **6**(1): 71–89.

Miyashita, K., E. Nara and T. Ota (1993). "Oxidative stability of polyunsaturated fatty-acids in an aqueous-solution." *Bioscience Biotechnology and Biochemistry* **57**(10): 1638–1640.

Mozafari, M. R., C. Johnson, S. Hatziantoniou and C. Demetzos (2008). "Nanoliposomes and their applications in food nanotechnology." *Journal of Liposome Research* **18**(4): 309–327.

Mozaffarian, D. and J. H. Y. Wu (2011). "Omega-3 fatty acids and cardiovascular disease effects on risk factors, molecular pathways, and clinical events." *Journal of the American College of Cardiology* **58**(20): 2047–2067.

Muller, R. H., S. Gohla and C. M. Keck (2011). "State of the art of nanocrystals—Special features, production, nanotoxicology aspects and intracellular delivery." *European Journal of Pharmaceutics and Biopharmaceutics* **78**(1): 1–9.

Mullertz, A., D. G. Fatouros, J. R. Smith, M. Vertzoni and C. Reppas (2012). "Insights into intermediate phases of human intestinal fluids visualized by atomic force microscopy and cryo-transmission electron microscopy ex vivo." *Molecular Pharmaceutics* **9**(2): 237–247.

Murrieta-Pazos, I., C. Gaiani, L. Galet, R. Calvet, B. Cuq and J. Scher (2012). "Food powders: Surface and form characterization revisited." *Journal of Food Engineering* **112**(1–2): 1–21.

Okuda, S., D. J. McClements and E. A. Decker (2005). "Impact of lipid physical state on the oxidation of methyl linolenate in oil-in-water emulsions." *Journal of Agricultural and Food Chemistry* **53**(24): 9624–9628.

Oliveira, A. C., T. S. Moretti, C. Boschini, J. C. C. Baliero, L. A. P. Freitas, O. Freitas and C. S. Favaro-Trindade (2007a). "Microencapsulation of B-lactis (BI 01) and L-acidophilus (LAC 4) by complex coacervation followed by spouted-bed drying." *Drying Technology* **25**(10): 1687–1693.

Oliveira, A. C., T. S. Moretti, C. Boschini, J. C. C. Baliero, O. Freitas and C. S. Favaro-Trindade (2007b). "Stability of microencapsulated B lactis (Bl 01) and L acidophilus (LAC 4) by complex coacervation followed by spray drying." *Journal of Microencapsulation* **24**(7): 685–693.

Parvez, S., K. A. Malik, S. A. Kang and H. Y. Kim (2006). "Probiotics and their fermented food products are beneficial for health." *Journal of Applied Microbiology* **100**(6): 1171–1185.

Patel, A. R., P. C. M. Heussen, J. Hazekamp, E. Drost and K. P. Velikov (2012). "Quercetin loaded biopolymeric colloidal particles prepared by simultaneous precipitation of quercetin with hydrophobic protein in aqueous medium." *Food Chemistry* **133**(2): 423–429.

Pawlik, A. K. and I. T. Norton (2012). "Encapsulation stability of duplex emulsions prepared with SPG cross-flow membrane, SPG rotating membrane and rotor-stator techniques— A comparison." *Journal of Membrane Science* **415**: 459–468.

Playne, M. J. (2002). "The health benefits of probiotics." *Food Australia* **54**(3): 71–74.

Porter, C. J. H., C. W. Pouton, J. F. Cuine and W. N. Charman (2006). Enhancing intestinal drug solubilisation using lipid-based delivery systems. *Annual Meeting of the American-Association-of-Pharmaceutical-Scientists*, San Antonio, TX.

Porter, C. J. H., N. L. Trevaskis and W. N. Charman (2007). "Lipids and lipid-based formulations: Optimizing the oral delivery of lipophilic drugs." *Nature Reviews Drug Discovery* **6**(3): 231–248.

Pouton, C. W. and C. J. Porter (2008). "Formulation of lipid-based delivery systems for oral administration: Materials, methods and strategies." *Advanced Drug Delivery Reviews* **60**(6): 625–637.

Poyato, C., I. Navarro-Blasco, M. I. Calvo, R. Y. Cavero, I. Astiasaran and D. Ansorena (2013). "Oxidative stability of O/W and W/O/W emulsions: Effect of lipid composition and antioxidant polarity." *Food Research International* **51**(1): 132–140.

Priya, A. J., S. P. Vijayalakshmi and A. M. Raichui (2011). "Enhanced survival of probiotic lactobacillus acidophilus by encapsulation with nanostructured polyelectrolyte layers through layer-by-layer approach." *Journal of Agricultural and Food Chemistry* **59**(21): 11838–11845.

Qian, C. and D. J. McClements (2011). "Formation of nanoemulsions stabilized by model food-grade emulsifiers using high-pressure homogenization: Factors affecting particle size." *Food Hydrocolloids* **25**(5): 1000–1008.

Qian, C., E. A. Decker, H. Xiao and D. J. McClements (2012). "Nanoemulsion delivery systems: Influence of carrier oil on beta-carotene bioaccessibility." *Food Chemistry* **135**(3): 1440–1447.

Ranadheera, R., S. K. Baines and M. C. Adams (2010). "Importance of food in probiotic efficacy." *Food Research International* **43**(1): 1–7.

Rao, J. J. and D. J. McClements (2010). "Stabilization of phase inversion temperature nano-emulsions by surfactant displacement." *Journal of Agricultural and Food Chemistry* **58**(11): 7059–7066.

Rao, J. J. and D. J. McClements (2011). "Formation of flavor oil microemulsions, nano-emulsions and emulsions: Influence of composition and preparation method." *Journal of Agricultural and Food Chemistry* **59**(9): 5026–5035.

Rein, M. J., M. Renouf, C. Cruz-Hernandez, L. Actis-Goretta, S. K. Thakkar and M. D. Pinto (2013). "Bioavailability of bioactive food compounds: A challenging journey to bioefficacy." *British Journal of Clinical Pharmacology* **75**(3): 588–602.

Riaz, Q. U. and T. Masud (2013). "Recent trends and applications of encapsulating materials for probiotic stability." *Critical Reviews in Food Science and Nutrition* **53**(3): 231–244.

Roberfroid, M., G. R. Gibson, L. Hoyles, A. L. McCartney, R. Rastall, I. Rowland, D. Wolvers, B. Watzl, H. Szajewska, B. Stahl, F. Guarner, F. Respondek, K. Whelan, V. Coxam, M. J. Davicco, L. Leotoing, Y. Wittrant, N. M. Delzenne, P. D. Cani, A. M. Neyrinck and A. Meheust (2010). "Prebiotic effects: Metabolic and health benefits." *British Journal of Nutrition* **104**: S1–S63.

Rozner, S., D. E. Shalev, A. I. Shames, M. F. Ottaviani, A. Aserin and N. Garti (2010). "Do food microemulsions and dietary mixed micelles interact?" *Colloids and Surfaces B-Biointerfaces* **77**(1): 22–30.

Saberi, A. H., Y. Fang and D. J. McClements (2013). "Fabrication of vitamin E-enriched nanoemulsions: Factors affecting particle size using spontaneous emulsification." *Journal of Colloid and Interface Science* **391**: 95–102.

Sagalowicz, L. and M. E. Leser (2010). "Delivery systems for liquid food products." *Current Opinion in Colloid and Interface Science* **15**(1–2): 61–72.

Salminen, H., K. Herrmann and J. Weiss (2013). "Oil-in-water emulsions as a delivery system for n-3 fatty acids in meat products." *Meat Science* **93**(3): 659–667.

Salvia-Trujillo, L., C. Qian, O. Martin-Belloso and D. J. McClements (2013). "Modulating beta-carotene bioaccessibility by controlling oil composition and concentration in edible nanoemulsions." *Food Chemistry* **139**(1–4): 878–884.

Sanders, M. E. and M. L. Marco (2010). Food formats for effective delivery of probiotics. In *Annual Review of Food Science and Technology*, M. P. Doyle and T. R. Klaenhammer (eds.). Palo Alto, CA, Annual Reviews, **1**: 65–85.

Sapei, L., M. A. Naqvi and D. Rousseau (2012). "Stability and release properties of double emulsions for food applications." *Food Hydrocolloids* **27**(2): 316–323.

Sikora, M., N. Badrie, A. K. Deisingh and S. Kowalski (2008). "Sauces and dressings: A review of properties and applications." *Critical Reviews in Food Science and Nutrition* **48**(1): 50–77.

Singh, H., A. Ye and D. Horne (2009). "Structuring food emulsions in the gastrointestinal tract to modify lipid digestion." *Progress in Lipid Research* **48**(2): 92–100.

Spernath, A. and A. Aserin (2006). "Microemulsions as carriers for drugs and nutraceuticals." *Advances in Colloid and Interface Science* **128**: 47–64.

Tamjidi, F., A. Nasirpour and M. Shahedi (2012). "Physicochemical and sensory properties of yogurt enriched with microencapsulated fish oil." *Food Science and Technology International* **18**(4): 381–390.

Taylor, T. M., P. M. Davidson, B. D. Bruce and J. Weiss (2005). "Liposomal nanocapsules in food science and agriculture." *Critical Reviews in Food Science and Nutrition* **45**(7–8): 587–605.

Teitelbaum, J. E. and W. A. Walker (2002). "Nutritional impact of pre- and probiotics as protective gastrointestinal organisms." *Annual Review of Nutrition* **22**: 107–138.

Teoh, P. L., H. Mirhosseini, S. Mustafa, A. S. M. Hussin and M. Y. A. Manap (2011). "Recent approaches in the development of encapsulated delivery systems for probiotics." *Food Biotechnology* **25**(1): 77–101.

Thies, C. (2012). Nanocapsules as delivery systems in the food, beverage and nutraceutical industries. In *Nanotechnology in the Food, Beverage and Nutraceutical Industries*, Q. Huang (ed.). Cambridge, UK, Woodhead Publ Ltd: 208–256.

Uhlemann, J. and I. Reiss (2010). "Product design and process engineering using the example of flavors." *Chemical Engineering and Technology* **33**(2): 199–212.

van Aken, G. A. and E. H. A. de Hoog (2009). Oral processing and perception of food emulsions: Relevence for fat reduction in foods. In *Designing Functional Foods: Measuring and Controlling Food Structure Breakdown and Nutrient Absorption*, D. J. McClements and E. A. Decker (eds.). Boca Raton, FL, CRC Press: 265–294.

van Aken, G. A., M. H. Vingerhoeds and E. H. A. de Hoog (2005). Colloidal behavior of food emulsions under oral conditions. In *Food Hydrocolloids, Interactions, Microstructure and Processing*, E. Dickinson (ed.). Cambridge, UK, The Royal Society of Chemistry: 356–366.

van Aken, G. A., M. H. Vingerhoeds and E. H. A. de Hoog (2007). "Food colloids under oral conditions." *Current Opinion in Colloid and Interface Science* **12**(4–5): 251–262.

Vega, C. and Y. H. Roos (2006). "Invited review: Spray-dried dairy and dairy-like— Emulsions compositional considerations." *Journal of Dairy Science* **89**(2): 383–401.

Velikov, K. P. and E. Pelan (2008). "Colloidal delivery systems for micronutrients and nutraceuticals." *Soft Matter* **4**(10): 1964–1980.

Waraho, T., D. J. McClements and E. A. Decker (2011). "Mechanisms of lipid oxidation in food dispersions." *Trends in Food Science and Technology* **22**(1): 3–13.

Weinbreck, F., M. Minor and C. G. De Kruif (2004). "Microencapsulation of oils using whey protein/gum arabic coacervates." *Journal of Microencapsulation* **21**(6): 667–679.

Yeap, Y. Y., N. L. Trevaskis, T. Quach, P. Tso, W. N. Charman and C. J. H. Porter (2013). "Intestinal bile secretion promotes drug absorption from lipid colloidal phases via induction of supersaturation." *Molecular Pharmaceutics* **10**(5): 1874–1889.

Zheng, M. Y., F. Liu, Z. W. Wang and J. H. Baoyindugurong (2011). "Formation and characterization of self-assembling fish oil microemulsions." *Colloid Journal* **73**(3): 319–326.

Ziani, K., Y. Fang and D. J. McClements (2012). "Encapsulation of functional lipophilic components in surfactant-based colloidal delivery systems: Vitamin E, vitamin D, and lemon oil." *Food Chemistry* **134**(2): 1106–1112.

# Chapter 10

## Key Physicochemical Concepts

## 10.1 Introduction

In this final chapter, a number of key physicochemical and mathematical concepts needed to facilitate the design, manufacture, and application of colloidal delivery systems for active ingredients are presented. This fundamental information is important for trying to establish a more quantitative basis for the development of successful delivery systems in the food and beverage industry.

## 10.2 Physical States and Phase Transitions

The active ingredients used in food and beverage products may exist in a number of different physical states, for example, gas, liquid, or solid (Walstra 2003). Similarly, the materials used to form the particle matrix that surrounds them may also exist in different physical states, typically either liquid or solid (McClements 2012). The physical state of a material within a delivery system may change when solution or environmental conditions such as solvent composition or temperature are altered, for example, because of melting, crystallization, evaporation, condensation, polymorphic changes, or glass transitions. Knowledge of the physical state and phase transitions of the various components used to fabricate a delivery system is therefore important for ensuring optimum functional performance. For this reason, a brief overview of some of the most important factors associated with the physical state of components in delivery systems is given.

## 10.2.1 Physical States

### 10.2.1.1 Thermodynamics

At thermodynamic equilibrium, a material predominately exists in the physical state (gas, liquid, solid) with the lowest free energy (McClements 2012). The free energy of a particular state is governed by enthalpy (energy stored in molecular interactions) and entropy (disorder) contributions. The free energy change ($\Delta G_T$) when a system undergoes a transition from one state to another can be conveniently represented by the following expression:

$$\Delta G_T = \Delta H_T - T \times \Delta S_T. \tag{10.1}$$

Here, $\Delta H_T$ is the change in enthalpy (molecular interaction strength) associated with the transition and $\Delta S_T$ is the change in entropy (disorder) associated with the transition. A phase transition is thermodynamically favorable when $\Delta G_T < 0$ but is thermodynamically unfavorable when $\Delta G_T > 0$. At the transition temperature ($T_m$), $\Delta G_T = 0$, and each state is equally favorable. The amount of energy stored in the molecular interactions in different states typically follows the order: solids > liquids > gas, whereas the entropy of different states typically follows the order: gas > liquids > solids. Thus, at relatively low temperatures (small $T$), the enthalpy contribution dominates (since strong attractive interactions means a more negative $\Delta H_T$), which favors the existence of the solid phase. Conversely, at high temperatures (large $T$), the entropy contribution dominates ($-T \times \Delta S_T$), which favors the existence of the gas phase. At intermediate temperatures, the balance of enthalpy and entropy effects favors the existence of the liquid phase.

### 10.2.1.2 Practical Implications

In practice, many materials can exist in a number of alternative forms within a particular physical state. For example, some solid polymer phases may exist in rubbery, glassy, or crystalline forms depending on environmental conditions and their thermal history (Bhandari and Howes 2000; Sablani et al. 2010). In addition, many crystalline fats can exist in a number of polymorphic forms that have different molecular packing, such as the α, β′, and β forms of long-chain triglycerides (Himawan et al. 2006; McClements 2012). Under a particular set of conditions, one of these forms usually has the lowest free energy and is therefore thermodynamically stable; however, the system may be trapped in a metastable state because of the presence of kinetic energy barriers (Figure 10.1). The precise nature of the physical state formed by a material often has important implications for the fabrication, stability, and functional performance of delivery systems. For example, the fate of an active ingredient dispersed within a liquid lipid particle that solidifies depends on the nature of the solid lipid matrix formed. The active ingredient may be expelled from the interior to the surface of

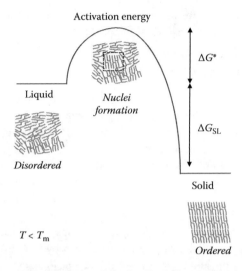

**Figure 10.1  Free energy changes (ΔG) associated with the solid–liquid phase transition. Below the melting temperature, the difference in free energy between the solid and liquid phases (ΔG$_{SL}$) is negative, thereby favoring the solid state, but there may be an activation energy (ΔG*) associated with nuclei formation that must be overcome before the transition occurs.**

the particle when the lipid solidifies into a highly organized crystalline structure that cannot easily accommodate it (Qian et al. 2013). On the other hand, the active ingredient may remain trapped within the solid matrix in the interior of the particle if the lipid solidifies into a less highly organized structure that can accommodate it (Hu et al. 2006). This has important implications for the release and chemical stability of some compounds; for example, flavor oils and carotenoids have been shown to be more susceptible to degradation after the crystallization of the lipid particle carrying them because of this reason (Mei et al. 2010; Qian et al. 2013).

The physical state of a material also has a number of other important consequences for the development of delivery systems (McClements 2010). The physical properties of materials depend strongly on their physical state, such as density, refractive index, and rheology. In turn, these physical properties influence the macroscopic properties of delivery systems, such as their stability, optical properties, mass transport properties, and textural attributes. In particular, the physical state of a particle matrix may have a major influence on the retention and release of active ingredients. The diffusion of a molecule through a fluid phase is primarily governed by the viscosity, whereas diffusion may be greatly inhibited in a solid phase. The physical state of an active ingredient may also have a pronounced influence on its bioavailability. Typically, the solid form of an active ingredient is less bioavailable than a solubilized form (Williams et al. 2013). The oral bioavailability also depends on the nature of the solid form, with amorphous forms being more bioavailable than crystalline forms (Brough and Williams 2013).

## 10.2.2 Crystallization and Melting

One or more components within a delivery system may undergo a change in its physical state in response to alterations in its environment, such as compositional or temperature changes (McClements 2012; Williams et al. 2013). In general, a number of different kinds of phase transitions may occur, such as solid to liquid (melting), liquid to solid (crystallization), gas to liquid (condensation), liquid to gas (evaporation), solid to gas (sublimation), rubber to glass (glass transition), and sol to gel (gelation) (McClements 2012). In this section, the main focus will be on crystallization and melting since this is one of the most important phase transitions utilized in the fabrication of food-grade delivery systems with specific encapsulation and release properties. It is usually convenient to divide the crystallization process into three sequential stages: supercooling, nucleation, and crystal growth. However, in reality, these three phenomena can occur simultaneously in different regions of a material.

### 10.2.2.1 Supercooling

In principle, a material should spontaneously undergo a liquid-to-solid transition when it is cooled below its melting point. In practice, a material may persist in a liquid state at temperatures well below its melting point, even though the solid state has the lowest free energy (Walstra 2003). The origin of this "supercooling" effect is the presence of a kinetic energy barrier that inhibits crystal formation even under circumstances where it is thermodynamically favorable (Figure 10.1). The height of the energy barrier depends on the ability of crystal nuclei to be formed that are stable enough to grow into crystals (Vekilov 2010a,b). The degree of supercooling of a material is defined as $\Delta T = T_m - T$, where $T_m$ is the melting point and $T$ is the temperature. The value of $\Delta T$ at which the formation of crystals is first observed depends on the nature of the material, the presence of any contaminating materials, the cooling rate, the structure of the material (e.g., particle size), and the application of external forces such as pressure, shear, or sonication (Hartel 2001; Walstra 2003). The phenomenon of supercooling is particularly important in colloidal systems, since each particle has a low probability of containing any impurities and so nucleation often proceeds by a homogeneous (rather than heterogeneous) nucleation. In this case, the crystallization temperature may be much lower than the melting temperature (McClements 2012).

### 10.2.2.2 Nucleation

Crystals can only grow after stable nuclei have been formed within a supercooled liquid (Kashchiev and van Rosmalen 2003; Leubner 2000; Lindfors et al. 2008;

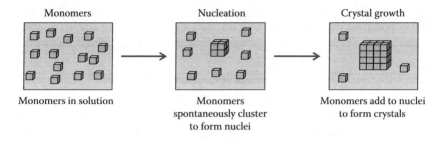

| Monomers | Nucleation | Crystal growth |
|---|---|---|
| Monomers in solution | Monomers spontaneously cluster to form nuclei | Monomers add to nuclei to form crystals |

**Figure 10.2    Schematic diagram of the classical nucleation theory developed to model nucleation and crystal growth.**

Maris 2006). Nuclei consist of small transient molecular clusters that form when a group of molecules encounter each other (Figure 10.2). After formation, nuclei may either dissociate back into the individual molecules or they may grow into crystals. The physical origin of supercooling can be explained in terms of the free energy changes associated with the formation of nuclei (Hartel 2001; Kashchiev and van Rosmalen 2003; Vekilov 2010; Walstra 2003). At temperatures below the melting point, the *bulk* crystalline state is thermodynamically more favorable than the *bulk* liquid state, and so there is a decrease in free energy when some of the molecules in the supercooled liquid associate with each other and form a nucleus. This negative (thermodynamically favorable) free energy ($\Delta G_V$) change is proportional to the volume of the nucleus formed. Conversely, the formation of a nucleus leads to the creation of a new interface between the solid phase and the liquid phase, which requires an input of free energy to overcome the interfacial tension (McClements 2012). This positive (thermodynamically unfavorable) free energy ($\Delta G_S$) change is proportional to the surface area of the nucleus formed. The total free energy change associated with nuclei formation is a combination of these volume and surface contributions (Herhold et al. 1999; Kashchiev et al. 2010):

$$\Delta G = \Delta G_V + \Delta G_S = \frac{4}{3}\pi r^3 \frac{\Delta H_f \Delta T}{T_m} + 4\pi r^2 \gamma_{SL}, \qquad (10.2)$$

where $r$ is nucleus radius, $\Delta H_f$ is the enthalpy change per unit volume associated with the liquid–solid transition, and $\gamma_{SL}$ is the solid–liquid interfacial tension. The volume contribution becomes increasingly negative as the radius of the nuclei increases, whereas the surface contribution becomes increasingly positive (Figure 10.3). The unfavorable (positive) surface contribution dominates for small nuclei, whereas the favorable (negative) volume term dominates for large nuclei. The overall free energy has a maximum value at a critical nucleus radius ($r^*$):

$$r^* = \frac{-2\gamma_{SL} T_m}{\Delta H_f \Delta T}. \qquad (10.3)$$

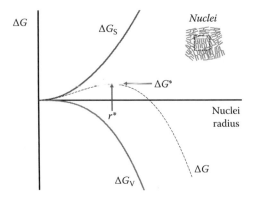

**Figure 10.3** **Schematic of free energy changes associated with formation of nuclei of different radii (r) in a supercooled melt. The overall free energy (ΔG) change depends on a volume contribution (ΔG_V) that favors nuclei formation and a surface contribution (ΔG_S) that opposes nuclei formation. There is a free energy maximum (ΔG*) and critical radius (r*) associated with nuclei formation. (Adapted from McClements [2005].)**

When a nucleus is formed in a supercooled liquid with a radius below this critical value ($r < r^*$), it tends to dissociate, but if it is formed with a radius above this critical value ($r > r^*$), then it tends to grow into a crystal. This equation predicts that the critical nuclei size ($r^*$) for the formation of stable nuclei decreases as the degree of supercooling increases ($\Delta T$), which accounts for the increase in nucleation rate with decreasing temperature typically observed experimentally. The free energy associated with the formation of a nucleus of the critical size ($\Delta G^*$) required for crystal growth is calculated by replacing $r$ in Equation 10.2 with the critical radius given in Equation 10.3 (Walstra 2003).

$$\Delta G^* = \frac{16\pi\gamma_{SL}^3 T_m^2}{3\Delta H_f^2 \Delta T^2} = \frac{4}{3}\pi r^{*2}\gamma_{SL} \qquad (10.4)$$

The rate of nuclei formation in a supercooled melt has been related to the activation energy ($\Delta G^*$) that must be overcome to form stable nuclei using classical nucleation theory (CNT):

$$J = J_0 \exp\left(-\frac{\Delta G^*}{k_B T}\right). \qquad (10.5)$$

The nucleation rate ($J$) is the number of stable nuclei formed per second per unit volume of material, $J_0$ is a preexponential factor, $k_B$ is Boltzmann's constant, and $T$ is the absolute temperature. Expressions for the preexponential factor have been calculated based on the assumption that there is one or more activation energies associated with the movement of molecules to the surface of the nuclei and with their incorporation into the nuclei (Walstra 2003). A schematic diagram of the dependence of the nucleation rate on the degree of supercooling is shown in Figure 10.4. The formation of stable nuclei is negligibly slow at temperatures just below the melting point owing to supercooling but increases dramatically as the degree of supercooling increases. In practice, measured nucleation rates may show a more complex dependence on temperature because of diffusional barriers associated with the retardation of molecular movement at low temperatures caused by the increase in viscosity (Hartel 2001). In these cases, there is often a maximum in the nucleation rate at a particular temperature. Nucleation rates predicted by CNT are often many orders of magnitude different from experimentally measured ones, but this theory does provide some valuable qualitative insights into the major factors influencing nucleation (Balibar and Caupin 2006; Hartel 2001; Walstra 2003).

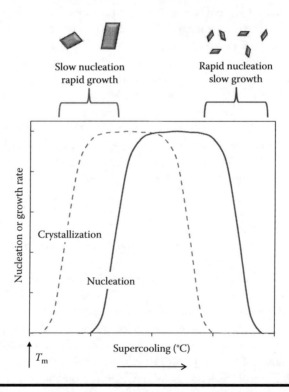

**Figure 10.4 Schematic representation of the dependence of the nucleation rate and crystal growth rate on temperature. Nucleation and crystal growth only occur once the material has been cooled below the melting point ($T_m$). (Adapted from McClements [2012].)**

The rate of nucleation in a supercooled liquid often depends on the nature of any surfaces that are in contact with the liquid that may promote nuclei formation, such as seed crystals, particulate impurities, air bubbles, or container walls. When the formation of nuclei in a pure supercooled liquid is unaffected by the presence of any surfaces, it is usually referred to as *homogeneous nucleation* and depends on the spontaneous clustering of the molecules in the liquid (Hartel 2001). On the other hand, when the supercooled liquid is in contact with surfaces that promote nuclei formation, then nucleation can occur at a considerably lower degree of supercooling (higher temperature) than for a pure system (Walstra 2003). Nucleation caused by the presence of surfaces is referred to as *heterogeneous nucleation* and can be divided into two types: primary and secondary (Hartel 2001; Kashchiev and van Rosmalen 2003). *Primary heterogeneous nucleation* occurs when the surfaces have a different chemical structure than the solute, for example, impurities or container surfaces. *Secondary heterogeneous nucleation* occurs when the impurities are seed crystals with the same chemical structure as the solute.

Heterogeneous nucleation occurs when the formation of nuclei in the presence of a surface is more thermodynamically favorable (lower $\Delta G^*$) than in a pure liquid (Kashchiev and van Rosmalen 2003). Consequently, the degree of supercooling required to initiate nucleation is reduced. Conversely, some types of impurities actually interfere with the formation of stable nuclei (higher $\Delta G^*$) and therefore decrease the nucleation rate and increase the degree of supercooling required (Hartel 2001; Walstra 2003). This phenomenon may occur because the impurities are incorporated into the surface of the nuclei and prevent any further molecules being incorporated. Whether an impurity acts as a catalyst or an inhibitor of nucleation depends on its molecular structure and interactions with the nuclei. This knowledge can be used to control the crystallization process in delivery systems and therefore alter their formation, stability, and functionality.

### 10.2.2.3 Crystal Growth

Once stable nuclei are formed in a supercooled liquid, they grow into crystals by incorporating additional molecules into the solid–liquid interface (Hartel 2001; Walstra 2003). Crystals have different faces that grow at different speeds leading to crystals with different sizes and shapes (e.g., ellipsoids, cubes, needles, spherulites, etc.). The overall growth rate depends on a variety of factors, including movement of crystallizing molecules to the crystal surface, movement of noncrystallizing molecules away from the crystal surface, incorporation of molecules into the crystal lattice, and removal of heat from the crystal surface. Any of these processes can be the rate-limiting step for crystal growth depending on the nature of the system. Consequently, a general theory that accurately models crystal growth is difficult to formulate. Typically, the crystal growth rate increases initially with supercooling, has a maximum value at a particular temperature, and then decreases upon further supercooling. The dependence of the crystal growth rate on temperature therefore follows a pattern similar to that of the nucleation rate, but the maximum rate of nuclei formation typically occurs at a temperature different from that of the maximum rate of crystal growth (Figure 10.4).

An important practical consequence of the different dependencies of the nucleation and growth rates on temperature is the possibility of controlling the size of the crystals produced (Herrera and Hartel 2000). A large number of small crystals will be formed when the system is supercooled to a temperature where the nucleation rate is much faster than the crystallization rate. Conversely, a small number of large crystals will be formed when the system is supercooled to a temperature where the nucleation rate is considerably slower than the crystallization rate (Figure 10.4). A number of different theories have been developed to theoretically model crystal growth (Hartel 2001; Kashchiev and van Rosmalen 2003), with the most appropriate model for a specific situation depending on the rate-limiting step (Hartel 2001).

## 10.2.2.4 Crystal Morphology and Structure

The precise nature of the crystals formed after crystallization is often important for the development of successful delivery systems, for example, the molecular packing, size, shape, and interactions of the crystals. As mentioned earlier, the crystal size can often be manipulated by controlling the thermal history of the liquid during cooling, for example, the cooling rate and holding temperature. A large number of small crystals tend to be formed when a liquid is rapidly cooled to a temperature well below its melting point, whereas a small number of large crystals tend to be formed when a liquid is slowly cooled to a temperature just below its melting point. This is because the nucleation rate tends to increase more rapidly than the crystallization rate with decreasing temperature (Figure 10.4). The size of the crystals produced has important implications for the formation, stability, texture, perception, and biological fate of some delivery systems. Sufficiently large crystals are often perceived as "grainy" or "sandy" within the mouth (Walstra 2003). The dissolution rate of crystalline particles within the gastrointestinal tract (GIT) tends to decrease as the particle size increases, which may influence the bioavailability of active ingredients (which may be crystalline themselves or trapped within crystalline matrices) (Williams et al. 2013).

The incorporation of active ingredients into crystalline particles may also depend on their thermal history. If a matrix material is cooled slowly, then there may be sufficient time for all the molecules to efficiently pack into the crystal structure, leading to a dense, highly ordered phase. On the other hand, if a matrix material is cooled rapidly, then there may not be sufficient time for the molecules to pack efficiently, leading to more dislocations and a less ordered structure (Timms 1991). The nature of the crystalline structure formed may influence its ability to encapsulate, retain, and release active ingredients; for example, some bioactive ingredients (such as carotenoids) have been shown to be expelled from highly ordered crystalline lipids (Qian et al. 2013).

Some materials are able to exist in more than one crystalline form ("polymorphs") that can be distinguished from each other because of differences in their molecular packing, morphology, and physicochemical properties (Caira 1998; Hartel 2001; Himawan et al. 2006; Rodriguez-Spong et al. 2004; Walstra 2003). The polymorphic form with the lowest free energy under a particular set of conditions is thermodynamically stable (Figure 10.5). However, a material may crystallize into a metastable polymorphic form and remain there for an extended period owing to the presence of kinetic energy barriers that separate different polymorphic forms. Knowledge of the different polymorphic forms that an active ingredient or matrix material can adopt, and the transition temperatures involved, is therefore important (Himawan et al. 2006; Sato 2001). Information about polymorphic transitions is particularly important for the formation of solid lipid nanoparticles and nanostructured lipid carriers, which are finding increasing use for the encapsulation and protection of bioactive lipophilic molecules (Bunjes 2011; Bunjes and Koch 2005).

Once the initial crystallization process is completed, there may still be changes in the nature of the crystals present over time as a result of postcrystallization events, such as crystal aggregation, sintering, or Ostwald ripening (Hartel 2001; Himawan et al. 2006; Walstra 2003). Aggregation occurs when two or more crystals are attracted to each other and form a physical connection. At a sufficiently high concentration, crystal aggregation may lead to the formation of a three-dimensional network of crystals that gives a material some solid-like characteristics. Sintering occurs when a number of crystals come into contact and fuse together, thereby forming larger, irregularly shaped particles. Ostwald ripening occurs when molecules migrate from smaller crystals to larger crystals through the intervening medium. Aggregation, sintering, and Ostwald ripening therefore change the size and morphology of the crystals present. Postcrystallization events are often

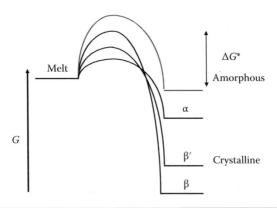

**Figure 10.5 Schematic representation of differences in free energy of a substance that can exist in a number of different physical states: liquid, crystalline (α, β′, and β polymorphs), and amorphous.**

undesirable because they adversely affect the physicochemical, functional, and sensory properties of the delivery system.

Some materials exist in an amorphous solid state (rather than a crystalline state) depending on their composition and the preparation conditions used (Williams et al. 2013). Amorphous solids tend to have higher solubilities in gastrointestinal fluids than their crystalline counterparts, which may be an advantage for increasing the release of some active ingredients in the GIT (Brough and Williams 2013; Rodriguez-Spong et al. 2004). Consequently, it may be important to select preparation conditions that favor the formation of an amorphous state, rather than a crystalline state.

### 10.2.2.5 Melting

Knowledge of the melting behavior of the different components used to fabricate delivery systems is often important, since it affects their production, stability, and performance (McClements 2012; Walstra 2003). The crystals in a material undergo a solid-to-liquid phase transition when they are heated above their melting points. The melting range is typically fairly sharp for pure substances, but it can be fairly broad for complex food ingredients that contain a mixture of different components. The temperature at which melting occurs is often considerably higher than that at which crystallization occurs because of supercooling effects. This is particularly true for solid–liquid phase transitions involving colloidal particles where the degree of supercooling can be much higher than that observed in bulk materials. For example, in emulsions and nanoemulsions, there may be a difference of 10°C to 20°C between the melting and crystallization temperatures depending on their particle size and the number of impurities present (Coupland 2002). This high degree of supercooling originated from the fact that the probability of finding an impurity that catalyzes nucleation within a fine particle is extremely small, and therefore nucleation tends to be homogeneous rather than heterogeneous (McClements 2012). The melting point may also depend on particle size owing to a thermodynamic effect: the melting point tends to decrease as particle curvature increases; that is, particle size decreases. This effect can be described by the Gibbs–Thomson equation (Sun and Simon 2007):

$$T_m(d) = T_m(\infty)\left(1 - \frac{4\gamma_{SL}}{\Delta H_{SL}\rho_S d}\right). \tag{10.6}$$

Here, $T_m(d)$ is the melting temperature of a particle with diameter $d$, $T_m(\infty)$ is the melting temperature of a particle with infinite diameter (bulk material), $\gamma_{SL}$ is the solid–liquid interfacial tension, $\Delta H_{SL}$ is the bulk heat of fusion, and $\rho_S$ is the density of the solid phase within the particle. This equation predicts that the melting point should decrease when the particles become very small, which is supported by experiments on small lipid particles (Schubert et al. 2005). Typically, the particle diameter has to be less than about 100 nm for this effect to become appreciable.

The small dimensions of the particles in colloidal delivery systems may also alter other aspects of the phase transitions of encapsulated active ingredients or particle matrices. The emulsification of lipids can cause an appreciable change in the rates of polymorphic transitions; for example, these transitions tend to occur much more quickly in emulsified fats than in bulk fats (Westesen et al. 1997). Lipid particles may also be highly unstable to aggregation at temperatures where the lipid phase is partly solid and partly liquid because of partial coalescence (Fredrick et al. 2010), which can have a pronounced effect on their shelf life, their rheology, and the functional performance of colloidal delivery systems.

## 10.2.3 Crystallization and Dissolution

Dissolution plays an important role in a number of physicochemical phenomena associated with the fabrication and application of colloidal delivery systems. The preparation of many delivery systems involves the dissolution of an active ingredient into a liquid carrier phase (McClements 2012). The release of active ingredients from certain types of delivery systems is the result of the particle matrix dissolving under certain environmental conditions, such as the mouth, stomach, or small intestine (Siepmann and Siepmann 2013). Ingested crystalline active ingredients may need to dissolve in the GIT before they can be absorbed by the body (Williams et al. 2013). In this section, the physicochemical basis for the dissolution of solutes in solvents is therefore considered. Initially, the solutes may be present in a liquid form (such as oil) or a solid form (such as a powder), and they may be bulk materials or particles.

### 10.2.3.1 Saturation

A solute typically has a finite solubility in a given solvent ($C_S^*$). At thermodynamic equilibrium, the solute is fully dissolved in the solvent below this concentration but is insoluble above this level and forms a separate phase (McClements 2012). Knowledge of the saturation concentration of a solute in the various phases that a delivery system is fabricated from is therefore crucial for designing the system to function properly. For example, the total amount of an active ingredient that can be incorporated into a delivery system in a soluble form depends on these saturation concentrations. In addition, knowledge of the saturation concentrations of the various materials that the delivery system comes into contact during its preparation, storage, and utilization may also be important, for example, containers, diluents, and gastrointestinal fluids. The saturation concentration depends on the nature of solute and solvent, as well as environmental conditions such as temperature and pressure. Ideally, one would like to predict the solubility of a solute in a particular solvent. The saturation concentrations of some substances can be related to the molecular characteristics of the solutes and solvents involved, with some success being reported through the use of various empirical, theoretical, and computational models (Delaney 2005; Modarresi et al. 2008; Wang and Hou 2011). For example, the relatively simple *general solubility equation* has been found to give a reasonable prediction of the water solubility of small solutes based on knowledge of a few physicochemical properties: $\log(S_W) = 0.5 - 0.01 (T_m - 25) - \log(K_{OW})$, where $S_W$ is the molar water solubility, $T_m$ is the melting point, and $K_{OW}$ is the oil–water partition coefficient of the solute (Ran et al. 2002; Yang et al. 2002). This model predicts that the water solubility of a solute should decrease as its melting point and $K_{OW}$ increase. Factors influencing the oil–water partition coefficient of solutes are discussed in more detail in a later section.

## 10.2.3.2 Supersaturation

In practice, it is often possible to dissolve a higher amount of a solute into a solvent than $C_S^*$ owing to *supersaturation* effects (Warren et al. 2010). The degree of supersaturation of a solution can be defined as $S = C_T / C_S^*$, where $C_T$ is the total solute concentration present. The physicochemical origin of supersaturation is similar to that of supercooling: there is an activation energy associated with the formation of nuclei that must be overcome before crystal growth can occur (Section 10.2.2.1). The degree of supersaturation that can be achieved before any crystals form on a reasonable timescale depends on storage temperature, the nature of the solute and solvent, the presence of any contaminating materials, and the application of external forces (Kashchiev and van Rosmalen 2003; Lindfors et al. 2008). The practical consequences of saturation and supersaturation can be highlighted by examining the form of the typical solubility profiles obtained experimentally when increasing amounts of solute are added to a solvent (Lindfors et al. 2008). At relatively low solute levels, the solute fully dissolves in the solvent and the samples appear clear, but at higher concentrations, some of the solute does not dissolve and the samples appear turbid or contain a layer of insoluble material (Figure 10.6). This insoluble matter may accumulate at either the top or the bottom of the sample depending on its density relative to that of the solvent. The insoluble material can be removed from the sample (e.g., by decanting, filtering, or centrifuging) and then the dissolved solute concentration present within the supernatant can be measured. A schematic plot of dissolved solute concentration ($C_D$) versus total solute concentration ($C_T$) is shown in Figure 10.6. There is initially a linear increase in $C_D$ with $C_T$ because all the added solute dissolves, but once a certain solute concentration is exceeded, there is a sharp decrease in $C_D$, after which $C_D$ remains fairly constant. The plateau value of $C_D$ observed at high solute concentrations is the saturation solute concentration ($C_S^*$). There is a range of solute concentrations where the amount of dissolved solute is higher than $C_S^*$, which is due to supersaturation. The degree of supersaturation in a system depends on solute type, solvent type, and environmental conditions such as temperature, mechanical forces, and time. Typically,

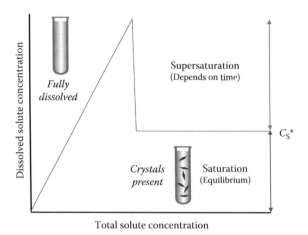

**Figure 10.6** **A representative plot from a typical solubility experiment showing the dissolved solute concentration ($C_D$) versus the total solute concentration added ($C_T$). At relatively low concentrations, the solute dissolves, but above a certain concentration, crystals form in the sample, and the solute concentration is at the saturation level ($C_S$).**

the degree of supersaturation decreases with time as more nucleation and crystal growth occurs in supersaturated solutions (Lindfors et al. 2008).

When carrying out solute dissolution experiments, it is particularly important to take into account that the rate of dissolution may vary considerably depending on the system (Lindfors et al. 2008). The dissolution of crystalline active ingredients can vary from a few seconds to a few days or even longer. In general, the rate of solute dissolution depends on solute properties (such as molecular packing and surface area), solvent properties (such as polarity and molecular weight), and environmental conditions (such as temperature, pressure, and applied mechanical forces). If one is trying to determine the saturation concentration of a solute, it is important to ensure that the system has come to equilibrium before carrying out any measurements. Information about dissolution kinetics can be achieved by measuring the change in dissolved solute concentration over time and establishing when equilibrium has been attained (Lindfors et al. 2008).

It is also important to establish the kinetics of solute dissolution when designing preparation methods for colloidal delivery systems containing solid active agents. For example, we have found that hydrophobic active ingredients (polymethoxyflavones) may be initially dissolved in nanoemulsion-based delivery systems owing to supersaturation effects, but that they form crystals during long-term storage that sediment to the bottom of the container (Li et al. 2012). This phenomenon would be undesirable for commercial applications because it would lead to adverse changes in product appearance, as well as to potential loss of biological activity (since the crystals may not be consumed or they may have a low bioavailability in the GIT). In general, it is important to ensure that any crystalline active ingredients dissolve in intestinal fluids within a reasonable time frame; otherwise, they may not be absorbed (Williams et al. 2013).

### 10.2.3.3 Dissolution Rate

The dissolution of a solute into a solvent can be considered to consist of two major steps: (i) solvation and detachment of the solute molecules from the solute–solvent interface, and (ii) movement of the solute molecules away from this interface and into the surrounding solution (Li and Jastic 2005). It is usually assumed that the mass transport of solute molecules away from the interface is the rate-limiting step, and so the dissolution process can be modeled using the Noyes–Whitney equation (Li and Jastic 2005):

$$\frac{dM}{dt} = A \times \frac{D}{h}\left(C_S^* - C_B\right). \tag{10.7}$$

Here, $M$ is the mass of solute leaving the particle, $t$ is the dissolution time, $A$ is the solute–solvent contact area, $D$ is the diffusion coefficient of the solute in the surrounding solution, $h$ is the thickness of the diffusion layer, $C_S^*$ is the saturation solubility, and $C_B$ is the concentration of solute in the surrounding solution. This equation predicts that a solute particle dissolves until the solute concentration in the surrounding solution reaches the solubility limit (i.e., $C_S^* = C_B$), and then any further dissolution is inhibited (i.e., supersaturation effects are ignored). The parameters $A$ and $h$ in the above equation are both time dependent, which means that this equation is difficult to solve for complex particle geometries. Nevertheless, a relatively simple expression can be derived to describe the dissolution of spherical particles (Li and Jastic 2005):

$$M^{1/3} = M_0^{1/3}\left(1 - \frac{DC_S^* t}{h r_0 \rho}\right). \tag{10.8}$$

Here, $M$ is the mass of solute remaining in the particles at time $t$, $M_0$ is the initial mass of solute in the particles, $\rho$ is the density of the solute in the particle, and $r_0$ is the initial radius of the solute particle. For small particles ($r < 25$ μm), $h$ can be taken to be approximately equal to the radius of the particle (i.e., $h = r_0$). The following simple expression has been derived to estimate the time required for small spherical solute particles to dissolve in well-agitated aqueous solutions (Li and Jastic 2005):

$$\tau = \frac{\rho r_0^2}{2DC_S^*}. \tag{10.9}$$

This equation predicts that solute particles should dissolve more rapidly as (i) the particle radius decreases, (ii) the diffusion coefficient of solute molecules in the surrounding solution increases, and (iii) the saturation solubility of the solute increases. The equations presented above were derived for spheres, whereas many solutes have much more complex structures, such as needles, cubes, and ellipsoids. Nevertheless, these equations still provide some useful qualitative insights into the major factors influencing the dissolution process. For example, the calculated dissolution times for solid particles dispersed in water are $2.5 \times 10^{-4}$, 0.025, 2.5, and 250 s, for particles with radii of 10, 100, 1000, and 10,000 nm, respectively (assuming $C_S = 0.01$ wt%, $r = 1000$ kg/m$^{-3}$ and $D = 2 \times 10^{-9}$ m$^2$/s). In reality, dissolution times may be considerably longer than these predicted values if the rate-limiting step is detachment of the solute molecules from the solid surface, rather than movement into the surrounding solution.

### 10.2.3.4 Temperature Dependence of Dissolution

In certain circumstances, it is important to know the temperature dependence of the saturation solubility of a solute when designing delivery systems (McClements and Xiao 2012). For example, a crystalline active ingredient may be dissolved within a solvent at high temperatures, but one wants to know whether it will remain in solution when the temperature is lowered. The temperature dependence of the saturation solubility of a solute in a solvent can be predicted assuming that the solvent and solute have widely differing melting points (>20°C) and form an ideal mixture (Walstra 2003):

$$x = \exp\left( \frac{\Delta H_{SL}}{R} \left[ \frac{1}{T_m} - \frac{1}{T} \right] \right). \tag{10.10}$$

Here, $x$ is the saturation solubility (mole fraction) of the higher melting point component (solute) in the lower melting point component (solvent), $T$ is the temperature, $T_m$ is the melting point, $R$ is the gas constant, and $\Delta H_{SL}$ is the molar heat of fusion (Walstra 2003). Above the melting point, $x = 1$; that is, the solute is completely liquid and miscible with the solvent. The mole fraction of a solute dissolved in a solvent at saturation can be converted into a mass fraction ($\Phi_S$) if the molar masses of the two components are known:

$$\Phi_S = \left[ 1 + \frac{M_S}{M_O}\left( \frac{1}{x} - 1 \right) \right]^{-1}. \tag{10.11}$$

Here, $M_S$ and $M_O$ are the molar masses of the solute and solvent, respectively. The above equations can be combined to predict the change in saturation

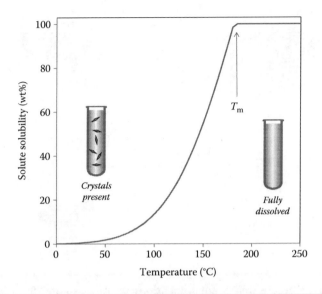

**Figure 10.7 Calculation of the temperature dependence of the solubility of a solid solute in a liquid carrier.** The calculations were carried out for $\beta$-carotene dissolved in a triacylglycerol oil: $\Delta H_f = 76$ J/g; $T_m = 181°C$; $M_S = 537$ g/mol; $M_O = 800$ g/mol.

solubility of a solute with temperature (Figure 10.7). The solubility increases as the temperature rises, and so it is possible to create a solution that contains a relatively high solute concentration at elevated temperatures. However, when this solution is cooled, the solubility decreases and so the system may become supersaturated or crystallize if the total amount of solute present exceeds $C_S^*$ at the final temperature. The rate of this process will depend on kinetic factors, such as nucleation and crystal growth rates as discussed earlier. In practice, the above equations have limited application because most real systems are nonideal mixtures, so their solubility behavior depends on the precise nature of the molecular interactions involved (Chebil et al. 2007). The above equations can be modified to take into account nonideal mixtures by dividing the right-hand side by the activity coefficient ($\gamma_i$). A number of theoretical models have been developed to predict the activity coefficients of solutes in solvents on the basis of their molecular characteristics (Modarresi et al. 2008).

### 10.2.3.5 Dissolution in Multiphase Systems

Most colloidal delivery systems are multiphase systems consisting of at least two phases, that is, the dispersed phase (particles) and the continuous phase (suspending fluid) (Figure 10.8). Solute solubility is typically different in different phases, and so it is important to establish the solubility of a solute in the overall system. The

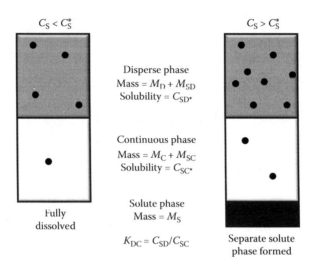

**Figure 10.8  A solute (black dots) partitions between solute (S), dispersed (D), and continuous (C) phases depending on equilibrium partition coefficients, solubilities, and concentrations.**

following expression can be used to predict the maximum amount of solute that can be dissolved in a two-phase system at equilibrium (McClements 2012):

$$\Phi_S^* = \frac{C_{SD}^* \left[ \Phi_D \left( K_{DC} - 1 \right) + \left( 1 - C_{SD}^* \right) \right]}{\left( 1 - C_{SD}^* \right) K_{DC}}.$$

(10.12)

Here, $C_{SD}^* \left( = M_{SD}^* / \left[ M_{SD}^* + M_D \right] \right)$ and $C_{SC}^* \left( = M_{SC}^* / \left[ M_{SC}^* + M_C \right] \right)$ are the equilibrium saturation concentrations of the solute in the dispersed and continuous phases (expressed as mass fractions), $\Phi_D$ is the mass fraction of dispersed phase in the overall system ($= M_D / [M_S + M_C + M_D]$), and $K_{DC}$ is the partition coefficient. $M_{SD}$ and $M_{SC}$ are the masses of the solute dissolved in the dispersed and continuous phases, while $M_S$, $M_D$, and $M_C$ refer to the masses of the solute, dispersed phase, and continuous phase, respectively. The superscript "*" refers to the condition of equilibrium saturation. This equation shows that the total amount of solute that can be dissolved in a two-phase delivery system depends on the amount of dispersed phase ($\Phi_D$), the saturation solubility of the solute in the disperse phase ($C_{SD}^*$), and the partition coefficient ($K_{DC}$). For systems that consist of oil and water (such as emulsions and nanoemulsions), the oil–water partition coefficient ($K_{OW}$) can be used as the appropriate $K_{DC}$. Predictions made using these equations show that the total amount of a nonpolar solute ($K_{OW} > 1$) that can be dissolved in an emulsion or nanoemulsion increases with increasing oil content (Figure 10.9). The above equations are useful for calculating the total amount

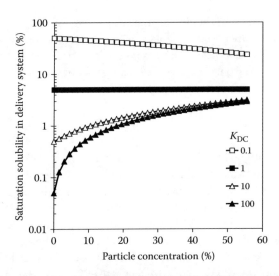

**Figure 10.9** **Influence of equilibrium partition coefficient ($K_{DC}$) and disperse phase (particle) concentration on the maximum amount of solute that can be solubilized within a delivery system. It is assumed that the saturation concentration of the solute in the dispersed phase ($C_{SD}^*$) is 5%.**

of an active ingredient that can be dissolved within a particular delivery system and to predict how the solubility will change if the system is diluted with other substances in a food product or in the GIT.

### 10.2.3.6 Nucleation and Crystal Growth

Nucleation and crystal growth in supersaturated solutions are closely related to those observed in supercooled melts (McClements 2012). However, nuclei are formed within a liquid that has a different molecular structure to the nucleating substance in the case of supersaturation, whereas they are formed in a liquid that has the same molecular structure in the case of supercooling (Carlert et al. 2010; Kashchiev and van Rosmalen 2003). An expression for the critical radius for the formation of stable nuclei in the case of supersaturation can be derived from classical nucleation theory:

$$r^* = \frac{2\gamma_{SL}(V_m/N_A)}{k_B T \ln(S)}. \tag{10.13}$$

The free energy associated with the formation of nuclei with this critical radius in a supersaturated solution is then given by:

$$\Delta G^* = \frac{16\pi\gamma_{SL}^3(V_m/N_A)^2}{3(k_B T \ln(S))^2}. \tag{10.14}$$

Here, $\gamma_{SL}$ is the solid–liquid interfacial tension, $V_m$ is the molar volume of the solute, $N_A$ is Avogadro's number, and $S$ is the degree of supersaturation ($S = C_B/C_S^*$). The rate at which nucleation proceeds is related to the activation energy ($\Delta G^*$) that must be overcome before stable nuclei persist in the supersaturated solution:

$$J = J_0 \exp\left( -\frac{\Delta G^*}{k_B T} \right). \tag{10.15}$$

The above equations highlight some of the major factors expected to influence nucleation in supersaturated solutions. First, the nucleation rate should increase as the degree of supersaturation increases: $S$ is always greater than 1, and so $\Delta G^*$ decreases as $\ln(S)$ increases. Second, the nucleation rate has a strong dependence on the interfacial tension, since $\Delta G^*$ is proportional to $\gamma_{SL}$ cubed. The interfacial tension depends on the nature of the solvent surrounding the solute molecules, which has important implications for situations where nucleation occurs in multiphase systems such as many delivery systems (McClements 2012). As the interfacial tension increases, the energy barrier for the formation of nuclei increases, which decreases the nucleation rate. Third, the nucleation rate increases as the diffusion coefficient of solute molecules in the surrounding solution increases, and hence one would expect faster nucleation in lower-viscosity solvents. Despite the classical nucleation theory being widely used to describe nucleation, it is based on a number of assumptions that severely limit its accuracy, and therefore more sophisticated theories have been developed (Chen et al. 2011; Erdemir et al. 2009; Vekilov 2010a,b,c).

## 10.3 Partitioning Phenomenon

The physical location of an active ingredient within a colloidal delivery system often has a major impact on its properties, such as the loading capacity, retention efficiency, release profile, and chemical stability. For example, some flavor molecules are stable to chemical degradation when they are located within the interior of oil droplets, but they are highly unstable when they are dispersed in the surrounding aqueous phase (Choi et al. 2009). In this section, some of the major factors influencing the partitioning of active ingredients in multiphase systems are therefore considered.

### 10.3.1 Equilibrium Partitioning Coefficients

The partitioning of a solute between two phases is described by an equilibrium partition coefficient (McClements 2012):

$$K_{12} = \frac{a_1}{a_2}, \tag{10.16}$$

where $a_1$ and $a_2$ are the activity coefficients of the solute molecule in Phase 1 and Phase 2, respectively. In cases where the solute concentrations in the two phases are relatively low, solute–solute interactions can be ignored and the activity coefficients can be approximated by concentrations:

$$K_{12} = \frac{c_1}{c_2}, \tag{10.17}$$

where $c_1$ and $c_2$ are the concentrations of the solute molecules in the two phases. The nature of the two phases involved can vary widely depending on the system in question, for example, gas–liquid, gas–solid, liquid–liquid, liquid–solid, and solid–solid. The magnitude of the equilibrium partition coefficient depends on the relative strength of the interactions between the solute molecules and their surroundings in the two different phases (Israelachvili 2011):

$$K_{12} = \exp\left(-\frac{\Delta G_{12}}{RT}\right), \tag{10.18}$$

where $\Delta G_{12}$ is the difference in free energy per mole of the solute in the two phases. This free energy term depends on the change in mixing enthalpy and entropy that occur when a solute molecule moves from one phase to another: $\Delta G_{12} = \Delta G_{int} - T\Delta S_{mix}$ (McClements 2005). The mixing entropy term ($\Delta S_{mix}$) favors the random distribution of solute molecules throughout the entire system, rather than their confinement to only one of the phases, and therefore it tends to drive the partition coefficient toward unity. The interaction energy change ($\Delta G_{int}$) is associated with the net alteration in the strength of the molecular interactions when solute molecules move from one phase to another. The value of $\Delta G_{int}$ associated with the movement of a solute from Phase 1 to Phase 2 may be negative (favorable), zero (neutral), or positive (unfavorable). The interaction energy change tends to favor a molecular distribution that optimizes the attractive forces between the solute molecules and the surrounding molecules, which may lead to preferential accumulation of the solute molecules in one phase or another. The equilibrium partition coefficient therefore depends on the molecular interactions of the solute molecules and of the solvent molecules in the two different phases. It therefore depends on the nature of the materials used to fabricate the particle matrix and the surrounding liquid in colloidal delivery systems. The overall free energy change ($\Delta G_{12}$) associated with partitioning of a solute between two phases depends on the net contribution of the interaction energy and entropy of mixing effects. When $\Delta G_{12}$ is highly negative ($K_{12} \gg 1$), the solute prefers to be in Phase 1; when $\Delta G_{12}$ is highly positive ($K_{12} \ll 1$), the solute prefers to be in Phase 2;

when $\Delta G_{12}$ is close to zero ($K_{12} \approx 1$), the solute prefers to be randomly distributed throughout the whole system.

## 10.3.2 Solute Partitioning in Delivery Systems

In this section, the equilibrium partitioning of a solute within a three-component system consisting of solute (S), dispersed phase (D), and continuous phase (C) is considered. Initially, a system is considered where the total amount of solute present may be comparable to that of the dispersed and continuous phases. In this case, the solute concentration may be below or above the equilibrium saturation concentration ($C_S^*$) for the overall system, and it may therefore be either fully dissolved or present as a separate phase (Figure 10.8). The overall saturation concentration for a delivery system can be calculated using the equation given earlier (Section 10.2.3.5) (McClements 2012).

### 10.3.2.1 Solute Partitioning below the Saturation Limit

At total solute concentrations below the saturation limit ($C_S < C_S^*$), the solute is fully dissolved in both the dispersed and continuous phases and its distribution is determined by the equilibrium partition coefficient ($K_{DC}$). The fraction of solute molecules present in the dispersed phase is given by the following equation (McClements 2012):

$$\phi_{SD} = \left( \frac{-b - \sqrt{b^2 - 4ac}}{2a} \right), \tag{10.19}$$

where $a = [1 - K_{DC}]$, $b = [K_{DC}(\Phi_S - \Phi_D) + (\Phi_D - 1)]$, and $c = K_{DC}\Phi_S\Phi_D$. This equation can be used to predict the distribution of solute molecules between the dispersed phase (particles) and continuous phase (surrounding fluid) as a function of particle concentration. A plot of the percentage of solute dissolved in the dispersed phase ($100 \times \Phi_{SD}/\Phi_S$) as a function of particle concentration is shown in Figure 10.10 for substances with different partition coefficients. In this example, it was assumed that the dispersed phase was oil and the continuous phase was water (i.e., an oil-in-water emulsion or nanoemulsion), so that $K_{DC} = K_{OW}$. As expected, the more hydrophobic a material (higher $K_{OW}$), the greater the percentage remaining in the oil droplets at a particular disperse phase concentration. For less hydrophobic materials (lower $K_{OW}$), a significant percentage of the solute may be present in the aqueous continuous phase, particularly when the oil concentration is low. This phenomenon has important consequences for the functional performance of many types of colloidal delivery systems. The chemical stability of many lipophilic components depends on their molecular environment; for example, some active

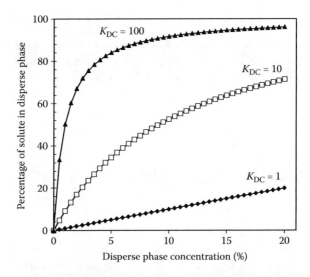

**Figure 10.10** **Influence of disperse phase (particle) concentration and equilibrium partition coefficient ($K_{DC}$) on the percentage of solute present within the disperse phase of a colloidal delivery system. In oil–water systems, for more nonpolar compounds, a greater percentage is present within the oil phase than for more hydrophilic compounds.**

substances undergo faster degradation when dissolved in water than in oil (Choi et al. 2009). Second, absorption of lipophilic components from the GIT depends on whether they are dissolved in the aqueous phase or in oil droplets (Porter et al. 2007; Williams et al. 2013). Third, the nucleation, growth, and ripening of crystalline solutes depend on their molecular environment and may therefore be altered by partitioning (McClements 2012).

A simplified equation can be developed if it is assumed that the concentration of the solute is much lower than that of either the dispersed or continuous phases (i.e., $\phi_S \ll \phi_D$ and $\phi_C$), so that the solute does not contribute significantly to the overall mass of the system. In this case,

$$\phi_{SD} = \phi_D K_{DC}[1 - \phi_D(1 - K_{DC})]^{-1}. \tag{10.20}$$

This equation shows that the amount of solute present within the particles increases with increasing particle concentration ($\phi_D$) and increasing $K_{DC}$ (for $K_{DC} > 1$). It is useful to use this equation to predict the distribution of an active ingredient in a colloidal delivery system under different conditions (e.g., dilution), since their molecular environment often plays an important role in determining their chemical stability, sensory perception, and biological function.

### 10.3.2.2 Solute Partitioning above the Saturation Level

When a sufficiently high amount of solute is added to a delivery system, it will no longer be soluble, and so it will form a separate phase (Figure 10.8). Above the saturation concentration for the delivery system ($\Phi_S > \Phi_S^*$), the solute will be saturated in both the disperse phase ($C_{SD}^*$) and the continuous phase ($C_{SC}^*$), and a third phase will be formed. The concentration (mass fraction) of the solute in the dispersed, continuous, and solute phases is then given by (McClements 2012):

$$\Phi_{SD}^* = \frac{C_{SD}^*\Phi_D}{1-C_{SD}^*} \quad \Phi_{SC}^* = \frac{C_{SD}^*\Phi_C}{K_{DC}-C_{SD}^*} \quad \Phi_{SS}^* = \Phi_S - \Phi_{SD}^* - \Phi_{SC}^* \quad (10.21)$$

Here, $\Phi_S$ is the total mass fraction of solute present in the overall delivery system. These equations can be used to determine the change in solute concentration with system composition ($\Phi_S$, $\Phi_D$, $\Phi_C$), saturation concentration ($C_{SD}^*$), and partition coefficient ($K_{DC} = C_{SD}^*/C_{SC}^*$). Calculations of the solute concentration in the three different phases with increasing solute concentration in the overall system are shown in Figure 10.11. Initially, the solute partitions between the dispersed and continuous phases, but once a critical concentration is exceeded, it forms a separate phase. This has important implications for the development of colloidal delivery systems. For example, if one is trying to disperse a crystalline

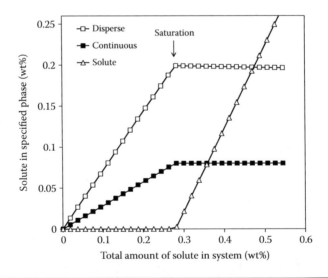

**Figure 10.11** Calculated amounts of solute in dispersed, continuous, and solute phases when the total amount of solute in the delivery system is increased. It was assumed that $C_{SD}^* = 0.01$, $K_{DC} = 10$, and $\phi_D = 0.2$.

active ingredient within an oil-in-water emulsion, then one may want to ensure that the solute concentration remains below the saturation concentration for the overall system; otherwise, crystals may form during storage (Li et al. 2012; Li et al. 2012). Crystal formation may adversely alter the physical stability and sensory perception of delivery systems, as well as reducing the bioavailability of the active ingredient.

### 10.3.2.3 Partitioning of Volatile Solutes

A number of active ingredients used in the food industry are volatile and therefore partition into the headspace above a product, for example, flavor molecules (McClements 2005). The influence of the properties of a delivery system on the headspace concentration of a volatile flavor is important since it influences the expected flavor profile. In this section, the partitioning of volatile solute molecules between the headspace and colloidal dispersions is considered (Figure 10.12). In this case, three equilibrium partition coefficients need to be defined:

$$K_{DC} = \frac{c_D}{c_C} \quad K_{GC} = \frac{c_G}{c_C} \quad K_{GD} = \frac{c_G}{c_D} \tag{10.22}$$

Here, the subscripts D, C, and G refer to the dispersed, continuous, and gas phases, respectively. Assuming that the amount of solute is much smaller than the

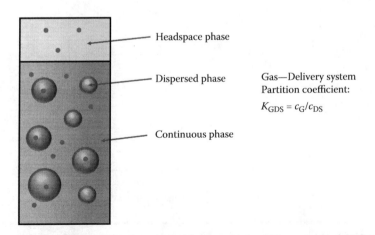

Headspace phase

Dispersed phase

Gas—Delivery system
Partition coefficient:

$K_{GDS} = c_G/c_{DS}$

Continuous phase

**Figure 10.12  Volatile molecules, such as flavors, may partition between the gaseous headspace and the delivery system by an amount that depends on the equilibrium partition coefficients and system composition.**

amount of continuous or dispersed phases, the partition coefficient between the headspace gas and the delivery system ($K_{GDS}$) is given by:

$$\frac{1}{K_{GDS}} = \frac{\phi_D}{K_{GD}} + \frac{\phi_C}{K_{GC}} = \frac{K_{DC}\phi_D}{K_{GC}} + \frac{(1-\phi_D)}{K_{GC}}. \tag{10.23}$$

Here, $\phi_D$ and $\phi_C$ are the volume fractions of the dispersed and continuous phases: $\phi_D + \phi_C = 1$. Thus, the partition coefficient between a delivery system and the headspace can be predicted from knowledge of $K_{GD}$ and $K_{GC}$ and the system composition.

The partitioning of volatile molecules above delivery systems has a major impact on the flavor profile of food products containing oil and water, such as oil-in-water emulsions and nanoemulsions (McClements 2005). In this case, the most appropriate partition coefficients are the gas–water ($K_{GC} = K_{GW}$) and oil–water ($K_{DC} = K_{OW}$) partition coefficients, which can be found in the literature for many common flavor molecules. Predictions of the amount of a volatile solute molecule in the headspace above oil-in-water emulsions with the same total flavor concentration but different disperse phase volume fractions are shown in Figure 10.13. For nonpolar flavors ($K_{OW} > 1$), the headspace flavor concentration decreases with increasing oil content. The steepness of this decrease increases as the flavor molecules become more

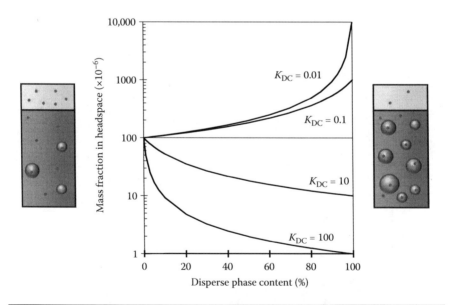

**Figure 10.13** The partitioning of volatile solute molecules above delivery systems depends on system composition and partition coefficients. For emulsions, the initial flavor intensity of systems with fixed flavor content is strongly dependent on oil content.

nonpolar, so that even a small increase in oil concentration in a dilute oil-in-water emulsion causes a large decrease in headspace concentration for highly nonpolar flavors ($K_{OW} \gg 1$). For polar flavors ($K_{OW} < 1$), the headspace flavor concentration increases slightly with increasing fat content at relatively low oil contents but more sharply at higher oil contents (Figure 10.13). The steepness of the changes at high oil contents increases as the flavor molecules become more hydrophilic. Practically, this means that the volatile nonpolar flavors in relatively dilute oil-in-water emulsions become more odorous as the fat content is decreased, whereas the volatile polar flavors remain relatively unchanged. This has important consequences when designing reduced-fat versions of food and beverage products.

### 10.3.2.4 Influence of Particle Size on Solubility

The above discussion does not take into account the potential influence of particle size on the partitioning of solute molecules. The assumption that solute partitioning is independent of particle size is likely to be valid for colloidal dispersions containing relatively large particles (i.e., $d > 500$ nm), because the influence of curvature on solubility is not very important. However, when the particle diameter falls below this value, there is a significant increase in the solubility of a solute within the surrounding liquid because of an increase in Laplace pressure (Van Eerdenbrugh et al. 2010). Theoretical calculations show that the equilibrium solubility of the material within a spherical particle increases as particle size decreases (Van Eerdenbrugh et al. 2010):

$$C(d) = C_S^* \exp\left( \frac{4\gamma V_m}{RTd} \right).$$
(10.24)

Here, $V_m$ is the molar volume of the solute, $\gamma$ is the interfacial tension at the solute–solvent interface, $C_S^*$ is the equilibrium solubility of the solute in the continuous phase, and $C(d)$ is the solubility of the solute when contained in a spherical particle of diameter $d$. If a solute makes up the entirety of a particle in a delivery system, then its partitioning between the dispersed and continuous phases depends on particle size. A plot of the change in relative solubility $(C(d))/C_S^*$ of a solute on particle diameter is shown in Figure 10.14 for a representative active ingredient (β-carotene). For relatively large particles ($d > 1000$ nm), the solubility of the solute in the surrounding liquid is close to the solubility of a bulk material. However, as the particle diameter decreases below about 200 nm, there is a steep increase in solubility. The effect of particle size on the water solubility of various lipophilic solutes has been demonstrated experimentally (Buckton and Beezer 1992; Van Eerdenbrugh et al. 2010). An increase in solute concentration in the aqueous phase would be expected to cause a corresponding increase in the headspace. Thus, the flavor profiles of food products may change as the size of the colloidal particles

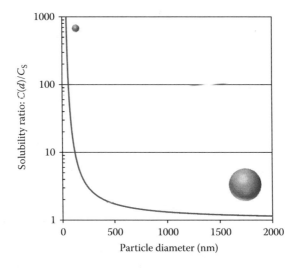

**Figure 10.14  Size dependence of the solubility of a solute. The solubility increases appreciably when the particle radius falls below about 200 nm. The calculations were made for β-carotene particles, with a molar volume of 570 cm³ mol⁻¹ and an assumed interfacial tension of 1 mJ m⁻².**

within them changes. Similarly, the solubility of a lipophilic bioactive compound contained within a colloidal particle in the surrounding gastrointestinal fluids will increase as the particle size decreases, which may increase its bioavailability.

## 10.4  Mass Transport Processes

The movement of molecules from one region to another within a colloidal delivery system is important for a number of reasons, including loading, retention, and release of active ingredients (Li and Jastic 2005). In this section, a brief overview of the major factors that may influence the mass transport of solute molecules through materials is given. In general, molecular movement is typically due to diffusion or convection. Diffusion occurs as a result of the thermal energy of the system, whereas convection occurs as a result of fluid flow. When studying the mass transport of molecules or particles within colloidal delivery systems, it is important to establish whether the dominant process is diffusion or convection.

### 10.4.1  Diffusion

In this section, the term *particle* is used to refer to both molecules and colloidal particles for the sake of convenience. The diffusion of a particle suspended in a fluid is a result of its thermal energy and takes place by a series of small steps that are random in direction and speed (Walstra 2003). Because of the random nature of the diffusion process, the average absolute distance $(x)$ that a particle moves from its original location is zero, but the square of the distance $(x^2)$ is finite. The root-mean-square distance $(\Delta\bar{x})$ that a particle moves in a particular direction is given by the following equation:

$$\Delta\bar{x} = \sqrt{2Dt}, \tag{10.25}$$

where $D$ is the translational diffusion coefficient of the particle and $t$ is the time. The diffusion coefficient can be related to the size of the particles and to the properties of the surrounding solvent (i.e., viscosity) by the following expression:

$$D = \frac{k_B T}{6\pi\eta r}. \tag{10.26}$$

Here, $k_B$ is Boltzmann's constant, $T$ is the absolute temperature, $\eta$ is the viscosity of the solvent surrounding the particle, and $r$ is the particle radius. This expression shows that the average distance moved by a particle in a certain time increases as the particle radius and viscosity decrease. For nonspherical particles, the above equation is modified by replacing $r$ with the hydrodynamic radius (Hunter 1986; Larson 1999).

Even though the movement of an individual particle in a fluid is random, there is a net movement of solute particles along activity gradients, from regions of high activity to regions of low activity (McClements 2012). For the sake of convenience, the solute concentration is often used rather than the solute activity. If there are any concentration gradients within a particular system, the solute particles will tend to move to make them disappear. Mathematical modeling of diffusion processes can be used to quantify the change in the solute concentration profile over time and distance in a system. An appreciation of the factors influencing diffusion can be obtained by examining mathematical expressions developed to describe the diffusion of solute particles across an infinite plane boundary separating two infinite phases (Li and Jastic 2005). Initially, the particle concentration on one side of the boundary is assumed to have a finite value $(C = C_i)$, whereas the particle concentration on the other side of the boundary is assumed to be zero $(C = 0)$. After they come into contact, particles diffuse from the solute-rich side to the solute-poor side because of the concentration gradient. Expressions have been derived to predict the change in solute concentration with time $(t)$ and distance $(x)$ from the boundary for this simple system and for more complex systems with other geometries (Li and Jastic 2005). To a first approximation, the amount of solute molecules released from the solute-rich side into the solute-poor side is given by Higuchi's equation (Siepmann and Peppas 2011):

$$M(t) = A\sqrt{C_S^*\left(2C_i - C_S^*\right)Dt}. \tag{10.27}$$

Here, $A$ is the surface area of the boundary between the two phases. This equation shows that a greater amount of solute is released as the surface area increases, the diffusion coefficient increases, and the concentration gradient $(C_i - C_S^*)$ increases. It also suggests that the amount of solute released should initially increase linearly with the square root of time. Equations similar to the one above can be derived for other kinds of geometries that are more applicable to delivery systems, such as release from spheres or coated spheres (see later).

## 10.4.2 Convection

In many situations, convection makes a major contribution to the mass transport of molecules or particles within colloidal delivery systems, for example, because of fluid flow in containers, the mouth, or the GIT. In this case, it is important to establish the precise nature of the fluid flow profile that the system experiences, for example, laminar or turbulent flow (Walstra 2003). In laminar flow, the fluid elements move in a smooth and regular manner, whereas in turbulent flow, they move in a chaotic fashion even though the average flow is in a particular direction. Numerous kinds of laminar flows are possible depending on the geometry of the system, for example, simple shear, rotational, and elongational. The degree of laminar flow can be quantified in terms of a *velocity gradient*. In turbulent flow, any fluid element is subjected to a random fluctuation in direction and speed. This leads to the formation of *eddies* within the fluid, which are responsible for highly efficient mixing. Whether the flow is laminar or turbulent in a particular system can be established using the Reynolds number (Re), which is a measure of the relative importance of inertial over frictional stresses. For flow in a cylindrical pipe, the Reynolds number is given by:

$$\mathrm{Re} = \frac{d\bar{v}\rho}{\eta}. \tag{10.28}$$

Here, $d$ is the diameter of the pipe, $\bar{v}$ is the flow rate, $\rho$ is the fluid density, and $\eta$ is the fluid viscosity. Turbulence begins when the Reynolds number exceeds a critical value (Re ≈ 2300). The mixing of different components within a fluid occurs much more effectively when the flow profile is turbulent rather than laminar. This has important consequences for the transport of delivery systems within the GIT and for the release of active ingredients from them. Mathematical equations developed to model the release of active ingredients from delivery systems are often based on the assumption that mass transport is due to diffusion, convection, or both. It is therefore important to be sure that the assumptions made in the development of a particular mathematical theory are appropriate for the practical situation that it is used to model.

# 10.5 Modeling Release Profiles from Colloidal Delivery Systems

One of the most important characteristics of colloidal delivery systems is their ability to release active ingredients (Cejkova and Stepanek 2013). It is therefore useful to have theoretical and mathematical models to describe and predict the release profiles of active ingredients from colloidal particles. As discussed earlier in this book, an active ingredient may be released from a colloidal particle through a number of mechanisms, including diffusion, dissolution, erosion, fragmentation, and swelling (Liechty et al. 2010; Peppas et al. 2000; Siepmann and Siepmann 2008) (Figure 1.4). In this section, a brief overview of some of the models that have been developed to describe the release of active ingredients from colloidal particles with different structures and release mechanisms is given.

## 10.5.1 Diffusion

An active ingredient (solute) may be released from a colloidal particle primarily owing to molecular diffusion (Siepmann and Siepmann 2012). In general, the rate at which an active ingredient is released depends on particle characteristics (such as initial size, shape, structure, and composition), solute characteristics (such as diffusion coefficient and interactions), environmental conditions (such as temperature and stirring), and the solute concentration gradient from the interior of the particle to the surrounding medium. Mathematical models have been developed to describe diffusion-limited release of active ingredients from particles with a variety of different shapes (Siepmann and Siepmann 2012). A number of different cases can be considered depending on the initial properties of the active ingredient and the colloidal particles, and some of the most common situations are highlighted below.

### 10.5.1.1 Release from Homogeneous Spheres

In the first situation, the release of an active ingredient initially dissolved within a homogeneous sphere is considered (Figure 10.15). In the pharmaceutical industry, this example is often referred to as a *monolithic solution* (Siepmann and Siepmann 2012). A mathematical theory, known as the Crank model, has been developed to describe the rate at which a solute is released from a homogeneous spherical particle surrounded by an infinite volume of a well-stirred liquid:

$$\frac{M(t)}{M_\infty} = 1 - \sum_{n=0}^{\infty} \frac{6}{\pi^2 n^2} \exp\left[-\frac{D\pi^2 n^2}{K_{DC} r^2} t\right]. \tag{10.29}$$

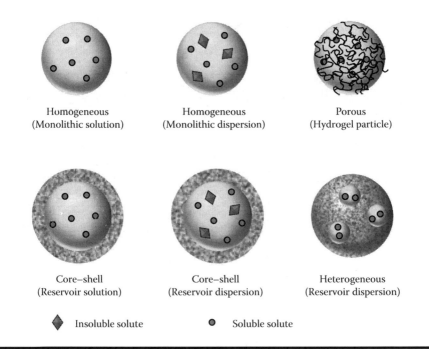

Homogeneous
(Monolithic solution)

Homogeneous
(Monolithic dispersion)

Porous
(Hydrogel particle)

Core–shell
(Reservoir solution)

Core–shell
(Reservoir dispersion)

Heterogeneous
(Reservoir dispersion)

◆ Insoluble solute    ● Soluble solute

**Figure 10.15    Schematic representation of different particle structures whose solute release rates can be modeled mathematically.**

Here, $M(t)$ is the total amount of solute that has diffused out of the sphere by time $t$, $M_\infty$ is the amount of solute that has diffused out of the particle after an infinite time (which is close to the initial amount present in the particle), $D$ is the diffusion coefficient of the solute within the particle matrix, $r$ is the particle radius, and $n$ is an integer. This equation assumes that the concentration of solute in the continuous phase is initially zero and therefore is only strictly applicable to systems with high $K_{DC}$ values. The above equation involves calculating an infinite series of terms, which makes it somewhat difficult to apply in practice and hence a number of simplifications have been developed. For example, release profiles can be more easily modeled using the following equation, which gives predictions that are in close agreement with the Crank model (Lian et al. 2004):

$$\frac{M(t)}{M_\infty} = 1 - \exp\left[-\frac{1.2 D \pi^2}{K_{DC} r^2} t\right]. \tag{10.30}$$

This model provides some useful insights into the factors that influence the rate of solute release from particles. For example, the release of solute molecules from oil droplets in oil-in-water emulsions can be modeled using these equations (Figure 10.16). There is a rapid initial release of solute molecules from the oil droplets into

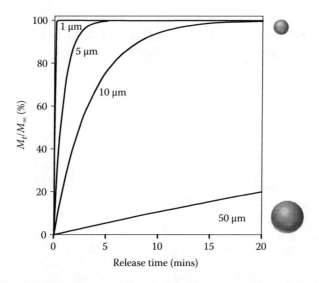

**Figure 10.16** **Influence of particle radius and equilibrium partition coefficient ($K_{DC}$) on the release profile of solute molecules from spherical particles. Predictions were made using the Crank model assuming a $K_{DC}$ of $10^3$ and a diffusion coefficient of $4 \times 10^{-10}$ m$^2$ s$^{-1}$.**

the aqueous phase, followed by a more gradual increase at longer times as the solute concentrations in the oil and aqueous phases approach their equilibrium values. The release rate increases as the particle size decreases, because the solute molecules have a shorter distance to diffuse out of the particles. Thus, the release rate of an active ingredient from a colloidal particle may be controlled by altering the particle size.

A convenient measure of the rate of solute release from a particle is the time required for half of the solute molecules to diffuse out, which is given by the following approximate expression derived from the Crank model (Lian et al. 2004):

$$t_{1/2} = \frac{0.0585 r^2 K_{DC}}{D}. \qquad (10.31)$$

The variation of $t_{1/2}$ with particle radius and equilibrium partition coefficient for solute molecules in oil-in-water emulsion or nanoemulsions (where $K_{DC} = K_{OW}$) is shown in Table 10.1 (ignoring the potential effects of particle size on $K_{OW}$). The time for half of the solute molecules contained within the oil droplets to be released is strongly dependent on the equilibrium partition coefficient ($K_{OW}$), with $t_{1/2}$ increasing with increasing $K_{OW}$. For polar solutes ($K_{OW} \leq 1$), release occurs extremely rapidly ($t_{1/2} < 5$ s) from particles with radii <100 μm. On the other hand, for nonpolar solutes ($K_{OW} \gg 1$), release may occur relatively slowly from larger particles. For example, $t_{1/2}$ is approximately 1.5 and 15 s for $K_{OW} = 100$ and 1000,

**Table 10.1  Predictions of the Influence of Particle Radius and Oil–Water Partition Coefficient on the Time Taken for Half of the Solute Molecules to Diffuse out of Spherical Oil Droplets (Crank Model)**

| | $K_{OW} = 1$ | $K_{OW} = 10$ | $K_{OW} = 100$ | $K_{OW} = 1000$ |
|---|---|---|---|---|
| $r$ ($\mu m$) | *Release Time:* $t_{1/2}$ *(s)* | | | |
| 0.1 | $1.46 \times 10^{-6}$ | $1.46 \times 10^{-5}$ | $1.46 \times 10^{-4}$ | $1.46 \times 10^{-3}$ |
| 0.2 | $5.84 \times 10^{-6}$ | $5.84 \times 10^{-5}$ | $5.84 \times 10^{-4}$ | $5.84 \times 10^{-3}$ |
| 0.5 | $3.65 \times 10^{-5}$ | $3.65 \times 10^{-4}$ | $3.65 \times 10^{-3}$ | $3.65 \times 10^{-2}$ |
| 1 | $1.46 \times 10^{-4}$ | $1.46 \times 10^{-3}$ | $1.46 \times 10^{-2}$ | $1.46 \times 10^{-1}$ |
| 2 | $5.84 \times 10^{-4}$ | $5.84 \times 10^{-3}$ | $5.84 \times 10^{-2}$ | $5.84 \times 10^{-1}$ |
| 5 | $3.65 \times 10^{-3}$ | $3.65 \times 10^{-2}$ | $3.65 \times 10^{-1}$ | 3.65 |
| 10 | $1.46 \times 10^{-2}$ | $1.46 \times 10^{-1}$ | 1.46 | 14.6 |
| 20 | $5.84 \times 10^{-2}$ | $5.84 \times 10^{-1}$ | 5.84 | 58.4 |
| 50 | $3.65 \times 10^{-1}$ | 3.65 | 36.5 | 365 |
| 100 | 1.46 | 14.6 | 146 | 1460 |

*Source:* Adapted from McClements (2005).

respectively, in emulsions containing droplets with a 10 μm radius. It is therefore possible to control the release profile of highly nonpolar solute molecules in oil-in-water emulsions by controlling their particle size.

The Crank model assumes that the rate-limiting step for release is the transport of solute molecules through the interior of the dispersed phase particles, that is, that there is no external resistance to mass transport owing to the surrounding continuous phase (Lian et al. 2004). Other models have been developed based on the assumption that the rate-limiting step is the mass transport of the solute into the continuous phase away from the particle surface, for example, the Sherwood equation (Arifin et al. 2006). General models have also been developed that take both of these effects into account (Arifin et al. 2006). These theories can be used to provide a more detailed analysis of the influence of particle size, diffusion coefficients, equilibrium partition coefficients, and fluid flow on solute release profiles.

The above equations assume that the release of the solute molecules occurs from a homogeneous sphere, whereas in reality, the release of solutes from colloidal delivery systems often occurs from more complicated particle structures, such as coated or embedded particles (Chapter 1). Theoretical or empirical models have been developed to describe a number of these more complex particle systems (Arifin et al. 2006; Siepmann and Siepmann 2008, 2011).

## 10.5.1.2 Release from Homogeneous Spheres Containing Insoluble Active Ingredients

In this case, it is assumed that the active ingredient is not fully soluble within the particle matrix. Instead, it is homogeneously distributed throughout the spherical particle in both insoluble and soluble forms (Figure 10.15). Only the soluble form of the drug is assumed to diffuse out of the particles, and it is assumed that the insoluble and soluble forms are in equilibrium. Thus, when some of the soluble form diffuses out of the particle, some of the insoluble form dissolves. In the pharmaceutical industry, this situation is often referred to as a *monolithic dispersion* (Siepmann and Siepmann 2012). The following equation has been derived to describe the release of an active ingredient from a monolithic dispersion containing spherical particles (Siepmann and Siepmann 2012):

$$\frac{M(t)}{M_\infty} - \frac{3}{2}\left[1 - \left(1 - \frac{M(t)}{M_\infty}\right)^{2/3}\right] = -\frac{3D}{r^2}\frac{c_S^*}{c_0}t. \qquad (10.32)$$

Here, $c_S^*$ is the saturation concentration of the active ingredient within the particle matrix and $c_0$ is the initial concentration of the active ingredient within the system. This expression must be solved numerically to determine the value of $M(t)$, since $M(t)$ cannot be isolated on only one side of the equation. However, this is relatively straightforward using modern mathematical software programs. This equation is based on a number of assumptions (Siepmann and Siepmann 2012): (i) transport of active ingredient within the particles is rate limiting (rather than their movement away from the particle surfaces); (ii) the dissolution of the insoluble form of the active ingredient is rapid; (iii) perfect sink conditions exist; (iv) the initial concentration of active ingredient in the particle is much higher than the saturation concentration; (v) the active ingredient is finely and homogeneously distributed throughout the particles; and (vi) the particles do not swell or dissolve (i.e., their size and shape remain constant). This equation predicts that the release rate should increase as the diffusion coefficient and saturation concentration increases and the particle size decreases.

## 10.5.1.3 Release from Porous Polymeric Spheres

A number of colloidal delivery systems consist of spherical particles that are composed of a porous polymer network, for example, hydrogel particles made from proteins or polysaccharide gels (Matalanis et al. 2011). These particles can be considered to consist of a network of polymers molecules with solute (dispersed phase) and solvent (continuous phase) trapped between them (Figure 10.15). In this case, the release of a solute depends on its ability to diffuse through the porous polymer network and depends on the size of the solute molecules relative to the pores in the network. The diffusion-controlled release models discussed above can still be used to predict the release of solutes from porous polymer particles (e.g., the Crank model), but the normal diffusion coefficient (*D*) is replaced by the following one (Chan and Neufeld 2009; Li et al. 2011b):

$$D_{gel} = D \exp\left(-\pi\left(\frac{r_H - r_f}{\zeta + 2r_f}\right)\right). \tag{10.33}$$

Here, *D* is the diffusion coefficient of the solute through the material in the pores, $r_H$ is the hydrodynamic radius of the solute, $r_f$ is the cross-sectional radius of the polymer chains in the gel network, and $\zeta$ is the pore diameter of the gel network. As mentioned earlier, the following parameters have been reported for alginate gels: pore diameter ($\zeta$) = 4 to 400 nm; alginate chain radius ($r_f$) = 0.83 nm (Chan and Neufeld 2009). As mentioned earlier, the diffusion coefficient of a solute in water is given by $D_w = k_B T / 6\pi\eta r_H$, where $k_B$ is Boltzmann's constant, *T* is the absolute temperature, and $\eta$ is the solvent viscosity. Predictions made using this equation show that release occurs more rapidly as the solute size decreases and the pore size increases (Figure 10.17). In reality, it may also be necessary to take into account specific molecular interactions of the solute molecules with the polymer network that may retard release, such as attractive electrostatic or hydrophobic interactions

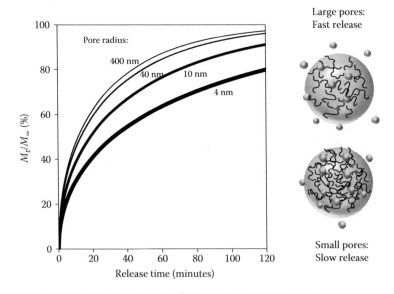

**Figure 10.17** Influence of particle network mesh diameter on the release profile of a model solute from spherical hydrogel particles. Predictions were made using the modified Crank model assuming a particle diameter of 2.4 mm, a polymer chain radius ($r_f$) of 0.83 nm, and a solute radius of 1.67 nm. (From Li et al. [2011b].)

## 10.5.1.4 Release from Coated Spheres

A number of colloidal delivery systems contain coated spherical particles; for example, they have core–shell structures (Figure 10.15). In the pharmaceutical industry, these kinds of systems are often referred to as *reservoir systems* (Siepmann and Siepmann 2008, 2012). The active ingredient may be completely soluble within the particle matrix (reservoir solutions) or it may be partly insoluble (reservoir dispersions). In addition, the active ingredient may be primarily dispersed within the core or shell of the particles. In these cases, the rate of release depends on the permeability ($P$) of the shell surrounding the core, which depends on its thickness ($h$), the shell–core partition coefficient ($K$), and the diffusion coefficient of the solute in the shell ($D_S$): $P = D_S K/h$. If the release rate is limited by the coating, then the rate of release increases as the diffusion coefficient and partition coefficient increase, and the thickness decreases. In this section, the influence of coatings on diffusion-controlled release of active ingredients from a number of model systems involving spherical particles is considered.

Initially, we consider release from core–shell spheres initially containing solute molecules evenly dissolved throughout the core (reservoir solutions), in which the rate-limiting step is diffusion of the solute across the shell (Siepmann and Siepmann 2012):

$$\frac{M(t)}{M_\infty} = 1 - \exp\left[-\frac{3r_O D_S K}{r_I^2 r_O - r_I^3} t\right].$$

(10.34)

Here, $K$ is the shell–core partition coefficient and $r_I$ and $r_O$ are the inner and outer radii of the shell. This equation shows that the release rate of an active ingredient should increase as its diffusion coefficient in the shell increases, as the shell–core partition coefficient increases, and as the thickness of the shell decreases. The above equation assumes that the shell–core partition coefficient ($K$) of the active ingredient is the same as the shell-continuous phase partition coefficient. In other words, it is assumed that the solubility of the solute in the particle matrix is similar to that in the continuous phase (which is often not the case in colloidal delivery systems). In addition, it is assumed that perfect sink conditions are maintained throughout the release process.

If an active ingredient is not completely soluble within the particle matrix ($C > C_S^*$), then it will be present in both insoluble and soluble forms. In the case where an insoluble active ingredient is homogeneously distributed throughout a spherical coated particle (reservoir dispersion) the release is given by (Siepmann and Siepmann 2012):

$$M(t) = \frac{4\pi D_S K C_S^* r_O r_I}{r_O - r_I} t.$$

(10.35)

Here, $C_S^*$ is the solubility of the active ingredient within the particle matrix. In this case, the dissolution of the insoluble fraction of the active ingredient within the particle is assumed to be much faster than the diffusion of the solute through the particle matrix. Thus, the chemical activity of the active ingredient remains constant throughout the duration of the release process as long as some of it remains insoluble. This equation predicts that the amount of active ingredient should increase as the diffusion coefficient in the shell increases, the shell–core partition coefficient increases, the saturation concentration of the solute in the core increases, and the shell thickness decreases. This knowledge may be used to design delivery systems with shells that can control the release of active ingredients.

### 10.5.1.5 Release from Heterogeneous Spheres

A number of delivery systems contain particles that contain a mixture of different phases, for example, multiple emulsions or filled hydrogel particles. In these cases, the active ingredient has to diffuse through a particle that contains different internal regions with different properties (Figure 5.15). Different regions may vary in their concentrations, sizes, structures, compositions, and physical properties (such as diffusion coefficients and partition coefficients). In this section, we consider diffusion of solute molecules out of large particles that contain a number of smaller particles embedded within them. One model that has been developed to predict release from multiphase particles uses an effective diffusion coefficient ($D_e$) in the release equations discussed earlier (Lian et al. 2004):

$$D_e = D_m \frac{2D_m(1-\phi_p) + K_{DC}D_p(1+\phi_p)}{D_m(2+\phi_p) + K_{DC}D_p(1-\phi_p)}. \tag{10.36}$$

Here $\phi_p$ is the volume fraction of the small particles within the larger particles and $D_p$ and $D_m$ are the diffusion coefficients of the solute molecules through the materials that make up the internal small particles and the surrounding matrix, respectively. This equation assumes perfect sink conditions, that the larger particles are spherical, that the inside of the large particle can be treated as an effective medium, and that the matrix surrounding the small particles does not affect $K_{DC}$ (Lian et al. 2004). The above equation indicates that the release of solute depends on the amount and diffusion coefficients of the different regions within the larger particle. This approach has been used to predict the influence of fat content on the release of flavor molecules from filled hydrogel particles (Lian et al. 2004).

## 10.5.2 Particle Dissolution

The release of active ingredients from some types of delivery systems is due to particle dissolution (Siepmann and Siepmann 2013). An active ingredient may make up the whole of the particles within a colloidal delivery system and is released when it experiences certain solution conditions and dissolves. Alternatively, an active ingredient may be trapped within a carrier matrix that dissolves under a particular set of conditions and releases the active ingredient. The initial composition and structure of a dissolution-based delivery system may vary depending on the application, for example, the solute may be initially dissolved or dispersed within a particle that has a homogeneous or core–shell structure. In the latter case, the active ingredient may be present in the core or shell of the particle. By incorporating solute molecules within different locations, it may be possible to design delivery systems with novel release characteristics; for example, one component is released rapidly, whereas another is released more slowly. For the sake of simplicity, we only consider the case where the solute makes up the entirety of a homogeneous spherical particle and is released by dissolution.

To a first approximation, the dissolution of a particle can be considered to be a multistep process involving wetting, solvation, and detachment of the solute molecules from the particle surface and then movement of the solute molecules away from the particle and into the surrounding continuous phase (Li and Jastic 2005; Siepmann and Siepmann 2013). This process is highly dependent on whether the surrounding solution is quiescent (nonstirred) or agitated (stirred). An excellent overview of the physical basis and mathematical description of dissolution-limited drug release in delivery systems has been given (Siepmann and Siepmann 2013).

The following expression has been derived to predict the release of a solute from a dissolving spherical particle over time (Li and Jastic 2005):

$$M(t) = M_0 \left( 1 - \left( 1 - \frac{DC_S^*}{hr_0\rho} \right)^3 \right). \tag{10.37}$$

Here, $M(t)$ is the mass of material released from the particles at time $t$, $M_0$ is the initial mass of the particles, $\rho$ is the density of the particles, $r_0$ is the initial particle radius, $D$ is the diffusion coefficient of the solute in the continuous phase surrounding the particle, $C_S^*$ is the solubility of the solute in the continuous phase, and $h$ is the thickness of the diffusion layer. For small particles ($r < 25\ \mu m$), $h$ is approximately equal to the particle radius ($h \approx r_0$). This equation assumes that the mass transport of solute molecules away from the particle surface is the rate-limiting step. The above equation predicts that the release rate should increase as the particle radius decreases, the diffusion coefficient of solute molecules in the surrounding medium increases (i.e., its viscosity decreases), and the solute solubility in the surrounding medium increases. It also predicts that a particle will continue to dissolve until the concentration of solute in the surrounding solution reaches the solubility limit ($C_S^*$).

If it is assumed that an active ingredient is initially dissolved within a particle matrix and is released into the surrounding fluid when the particle matrix is dissolved, then the amount released over time can still be described by the same equation, but the right-hand side should be multiplied by a factor $\Phi$, which represents the mass fraction of solute within the particle.

A number of other theoretical and empirical models that have been developed to describe the release of active ingredients owing to dissolution or erosion from different types of particles under various conditions have been reviewed elsewhere (Siepmann and Siepmann 2008, 2013).

### 10.5.3 Particle Swelling

In this case, an active ingredient is initially trapped within a polymer matrix inside a particle (Figure 10.18). The active ingredient is then released from the polymer matrix when it absorbs solvent molecules from the continuous phase surrounding the particle (Li and Jastic 2005; Oh et al. 2008). This process causes the particles to swell, which increases the pore size and facilitates the diffusion of active ingredients into the surrounding continuous phase. Swelling is an important process in a number of pharmaceutical applications, such as drug release from tablets fabricated from polymers, for example, hydroxypropyl methylcellulose or HPMC (Siepmann and Siepmann 2008). A number of physicochemical processes have to be considered when developing mathematical models to describe solute release by this mechanism: diffusion of solvent molecules into the particles, solvation of the

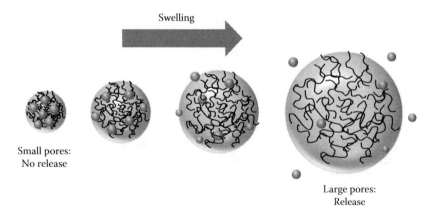

**Figure 10.18  A delivery system containing polymeric particles may release a solute because of a swelling mechanism.**

polymer molecules, glass–rubbery transitions of the polymer matrix, diffusion of solute molecules out of the particles, and dissolution of the solute if it is initially in an insoluble form (Li and Jastic 2005, Siepmann and Siepmann 2008). In practice, particle swelling may either decrease or increase the rate of solute release. A decrease in solute release may occur because of the increase in particle size (increased diffusion path length) and decrease in solute concentration gradient. On the other hand, an increase in solute release may occur because of the increase in polymer mobility and pore size. Because of the complexity of the physical processes involved, the mathematical models developed to describe this process are usually quite complex and often have to be solved numerically (Siepmann and Siepmann 2008). The following simple semiempirical model (the "Peppas equation") has been used to describe the release of active ingredients from spherical polymeric particles caused by swelling:

$$M(t) = M_\infty k t^{0.85}. \tag{10.38}$$

Here, $k$ is a constant that depends on the structure and geometry of the system and which can be determined experimentally by measuring the amount of active ingredient released from the particles over time. In food applications, it may be useful to encapsulate an active ingredient in a polymeric particle that is designed to swell under a particular set of conditions, for example, a dried particle that swells when it is introduced into an aqueous solution or a wet particle that swells when solution conditions are altered (such as pH, ionic strength, or temperature). In these cases, the active ingredient would initially be encapsulated within a particle with a pore size small enough to prevent it from escaping. However, it would be released once the particle swells in response to a change in solution or environmental conditions (Figure 10.18). Conversely, swelling/shrinking of particles may be used to load active ingredients into delivery systems. An active ingredient could be loaded into a particle by swelling an empty particle first, then mixing it with the active ingredient, and then changing the environmental conditions so that the particle shrinks and traps the ingredient.

## 10.5.4 Particle Matrix Degradation

In this case, it is assumed that an active ingredient is released from a particle owing to chemical degradation of the molecules within the particle matrix (Jain et al. 2010; Li and Jastic 2005). Initially, the active ingredient is dissolved or dispersed within a particle matrix. When the delivery system encounters particular environmental conditions (such as pH, ionic strength, or enzyme activity), the molecules within the particle matrix are chemically degraded, for example, owing to hydrolysis of biopolymer chains or lipid molecules (Chen et al. 2006; McClements et al. 2009b; McClements and Li 2010). Consequently, the active ingredients may be released into the continuous phase because of disintegration of the particle matrix or because of an increase

**Table 10.2  Examples of Materials Used to Fabricate Particle Matrices and Their Potential Enzymatic Degradation Mechanisms**

| Matrix Material | Degradation Mechanism | Release Location |
|---|---|---|
| Triacylglycerols | Gastric and pancreatic lipases | Stomach and small intestine |
| Proteins | Pepsin, trypsin, and chymotrypsin | Stomach and small intestine |
| Starch | Amylase and α-glycosidase | Mouth, stomach, and small intestine |
| Dietary fibers | Colonic bacterial enzymes | Colon |

in the pore size of the particle matrix. Degradation may occur throughout the entire particle, or it may start from the particle exterior and move inward. A number of physicochemical events may be involved in this kind of release mechanism: diffusion of degrading substances (e.g., strong acids, strong bases, or enzymes) into the particle, chemical degradation kinetics of the molecules within the matrix, diffusion of the active ingredient through the remaining particle matrix, and dissolution of the active ingredient in the case where it is initially insoluble. Degradation-based release is highly dependent on the nature of the molecules that make up the particle matrix and is therefore more difficult to model mathematically than other forms of solute release. Nevertheless, this mechanism is particularly useful for developing delivery systems with specific release characteristics, for example, degradation when a particular location of the GIT is reached. A number of particle matrix materials and gastrointestinal enzymes that may stimulate their degradation are highlighted in Table 10.2.

## 10.5.5  Particle Matrix Fragmentation

In this case, it is assumed that an active ingredient is released from a particle owing to physical fragmentation of the particle matrix (Li and Jastic 2005). The active ingredient is initially dissolved or dispersed within the particle matrix. The active ingredient is released when the matrix material is broken down into fragments as a result of physical disruption, for example, shearing, grinding, or mastication. The rate of release depends on the fracture properties of the particle matrix, such as the fracture stress and strain, as well as the size and shape of the fragments formed. The active component may still diffuse out of the fragments formed, but it will be released at a faster rate because of the increased surface area and decreased diffusion path associated with smaller particles.

## 10.5.6 Establishing Release Mechanisms

The design of an effective colloidal delivery system for an active ingredient often depends on establishing the most appropriate release mechanism for the application involved. The nature of the release mechanism will largely determine the release profile (e.g., burst, sustained, triggered, or targeted). It is usually necessary to carry out a detailed series of systematic experiments using a variety of analytical tools to establish the nature of the release mechanism. Chemical, spectroscopic, or chromatographic methods may be needed to quantify the release profile of the active ingredient from the delivery system. Particle sizing or microscopy methods may be required to monitor changes in particle dimensions after they are introduced into a particular set of environmental conditions, so as to establish whether particle dissolution, erosion, fragmentation, or swelling is occurring. Labeling methods (such as radioactive or fluorescence) may be needed to measure the location and transport of active ingredients within delivery systems. Calorimetry, spectroscopy, or scattering methods may be used to monitor changes in the physical state of the bioactive ingredient or particle matrix, such as dissolution, sol–gel, melting, crystallization, or polymorphism. Once the physical basis of the release mechanism has been established, it is possible to select an appropriate mathematical theory to model it (Siepmann and Siepmann 2008, 2013). This theory can then be used to quantify the major factors that would be expected to influence the release rate, which will facilitate the rational design of a delivery system with the desired release profile.

# 10.6 Modeling Particle Aggregation

One of the most important characteristics of colloidal delivery systems is the stability of the particles to aggregation. Particle aggregation depends on the relative strength and range of the attractive and repulsive interactions operating in the system, as well as the mechanism responsible for particle–particle encounters, for example, Brownian motion, mechanical agitation, or gravitational forces (Israelachvili 2011; McClements 2005). A brief overview of the major types of colloidal interactions and their influence of the aggregation stability of colloidal dispersions is therefore given in this section.

## 10.6.1 Colloidal Interactions

There are numerous types of attractive and repulsive interactions that may operate in a colloidal dispersion, with the most important ones depending on the specific system. To a first approximation, the overall interaction potential $w(h)$ between a pair of colloidal particles separated by a surface-to-surface distance $h$ can be represented as a sum of van der Waals, steric, electrostatic, hydrophobic, and depletion interactions, respectively:

$$w(h) = w_V(h) + w_S(h) + w_E(h) + w_H(h) + w_D(h). \tag{10.39}$$

The sign, magnitude, and range of each of these colloidal interactions can be predicted using mathematical models (see below). When designing a colloidal delivery system, it is important to be aware of the dominant colloidal interactions that operate, and to establish the major factors that influence them.

### 10.6.1.1 van der Waals Interactions

van der Waals interactions are relatively long-range, medium-strength attractive interactions that operate between all kinds of colloidal particles. If there were no repulsive colloidal interactions acting in a colloidal dispersion, then the particles would always have a tendency to aggregate because of these attractive interactions. To a first approximation, the van der Waals interaction between two spheres suspended in a liquid is given by the following equation (McClements 2005):

$$w_V(h) = -\frac{A_H r}{12h}. \tag{10.40}$$

Here, $r$ is the particle radius, $h$ is particle surface-to-surface separation, and $A_H$ is the Hamaker function ($\approx 0.75 \times 10^{-20}$ J for oil–water systems). The value of the Hamaker function depends on the physicochemical properties of the particle and surrounding medium, such as their dielectric constants and refractive indices. In practice, the Hamaker function should be modified to take into account retardation effects and electrostatic screening effects, which can cause an appreciable reduction in its value (Israelachvili 2011; McClements 2005). This equation predicts that the van der Waals attraction increases with increasing Hamaker function, increasing radius, and decreasing particle separation.

### 10.6.1.2 Steric Interactions

Steric interactions are extremely strong, short-range repulsive interactions that also operate between all types of colloidal particles (Israelachvili 2011). When two colloidal particles approach each other closely, their interfacial layers overlap, which generates a strong repulsion because of entropy effects (two molecules cannot occupy the same space). To a first approximation, the steric repulsion can be described by the following equation:

$$w_S(h) = \left(\frac{2\delta}{h}\right)^{12} kT. \tag{10.41}$$

Here, δ is the thickness of the interfacial layer of the particles, $k$ is Boltzmann's constant ($1.381 \times 10^{-23}$ J K$^{-1}$), and $T$ is the absolute temperature. This equation predicts that the steric interaction is close to zero where the distance between the particles is greater than 2δ but is strongly repulsive at closer separations when the interfacial layers overlap. A large steric repulsion is typically what stops colloidal particles from merging together when they collide with each other.

### 10.6.1.3 Electrostatic Interactions

Electrostatic interactions are important in colloidal dispersions that contain electrically charged particles (Israelachvili 2011). They may be either attractive or repulsive depending on the relative signs of the charges on different types of particles. When all the particles have the same charge sign (++ or −−), the interaction is repulsive, but when they have different charge signs (+− or −+), the interaction is attractive. The strength and range of electrostatic interactions depend on the magnitude of the electrical charge on the particle surfaces and the composition of the surrounding liquid (i.e., pH, ionic strength, and dielectric constant). For this reason, they are one of the colloidal interactions that are easiest to manipulate when designing colloidal delivery systems. The electrostatic interaction between two similarly sized spherical particles suspended in a liquid is given by the following equation (McClements 2005):

$$w_E(h) = -2\pi\varepsilon_0\varepsilon_R r\Psi_1\Psi_2 \ln(1 - e^{-\kappa h}). \qquad (10.42)$$

Here, $\varepsilon_0$ is the permittivity of a vacuum ($8.8542 \times 10^{-12}$ C J$^{-1}$ m), $\varepsilon_R$ is the relative permittivity of the continuous phase (≈80 for water), $\Psi_1$ and $\Psi_2$ are the surface potentials of the two types of particles (which is often assumed to be equal to the ζ-potential), and $\kappa^{-1}$ is the Debye screening length. If $\Psi_1$ and $\Psi_2$ have the same sign, the interaction is repulsive, but if they have opposite signs, the interaction is attractive. The Debye screening length is given by the following expression for aqueous solutions at ambient temperature: $\kappa^{-1} \approx 0.304/\sqrt{I}$ nm, where $I$ is the ionic strength of the continuous phase expressed in moles per liter. This equation assumes that the magnitude of the electrical potential on the particles is relatively small (<25 mM) and that the electrostatic interaction occurs under constant surface charge density (σ) conditions (Israelachvili 2011). It predicts that the strength of the electrostatic interaction should increase with increasing particle charge, decreasing salt concentration, decreasing particle separation, and increasing particle size.

Electrostatic interactions play an important role in determining the stability and functional performance of colloidal delivery systems that contain electrically charged particles. Electrostatic repulsion between similarly charged particles is commonly used to prevent aggregation in colloidal dispersions (Israelachvili 2011). In this case, it is important that the particles have a sufficiently high electrical charge (ζ-potential), that the ionic strength is not too high, and that there are no oppositely charged species that promote bridging flocculation (McClements 2005). Electrostatic attraction between oppositely charged particles may be utilized to create microclusters through heteroaggregation, which can be used to modulate textural or release characteristics of delivery systems (Mao and McClements 2012, 2013). Electrostatic interactions are also important in building other types of colloidal delivery systems, such as multilayer emulsions (Guzey and McClements 2006) and coacervates (Schmitt and Turgeon 2011; Turgeon et al. 2007). Electrostatic interactions between colloidal particles and surfaces are important for certain applications of delivery systems. A charged colloidal particle may adhere to an oppositely charged surface in a container (such as a bottle), which may lead to some of the encapsulated active ingredient being lost (Ziani et al. 2011). A charged colloidal particle in a delivery system may also be attracted to the lining of the GIT, which will alter its retention and release characteristics. Cationic particles are often used in drug delivery applications because of their ability to adhere to the mucus layer (Durrer et al. 1999; Teng et al. 2013).

### 10.6.1.4 Hydrophobic Interactions

Hydrophobic interactions are important in colloidal dispersions that contain particles that have an appreciable surface hydrophobicity. A semiempirical equation to describe the dependence of the hydrophobic attraction on particle separation has been developed from experimental measurements of the hydrophobic attraction between planar surfaces (Israelachvili 2011; McClements 2005):

$$w_H(h) = -2\pi r\gamma\Phi\lambda e^{-h/\lambda}. \tag{10.43}$$

Here, $\gamma$ is the interfacial tension at the boundary between the nonpolar groups and water ($\approx 50 \times 10^{-3}$ J m$^{-2}$), $\Phi$ is the fractional hydrophobicity of the particle surfaces (which varies from 0 for completely polar surfaces to 1 for completely nonpolar surfaces), and $\lambda$ is the decay length of the hydrophobic interactions ($\lambda \approx 1$ nm). This equation predicts that the strength of the hydrophobic attraction between colloidal particles increases as their surface hydrophobicity increases, particle size increases, and particle separation decreases.

There are a number of examples where hydrophobic interactions are important in the development of colloidal delivery systems. Fat droplets covered with emulsifiers that have some exposed nonpolar groups, such as surface or thermally denatured proteins, may have an appreciable surface hydrophobicity that leads to aggregation (Kim et al. 2002a,b). Biopolymer nanoparticles or microparticles formed by antisolvent precipitation of hydrophobic proteins, such as zein, are susceptible to aggregation because they have an appreciable surface hydrophobicity (Patel et al. 2010). Their stability can often be improved by adding emulsifiers that adsorb to nonpolar patches on the biopolymer particle surfaces, thereby reducing the surface hydrophobicity and attractive hydrophobic interactions. Fat droplets that are not fully covered with emulsifiers may also have an appreciable surface hydrophobicity, which makes them prone to droplet aggregation. This phenomenon occurs during homogenization under conditions where there is insufficient emulsifier present to cover the fat droplet surfaces (Jafari et al. 2007; Walstra 2003).

### 10.6.1.5 Depletion Interactions

Depletion interactions occur in colloidal dispersions that contain a mixture of different sized particles, for example, larger particles and smaller particles. The larger particles may be lipid droplets, liposomes, or hydrogel particles, whereas the smaller particles may be biopolymers (such as proteins or polysaccharides), surfactant micelles, or fine solid particles. The depletion interactions originate from the fact that the smaller particles are excluded from a narrow region surrounding the surfaces of the larger particles (Israelachvili 2011; McClements 2005). The thickness of this excluded region is approximately equal to the radius of the smaller particles. The concentration of the smaller particles in the excluded region is effectively zero, while it is finite in the surrounding continuous phase. As a consequence, there is an osmotic driving force that favors movement of solvent molecules from the excluded zone into the continuous phase, so as to dilute the smaller particles and thus reduce the concentration gradient. This situation can be achieved by two larger particles aggregating with each other, thereby reducing the volume of the exclusion region, which manifests itself as an attractive force between the larger particles. Thus, there is an osmotic driving force that favors particle aggregation and which increases as the concentration of smaller particles in the aqueous phase increases. To a first approximation, the interaction potential between two particles as a result of depletion interactions is given by (McClements 2005):

$$w_{\text{depletion}}(h) = -\frac{2}{3}\pi r^3 P_{\text{OSM}}\left(2\left(1+\frac{r_c}{r}\right)^3 + \left(1+\frac{h}{2r}\right)^3 - 3\left(1+\frac{r_c}{r}\right)^2\left(1+\frac{h}{2r}\right)\right). \quad (10.44)$$

Here, $P_{\text{OSM}}$ is the osmotic pressure arising from the exclusion of the smaller particles from the excluded region and $r_c$ is the radius of the smaller particles. This interaction is attractive and has a range approximately equal to the diameter of the smaller particles. Depletion interactions are therefore important in colloidal dispersions that contain a relatively high number concentration of smaller particles, such as polysaccharide molecules, protein clusters (such as casein micelles), or surfactant micelles.

### 10.6.1.6 Overall Interactions

The above equations can be used to predict the colloidal interactions between the particles in colloidal delivery systems. The colloidal interactions that need to be included in these calculations depend on the precise nature of the delivery system, for example, particle characteristics (such as charge, hydrophobicity, and interfacial thickness) and solution properties (such as pH, ionic strength, and composition). For example, electrostatic interactions only need to be included in the calculations for electrically charged particles, depletion interactions only need to be included when there are appreciable levels of nonadsorbed particles or polymers, and hydrophobic interactions only need to be included if the particles have appreciable surface hydrophobicity. An example of the utility of this approach is given in Figure 10.19, which shows the influence of ionic strength on the colloidal interactions between protein-coated lipid droplets. In this system, there is an attractive van der Waals interaction between the droplets and repulsive electrostatic and steric interactions. At low ionic strengths, the electrostatic repulsion is relatively strong and leads to a large energy barrier that prevents particle aggregation. At high ionic strengths, the electrostatic repulsion is screened and particle aggregation occurs because the attractive interactions (van der Waals) dominate the repulsive interactions (electrostatic).

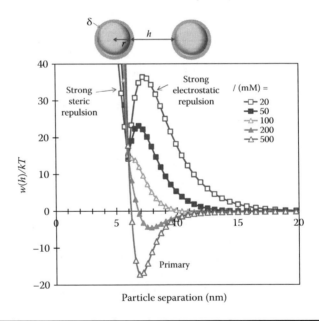

**Figure 10.19   The colloidal interactions between two particles can be calculated using mathematical models. These models can be used to predict the stability of colloidal particles to aggregation. In this case, the influence of ionic strength on the colloidal interactions between two protein-stabilized fat droplets is calculated. (Adapted from McClements [2010].)**

## 10.6.2 Calculation of Aggregation Kinetics

The kinetics of particle aggregation in dilute colloidal dispersions can be described by the following equations (McClements 2005):

$$-\frac{dn}{dt} = \frac{1}{2}k_B n^2 E = \frac{4kTn^2 E}{3\eta_1} \tag{10.45}$$

$$E = \left(2r \int_0^\infty \frac{\exp[w(h)/kT]}{h^2} dh\right)^{-1} \tag{10.46}$$

Here, $n$ is the total number of particles per unit volume at time $t$, $k_B$ is a second-order rate constant ($m^3\ s^{-1}$) determined by the number of collisions per second per unit volume owing to Brownian motion of the particles, $\eta_1$ is the viscosity of the continuous phase, and $E$ is the collision efficiency (i.e., the fraction of collisions leading to aggregation). Equation 10.45 assumes that the collisions are due to Brownian motion (rather than gravity or stirring), that the rate constant is independent of aggregate size, and that the collision efficiency is governed by colloidal (rather than hydrodynamic) interactions. Equation 10.45 can be integrated to give the following expression for the change in the total number of particles within the colloidal dispersion ($n_T$) with time (McClements 2005):

$$n_T(t) = \frac{n_0}{1 + \frac{1}{2}k_B E n_0 t}. \tag{10.47}$$

Here, $n_0$ is the initial number of particles per unit volume. The change in particle size with time owing to aggregation can be predicted assuming that at any given time the particles are monodisperse with an "effective" particle radius given by (McClements 2005):

$$r(t) = \sqrt[3]{\frac{3\phi}{4\pi n_T(t)}} = \sqrt[3]{r_0^3 + \left(\frac{\phi kTE}{\eta\pi}\right)t}. \tag{10.48}$$

This equation is only suitable for prediction of particle growth in the initial stages of particle aggregation. It predicts that the particle size should increase faster as the particle concentration ($\phi$) increases, as the particle radius ($r$) decreases, as the collision efficiency ($E$) increases, and as the viscosity ($\eta$) decreases. The above equations are useful for predicting the factors that influence the aggregation stability of colloidal dispersions. For example, they can be used to explore the influence of parameters such as particle characteristics (e.g., size, charge, hydrophobicity, and interfacial thickness) and solution conditions (e.g., pH or ionic strength) on aggregation stability (McClements 2005).

# 10.7 Modeling Bioavailability of Bioactives in the GIT

There is increasing interest in the utilization of colloidal delivery systems to release bioactive components (such as vitamins, minerals, and nutraceuticals) within the GIT (McClements et al. 2009b; McClements and Li 2010; Velikov and Pelan 2008). In this case, it is important to understand the physicochemical and physiological processes that occur as a delivery system passes through the various stages of the GIT. A number of computational models have been developed to model the release and absorption of drugs within the GIT, and some of these are commercially available and widely used in the pharmaceutical industry to optimize the performance of drug delivery systems (Iacocca et al. 2010; Mathias and Crison 2012; Sugano 2009a,b). Modeling the release and absorption of food-grade bioactive components after ingestion is much more difficult because of the variability in the food matrices in which they are contained. For example, a nutraceutical may be delivered in a low-viscosity beverage emulsion, a viscoelastic yogurt, or a solid granola bar, which have different compositions and structures and so may behave very differently within the GIT. Thus, the initial environment of the bioactive components within a delivery system or food matrix may have a strong influence on their subsequent biological fate. The main aims of this section are to highlight some of the most important physicochemical and physiological processes that influence the release and absorption of bioactive components from foods and to provide mathematical expressions to describe these processes where possible.

## 10.7.1 Release of Bioactive Components

Prior to ingestion, the bioactive components within a food or beverage product are typically contained within some kind of structure, such as a lipid or biopolymer particle. Before they can be absorbed by the human body, they usually have to be released from this initial location. In this section, the breakdown of food structures and the release of bioactive components within the GIT are therefore considered. Ingested foods experience a complex series of physicochemical and physiological processes as they pass through the different regions of the human GIT (Figure 10.20) (McClements et al. 2009a; Singh et al. 2009).

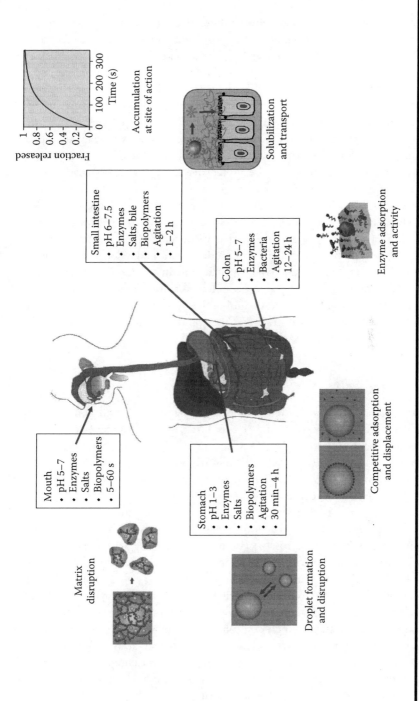

**Figure 10.20** Schematic representation of some of the physicochemical and physiological processes occurring when colloidal delivery systems pass through the human GIT.

## 10.7.1.1 Mouth

The first environment a colloidal delivery system experiences after ingestion is the mouth (Figure 10.20) (McClements et al. 2009a,b; Salles et al. 2011; Singh and Sarkar 2011; Singh et al. 2009; Stieger and van de Velde 2013; van Aken 2010). An ingested delivery system undergoes a series of changes in composition and structure as a result of oral processing within the mouth: it is mixed with saliva; it may be diluted, dissolved, or dispersed; its pH, ionic strength, and temperature may change; it may undergo phase changes (such as solid–liquid, sol–gel, or helix–coil transitions); it may be acted upon by digestive enzymes (such as amylase and lipase); it may interact with the surfaces of the mouth; and it experiences a complex flow/force pattern. As a result, there may be appreciable changes in the composition and structure of a delivery system after oral processing. Because of the complexity of these processes, it is difficult to develop mathematical models that can be used to predict the behavior of all different kinds of delivery systems within the mouth. Instead, it is often better to develop models for specific types of delivery systems and foods. For example, a mathematical model could be developed to predict the erosion or fragmentation of colloidal particles within the mouth on the basis of their specific physicochemical properties. However, it is often more practical to experimentally measure the breakdown of delivery systems within the oral cavity, rather than trying to predict it, owing to the complexity of the processes involved (Hutchings et al. 2011; Koc et al. 2013).

## 10.7.1.2 Stomach

After an ingested delivery system is lubricated with saliva and converted into a form suitable for swallowing (the "bolus"), it passes down the esophagus and enters the stomach (Figure 10.18) (McClements et al. 2009a,b; Singh and Sarkar 2011; Singh et al. 2009; van Aken 2010). Within the gastric cavity, the ingested material is exposed to a variety of physicochemical and physiological conditions that may further alter its composition and structure: strong acid (pH 1 to 3), high ionic strength (e.g., calcium and sodium salts), high enzyme activity (e.g., proteases and lipases), surface-active substances (e.g., phospholipids and proteins), and complex flow/force patterns (gastric motility). Strong acids may initiate the chemical degradation of some of the components within a colloidal delivery system or food matrix, for example, surfactant, protein, or polysaccharide hydrolysis. Alterations in pH and ionic strength may change the electrical characteristics of ionizable groups, thereby altering the magnitude and range of any electrostatic interactions in the system, which could lead to changes in the integrity, permeability, or aggregation state of colloidal particles. Surface-active substances in the gastric juices may adsorb to particle surfaces because of competitive adsorption or coadsorption processes, thereby altering their surface characteristics. Gastric lipases initiate lipid digestion, whereas gastric proteases begin protein digestion.

In addition, amylases from the mouth may continue to break down starches within the stomach.

Studies have shown that there may be appreciable changes in the composition, structure, dimensions, electrical charge, and physical state of the components in colloidal delivery systems within the stomach (Golding et al. 2011; McClements and Xiao 2012; Singh and Sarkar 2011; Singh et al. 2009). The degree and nature of these changes depend on the precise nature of the delivery system and food matrix. Studies have shown that certain types of protein-coated fat droplets are highly prone to droplet flocculation and coalescence within the stomach, whereas some types of surfactant-coated oil droplets are stable within the gastric environment. The degree of particle aggregation and gravitational separation within the stomach may have important implications for the biological response of humans to consumed foods, for example, the feelings of satiety and satiation (Halford and Harrold 2012; Lundin et al. 2008). Acid-stable emulsions that remain evenly distributed throughout the stomach reduce gastric emptying and increase satiety more effectively than acid-unstable emulsions that aggregate and cream within the stomach. This may have important implications for the development of functional foods to tackle overweight and obesity (Marciani et al. 2009; Marciani et al. 2006, 2007). Delivery systems can be designed so that they increase the feeling of satiety and satiation and therefore lead to lower overall food consumption.

Mathematical modeling of the release of bioactive components from colloidal delivery systems or food matrices within the stomach is usually difficult because of their compositional and structural complexity. In addition, it is important to take into account the actual structure of the delivery system within the stomach, which may be very different from that which was originally ingested. For certain types of colloidal delivery systems, the mathematical models developed to model gastric release in the pharmaceutical industry may also be suitable for application in the food industry. If the release of a bioactive component from a colloidal particle occurs through diffusion-, dissolution-, or erosion-limited processes, then models similar to those presented in Section 10.5 may be adopted (Sugano 2009a–c). However, it is important to modify the models to take into account the actual conditions occurring within the stomach, such as a confined volume, gastric motility, dynamic compositional changes, and the fact that some of the released bioactive components may be absorbed during the process (Turner et al. 2012).

## 10.7.1.3 Small Intestine

The material leaving the stomach (the "chyme") is forced through a narrow muscular valve (the "pylorus sphincter") that controls the amount of food entering the small intestine (Figure 10.20) (Barrett 2006). The chyme is mixed with alkaline small intestinal fluids (which contain bile salts, phospholipids, pancreatic lipase, colipase, bicarbonate, and various other salts) within the small intestine, which causes the pH to increase closer to neutral (typically within the range of pH 5.4 to

7.4). Small intestinal fluids are designed by nature to further digest any macronutrients remaining within the food so that they can be released into the GIT and absorbed by the epithelium cells (Barrett 2006). Proteins are converted to peptides and amino acids by proteases, triacylglycerols are converted into free fatty acids (FFAs) and monoacylglycerols (MAGs) by lipases, and starches are converted into oligosaccharides and glucose by amylases. An important factor influencing the absorption of certain bioactive components (particularly lipophilic ones) is the formation of colloidal structures within the small intestine that enhance the solubilization, transport, and absorption of micronutrients and digested macronutrients (van Aken 2010). For example, mixed micelles and vesicles that are capable of solubilizing FFAs, MAGs, oil-soluble vitamins, and nutraceuticals are formed (Porter et al. 2007). These mixed micelles transport the solubilized components through the mucus layer until they reach the surfaces of the enterocytes where they are absorbed (Figure 10.21).

The release of bioactive components from colloidal delivery systems within the GIT is highly dependent on its composition and structure. Numerous mathematical models have been developed in the pharmaceutical industry to model the release of drugs within the small intestine (Sjogren et al. 2013; Sugano 2009, 2011), and many of these are also suitable for modeling release of bioactive components from food-based colloidal delivery systems. For lipophilic bioactive components that are coingested with lipids, it is necessary to include the influence of bile salt solubilization and transport within the mathematical theories used to model release (Sugano 2009a–c; Sugano et al. 2010; Turner et al. 2012). As with the stomach, it is important to include factors such

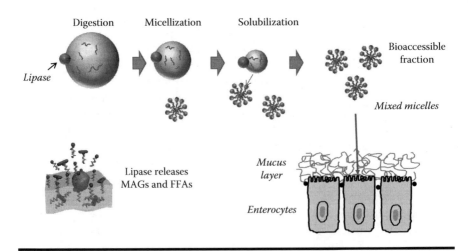

**Figure 10.21 Schematic representation of some of the processes that influence the release of lipophilic bioactive agents in the GIT: lipid digestion, micelle formation, solubilization, and transport through the mucus layer.**

as confined volume, intestinal mobility, dynamic compositional changes, and bioactive absorption.

### 10.7.1.4 Colon

The majority of an ingested food is normally broken down and absorbed in the stomach and small intestine, but a fraction of it may reach the colon (Figure 10.20). If all the components used to fabricate a colloidal delivery system were fully digestible, then one would expect it to be largely digested and absorbed within the stomach and small intestine. For example, one would expect this to happen for lipid particles with a triacylglycerol core and a protein shell, for example, β-lactoglobulin stabilized corn oil-in-water nanoemulsions (Li et al. 2011). On the other hand, if one or more of the components used to fabricate a colloidal delivery system were indigestible, then it may be able to reach the colon without being absorbed. For example, this might occur for indigestible oil droplets or for digestible oil droplets surrounded by dietary fiber shells or matrices (McClements and Li 2010). The colon contains a broad range of bacteria that are capable of breaking down and utilizing various food components that reach it (Basit 2005). For example, many dietary fibers are broken down by digestive enzymes released from colonic bacteria, leading to the formation of short chain fatty acids and other products that may be beneficial to human health (Kumar et al. 2012; O'Keefe 2008). This phenomenon may also be utilized in the fabrication of colloidal delivery systems specifically designed to release bioactive components within the colon (Kosaraju 2005; Patten et al. 2009). On the other hand, there may be some adverse effects associated with altering the amount and type of undigested food materials reaching the colon (e.g., lipids or proteins), and further work is needed in this area.

## 10.7.2 Absorption of Bioactive Components

After the bioactive components within an ingested colloidal delivery system have been released from a food or beverage matrix, then they need to be absorbed in an appropriate location within the GIT (Barrett 2006; McClements et al. 2009a; Singh et al. 2008). The majority of absorption of bioactive food components typically occurs within the small intestine, although some may occur within the mouth, esophagus, and stomach prior to reaching the small intestine. In addition, any bioactive components that pass through the upper intestine may be absorbed in the colon in either their native or metabolized forms. After absorption, a bioactive component is typically transported to the systemic circulation where it is then distributed to different tissues, metabolized, stored, or excreted. There are two important factors to consider in the case of colloidal delivery systems containing bioactive components encapsulated within carrier particles: (i) the absorption of the bioactive components themselves and (ii) the potential absorption of the carrier particles (McClements 2013). In many cases, carrier particles may influence the absorption of bioactive compounds, without actually being absorbed themselves; for example, digestible lipid carrier particles increase the amount of mixed micelles formed in the small intestine that can solubilize and transport bioactive components (Porter et al. 2007). Toxicologically, it is important to establish the influence of any colloidal delivery system on human health, as the colloidal particles may be absorbed or they may influence the normal absorption of bioactive components (Bouwmeester et al. 2009; Magnuson et al. 2011).

When considering the biological fate of bioactive components or carrier particles, it is important to establish the region where their absorption occurs within the GIT, as well as their state when they reach the absorption site (McClements 2013). If the original carrier particle was fully digested before reaching the absorption site, then one only needs to consider absorption of the bioactive component. On the other hand, if the ingested carrier particle had an indigestible core or shell, then it may remain intact at the absorption site, and so one has to consider its absorption too.

## 10.7.2.1 Absorption of Particles

In this section, the potential absorption of entire colloidal particles by the epithelium cells within the GIT is considered. The particles in colloidal delivery systems are typically composed of bioactive components encapsulated within some kind of structured carrier matrix. As mentioned earlier, when the colloidal particles reach the small intestine (or another location where absorption occurs), their composition and structure may be very different from those of the originally ingested particles. The nature of the particles at the absorption site will influence the rate and extent of their absorption. There are two major barriers to the absorption of colloidal particles in the GIT: (i) they must first penetrate the mucus layer that coats the epithelium cells, and (ii) they must be transported across the epithelium cells and into the body (Ensign et al. 2012). The mucus layer is a compositionally and structurally complex porous gel-like material that coats the epithelium cells (Acosta 2009; Cone 2009; Ensign et al. 2012). The thickness, structure, and composition of the mucus layer are highly dynamic and depend on GIT location, as well as on an individual's diet, health status, and genetics. The mucus layer mainly consists of water, biopolymers, lipids, and minerals, and serves a number of important biological functions, such as facilitation of food transport through the GIT owing to lubrication and protection of the GIT from ingested pathogens and foreign particles. The mucus layer normally restricts particle movement from the lumen to the epithelium cells by two main mechanisms (Cone 2009; Ensign et al. 2012):

■ *Mesh size*—The mucus layer consists of a physically and chemically crosslinked and entangled biopolymer network. In this case, the rate of particle transport through the mucus layer depends on pore dimensions relative to particle dimensions. The pore size of the biopolymer network in the mucus layer is polydisperse, but a critical cutoff point for particle penetration of around 400 nm has been reported (Cone 2009). Particles considerably smaller than the mesh size can easily penetrate through the mucus layer and reach the epithelium cells, whereas those with similar or larger dimensions will be hindered.

- *Molecular interactions*—The mucus layer contains chemically complex biopolymer molecules that contain a mixture of nonpolar, polar, anionic, and cationic functional groups (Cone 2009). These biopolymer molecules are therefore capable of interacting with colloidal particles through a variety of molecular interactions, such as van der Waals, electrostatic, hydrophobic, and hydrogen bonding. The nature of the interactions between colloidal particles and the mucus layer influences their ability to travel through and reach the epithelium cells. If a colloidal particle is strongly attracted to the biopolymer molecules in the mucus layer, then it may become trapped rather than being transported across it. This mucoadhesion process increases the residence time of colloidal particles within the GIT, which may lead to a greater degree of digestion and absorption. On the other hand, it means that the particle will not reach the epithelium cells and be absorbed. If a particle is not strongly attracted to the mucus layer, and it is small enough, then it may penetrate through the mucus layer and therefore reach the epithelium cells.

After a colloidal particle has traveled through the mucus layer, it may be absorbed by the epithelium cells depending on its dimensions, shape, and surface characteristics (Cone 2009; Ensign et al. 2012). There are two major kinds of epithelium cells within the GIT where particle absorption may occur: enterocytes and M-cells (des Rieux et al. 2006; Ensign et al. 2012). Enterocytes are the most numerous type of cell lining the GIT, but they are not highly efficient at absorbing particles. Conversely, M-cells are much less numerous (<1% of epithelium surface), but they are more efficient at absorbing particles. M-cells are primarily located in specialized regions (called "Peyer's patches") responsible for absorbing and sampling ingested antigens, such as microorganisms, macromolecules, and certain types of colloidal particles (des Rieux et al. 2006; Frohlich and Roblegg 2012). After absorption, these substances are delivered to the underlying lymphoid system, where they may induce an immune response. Colloidal particles may be absorbed by epithelium cells through various translocation mechanisms (Frohlich and Roblegg 2012; Powell et al. 2010) (Figure 10.22):

*Paracellular*: Small colloidal particles may pass between the narrow gaps ("tight junctions") separating neighboring epithelial cells (Yuan et al. 2007). The critical cutoff point for this process is relatively small (a few nanometers) under normal circumstances, but certain types of ingested substances may increase the dimensions of the tight junctions, such as some surfactants, polymers, and chelating agents, thereby allowing bigger particles to pass through.

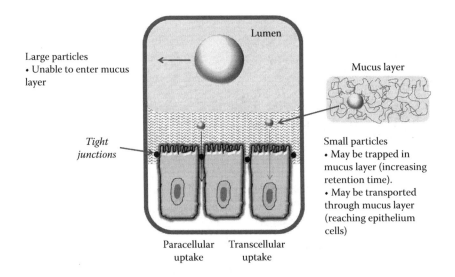

**Figure 10.22    Small particles may be trapped within the mucus layer or may be transported across it and can be adsorbed directly, which could alter the bioavailability of lipophilic substances.**

*Transcellular*: Some colloidal particles are small enough to be directly absorbed through epithelial cell walls by either passive or active transport mechanisms. Absorption typically occurs by an "endocytosis" process that involves the colloidal particle encountering the cell membrane, a portion of the cell membrane wrapping itself around the particle, and then part of the cell membrane budding off to form a vesicle with the colloidal particle trapped inside. The vesicle then moves into the interior of the cell where it may be further processed. This process can occur in both enterocytes and M-cells, with the latter cell type being more efficient at particle absorption. The critical cut-off point for this process depends on the precise nature of the transcellular absorption mechanism but has been estimated to be around 50 to 100 nm for enterocytes and around 20 to 500 nm for M-cells (Frohlich and Roblegg 2012; Powell et al. 2010).

*Persorption*: Some colloidal particles are able to be absorbed through small temporary holes formed within the lining of the GIT when some of the epithelium cells are shed and replaced (Powell et al. 2010).

The precise pathway taken, as well as the ultimate rate and extent of particle uptake, depends on particle properties, such as size, shape, charge, and surface chemistry (des Rieux et al. 2006; Hu et al. 2009). At present, there is a relatively poor understanding of the relationship between the properties of colloidal particles and their absorption route and extent. Nevertheless, some studies have shown that certain kinds of colloidal particles are directly absorbed by an amount that depends on their characteristics. For example, cationic nanoparticles have been reported to be absorbed more readily than anionic ones (Bouwmeester et al. 2009; Hagens et al. 2007): smaller particles are often absorbed more readily than larger ones (but there may be an optimum size for particle absorption) (Hu et al. 2009); hydrophobic particles are absorbed more readily than hydrophilic ones (des Rieux et al. 2006; Ensign et al. 2012). In practice, the role of particle characteristics is likely to depend on the precise nature of the colloidal particles and translocation mechanisms involved, as well as any changes in the character of the colloidal particles as they pass through the GIT prior to absorption. Clearly, further work is needed in this area to fully understand the role of specific particle characteristics on their absorption.

The physicochemical characteristics of colloidal particles also determine their fate after they have entered the epithelium cells: (i) they may be digested by cellular enzymes; (ii) they may be transported out of the cells into the blood or lymphatic systems; or (iii) they may accumulate within cells (Brigger et al. 2012; Hu et al. 2009). Some types of colloidal particles have been shown to promote cellular damage if they accumulate within cells (Hu, Mao et al. 2009; Unfried et al. 2007). Particles that are transported out of the epithelial cells via the portal vein or lymphatic system will circulate through the human body, where they may be metabolized, may be excreted, or may accumulate within certain tissues (Bouwmeester et al. 2009). The biological fate of an ingested particle depends on its physicochemical characteristics; for example, smaller (inorganic) particles have been reported to accumulate within a wider range of tissues than larger ones (De Jong et al. 2008). The development of quantitative theories to predict the absorption of colloidal particles by epithelium cells is difficult because of the complex biological processes involved. Instead, it is usually better to carry out systematic experiments to identify the physiological and physicochemical mechanisms involved.

## 10.7.2.2 Bioactive Absorption

The bioavailability of an ingested bioactive component is often defined as the fraction that eventually reaches the systemic circulation or the appropriate site of action (Versantvoort et al. 2004). The overall bioavailability ($F$) of a bioactive component can be conceptually divided into a number of different factors (Arnott and Planey 2012):

$$F = F_L \times F_A \times F_D \times F_M \times F_E. \tag{10.49}$$

$F_L$ is the fraction of bioactive component *liberated* from the food matrix and released into the lumen of the GIT in a bioaccessible form. $F_A$ is the fraction of the released bioactive component that is *absorbed* by the epithelial cells. $F_D$ is the fraction of the absorbed bioactive component that reaches the site of action after *distribution* within the body. $F_M$ is the fraction of the absorbed bioactive component that reaches the site of action in a *metabolically* active form and depends on any chemical or enzymatic modifications that take place before and after ingestion. $F_E$ is the fraction of the absorbed bioactive component that has not been *excreted* by the body. In reality, each of these different parameters varies after ingestion of a bioactive component, which leads to a profile of $F$ versus time at the specified site of action.

Many of the most important processes that influence these factors have been identified (Bauer et al. 2005; Bermudez et al. 2004; Fave et al. 2004):

*Liberation*: Bioactive components may have to be released from a delivery system or food matrix before they can be absorbed. In the pharmaceutical industry, the solubility of a bioactive agent within the gastrointestinal fluids is usually taken as a measure of its ability to be released from a delivery system (Buckley et al. 2013; Dahan et al. 2009). In this case, different bioactive components are classified according to their solubilities in intestinal fluids. It is often assumed that only those bioactive components that can be solubilized within the GIT fluids will be effectively absorbed (although insoluble particles may also be absorbed by the translocation mechanisms discussed in the previous section). For foods and beverages, the bioactive components and the delivery systems containing them often have to be released from complex food matrices. In this case, the release mechanism depends on the nature of the bioactive components, the delivery system, and the food matrix. For encapsulated lipophilic bioactives, this may involve digestion of carrier lipids, followed by solubilization and transport by mixed micelles within the small intestine. For encapsulated hydrophilic bioactives, this may involve digestion of lipid, protein, or starch matrices, leading to release of the bioactives into the aqueous intestinal fluids. For crystalline or amorphous solids, this may involve the dissolution of the material into the intestinal fluids. The amount of a bioactive component released into the GIT can be used as a measure of the fraction liberated ($F_L$) in a bioaccessible form (i.e., a form suitable for absorption).

*Absorption*: After a bioactive component has been liberated, it needs to be absorbed by the epithelium cells that line the GIT. The precise absorption mechanism involved depends on the nature of the bioactive component, for example, its molecular weight, polarity, and physical state. Absorption may take place by either active or passive transport mechanisms involving transcellular, paracellular, or persorption processes (see the last section). One of the key factors influencing the efficiency of the absorption process is the permeability of the mucus layer and epithelium cells for the bioactive component. In the pharmaceutical industry, drugs are often classified according to their permeability in epithelium cells using the biopharmaceutical classification scheme, that is, BCS (Buckley et al. 2013; Dahan et al. 2009). It is assumed that only those liberated bioactive components that have high *permeability* will be effectively absorbed. In practice, a bioactive component should have a high solubility and permeability to be easily absorbed. The uptake of bioactive components by the human body may therefore be limited owing to their poor solubility or their low permeability. There are four major categories of bioactive components according to the BCS system: *Class I*—high solubility and high permeability, *Class II*—low solubility and high permeability, *Class III*—high solubility and low permeability, and *Class IV*—low solubility and low permeability. Class I compounds are readily absorbed within the GIT, whereas there may be limits to absorption for compounds in the other three classes. Many lipophilic bioactive compounds fall into Class II since they naturally have a low solubility in intestinal fluids, but they have a high permeability through epithelium cell membranes. In this case, the lipophilic bioactives would be solubilized in mixed micelles that travel through the mucus layer to the intestinal enterocyte cells where they are absorbed by passive or active transfer mechanisms (Porter et al. 2007; Singh et al. 2008). This cell permeation step determines the fraction of the liberated lipophilic component that is eventually absorbed ($F_A$) by the body.

*Distribution*: After absorption, bioactive components are distributed among different sites within the human body, such as the lymphatic system, systemic circulation, liver, kidney, adipose tissue, heart, lungs, brain, and so on (Roberts and Hall 2013). The molecular and physicochemical characteristics of the bioactive components, as well as those of any coingested food components, may potentially alter the distribution of bioactive components. The target tissue for a bioactive component depends on the nature of the biological response required, such as enhanced performance, maintenance of general well-being, prevention of chronic disease, or treatment of specific acute diseases.

*Metabolism*: Bioactive components may be transformed after ingestion owing to chemical (such as acid hydrolysis) or biochemical (such as enzyme activity) processes occurring within the human body (Laparra and Sanz 2010; Spencer et al. 2006). Many bioactive agents are modified by specific enzymes associated with the epithelium cells or microorganisms that inhabit the GIT, for example, reduction, conjugation, dehydroxylation, cyclization, and methylation (Wang and Qiu 2013). Changes in the chemical structure of a bioactive component may alter its solubility, absorption, and bioactivity. The metabolism of lipophilic bioactives is strongly influenced by the route that they are transported to the systemic circulation. Large lipophilic agents tend to be transported via the lymphatic system, whereas smaller ones tend to be transported via the portal vein and liver (Iqbal and Hussain 2009; Trevaskis et al. 2006). Lipophilic bioactives that pass through the liver may be highly metabolized before reaching the systemic circulation, thereby altering their biological activity. In some cases, molecular transformations increase bioactivity, whereas in other cases, they decrease it. The transformations of a bioactive component as it passes through the different regions of the body therefore determine the fraction that arrives at the site of action in a metabolically active state ($F_M$).

*Excretion*: Bioactive components and their metabolites are eventually excreted from the human body through a variety of mechanisms and may end up in the feces, urine, sweat, or breath (Manach et al. 2009; Tomas-Barberan et al. 2009). Obviously, only the fraction of a bioactive component that remains within the human body ($F_E$) is able to have a biological effect.

The oral bioavailability of bioactive components can be improved by designing colloidal delivery systems that increase the fraction: liberated ($F_L$), absorbed ($F_A$), reaching the site of action ($F_D$), remaining in a metabolically active form ($F_M$), or not excreted ($F_E$). This can be achieved by manipulating the composition and structure of the delivery system and food matrix based on knowledge of the impact of the factors influencing the biological fate of specific bioactives.

## 10.8 Summary

This chapter has highlighted some of the physical and mathematical models that can be used to aid the design of effective colloidal delivery systems for utilization within the food and other industries. These models are particularly important because they help to identify the most important factors expected to influence the functional performance of a particular delivery system, and they enable one to quantify the influence of specific compositional and structural features on encapsulation, retention, and release properties. For example, the effects of particle composition, size, or interfacial properties on the stability or release characteristics of a colloidal delivery system could be explored using a suitable mathematical model. Ultimately, it would be useful to have a general model to predict the release characteristics of specific delivery systems on the basis of their initial composition and structure, analogous to those commercially available for predicting drug release in the pharmaceutical industry. Nevertheless, the composition and structural complexity of most food and beverage products, as well as the complexity of the events occurring within the GIT after ingestion, make this particularly difficult.

## References

Acosta, E. (2009). "Bioavailability of nanoparticles in nutrient and nutraceutical delivery." *Current Opinion in Colloid and Interface Science* **14**(1): 3–15.

Arifin, D. Y., L. Y. Lee and C. H. Wang (2006). "Mathematical modeling and simulation of drug release from microspheres: Implications to drug delivery systems." *Advanced Drug Delivery Reviews* **58**(12–13): 1274–1325.

Arnott, J. A. and S. L. Planey (2012). "The influence of lipophilicity in drug discovery and design." *Expert Opinion on Drug Discovery* **7**(10): 863–875.

Balibar, S. and F. Caupin (2006). "Nucleation of crystals from their liquid phase." *Comptes Rendus Physique* **7**(9–10): 988–999.

Barrett, K. E. (2006). *Gastrointestinal Physiology.* New York, McGraw-Hill.

Basit, A. W. (2005). "Advances in colonic drug delivery." *Drugs* **65**(14): 1991–2007.

Bauer, E., S. Jakob and R. Mosenthin (2005). "Principles of physiology of lipid digestion." *Asian-Australasian Journal of Animal Sciences* **18**(2): 282–295.

Bermudez, B., Y. M. Pacheco, S. Lopez, R. Abia and F. J. G. Muriana (2004). "Digestion and absorption of olive oil." *Grasas Y Aceites* **55**(1): 1–10.

Bhandari, B. R. and T. Howes (2000). "Glass transition in processing and stability of food." *Food Australia* **52**(12): 579–585.

Bouwmeester, H., S. Dekkers, M. Y. Noordam, W. I. Hagens, A. S. Bulder, C. de Heer, S. ten Voorde, S. W. P. Wijnhoven, H. J. P. Marvin and A. Sips (2009). "Review of health safety aspects of nanotechnologies in food production." *Regulatory Toxicology and Pharmacology* **53**(1): 52–62.

Brigger, I., C. Dubernet and P. Couvreur (2012). "Nanoparticles in cancer therapy and diagnosis." *Advanced Drug Delivery Reviews* **64**: 24–36.

Brough, C. and R. O. Williams (2013). "Amorphous solid dispersions and nano-crystal technologies for poorly water-soluble drug delivery." *International Journal of Pharmaceutics* **453**(1): 157–166.

Buckley, S. T., K. J. Frank, G. Fricker and M. Brandl (2013). "Biopharmaceutical classification of poorly soluble drugs with respect to enabling formulations." *European Journal of Pharmaceutical Sciences* **50**(1): 8–16.

Buckton, G. and A. E. Beezer (1992). "The relationship between particle-size and solubility." *International Journal of Pharmaceutics* **82**(3): R7–R10.

Bunjes, H. (2011). "Structural properties of solid lipid based colloidal drug delivery systems." *Current Opinion in Colloid and Interface Science* **16**(5): 405–411.

Bunjes, H. and M. H. J. Koch (2005). "Saturated phospholipids promote crystallization but slow down polymorphic transitions in triglyceride nanoparticles." *Journal of Controlled Release* **107**(2): 229–243.

Caira, M. R. (1998). Crystalline polymorphism of organic compounds. In *Design of Organic Solids*, E. Weber (ed.). Berlin 33, Springer-Verlag Berlin **198**: 163–208.

Carlert, S., A. Palsson, G. Hanisch, C. von Corswant, C. Nilsson, L. Lindfors, H. Lennernas and B. Abrahamsson (2010). "Predicting intestinal precipitation-A case example for a basic BCS class II drug." *Pharmaceutical Research* **27**(10): 2119–2130.

Cejkova, J. and F. Stepanek (2013). "Compartmentalized and internally structured particles for drug delivery—A review." *Current Pharmaceutical Design* **19**(35): 6298–6314.

Chan, A. W. and R. J. Neufeld (2009). "Modeling the controllable pH-responsive swelling and pore size of networked alginate based biomaterials." *Biomaterials* **30**(30): 6119–6129.

Chebil, L., C. Humeau, J. Anthoni, F. Dehez, J. M. Engasser and M. Ghoul (2007). "Solubility of flavonoids in organic solvents." *Journal of Chemical and Engineering Data* **52**(5): 1552–1556.

Chen, J., B. Sarma, J. M. B. Evans and A. S. Myerson (2011). "Pharmaceutical crystallization." *Crystal Growth and Design* **11**(4): 887–895.

Chen, L. Y., G. E. Remondetto and M. Subirade (2006). "Food protein-based materials as nutraceutical delivery systems." *Trends in Food Science and Technology* **17**(5): 272–283.

Choi, S. J., E. A. Decker, L. Henson, L. M. Popplewell and D. J. McClements (2009). "Stability of citral in oil-in-water emulsions prepared with medium-chain triacylglycerols and triacetin." *Journal of Agricultural and Food Chemistry* **57**(23): 11349–11353.

Cone, R. A. (2009). "Barrier properties of mucus." *Advanced Drug Delivery Reviews* **61**(2): 75–85.

Coupland, J. N. (2002). "Crystallization in emulsions." *Current Opinion in Colloid and Interface Science* **7**(5–6): 445–450.

Dahan, A., J. M. Miller and G. L. Amidon (2009). "Prediction of solubility and permeability class membership: Provisional BCS classification of the world's top oral drugs." *AAPS Journal* **11**(4): 740–746.

De Jong, W. H., W. I. Hagens, P. Krystek, M. C. Burger, A. Sips and R. E. Geertsma (2008). "Particle size-dependent organ distribution of gold nanoparticles after intravenous administration." *Biomaterials* **29**(12): 1912–1919.

Delaney, J. S. (2005). "Predicting aqueous solubility from structure." *Drug Discovery Today* **10**(4): 289–295.

des Rieux, A., V. Fievez, M. Garinot, Y. J. Schneider and V. Preat (2006). "Nanoparticles as potential oral delivery systems of proteins and vaccines: A mechanistic approach." *Journal of Controlled Release* **116**(1): 1–27.

Durrer, C., J. M. Irache, D. Duchene and G. Ponchel (1999). "Influence of colloid stabilizing agents on interactions between nanoparticles and mucus." *Stp Pharma Sciences* **9**(5): 437–441.

Ensign, L. M., R. Cone and J. Hanes (2012). "Oral drug delivery with polymeric nanoparticles: The gastrointestinal mucus barriers." *Advanced Drug Delivery Reviews* **64**(6): 557–570.

Erdemir, D., A. Y. Lee and A. S. Myerson (2009). "Nucleation of crystals from solution: Classical and two-step models." *Accounts of Chemical Research* **42**(5): 621–629.

Fave, G., T. C. Coste and M. Armand (2004). "Physicochemical properties of lipids: New strategies to manage fatty acid bioavailability." *Cellular and Molecular Biology* **50**(7): 815–831.

Fredrick, E., P. Walstra and K. Dewettinck (2010). "Factors governing partial coalescence in oil-in-water emulsions." *Advances in Colloid and Interface Science* **153**(1–2): 30–42.

Frohlich, E. and E. Roblegg (2012). "Models for oral uptake of nanoparticles in consumer products." *Toxicology* **291**(1–3): 10–17.

Golding, M., T. J. Wooster, L. Day, M. Xu, L. Lundin, J. Keogh and P. Clifton (2011). "Impact of gastric structuring on the lipolysis of emulsified lipids." *Soft Matter* **7**(7): 3513–3523.

Guzey, D. and D. J. McClements (2006). "Formation, stability and properties of multilayer emulsions for application in the food industry." *Advances in Colloid and Interface Science* **128**: 227–248.

Hagens, W. I., A. G. Oomen, W. H. de Jong, F. R. Cassee and A. Sips (2007). "What do we (need to) know about the kinetic properties of nanoparticles in the body?" *Regulatory Toxicology and Pharmacology* **49**(3): 217–229.

Halford, J. C. G. and J. A. Harrold (2012). "Satiety-enhancing products for appetite control: Science and regulation of functional foods for weight management." *Proceedings of the Nutrition Society* **71**(2): 350–362.

Hartel, R. W. (2001). *Crystallization in Foods.* Gaithersburg, MD, Aspen Publishers.

Herhold, A. B., D. Ertas, A. J. Levine and H. E. King (1999). "Impurity mediated nucleation in hexadecane-in-water emulsions." *Physical Review E* **59**(6): 6946–6955.

Herrera, M. L. and R. W. Hartel (2000). "Effect of processing conditions on crystallization kinetics of a milk fat model system." *Journal of the American Oil Chemists Society* **77**(11): 1177–1187.

Himawan, C., V. M. Starov and A. G. F. Stapley (2006). "Thermodynamic and kinetic aspects of fat crystallization." *Advances in Colloid and Interface Science* **122**(1–3): 3–33.

Hu, F. Q., S. P. Jiang, Y. Z. Du, H. Yuan, Y. Q. Ye and S. Zeng (2006). "Preparation and characteristics of monostearin nanostructured lipid carriers." *International Journal of Pharmaceutics* **314**(1): 83–89.

Hu, L., Z. W. Mao and C. Y. Gao (2009). "Colloidal particles for cellular uptake and delivery." *Journal of Materials Chemistry* **19**(20): 3108–3115.

Hunter, R. J. (1986). *Foundations of Colloid Science*. Oxford, UK, Oxford University Press.

Hutchings, S. C., K. D. Foster, J. E. Bronlund, R. G. Lentle, J. R. Jones and M. P. Morgenstern (2011). "Mastication of heterogeneous foods: Peanuts inside two different food matrices." *Food Quality and Preference* **22**(4): 332–339.

Iacocca, R. G., C. L. Burcham and L. R. Hilden (2010). "Particle engineering: A strategy for establishing drug substance physical property specifications during small molecule development." *Journal of Pharmaceutical Sciences* **99**(1): 51–75.

Iqbal, J. and M. M. Hussain (2009). "Intestinal lipid absorption." *American Journal of Physiology-Endocrinology and Metabolism* **296**(6): E1183–E1194.

Israelachvili, J. (2011). *Intermolecular and Surface Forces*, 3rd Edition. London, Academic Press.

Jafari, S. M., Y. H. He and B. Bhandari (2007). "Production of sub-micron emulsions by ultrasound and microfluidization techniques." *Journal of Food Engineering* **82**(4): 478–488.

Jain, G. K., S. A. Pathan, S. Akhter, N. Ahmad, N. Jain, S. Talegaonkar, R. K. Khar and F. J. Ahmad (2010). "Mechanistic study of hydrolytic erosion and drug release behaviour of PLGA nanoparticles: Influence of chitosan." *Polymer Degradation and Stability* **95**(12): 2360–2366.

Kashchiev, D., A. Borissova, R. B. Hammond and K. J. Roberts (2010). "Effect of cooling rate on the critical undercooling for crystallization." *Journal of Crystal Growth* **312**(5): 698–704.

Kashchiev, D. and G. M. van Rosmalen (2003). "Review: Nucleation in solutions revisited." *Crystal Research and Technology* **38**(7–8): 555–574.

Kim, H. J., E. A. Decker and D. J. McClements (2002a). "Impact of protein surface denaturation on droplet flocculation in hexadecane oil-in-water emulsions stabilized by beta-lactoglobulin." *Journal of Agricultural and Food Chemistry* **50**(24): 7131–7137.

Kim, H. J., E. A. Decker and D. J. McClements (2002b). "Role of postadsorption conformation changes of beta-lactoglobulin on its ability to stabilize oil droplets against flocculation during heating at neutral pH." *Langmuir* **18**(20): 7577–7583.

Koc, H., C. J. Vinyard, G. K. Essick and E. A. Foegeding (2013). Food oral processing: Conversion of food structure to textural perception. In *Annual Review of Food Science and Technology*, M. P. Doyle and T. R. Klaenhammer (eds.). Palo Alto, CA, Annual Reviews **4**: 237–266.

Kosaraju, S. L. (2005). "Colon targeted delivery systems: Review of polysaccharides for encapsulation and delivery." *Critical Reviews in Food Science and Nutrition* **45**(4): 251–258.

Kumar, V., A. K. Sinha, H. P. S. Makkar, G. de Boeck and K. Becker (2012). "Dietary roles of non-starch polysachharides in human nutrition: A review." *Critical Reviews in Food Science and Nutrition* **52**(10): 899–935.

Laparra, J. M. and Y. Sanz (2010). "Interactions of gut microbiota with functional food components and nutraceuticals." *Pharmacological Research* **61**(3): 219–225.

Larson, R. G. (1999). *The Structure and Rheology of Complex Fluids*. Oxford, UK, Oxford University Press.

Leubner, I. H. (2000). "Particle nucleation and growth models." *Current Opinion in Colloid and Interface Science* **5**(1–2): 151–159.

Li, X. and B. R. Jastic (2005). *Design of Controlled Release Drug Delivery Systems*. New York, McGraw-Hill Professional.

Li, Y., M. Hu, Y. M. Du and D. J. McClements (2011a). "Controlling lipid nanoemulsion digestion using nanolaminated biopolymer coatings." *Journal of Microencapsulation* **28**(3): 166–175.

Li, Y., M. Hu, Y. M. Du, H. Xiao and D. J. McClements (2011b). "Control of lipase digestibility of emulsified lipids by encapsulation within calcium alginate beads." *Food Hydrocolloids* **25**(1): 122–130.

Li, Y., H. Xiao and D. J. McClements (2012). "Encapsulation and delivery of crystalline hydrophobic nutraceuticals using nanoemulsions: Factors affecting polymethoxyflavone solubility." *Food Biophysics* **7**(4): 341–353.

Li, Y., J. K. Zheng, H. Xiao and D. J. McClements (2012). "Nanoemulsion-based delivery systems for poorly water-soluble bioactive compounds: Influence of formulation parameters on polymethoxyflavone crystallization." *Food Hydrocolloids* **27**(2): 517–528.

Lian, G. P., M. E. Malone, J. E. Homan and I. T. Norton (2004). "A mathematical model of volatile release in mouth from the dispersion of gelled emulsion particles." *Journal of Controlled Release* **98**(1): 139–155.

Liechty, W. B., D. R. Kryscio, B. V. Slaughter and N. A. Peppas (2010). Polymers for drug delivery systems. In *Annual Review of Chemical and Biomolecular Engineering*, J. M. Prausnitz, M. F. Doherty and M. A. Segalman (eds.). Palo Alto, CA, Annual Reviews **1**: 149–173.

Lindfors, L., S. Forssen, J. Westergren and U. Olsson (2008). "Nucleation and crystal growth in supersaturated solutions of a model drug." *Journal of Colloid and Interface Science* **325**(2): 404–413.

Lundin, L., M. Golding and T. J. Wooster (2008). "Understanding food structure and function in developing food for appetite control." *Nutrition and Dietetics* **65**: S79–S85.

Magnuson, B. A., T. S. Jonaitis and J. W. Card (2011). "A brief review of the occurrence, use, and safety of food-related nanomaterials." *Journal of Food Science* **76**(6): R126–R133.

Manach, C., J. Hubert, R. Llorach and A. Scalbert (2009). "The complex links between dietary phytochemicals and human health deciphered by metabolomics." *Molecular Nutrition and Food Research* **53**(10): 1303–1315.

Mao, Y. Y. and D. J. McClements (2012). "Fabrication of viscous and paste-like materials by controlled heteroaggregation of oppositely charged lipid droplets." *Food Chemistry* **134**(2): 872–879.

Mao, Y. Y. and D. J. McClements (2013). "Modification of emulsion properties by heteroaggregation of oppositely charged starch-coated and protein-coated fat droplets." *Food Hydrocolloids* **33**(2): 320–326.

Marciani, L., R. Faulks, M. S. J. Wickham, D. Bush, B. Pick, J. Wright, E. F. Cox, A. Fillery-Travis, P. A. Gowland and R. C. Spiller (2009). "Effect of intragastric acid stability of fat emulsions on gastric emptying, plasma lipid profile and postprandial satiety." *British Journal of Nutrition* **101**(6): 919–928.

Marciani, L., M. Wickham, G. Singh, D. Bush, B. Pick, E. Cox, A. Fillery-Travis et al. (2006). "Delaying gastric emptying and enhancing cholecystokinin release and satiety by using acid stable fat emulsions." *Gastroenterology* **130**(4): A227.

Marciani, L., M. Wickham, G. Singh, D. Bush, B. Pick, E. Cox, A. Fillery-Travis et al. (2007). "Enhancement of intragastric acid stability of a fat emulsion meal delays gastric emptying and increases cholecystokinin release and gallbladder contraction." *American Journal of Physiology-Gastrointestinal and Liver Physiology* **292**(6): G1607–G1613.

Maris, H. J. (2006). "Introduction to the physics of nucleation." *Comptes Rendus Physique* **7**(9–10): 946–958.

Matalanis, A., O. G. Jones and D. J. McClements (2011). "Structured biopolymer-based delivery systems for encapsulation, protection, and release of lipophilic compounds." *Food Hydrocolloids* **25**(8): 1865–1880.

Mathias, N. R. and J. Crison (2012). "The use of modeling tools to drive efficient oral product design." *AAPS Journal* **14**(3): 591–600.

McClements, D. J. (2005). *Food Emulsions: Principles, Practice, and Techniques*. Boca Raton, FL, CRC Press.

McClements, D. J. (2010). "Emulsion design to improve the delivery of functional lipophilic components." *Annual Review of Food Science and Technology* **1**: 241–269.

McClements, D. J. (2012). "Crystals and crystallization in oil-in-water emulsions: Implications for emulsion-based delivery systems." *Advances in Colloid and Interface Science* **174**: 1–30.

McClements, D. J. (2013). "Edible lipid nanoparticles: Digestion, absorption, and potential toxicity." *Progress in Lipid Research* **52**(4): 409–423.

McClements, D. J., E. A. Decker and Y. Park (2009a). "Controlling lipid bioavailability through physicochemical and structural approaches." *Critical Reviews in Food Science and Nutrition* **49**(1): 48–67.

McClements, D. J., E. A. Decker, Y. Park and J. Weiss (2009b). "Structural design principles for delivery of bioactive components in nutraceuticals and functional foods." *Critical Reviews in Food Science and Nutrition* **49**(6): 577–606.

McClements, D. J. and Y. Li (2010). "Structured emulsion-based delivery systems: Controlling the digestion and release of lipophilic food components." *Advances in Colloid and Interface Science* **159**(2): 213–228.

McClements, D. J. and H. Xiao (2012). "Potential biological fate of ingested nanoemulsions: Influence of particle characteristics." *Food and Function* **3**(3): 202–220.

Mei, L. Y., S. J. Choi, J. Alamed, L. Henson, M. Popplewell, D. J. McClements and E. A. Decker (2010). "Citral stability in oil-in-water emulsions with solid or liquid octadecane." *Journal of Agricultural and Food Chemistry* **58**(1): 533–536.

Modarresi, H., E. Conte, J. Abildskov, R. Gani and P. Crafts (2008). "Model-based calculation of solid solubility for solvent selection—A review." *Industrial and Engineering Chemistry Research* **47**(15): 5234–5242.

O'Keefe, S. J. D. (2008). "Nutrition and colonic health: The critical role of the microbiota." *Current Opinion in Gastroenterology* **24**(1): 51–58.

Oh, J. K., R. Drumright, D. J. Siegwart and K. Matyjaszewski (2008). "The development of microgels/nanogels for drug delivery applications." *Progress in Polymer Science* **33**(4): 448–477.

Patel, A., Y. C. Hu, J. K. Tiwari and K. P. Velikov (2010). "Synthesis and characterisation of zein-curcumin colloidal particles." *Soft Matter* **6**(24): 6192–6199.

Patten, G. S., M. A. Augustin, L. Sanguansri, R. J. Head and M. Y. Abeywardena (2009). "Site specific delivery of microencapsulated fish oil to the gastrointestinal tract of the rat." *Digestive Diseases and Sciences* **54**(3): 511–521.

Peppas, N. A., P. Bures, W. Leobandung and H. Ichikawa (2000). "Hydrogels in pharmaceutical formulations." *European Journal of Pharmaceutics and Biopharmaceutics* **50**(1): 27–46.

Porter, C. J. H., N. L. Trevaskis and W. N. Charman (2007). "Lipids and lipid-based formulations: Optimizing the oral delivery of lipophilic drugs." *Nature Reviews Drug Discovery* **6**(3): 231–248.

Powell, J. J., N. Faria, E. Thomas-McKay and L. C. Pele (2010). "Origin and fate of dietary nanoparticles and microparticles in the gastrointestinal tract." *Journal of Autoimmunity* **34**(3): J226–J233.

Qian, C., E. A. Decker, H. Xiao and D. J. McClements (2013). "Impact of lipid nanoparticle physical state on particle aggregation and beta-carotene degradation: Potential limitations of solid lipid nanoparticles." *Food Research International* **52**(1): 342–349.

Ran, Y. Q., Y. He, G. Yang, J. L. H. Johnson and S. H. Yalkowsky (2002). "Estimation of aqueous solubility of organic compounds by using the general solubility equation." *Chemosphere* **48**(5): 487–509.

Roberts, D. J. and R. I. Hall (2013). "Drug absorption, distribution, metabolism and excretion considerations in critically ill adults." *Expert Opinion on Drug Metabolism and Toxicology* **9**(9): 1067–1084.

Rodriguez-Spong, B., C. P. Price, A. Jayasankar, A. J. Matzger and N. Rodriguez-Hornedo (2004). "General principles of pharmaceutical solid polymorphism: A supramolecular perspective." *Advanced Drug Delivery Reviews* **56**(3): 241–274.

Sablani, S. S., R. M. Syamaladevi and B. G. Swanson (2010). "A review of methods, data and applications of state diagrams of food systems." *Food Engineering Reviews* **2**(3): 168–203.

Salles, C., M. C. Chagnon, G. Feron, E. Guichard, H. Laboure, M. Morzel, E. Semon, A. Tarrega and C. Yven (2011). "In-mouth mechanisms leading to flavor release and perception." *Critical Reviews in Food Science and Nutrition* **51**(1): 67–90.

Sato, K. (2001). "Crystallization behaviour of fats and lipids—A review." *Chemical Engineering Science* **56**(7): 2255–2265.

Schmitt, C. and S. L. Turgeon (2011). "Protein/polysaccharide complexes and coacervates in food systems." *Advances in Colloid and Interface Science* **167**(1–2): 63–70.

Schubert, M. A., B. C. Schicke and C. C. Muller-Goymann (2005). "Thermal analysis of the crystallization and melting behavior of lipid matrices and lipid nanoparticles containing high amounts of lecithin." *International Journal of Pharmaceutics* **298**(1): 242–254.

Siepmann, J. and N. A. Peppas (2011). "Higuchi equation: Derivation, applications, use and misuse." *International Journal of Pharmaceutics* **418**(1): 6–12.

Siepmann, J. and F. Siepmann (2008). "Mathematical modeling of drug delivery." *International Journal of Pharmaceutics* **364**(2): 328–343.

Siepmann, J. and F. Siepmann (2011). "Mathematical modeling of drug release from lipid dosage forms." *International Journal of Pharmaceutics* **418**(1): 42–53.

Siepmann, J. and F. Siepmann (2012). "Modeling of diffusion controlled drug delivery." *Journal of Controlled Release* **161**(2): 351–362.

Siepmann, J. and F. Siepmann (2013). "Mathematical modeling of drug dissolution." *International Journal of Pharmaceutics* **453**(1): 12–24.

Singh, H. and A. Sarkar (2011). "Behaviour of protein-stabilised emulsions under various physiological conditions." *Advances in Colloid and Interface Science* **165**(1): 47–57.

Singh, H., A. Q. Ye and D. Horne (2009). "Structuring food emulsions in the gastrointestinal tract to modify lipid digestion." *Progress in Lipid Research* **48**(2): 92–100.

Sjogren, E., J. Westergren, I. Grant, G. Hanisch, L. Lindfors, H. Lennernas, B. Abrahamsson and C. Tannergren (2013). "In silico predictions of gastrointestinal drug absorption in pharmaceutical product development: Application of the mechanistic absorption model GI-Sim." *European Journal of Pharmaceutical Sciences* **49**(4): 679–698.

Spencer, J. P. E., M. Abd El Mohsen and A. M. Minihane (2006). "Metabolism of dietary phytochemicals: A review of the metabolic forms identified in humans." *Current Topics in Nutraceutical Research* **4**(3–4): 187–203.

Stieger, M. and F. van de Velde (2013). "Microstructure, texture and oral processing: New ways to reduce sugar and salt in foods." *Current Opinion in Colloid and Interface Science* **18**(4): 334–348.

Sugano, K. (2009a). "Computational oral absorption simulation for low-solubility compounds." *Chemistry and Biodiversity* **6**(11): 2014–2029.

Sugano, K. (2009b). "Estimation of effective intestinal membrane permeability considering bile micelle solubilisation." *International Journal of Pharmaceutics* **368**(1–2): 116–122.

Sugano, K. (2009c). "Introduction to computational oral absorption simulation." *Expert Opinion on Drug Metabolism and Toxicology* **5**(3): 259–293.

Sugano, K. (2011). "Fraction of a dose absorbed estimation for structurally diverse low solubility compounds." *International Journal of Pharmaceutics* **405**(1–2): 79–89.

Sugano, K., M. Kataoka, C. D. Mathews and S. Yamashita (2010). "Prediction of food effect by bile micelles on oral drug absorption considering free fraction in intestinal fluid." *European Journal of Pharmaceutical Sciences* **40**(2): 118–124.

Sun, J. and S. L. Simon (2007). "The melting behavior of aluminum nanoparticles." *Thermochimica Acta* **463**(1–2): 32–40.

Teng, Z., Y. Li, Y. C. Luo, B. C. Zhang and Q. Wang (2013). "Cationic beta-lactoglobulin nanoparticles as a bioavailability enhancer: Protein characterization and particle formation." *Biomacromolecules* **14**(8): 2848–2856.

Timms, R. E. (1991). "Crystallization of fats." *Chemistry and Industry* 342.

Tomas-Barberan, F. A., A. Gil-Izquierdo and D. A. Moreno (2009). Bioavailability and metabolism of phenolic compounds and glucosinolates. In *Designing Functional Foods: Measuring and Controlling Food Structure Breakdown and Nutrient Absorption*, D. J. McClements and E. A. Decker (eds.). Cambridge, UK, Woodhead Publ Ltd: 194–229.

Trevaskis, N. L., W. N. Charman and C. J. H. Porter (2006). Lipid-based delivery systems and intestinal lymphatic drug transport: A mechanistic update. *Annual Meeting of the American-Association-of-Pharmaceutical-Scientists*, San Antonio, TX.

Turgeon, S. L., C. Schmitt and C. Sanchez (2007). "Protein-polysaccharide complexes and coacervates." *Current Opinion in Colloid and Interface Science* **12**(4–5): 166–178.

Turner, D. C., F. C. Yin, J. T. Kindt and H. L. Zhang (2012). "Understanding pharmacokinetic food effects using molecular dynamics simulation coupled with physiologically based pharmacokinetic modeling." *Biopharmaceutics and Drug Disposition* **33**(9): 510–521.

Unfried, K., C. Albrecht, L. O. Klotz, A. Von Mikecz, S. Grether-Beck and R. P. F. Schins (2007). "Cellular responses to nanoparticles: Target structures and mechanisms." *Nanotoxicology* **1**(1): 52–71.

van Aken, G. A. (2010). "Relating food emulsion structure and composition to the way it is processed in the gastrointestinal tract and physiological responses: What are the opportunities?" *Food Biophysics* **5**(4): 258–283.

Van Eerdenbrugh, B., J. Vermant, J. A. Martens, L. Froyen, J. Van Humbeeck, G. Van den Monter and P. Augustijns (2010). "Solubility increases associated with crystalline drug nanoparticles: Methodologies and significance." *Molecular Pharmaceutics* **7**(5): 1858–1870.

Vekilov, P. G. (2010a). "Nucleation." *Crystal Growth and Design* **10**(12): 5007–5019.

Vekilov, P. G. (2010b). Nucleation of crystals in solution. In *Selected Topics on Crystal Growth*, M. Wang, K. Tsukamoto and D. Wu (eds.). College Park, MD, American Institute of Physics **1270**: 60–77.

Vekilov, P. G. (2010c). "The two-step mechanism of nucleation of crystals in solution." *Nanoscale* **2**(11): 2346–2357.

Velikov, K. P. and E. Pelan (2008). "Colloidal delivery systems for micronutrients and nutraceuticals." *Soft Matter* **4**(10): 1964–1980.

Versantvoort, C. H. M., E. Van de Kamp and C. J. M. Rompelberg (2004). Development and applicability of an in vitro digestion model in assessing the bioaccessibility of contaminants from food. RIVM report 320102002/2004, Bilthoven, Inspectorate of Health Inspection: 1–87.

Walstra, P. (2003). *Physical Chemistry of Foods*. New York, Marcel Decker.

Wang, J. M. and T. J. Hou (2011). "Recent advances on aqueous solubility prediction." *Combinatorial Chemistry and High Throughput Screening* **14**(5): 328–338.

Wang, K. and F. Qiu (2013). "Curcuminoid metabolism and its contribution to the pharmacological effects." *Current Drug Metabolism* **14**(7): 791–806.

Warren, D. B., H. Benameur, C. J. H. Porter and C. W. Pouton (2010). "Using polymeric precipitation inhibitors to improve the absorption of poorly water-soluble drugs: A mechanistic basis for utility." *Journal of Drug Targeting* **18**(10): 704–731.

Westesen, K., H. Bunjes and M. H. J. Koch (1997). "Physicochemical characterization of lipid nanoparticles and evaluation of their drug loading capacity and sustained release potential." *Journal of Controlled Release* **48**(2–3): 223–236.

Williams, H. D., N. L. Trevaskis, S. A. Charman, R. M. Shanker, W. N. Charman, C. W. Pouton and C. J. H. Porter (2013). "Strategies to address low drug solubility in discovery and development." *Pharmacological Reviews* **65**(1): 315–499.

Yang, G., Y. Q. Ran and S. H. Yalkowsky (2002). "Prediction of the aqueous solubility: Comparison of the general solubility equation and the method using an amended solvation energy relationship." *Journal of Pharmaceutical Sciences* **91**(2): 517–533.

Yuan, H., J. Chen, Y. Z. Du, F. Q. Hu, S. Zeng and H. L. Zhao (2007). "Studies on oral absorption of stearic acid SLN by a novel fluorometric method." *Colloids and Surfaces B-Biointerfaces* **58**(2): 157–164.

Ziani, K., J. A. Barish, D. J. McClements and J. M. Goddard (2011). "Manipulating interactions between functional colloidal particles and polyethylene surfaces using interfacial engineering." *Journal of Colloid and Interface Science* **360**(1): 31–38.

# Index

Page numbers followed by f and t indicate figures and tables, respectively.